Essential Microbiology

Stuart Hogg

The University of Glamorgan, UK

John Wiley & Sons, Ltd

Telephone (+44) 1243 779777

Email (for orders and customer service enquiries): cs-books@wiley.co.uk
Visit our Home Page on www.wileyeurope.com or www.wiley.com

Reprinted with corrections October and December 2005

Other Wiley Editorial Offices

John Wiley & Sons Inc., 111 River Street, Hoboken, NJ 07030, USA

Jossey-Bass, 989 Market Street, San Francisco, CA 94103-1741, USA

Wiley-VCH Verlag GmbH, Boschstr. 12, D-69469 Weinheim, Germany

John Wiley & Sons Australia Ltd, 33 Park Road, Milton, Queensland 4064, Australia

John Wiley & Sons (Asia) Pte Ltd, 2 Clementi Loop #02-01, Jin Xing Distripark, Singapore 129809

John Wiley & Sons Canada Ltd, 22 Worcester Road, Etobicoke, Ontario, Canada M9W 1L1

Wiley also publishes its books in a variety of electronic formats. Some content that appears in print may not be available in electronic books.

Library of Congress Cataloguing-in-Publication Data

British Library Cataloguing in Publication Data

A catalogue record for this book is available from the British Library

ISBN 10: 0-471-49753-3 (H/B) ISBN 13: 9780-471-49753-0 (H/B)
ISBN 10: 0-471-49754-1 (P/B) ISBN 13: 978-0-471-49754-7 (P/B)

Typeset in 10/12pt Sabon by TechBooks, New Delhi, India
Printed and bound in Great Britain by Antony Rowe, Ltd, Chippenham, Wiltshire
This book is printed on acid-free paper responsibly manufactured from sustainable forestry in which at least two trees are planted for each one used for paper production.

Contents

Preface

Every year, in UK universities alone, many hundreds of students study microbiology as part of an undergraduate course. For some, the subject will form the major part of their studies, leading to a BSc degree in Microbiology, or a related subject such as Bacteriology or Biotechnology. For the majority, however, the study of microbiology will be a brief encounter, forming only a minor part of their course content.

A number of excellent and well-established textbooks are available to support the study of microbiology; such titles are mostly over 1000 pages in length, beautifully illustrated in colour, and rather expensive. This book in no way seeks to replace or compete with such texts, which will serve specialist students well throughout their three years of study, and represent a sound investment. It is directed rather towards the second group of students, who require a text that is less detailed, less comprehensive, and less expensive! The majority of the students in my own classes are enrolled on BSc degrees in Biology, Human Biology and Forensic Science; I have felt increasingly uncomfortable about recommending that they invest a substantial sum of money on a book much of whose content is irrelevant to their needs. Alternative recommendations, however, are not thick on the ground. This, then, was my initial stimulus to write a book of 'microbiology for the non-microbiologist'.

The facts and principles you will find here are no different from those described elsewhere, but I have tried to select those topics that one might expect to encounter in years 1 and 2 of a typical non-specialist degree in the life sciences or related disciplines. Above all, I have tried to *explain* concepts or mechanisms; one thing my research for this book has taught me is that textbooks are *not* always right, and they certainly don't always explain things as clearly as they might. It is my wish that the present text will give the attentive reader a clear understanding of sometimes complex issues, whilst avoiding over-simplification.

The book is arranged into seven sections, the fourth of which, Microbial Genetics, acts as a pivot, leading from principles to applications of microbiology. Depending on their starting knowledge, readers may 'dip into' the book at specific topics, but those whose biological and chemical knowledge is limited are strongly recommended to read Chapters 2 and 3 for the foundation necessary for the understanding of later chapters. Occasional boxes are inserted into the text, which provide some further enlightenment on the topic being discussed, or offer supplementary information for the inquisitive reader. As far as possible, diagrams are limited to simple line drawings, most of which could be memorised for reproduction in an examination setting. Although a Glossary is provided at the end of the book, new words are also defined in the text at the point of their first introduction, to facilitate uninterrupted reading. All chapters except the first are followed by a self-test section in which readers may review their knowledge and understanding by 'filling in the gaps' in incomplete sentences; the answers are all to be found in the text, and so are not provided separately. The only exceptions to this are two numerical questions, the solutions to which are to be found at the back of the book. By completing the self-test questions, the reader effectively provides a summary for the chapter.

A book such as this stands or falls by the reception it receives from its target readership. I should be pleased to receive any comments on the content and style of *Essential Microbiology* from students and their tutors, all of which will be given serious consideration for inclusion in any further editions.

Stuart Hogg
January 2005

Acknowledgements

I would like to thank those colleagues who took the time to read over individual chapters of this book, and those who reviewed the entire manuscript. Their comments have been gratefully received, and in some cases spared me from the embarrassment of seeing my mistakes perpetuated in print.

Thanks are also due to my editorial team at John Wiley, Rachael Ballard and Andy Slade, and production editor Robert Hambrook for ensuring smooth production of this book.

I am grateful to those publishers and individuals who have granted permission to reproduce diagrams. Every effort has been made to trace holders of copyright; any inadvertent omissions will gladly be rectified in any future editions of this book.

Finally, I would like to express my gratitude to my family for allowing me to devote so many weekends to 'the book'.

Part I

Introduction

1

Microbiology: What, Why and How?

As you begin to explore the world of microorganisms, one of the first things you'll notice is their extraordinary diversity – of structure, function, habitat and applications. Microorganisms (or microbes) inhabit every corner of the globe, are indispensable to life on Earth, are responsible for some of the most deadly human diseases and form the basis of many industrial processes. Yet until a few hundred years ago, nobody knew they existed!

In this opening chapter, we offer some answers to three questions:

- *What* is microbiology?
- *Why* is it such an important subject?
- *How* have we gained our present knowledge of microbiology?

What is microbiology?

Things aren't always the way they seem. On the face of it, 'microbiology' should be an easy word to define: the science (*logos*) of small (*micro*) life (*bios*), or to put it another way, the study of living things so small that they cannot be seen with the naked eye. Bacteria neatly fit this definition, but what about fungi and algae? These two groups each contain members that are far from microscopic. On the other hand, certain animals, such as nematode worms, can be microscopic, yet are not considered to be the domain of the microbiologist. Viruses represent another special case; they are most certainly microscopic (indeed, most are submicroscopic), but by most accepted definitions they are not living. Nevertheless, these too fall within the remit of the microbiologist.

In the central section of this book you can read about the thorny issue of microbial classification and gain some understanding of just what is and what is not regarded as a microorganism.

Why is microbiology important?

To the lay person, microbiology means the study of sinister, invisible 'bugs' that cause disease. As a subject, it generally only impinges on the popular consciousness in news coverage of the latest 'health scare'. It may come as something of a surprise therefore to learn that the vast majority of microorganisms coexist alongside us without causing any harm. Indeed, many perform vital tasks such as the recycling of essential elements, without which life on our planet could not continue, as we will examine in Chapter 16. Other microorganisms have been exploited by humans for our own benefit, for instance in the manufacture of antibiotics (Chapter 14) and foodstuffs (Chapter 17). To get some idea of the importance of microbiology in the world today, just consider the following list of some of the general areas in which the expertise of a microbiologist might be used:

- medicine
- environmental science
- food and drink production
- fundamental research
- agriculture
- pharmaceutical industry
- genetic engineering.

The popular perception among the general public, however, remains one of infections and plagues. Think back to the first time you ever heard about microorganisms; almost certainly, it was when you were a child and your parents impressed on you the dangers of 'germs' from dirty hands or eating things after they'd been on the floor. In reality, only a couple of hundred out of the half million or so known bacterial species give rise to infections in humans; these are termed *pathogens*, and have tended to dominate our view of the microbial world.

A *pathogen* is an organism with the potential to cause disease.

In the next few pages we shall review some of the landmark developments in the history of microbiology, and see how the main driving force throughout this time, but particularly in the early days, has been the desire to understand the nature and cause of infectious diseases in humans.

How do we know? Microbiology in perspective: to the 'golden age' and beyond

We have learnt an astonishing amount about the invisible world of microorganisms, particularly over the last century and a half. How has this happened? The penetrating insights of brilliant individuals are rightly celebrated, but a great many 'breakthroughs' or 'discoveries' have only been made possible thanks to some (frequently unsung) development in microbiological methodology. For example, on the basis that 'seeing is believing', it was only when we had the means to *see* microorganisms under a microscope that we could prove their existence.

Microorganisms had been on the Earth for some 4000 million years, when Antoni van Leeuwenhoek started out on his pioneering microscope work in 1673. Leeuwenhoek was an amateur scientist who spent much of his spare time grinding glass lenses

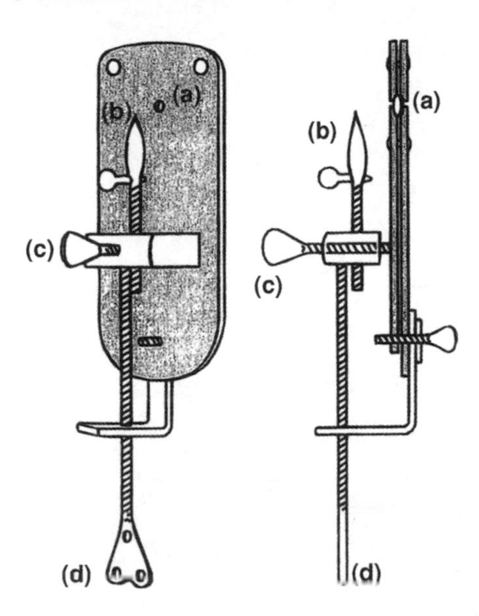

Figure 1.1 Leeuwenhoek's microscope. The lens (a) was held between two brass plates and used to view the specimen, which was placed on the mounting pin (b). Focusing was achieved by means of two screws (c) and (d). Some of Leeuwenhoek's microscopes could magnify up to 300 times. Original source: *Antony van Leeuwenhoek and his little animals* by CE Dobell (1932)

to produce simple microscopes (Figure 1.1). His detailed drawings make it clear that the 'animalcules' he observed from a variety of sources included representatives of what later became known as protozoa, bacteria and fungi. Where did these creatures come from? Arguments about the origin of living things revolved around the long held belief in spontaneous generation, the idea that living organisms could arise from non-living matter. In an elegant experiment, the Italian Francesco Redi (1626–1697) showed that the larvae found on putrefying meat arose from eggs deposited by flies, and not spontaneously as a result of the decay process. This can be seen as the beginning of the end for the spontaneous generation theory, but many still clung to the idea, claiming that while it may not have been true for larger organisms, it must surely be so for minute creatures such as those demonstrated by Leeuwenhoek. Despite mounting evidence against the theory, as late as 1859, fresh 'proof' was still being brought forward in its support. Enter onto the scene Louis Pasteur (1822–1895), still arguably the most famous figure in the history of microbiology. Pasteur trained as a chemist, and made a lasting contribution to the science of stereochemistry before turning his attention to spoilage problems in the wine industry. He noticed that when lactic acid was produced in wine instead of alcohol, rod-shaped bacteria were always present, as well as the expected yeast cells. This led him to believe that while the yeast produced the alcohol, the bacteria were responsible for the spoilage, and that both types of organism had originated in the environment. Exasperated by continued efforts to substantiate the theory of spontaneous generation, he set out to disprove it once and for all. In response to a call from the French Academy of Science, he carried out a series of experiments that led to the acceptance of *biogenesis*, the idea that life arises only from already existing life. Using his famous swan-necked flasks (Figure 1.2), he demonstrated in 1861 that as long as dust

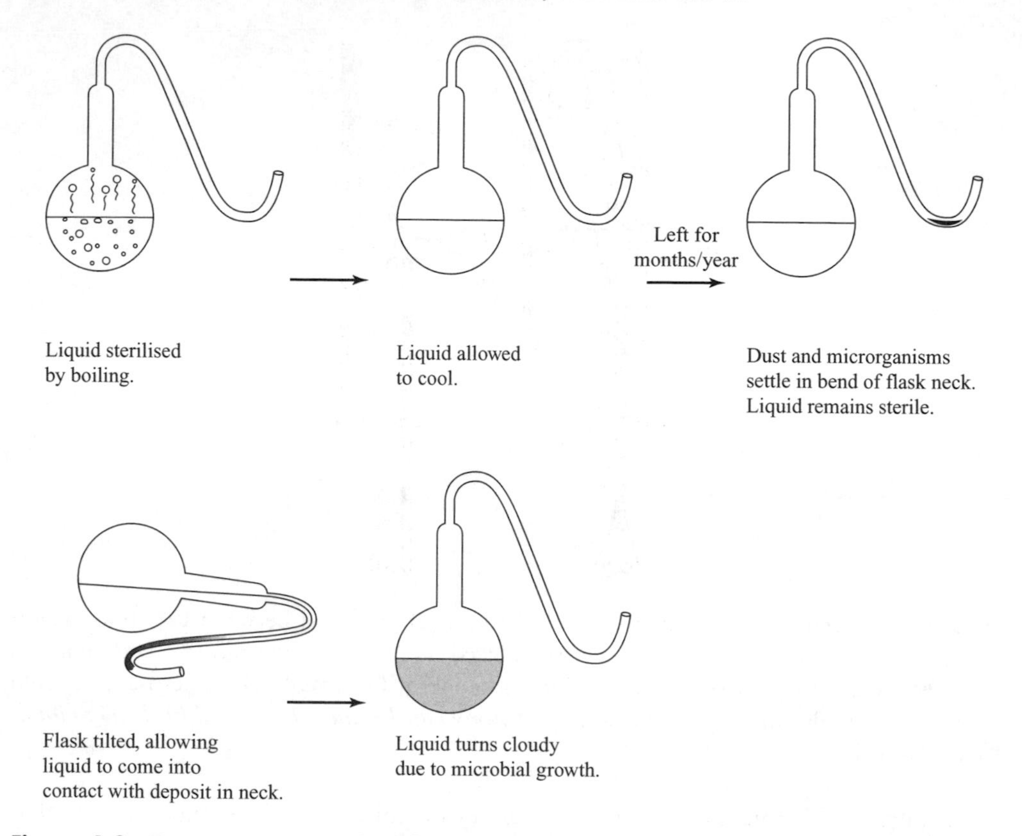
Left for
months/year
Liquid sterilised
by boiling.
Liquid allowed
to cool.
Dust and microrganisms
settle in bend of flask neck.
Liquid remains sterile.
Flask tilted, allowing
liquid to come into
contact with deposit in neck.
Liquid turns cloudy
due to microbial growth.

Figure 1.2 Pasteur's swan-necked flasks. Broth solutions rich in nutrients were placed in flasks and boiled. The necks of the flasks were heated and drawn out into a curve, but kept open to the atmosphere. Pasteur showed that the broth remained sterile because any contaminating dust and microorganisms remained trapped in the neck of the flask as long as it remained upright

particles (and the microorganisms carried on them) were excluded, the contents would remain sterile. This also disproved the idea held by many that there was some element in the air itself that was capable of initiating microbial growth. In Pasteur's words '.... the doctrine of spontaneous generation will never recover from this mortal blow. *There is no known circumstance in which it can be affirmed that microscopic beings came into the world without germs, without parents similar to themselves.*' Pasteur's findings on wine contamination led inevitably to the idea that microorganisms may be also be responsible for diseases in humans, animals and plants.

The notion that some invisible (and therefore, presumably, extremely small) living creatures were responsible for certain diseases was not a new one. Long before microorganisms had been shown to exist, the Roman philosopher Lucretius (~98–55 BC) and much later the physician Girolamo Fracastoro (1478–1553) had supported the idea. Fracastoro wrote 'Contagion is an infection that passes from one thing to another' and recognised three forms of transmission: by direct contact, through inanimate objects and via the air. We still class transmissibility of infectious disease in much the same way today. The prevailing belief at the time, however, was that an infectious disease was due to something called a *miasma,* a poisonous vapour arising from dead or diseased bodies, or to an imbalance between the four humours of the body (blood, phlegm, yellow bile and black bile). During the 19th century, many diseases were shown, one by one, to be caused by microorganisms. In 1835, Agostino Bassi showed that a disease of silkworms was due to a fungal infection, and 10 years later, Miles Berkeley demonstrated that a fungus was also responsible for the great Irish potato blight. Joseph Lister's pioneering work on antiseptic surgery provided strong, albeit indirect, evidence of the involvement of microorganisms in infections of humans. The use of heat-treated instruments and of phenol both on dressings and actually sprayed in a mist over the surgical area, was found greatly to reduce the number of fatalities following surgery. Around the same time, in the 1860s, the indefatigable Pasteur had shown that a parasitic protozoan was the cause of another disease of silkworms called *pébrine*, which had devastated the French silk industry.

The first proof of the involvement of bacteria in disease and the definitive proof of the germ theory of disease came from the German Robert Koch. In 1876 Koch showed the relationship between the cattle disease *anthrax* and a bacillus which we now know as *Bacillus anthracis*. Koch infected healthy mice with blood from diseased cattle and sheep, and noted that the symptoms of the disease appeared in the mice, and that rod shaped bacteria could be isolated from their blood. These could be grown in culture, where they multiplied and produced spores. Injection of healthy mice with these spores (or more bacilli) led them too to develop anthrax and once again the bacteria were isolated from their blood. These results led Koch to formalise the criteria necessary to prove a causal relationship between a specific disease condition and a particular microorganism. These criteria became known as *Koch's postulates* (Box 1.1), and are still in use today.

A *bacillus* is a rod-shaped bacterium.

Box 1.1 Koch's postulates

1 The microorganism must be present in every instance of the disease and absent from healthy individuals.

2 The microorganism must be capable of being isolated and grown in pure culture.

3 When the microorganism is inoculated into a healthy host, the same disease condition must result.

4 The same microorganism must be re-isolated from the experimentally infected host.

Despite their value, it is now realised that Koch's postulates do have certain limitations. It is known for example that certain agents responsible for causing disease (e.g. viruses, prions: see Chapter 10) can't be grown *in vitro,* but only in host cells. Also, the healthy animal in Postulate 3 is seldom human, so a degree of extrapolation is necessary – if agent X does not cause disease in

The term *in vitro* (= 'in glass') is used to describe procedures performed outside of the living organism in test tubes, etc. (c.f *in vivo*).

Table 1.1 The discovery of some major human pathogens

Year	Disease	Causative agent	Discoverer
1876	Anthrax	*Bacillus anthracis*	Koch
1879	Gonorrhoea	*Neisseria gonorrhoeae*	Neisser
1880	Typhoid fever	*Salmonella typhi*	Gaffky
1880	Malaria	*Plasmodium* sp	Laveran
1882	Tuberculosis	*Mycobacterium tuberculosis*	Koch
1883	Cholera	*Vibrio cholerae*	Koch
1883/4	Diphtheria	*Corynebacterium diphtheriae*	Klebs & Loeffler
1885	Tetanus	*Clostridium tetani*	Nicoaier & Kitasato
1886	Pneumonia (bacterial)	*Streptococcus pneumoniae*	Fraenkel
1892	Gas gangrene	*Clostridium perfringens*	Welch & Nuttall
1894	Plague	*Yersinia pestis*	Kitasato & Yersin
1896	Botulism	*Clostridium botulinum*	Van Ermengem
1898	Dysentery	*Shigella dysenteriae*	Shiga
1901	Yellow fever	Flavivirus	Reed
1905	Syphilis	*Treponema pallidum*	Schaudinn & Hoffman
1906	Whooping cough	*Bordetella pertussis*	Bordet & Gengou
1909	Rocky Mountain spotted fever	*Rickettsia rickettsii*	Ricketts

a laboratory animal, can we be sure it won't in humans? Furthermore, some diseases are caused by more than one organism, and some organisms are responsible for more than one disease. On the other hand, the value of Koch's postulates goes beyond just defining the causative agent of a particular disease, and allows us to ascribe a specific effect (of whatever kind) to a given microorganism.

Critical to the development of Koch's postulates was the advance in culturing techniques, enabling the isolation and pure culture of specific microorganisms. These are discussed in more detail in Chapter 4. The development of pure cultures revolutionised microbiology, and within the next 30 years or so, the pathogens responsible for the majority of common human bacterial diseases had been isolated and identified. Not without just cause is this period known as the 'golden age' of microbiology! Table 1.1 summarises the discovery of some major human pathogens.

A *pure* or *axenic* culture contains one type of organism only, and is completely free from contaminants.

At around the same time, Charles Chamberland, a pupil of Pasteur's, invented the autoclave, contributing greatly to the development of pure cultures.

Koch's greatest achievement was in using the advances in methodology and the principles of his own postulates to demonstrate the identity of the causative agent of tuberculosis, which at the time was responsible for around one in every seven human deaths in Europe. Although it was believed by many to have a microbial cause, the causative agent had never been observed, either in culture or in the affected tissues. We now know that *Mycobacterium tuberculosis* (the tubercle bacillus) is very difficult to stain by conventional methods due to the high lipid content of the cell wall surface. Koch developed a staining technique that enabled it to be seen, but realised that in order to satisfy his own postulates, he must isolate the organism and grow it in culture. Again, there were technical difficulties, since even under favourable conditions, *M. tuberculosis* grows slowly, but eventually Koch was able to demonstrate the infectivity of the cultured organisms towards guinea pigs. He was then able to isolate them again from the diseased animal and use them to cause disease in uninfected animals, thus satisfying the remainder of his postulates.

Although most bacterial diseases of humans and their aetiological agents have now been identified, important variants continue to evolve and emerge. Notable examples in recent times include Legionnaires' disease, an acute respiratory infection caused by the previously unrecognised genus, *Legionella*, and Lyme disease, a tickborne infection first described in Connecticut, USA in the mid-1970s. Also, a newly recognised pathogen, *Helicobacter pylori,* has been shown to play an important (and previously unsuspected) role in the development of peptic ulcers. There still remain a few diseases that some investigators suspect are caused by bacteria, but for which no pathogen has been identified.

> *Aetiology* is the cause or origin of a disease.

Following the discovery of viruses during the last decade of the 19th century (see Chapter 10), it was soon established that many diseases of plants, animals and humans were caused by these minute, non-cellular agents.

The major achievement of the first half of the 20th century was the development of antibiotics and other antimicrobial agents, a topic discussed in some detail in Chapter 14. Infectious diseases that previously accounted for millions of deaths became treatable by a simple course of therapy, at least in the affluent West, where such medications were readily available.

If the decades either side of 1900 have become known as the golden age of microbiology, the second half of the twentieth century will surely be remembered as the golden age of molecular genetics. Following on from the achievements of others such as Griffith and Avery, the publication of Watson and Crick's structure for DNA in 1953 heralded an extraordinary 50 years of achievement in this area, culminating at the turn of the 21st century in the completion of the Human Genome Project.

> The *Human Genome Project* is an international effort to map and sequence all the DNA in the human genome. The project has also involved sequencing the genomes of several other organisms.

What, you might ask, has this genetic revolution to do with microbiology? Well, all the early work in molecular genetics was carried out on bacteria and viruses, as you'll learn in Chapter 11, and microbial systems have also been absolutely central to the development of genetic engineering over the last three decades (Chapter 12). Also, as part of the Human Genome Project, the genomes of several microorganisms have been decoded, and it will become increasingly easy to do the same for others in the future, thanks to methodological advances made during the project. Having this information will help us to understand in greater detail the disease strategies of microorganisms, and to devise ways of countering them.

As we have seen, a recurring theme in the history of microbiology has been the way that advances in knowledge have followed on from methodological or technological developments, and we shall refer to a number of such developments during the course of this book. To conclude this introduction to microbiology, we shall return to the instrument that, in some respects, started it all. In any microbiology course, you are sure to spend some time looking down a microscope, and to get the most out of the instrument it is essential that you understand the principles of how it works. The following pages attempt to explain these principles.

Light microscopy

Try this simple experiment. Fill a glass with water, then partly immerse a pencil and observe from one side; what do you see? The apparent 'bending' of the pencil is due to rays of light being slowed down as they enter the water, because air and water have different *refractive indices*. Light rays are similarly retarded as they enter glass and all optical instruments are based on this phenomenon. The compound light microscope consists of three sets of lenses (Figure 1.3):

> The *refractive index* of a substance is the ratio between the velocity of light as it passes through that substance and its velocity in a vacuum. It is a measure of how much the substance slows down and therefore refracts the light.

- the *condenser* focuses light onto the specimen to give optimum illumination
- the *objective* provides a magnified and inverted image of the specimen
- the *eyepiece* adds further magnification

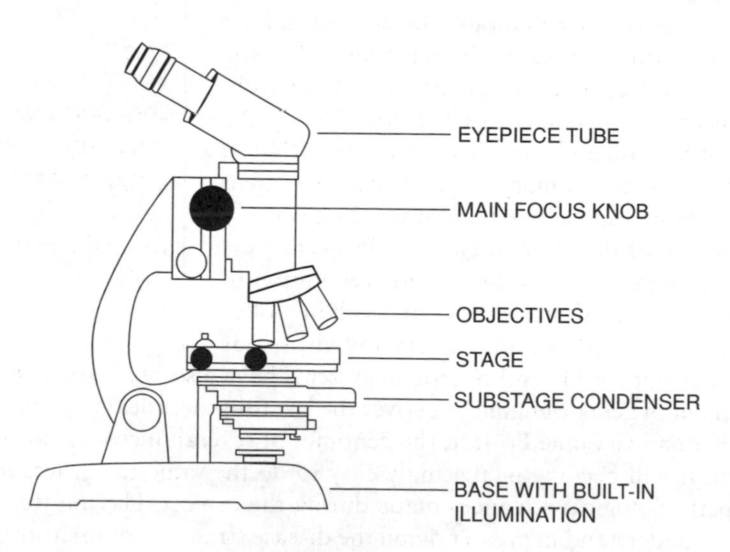

Figure 1.3 The compound light microscope. Modern microscopes have a built-in light source. The light is focused onto the specimen by the condenser lens, and then passes into the body of the microscope via the objective lens. Rotating the objective nosepiece allows different magnifications to be selected. The amount of light entering the microscope is controlled by an iris diaphragm. Light microscopy allows meaningful magnification of up to around 1000×

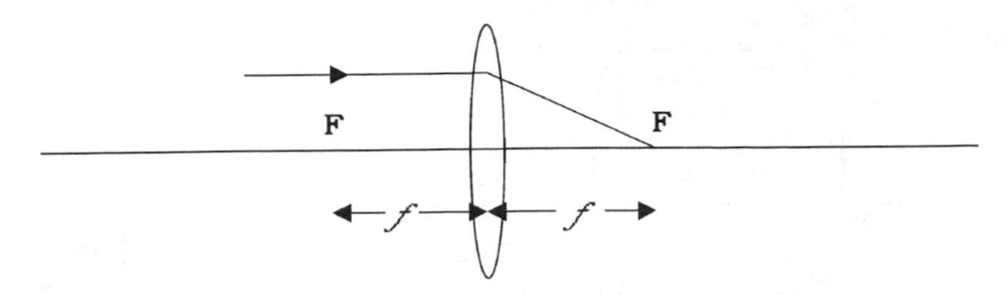

Figure 1.4 Light rays parallel to the axis of a convex lens pass through the focal point. The distance from the centre of the lens to the focal point is called the focal length (*f*) of the lens

Most microscopes have three or four different objectives, giving a range of magnifications, typically from 10× to 100×. The total magnification is obtained by multiplying this by the eyepiece value (usually 10×), thus giving a maximum magnification of 1000×.

In order to appreciate how this magnification is achieved, we need to understand the behaviour of light passing through a convex lens:

- rays parallel to the axis of the lens are brought to a focus at the *focal point* of the lens (Figure 1.4)
- similarly, rays entering the lens from the focal point emerge parallel to the axis
- rays passing through the centre of the lens from any angle are undeviated.

Because the condenser is not involved in magnification, it need not concern us here. Consider now what happens when light passes through an objective lens from an object AB situated slightly beyond its focal point (Figure 1.5a). Starting at the tip of the object, a ray parallel to the axis will leave the lens and pass through the focal point; a ray leaving the same point and passing through the centre of the lens will be undeviated. The point at which the two rays converge is an image of the original point formed by the lens. The same thing happens at an infinite number of points along the object's length, resulting in a primary image of the specimen, A′B′. What can we say about this image, compared to the original specimen AB? It is magnified and it is inverted (i.e. it appears upside down).

The primary image now serves as an object for a second lens, the eyepiece, and is magnified further (Figure 1.5b); this time the object is situated within the focal length. Using the same principles as before, we can construct a ray diagram, but this time we find that the two lines drawn from a point do not converge on the other side of the lens, but actually get further apart. The point at which the lines do eventually converge is actually 'further back' than the original object! What does this mean? The secondary image only *appears* to be coming from A″ B″, and isn't actually there. An image such as this is

A *real* image is one that can be projected onto a flat surface such as a screen. A virtual image does not exist in space and cannot be projected in this way. A familiar example is the image seen in a mirror.

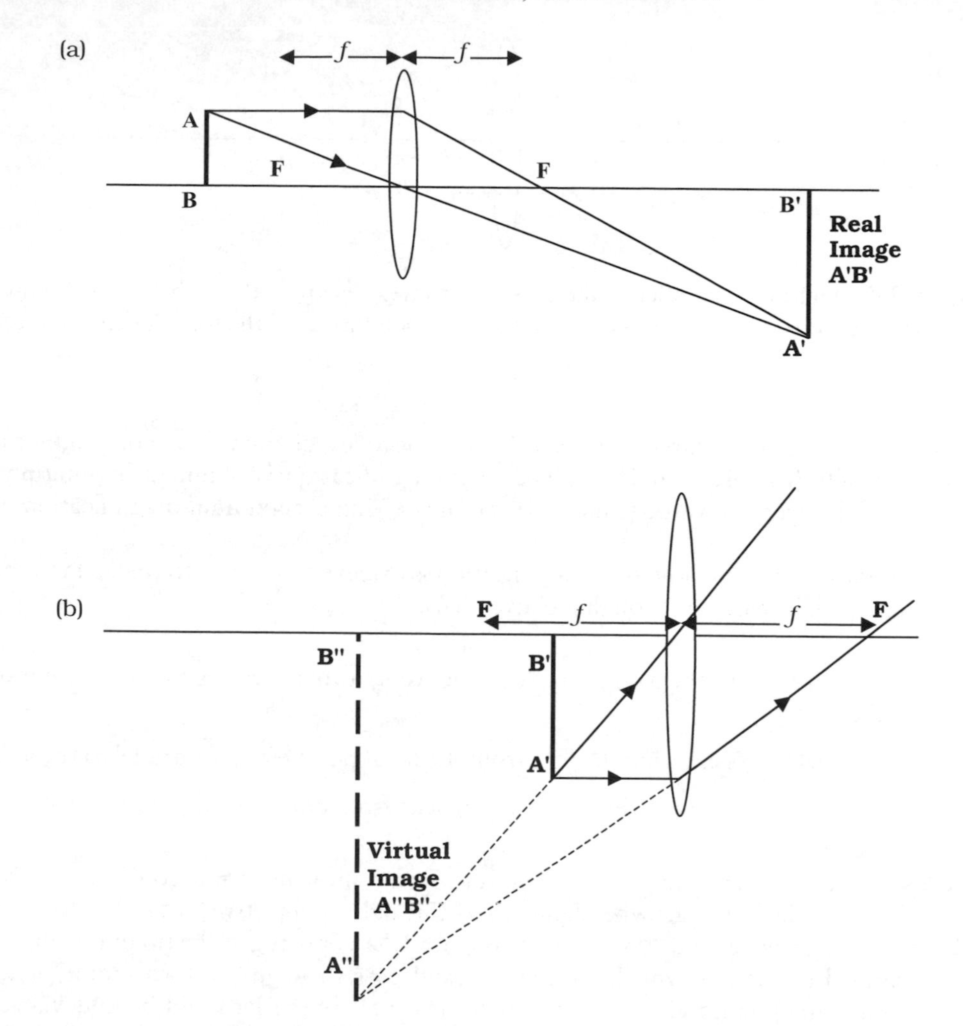

Figure 1.5 The objective lens and eyepiece lens combine to produce a magnified image of the specimen. (a) Light rays from the specimen AB pass through the objective lens to give a magnified, inverted and real primary image. (b) The eyepiece lens magnifies this further to produce a virtual image of the specimen

called a *virtual* image. Today's reader, familiar with the concept of virtual reality, will probably find it easier to come to terms with this than have students of earlier generations! The primary image A′B′, on the other hand, is a *real* image; if a screen was placed at that position, the image would be projected onto it. If we compare A″B″ with A′B′, we can see that it has been further magnified, but not further inverted, so it is still upside down compared with the original. One of the most difficult things to get used to when you first use a microscope is that everything appears 'wrong way around'. The rays of light emerging from the eyepiece lens are focussed by the lens of the observer's eye to form a real image on the retina of the viewer's eye.

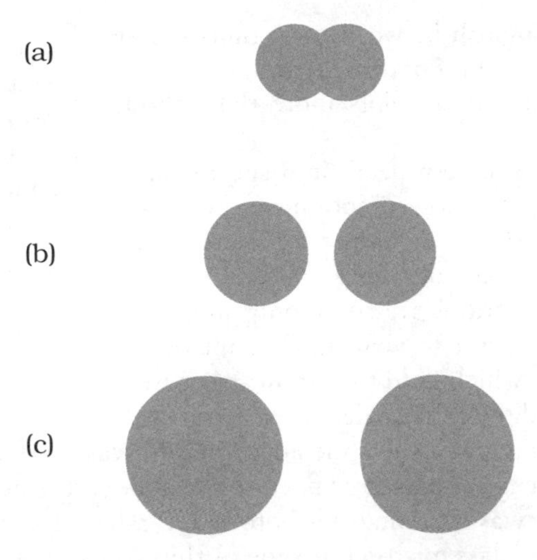

Figure 1.6 Magnification must be accompanied by improved resolution. Compared to (a), the image in (b) is magnified, but also provides improved detail; there are two objects, not just one. Further magnification, as seen in (c), provides no further information (empty magnification)

So, a combination of two lens systems allows us to see a considerably magnified image of our specimen. To continue magnifying an image beyond a certain point, however, serves little purpose, if it is not accompanied by an increase in detail (Figure 1.6). This is termed *empty magnification*. The *resolution* (resolving power, d) of a microscope is its capacity for discerning detail. More specifically, it is the ability to distinguish between two points a short distance apart, and is determined by the equation:

$$d = \frac{0.61\lambda}{n \sin\theta}$$

where λ is the wavelength of the light source, n is the refractive index of the air or liquid between the objective lens and the specimen and θ is the aperture angle (a measure of the light-gathering ability of the lens).

> Immersion oil is used to improve the resolution of a light microscope at high power. It has the same refractive index as glass and is placed between the high power objective and the glass slide. With no layer of air, more light from the specimen enters the objective lens instead of being refracted outside of it, resulting in a sharper image.

The expression $n \sin\theta$ is called the *numerical aperture* and for high quality lenses has a value of around 1.4. The lowest wavelength of light visible to the human eye is approximately 400 nm, so the maximum resolving power for a light microscope is approximately

$$d = \frac{0.61 \times 400}{1.4} = 0.17\mu\text{m}$$

that is, it cannot distinguish between two points closer together than about 0.2 μm. For comparison, the naked eye is unable to resolve two points more than about 0.2 mm apart.

A nanometre (nm) is 1 millionth of a millimetre. There are 1000 nm in one micron (μm), which is therefore one thousandth of a millimetre.
1 mm = 10^{-3} m
1 μm = 10^{-6} m
1 nm = 10^{-9} m

For us to be able to discern detail in a specimen, it must have contrast; most biological specimens, however, are more or less colourless, so unless a structure is appreciably denser than its surroundings, it will not stand out. This is why preparations are commonly subjected to staining procedures prior to viewing. The introduction of coloured dyes, which bind to certain structures, enables the viewer to discern more detail.

Since staining procedures involve the addition and washing off of liquid stains, the sample must clearly be immobilised or fixed to the slide if it is not to end up down the sink. The commonest way of doing this is to make a heat-fixed smear; this kills and attaches the cells to the glass microscope slide. A thin aqueous suspension of the cells is spread across the slide, allowed to dry, then passed (sample side up!) through a flame a few times. Excessive heating must be avoided, as it would distort the natural structure of the cells.

Using simple stains, such as methylene blue, we can see the size and shape of bacterial cells, for example, and their arrangement, while the binding properties of differential stains react with specific structures, helping us to differentiate between bacterial types. Probably the most widely used bacterial stain is the Gram stain (see Box 1.2), which for more than 100 years has been an invaluable first step in the identification of unknown bacteria.

Box 1.2 The Gram stain

The Gram stain involves the sequential use of two stains (see below). The critical stage is step 3; some cells will resist the alcohol treatment and retain the crystal violet, while others become decolorised. The counterstain (safranin or neutral red) is weaker than the crystal violet, and will only be apparent in those cells that have been decolorised.

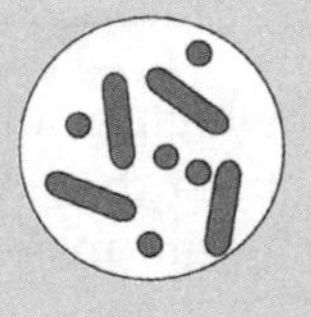

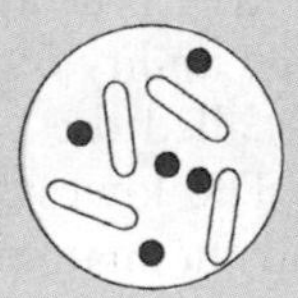

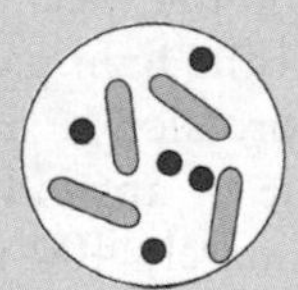

Add crystal violet (primary stain)	Add iodine (mordant)	Alcohol wash (decolourisation)	Add counterstain

The Gram stain is a differential stain, which only takes a few minutes to carry out, and which enables us to place a bacterial specimen into one of two groups, Gram-positive or Gram-negative. The reason for this differential reaction to the stain was not understood for many years, but is now seen to be a reflection of differences in cell wall structure, discussed in more detail in Chapter 3.

Specialised forms of microscopy have been developed to allow the viewer to discern detail in living, unstained specimens; these include phase contrast and dark-field microscopy. We can also gain an estimate of the number of microorganisms in a sample by directly counting them under the microscope. This is discussed along with other enumeration methods in Chapter 5.

Phase contrast microscopy exploits differences in thickness and refractive index of transparent objects such as living cells to give improved contrast.

Dark field microscopy employs a modified condenser. It works by blocking out direct light, and viewing the object only by the light it diffracts.

Electron microscopy

From the equation shown above, you can see that if it were possible to use a shorter wavelength of light, we could improve the resolving power of a microscope. However, because we are limited by the wavelength of light visible to the human eye, we are not able to do this with the light microscope. The electron microscope is able to achieve greater magnification and resolution because it uses a high voltage beam of electrons, whose wavelength is very much shorter than that of visible light. Consequently we are able to resolve points that are much closer together than is possible even with the very best light microscope. The resolving power of an electron microscope may be as low as 1–2 nm, enabling us to see viruses, for example, and the internal structure of cells. The greatly improved resolution means that specimens can be meaningfully magnified over 100 000×.

Electron microscopes, which were first developed in the 1930s and 1940s, use ring-shaped electromagnets as 'lenses' to focus the beam of electrons onto the specimen. Because the electrons would collide with, and be deflected by, molecules in the air, electron microscopes require a pump to maintain a vacuum in the column of the instrument. There are two principal types of electron microscope, the transmission electron microscope (TEM) and the scanning electron microscope (SEM).

Figure 1.7 shows the main features of a TEM. As the name suggests, the electron beam passes *through* the specimen and is scattered according to the density of the different parts. Due to the limited penetrating power of the electrons, extremely thin sections (<100 nm, or less than one-tenth of the diameter of a bacterial cell) must be cut, using a diamond knife. To allow this, the specimen must be fixed and dehydrated, a process that can introduce shrinkage and distortion to its structure if not correctly performed.

After being magnified by an objective 'lens', an image of the specimen is projected onto a fluorescent screen or photographic plate. More dense areas, which scatter the beam, appear dark, and those where it has passed through are light. It is often necessary to enhance contrast artificially, by means of 'staining' techniques that involve coating the specimen with a thin layer of a compound containing a heavy metal, such as osmium or palladium. It will be evident from the foregoing description of sample preparation and use of a vacuum that electron microscopy cannot be used to study living specimens.

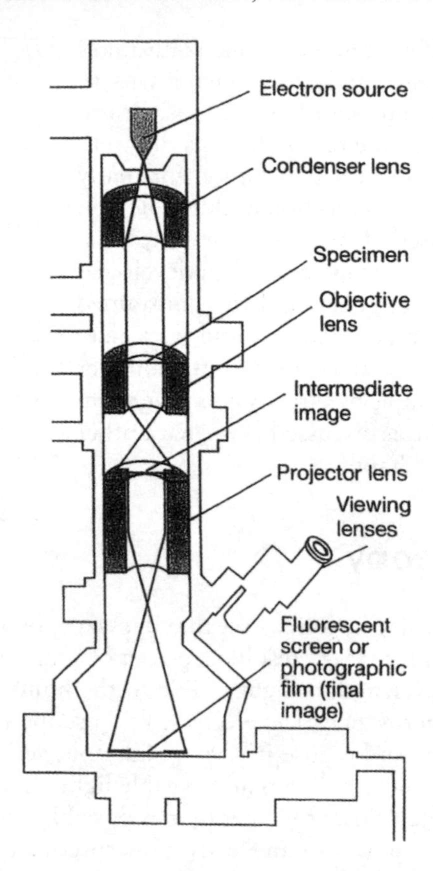

Figure 1.7 The transmission electron microscope. Electrons from a tungsten filament pass through a vacuum chamber and are focused by powerful electromagnets. Passage through the specimen causes a scattering of the electrons to form an image that is captured on a fluorescent screen. From Black, JG: Microbiology: Principles and Explorations, 4th edn, John Wiley & Sons Inc., 1999. Reproduced by permission of the publishers

The TEM has been invaluable in advancing our knowledge of the fine structure of cells, microbial or otherwise. The resulting image is, however, a flat, two-dimensional one, and of limited use if we wish to learn about the surface of a cell or a virus. For this, we turn to SEM. The scanning electron microscope was developed in the 1960s and provides vivid, sometimes startling, three-dimensional images of surface structure. Samples are dehydrated and coated with gold to give a layer a few nanometres thick. A fine beam of electrons probes back and forth across the surface of the specimen and causes secondary electrons to be given off. The number of these, and the angle at which they are emitted, depends on the topography of the specimen's surface. SEM does not have quite the resolving power of the TEM, and therefore does not operate at such high magnifications. Between them, SEM and TEM have opened up a whole new world to microbiologists, allowing us to put advances in our knowledge of microbial biochemistry and genetics into a structural context.

2
Biochemical Principles

All matter, whether living or non-living, is made up of *atoms*; the atom is the smallest unit of matter capable of entering into a chemical reaction. Atoms can combine together by *bonding*, to form *molecules*, which range from the small and simple to the large and complex. The latter are known as *macromolecules*; major cellular constituents such as carbohydrates and proteins belong to this group and it is with these that this chapter is mainly concerned (Table 2.1). In order to appreciate how these macromolecules operate in the structure and function of microbial cells however, we need to review the basic principles of how atoms are constructed and how they interact with one other.

Atomic structure

All atoms have a central, positively charged *nucleus*, which is very dense, and makes up most of the mass of the atom. The nucleus is made up of two types of particle, *protons* and *neutrons*. Protons carry a positive charge, and neutrons are uncharged, hence the nucleus overall is positively charged. It is surrounded by much lighter, and rapidly orbiting, *electrons* (Figure 2.1). These are negatively charged, the charge being equal (but of course opposite) to that of the protons, but they have only 1/1840 of the mass of either protons or neutrons. The attractive force between the positively charged protons and the negatively charged electrons holds the atom together.

The number of protons in the nucleus is called the *atomic number*, and ranges from 1 to over 100. The combined total of protons and neutrons is known as the *mass number*. All atoms have an equal number of protons and electrons, so regardless of the atomic number, the overall charge on the atom will always be zero.

Atoms having the same atomic number have the same chemical properties; such atoms all belong to the same *element*. An element is made up of one type of atom only and cannot be chemically broken down into simpler substances; thus pure copper for example is made up entirely of copper atoms. There are 92 of these elements occurring naturally, 26 of which commonly occur in living things. Each element has been given a universally agreed symbol; examples which we shall encounter in biological macromolecules include carbon (C), hydrogen (H) and oxygen (O). The atomic numbers of selected elements are shown in Table 2.2.

The relationship between neutrons, protons, atomic number, and mass number is illustrated in Table 2.3, using carbon as an example, since all living matter is based

Table 2.1 Biological macromolecules

Proteins	Carbohydrates	Lipids	Nucleic acids
Enzymes	Sugars	Triacylglycerols (fats)	DNA
Receptors	Cellulose	Phospholipids	RNA
Antibodies	Starch	Waxes	
Structural proteins		Sterols	

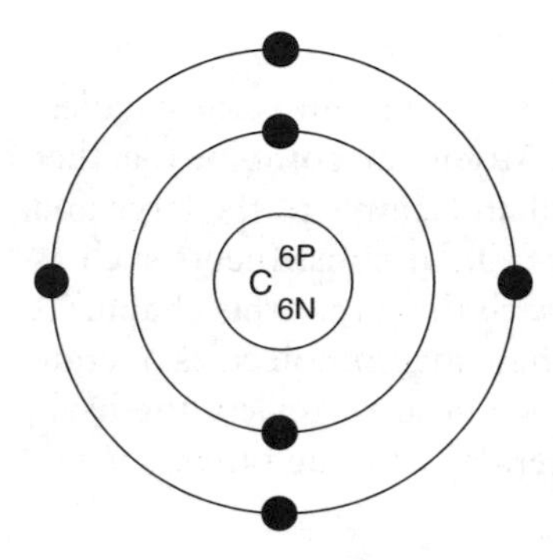

Figure 2.1 Atomic structure. The nucleus of a carbon atom contains six protons and six neutrons, surrounded by six electrons. Note how these are distributed between inner (2) and outer (4) electron shells

Table 2.2 Symbols and atomic numbers of some elements occurring in living systems

Element	Symbol	Atomic no.
Hydrogen	H	1
Carbon	C	6
Nitrogen	N	7
Oxygen	O	8
Sodium	Na	11
Magnesium	Mg	12
Phosphorus	P	15
Sulphur	S	16
Chlorine	Cl	17
Potassium	K	19
Iron	Fe	26

Table 2.3 The vital statistics of carbon

No. of Protons	No. of Neutrons	Atomic number	Mass number	Atomic mass
6	6	6	12	12.011

upon this element. The carbon represented can be expressed in the form:

$$^{12}_{6}C = \begin{array}{l}\text{(mass number)} \\ \text{(Element symbol)} \\ \text{(atomic number)}\end{array}$$

The number of neutrons in an atom can be deduced by subtracting the atomic number from the mass number. In the case of carbon, this is the same as the number of protons (6), but this is not always so. Phosphorus for example has 15 protons and 16 neutrons, giving it an atomic number of 15 and a mass number of 31.

Isotopes

Although the number of protons in the nucleus of a given element is always the same, the number of neutrons can vary, to give different forms or *isotopes* of that element. Carbon-14 (^{14}C) is a naturally occurring but rare isotope of carbon that has eight neutrons instead of six, hence the atomic mass of 14. Carbon-13 (^{13}C) is a rather more common isotope, making up around 1 per cent of naturally occurring carbon; it has seven neutrons per atomic nucleus. The *atomic mass* (or atomic weight) of an element is the average of the mass numbers of an element's different isotopes, taking into account the proportions in which they occur. (Box 2.1 shows how atomic weight is used to quantify amounts of compounds using moles.) Carbon-12 is by far the predominant form of the element in nature, but the existence of small amounts of the other forms means that the atomic mass is 12.011. Some isotopes are stable, while others decay spontaneously, with the release of subatomic particles. The latter are called *radioisotopes*; ^{14}C is a radioisotope, while the other two forms of carbon are stable isotopes. Radioisotopes have been an extremely useful research tool in a number of areas of molecular biology.

The electrons that orbit around the nucleus do not do so randomly, but are arranged in a series of electron shells, radiating out from the nucleus (Figure 2.1). These layers correspond to different energy levels, with the highest energy levels being located furthest away from the nucleus. Each shell can accommodate a maximum number of electrons, and electrons always fill up the shells starting at the innermost one, that is, the one with the lowest energy level. In our example, carbon has filled the first shell with two electrons, and occupied four of the eight available spaces on the second.

The chemical properties of atoms are determined by the number of electrons in the outermost occupied shell. Neon, one of the 'noble' gases, has an atomic number of 10, completely filling the first two shells, and is chemically unreactive or *inert*. Atoms that do not achieve a similar configuration are unstable, or *reactive*. Reactions take place between atoms that attempt to achieve stability by attaining a full outer shell. These reactions may involve atoms of the same element or ones of different elements; the result in either case is a *molecule* or *ion* (see below). Figure 2.2 shows how atoms combine to form a molecule. A substance made up of molecules containing two or more different elements is called a *compound*. In each example, the product of the reaction has a full outer electron shell; note that some atoms are donating electrons, while others are accepting them.

The number of unfilled spaces in the outermost electron shell determines the reactivity of an atom. If most of the spaces in the outermost shell are full, or if most are empty, atoms tend to strive for stability by gaining or losing electrons, as shown in Figure 2.3.

Box 2.1 How heavy is a mole?

When you work in a laboratory, something you'll need to come to grips with sooner or later is the matter of quantifying the amounts and concentrations of substances used. Central to this is the *mole*, so before we go any further, let's define this:

A mole is the molecular mass of a compound expressed in grams.

(The *molecular mass* is simply the sum of the atomic mass of all the atoms in a compound.)

So, to take sodium chloride as an example:

Molecular formula	=	NaCl (one atom each of sodium and chlorine)
Atomic mass of sodium	=	22.99
Atomic mass of chlorine	=	35.45
∴ Molecular mass	=	58.44

Thus *one mole of sodium chloride equals 58.44 grams (58.44 g)*

Concentrations are expressed in terms of mass per volume, so here we introduce the idea of the *molar solution*. This is a solution containing one mole dissolved in a final volume of 1 litre of an appropriate solvent (usually water).

Molar solution = one mole per litre

A one molar (1 M) solution of sodium chloride therefore contains 58.44 g dissolved in water and made up to 1 litre. A 2 M solution would contain 116.88 g in a litre, and so on.

In biological systems, a molar solution of anything is actually rather concentrated, so we tend to deal in solutions which are so many millimolar (mM, one thousandth of a mole per litre) or micromolar (μM, one millionth of a mole per litre).

Why bother with moles?

So far, so good, but why can't we just deal in grams, or grams per litre? Consider the following example. You've been let loose in the laboratory, and been asked to compare the effects of supplementing the growth medium of a bacterial culture with several different amino acids. 'Easy', you think. 'Add X milligrams of each to the normal growth medium, and see which stimulates growth the most'. The problem is that although you may be adding the same *weight* of each amino acid, you're not adding the same number of *molecules*, because each has a different molecular mass. If you add the same number of moles (or millimoles or micromoles) of each instead, you would be comparing the effect of the same number of molecules of each, and thus obtain a much more meaningful comparison. This is because *1 mole of one compound contains the same number of molecules as a mole of any other compound*. This number is called *Avogadro's Number*, and is 6.023×10^{23} molecules per mole.

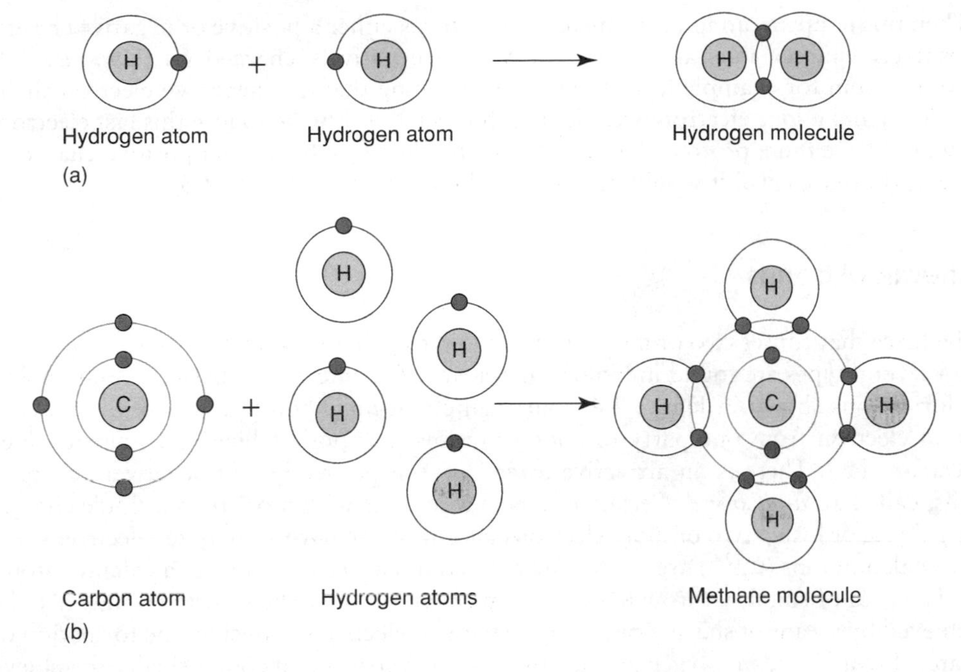

Figure 2.2 The formation of molecules of (a) hydrogen and (b) methane by covalent bonding. Each atom achieves a full set of electrons in its outer shell by sharing with another atom. A shared pair of electrons constitutes a covalent bond

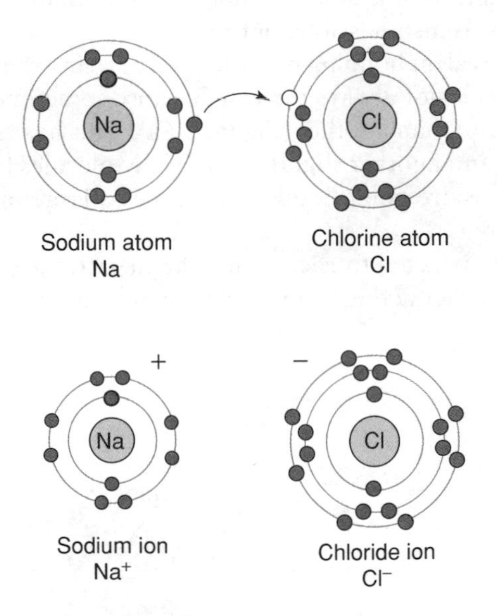

Figure 2.3 Ion formation. Sodium achieves stability by losing the lone electron from its outermost shell. The resulting sodium ion Na^+ has 11 protons and 10 electrons, hence it carries a single positive charge. Chlorine becomes ionised to chloride (Cl^-) when it gains an electron to complete its outer shell

When this happens, an ion is formed, which carries either a positive or negative charge. Positively charged ions are called *cations* and negatively charged ones *anions*. The sodium atom for example has 11 electrons, meaning that the inner two electron shells are filled and a lone electron occupies the third shell. If it were to lose this last electron, it would have more protons than electrons, and therefore have a net positive charge of one; if this happened, it would become a sodium ion, Na^+ (Figure 2.3).

Chemical bonds

The force that causes two or more atoms to join together is known as a *chemical bond*, and several types are found in biological systems. The interaction between sodium and chloride ions shown in Figure 2.4 is an example of *ionic* bonding, where the transfer of an electron from one party to another means that both achieve a complete outer electron shell. There is an attractive force between positively and negatively charged ions, called an *ionic bond*. Certain elements form ions with more than a single charge, by gaining or losing two or more electrons in order to achieve a full outer electron shell; thus calcium ions (Ca^{2+}) are formed by the loss of two electrons from a calcium atom.

The goal of stability through a full complement of outer shell electrons may also be achieved by means of sharing one or more pairs of electrons. Consider the formation of water (Figure 2.2); an oxygen atom, which has two spaces in its outer shell, can achieve a full complement by sharing electrons from two separate hydrogen atoms. This type of bond is a *covalent bond*.

Sometimes, a pair of atoms share not one but two pairs of electrons (see Figure 2.5). This involves the formation of a double bond. Triple bonding, through the sharing of three pairs of electrons, is also possible, but rare.

In the examples of covalent bonding we've looked at so far, the sharing of the electrons has been equal, but this is not always the case because sometimes the electrons may be drawn closer to one atom than another (Figure 2.6a). This has the effect of making one atom slightly negative and another slightly positive. Molecules like this are called *polar* molecules and the bonds are polar bonds. Sometimes a large molecule may have both polar and non-polar areas.

Polar molecules are attracted to each other, the negative areas of one molecule and the positive areas of another acting as magnets for one another (Figure 2.6b). In water,

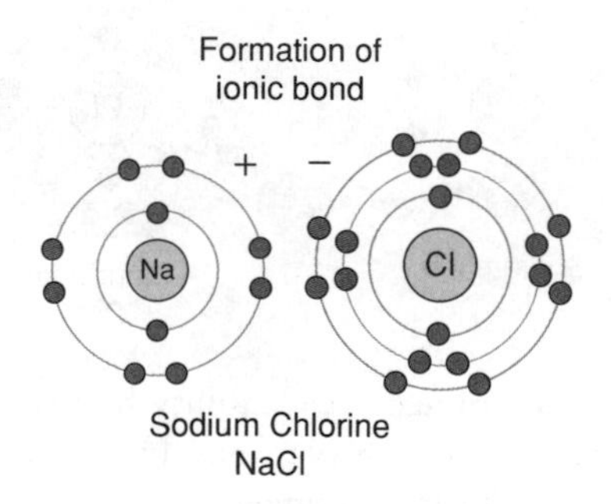

Figure 2.4 A positively charged Na^+ and negatively charged Cl^- attract each other, and an ionic bond is formed. The result is a molecule of sodium chloride

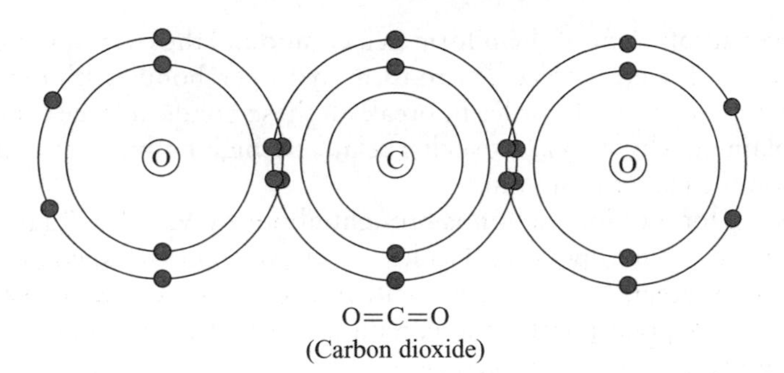

Figure 2.5 Double bond formation. In the formation of carbon dioxide, the carbon atom shares *two* pairs of electrons with each oxygen atom

hydrogen atoms bearing a positive charge are drawn to the negatively charged oxygens. You only have to look at raindrops on a window pane fusing together to see how this bonding is reflected in the physical properties of the compound.

This attraction between polar atoms is called *hydrogen bonding*, and can take place between covalently bonded hydrogen and any electronegative atom, most commonly oxygen or nitrogen. Hydrogen bonds are much weaker than either ionic or covalent

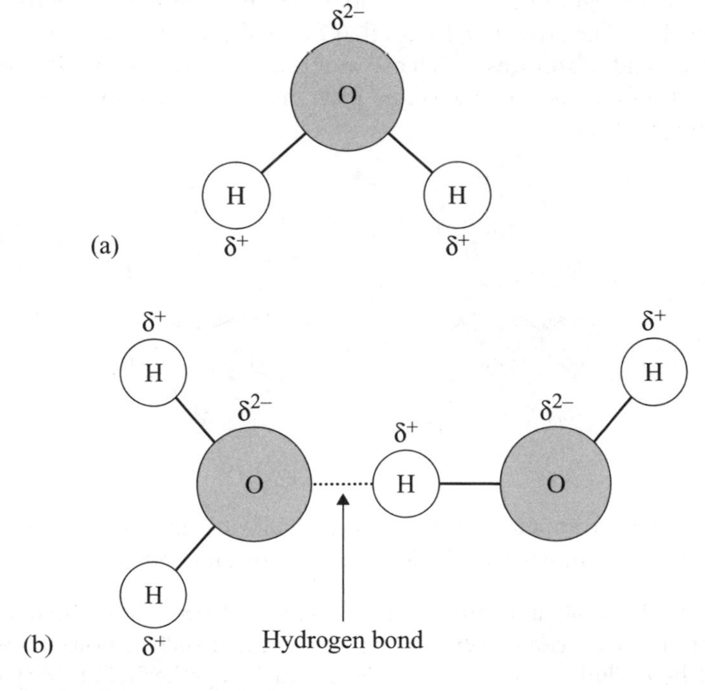

Figure 2.6 (a) The electrons of the hydrogen atoms are strongly attracted to the oxygen atom, causing this part of the water molecule to carry a slightly negative charge, and the hydrogen part a slightly positive one. (b) Because of their polar nature, water molecules are attracted to each other by hydrogen bonding. Hydrogen bonding is much weaker than ionic or covalent bonding, but plays an important role in the structure of macromolecules such as proteins and nucleic acids

bonds; however, if sufficient of them form in a compound the overall bonding force can be appreciable. Each water molecule can form hydrogen bonds with others of its kind in four places (Figure 2.6b). In order to break all these bonds, a large input of energy is required, explaining why water has such a relatively high boiling point, and why most of the water on the planet is in liquid form.

Another weak form of interaction is brought about by Van der Waals forces, which occur briefly when two non-polar molecules (or parts of molecules) come into very close contact with one another. Although transient, and generally even weaker than hydrogen bonds, they occur in great numbers in certain macromolecules and play an important role in holding proteins together (see below).

Water is essential for living things, both in the composition of their cells and in the environment surrounding them. Organisms are made up of between 60 and 95 per cent water by weight, and even inert, dormant forms like spores and seeds have a significant water component. This dependence on water is a function of its unique properties, which in turn derive from its polar nature.

Water is the medium in which most biochemical reactions take place; it is a highly efficient *solvent*, indeed more substances will dissolve in water than in any other solvent. Substances held together by ionic bonds tend to dissociate into anions and cations in water, because as individual solute molecules become surrounded by molecules of water, *hydration shells* are formed, in which the negatively charged parts of the solute attract the positive region of the water molecule, and the positive parts the negative region (Figure 2.7). The attractive forces that allow the solute to dissolve are called *hydrophilic* forces, and substances which are water-soluble are hydrophilic (water-loving). Other polar substances such as sugars and proteins are also soluble in water by forming hydrophilic interactions.

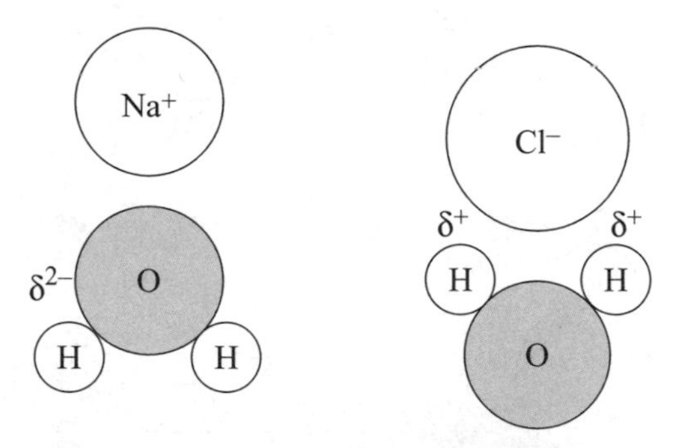

Figure 2.7 An ionic compound such as sodium chloride dissociates in water to its constituent ions. Water molecules form hydration shells around both Na^+ and Cl^- ions

Molecules such as oils and fats are non-polar, and because of their non-reactivity with water are termed *hydrophobic* ('water-fearing'). If such a molecule is mixed with water, it will be excluded, as water molecules 'stick together'. This very exclusion by water can act as a cohesive force among hydrophobic molecules (or hydrophobic areas of large molecules). This is often called hydrophobic bonding, but is not really bonding as such, rather a shared avoidance of water. All living cells have a hydrophilic interior surrounded by a hydrophobic membrane, as we will see in Chapter 3.

An *amphipathic* substance is one which is part polar and part non-polar. When such a substance is mixed with water, *micelles* are formed (Figure 2.8); the non-polar

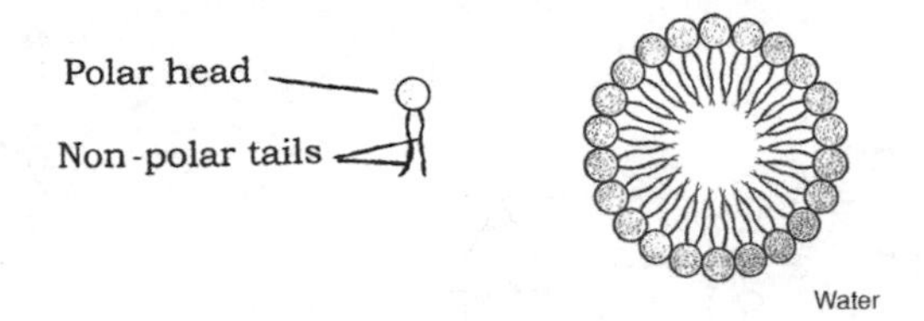

Figure 2.8 In an aqueous environment, amphipathic substances align their molecules so that the non-polar parts are hidden away from the water. From Black, JG: Microbiology: Principles and Explorations, 4th edn, John Wiley & Sons Inc., 1999. Reproduced by permission of the publishers

parts are excluded by the water and group together as described above, leaving the polar groups pointing outwards into the water, where they are attracted by hydrophilic forces. Detergents exert their action by trapping insoluble grease inside the centre of a micelle, while interaction with water allows them to be rinsed away.

Water takes part in many essential metabolic reactions, and its polar nature allows for the breakdown to hydrogen and hydroxyl ions (H^+ and OH^-), and re-synthesis as water. Water acts as a reactant in hydrolysis reactions such as:

$$A\text{—}B + H_2O \rightarrow A\text{—}H + B\text{—}OH$$

and as a product in certain synthetic reactions, such as:

$$A\text{—}H + B\text{—}OH \rightarrow A\text{—}B + H_2O$$

Acids, bases, and pH

Only a minute proportion of water molecules, something like one in every 5×10^8, is present in its dissociated form, but as we have already seen, the H^+ and OH^- ions play an important part in cellular reactions. A solution becomes acid or alkaline if there is an imbalance in the amount of these ions present. If there is an excess of H^+, the solution becomes *acid*, whilst if OH^- predominates, it becomes *alkaline*. The *pH* of a solution is an expression of the molar concentration of hydrogen ions:

$$\text{pH} = -\log_{10}[H^+]$$

In pure water, hydrogen ions are present at a concentration of 10^{-7} M, thus the pH is 7.0. This is called neutrality, where the solution is neither acid or alkaline. At higher concentrations of H^+, such as 10^{-3} M (1 millimolar), the pH value is lower, in this case 3.0, so acid solutions have a value below 7. Conversely, alkaline solutions have a pH above 7. You will see from this example that an increase of 10^4 (10 000)-fold in the $[H^+]$ leads to a change of only four points on the pH scale. This is because it is a logarithmic scale; thus a solution of pH 10 is 10 times more alkaline than one of pH 9, and 100 times more than one of pH 8. Figure 2.9 shows the pH value of a number of familiar substances.

Most microorganisms live in an aqueous environment, and the pH of this is very important. Most will only tolerate a small range of pH, and the majority occupy a range around neutrality, although as we shall see later on in this book, there are some

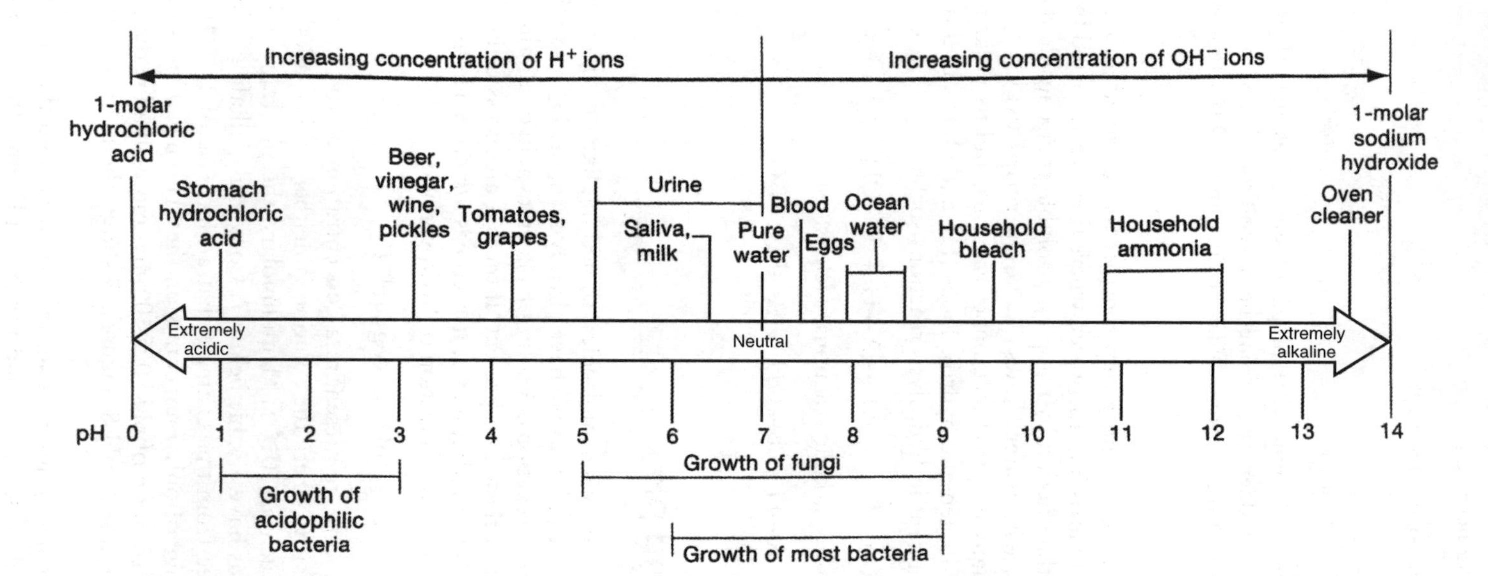

Figure 2.9 The pH value of some common substances. Most microorganisms exist at pH values around neutrality, but representatives are found at extremes of both acidity and alkality. From Black, JG: Microbiology: Principles and Explorations, 4th edn, John Wiley & Sons Inc., 1999. Reproduced by permission of the publishers

Table 2.4 Occurrence and characteristics of some functional groups

Functional Group	Formula	Type of molecule	Found in:	Remarks
Hydroxyl	-OH	Alcohols	Sugars	Polar group, making organic molecules more water soluble
Carbonyl	$-C(=O)H$	Aldehydes	Sugars	Carbonyl at end of chain
	$>C=O$	Ketones	Sugars	Carbonyl elsewhere in chain
Carboxyl	-COOH	Carboxylic acids	Sugars, fats, amino acids	
Amino	$-NH_2$	Amines	Amino acids, proteins	Can gain H^+ to become NH_3^+
Sulphhydryl	-SH	Thiols	Amino acids, proteins	Oxidises to give S=S bonds
Phosphate	$-O-P(=O)(O^-)-O^-$		Phospholipids nucleic acids	Involved in energy transfer

startling exceptions to this. Most of the important molecules involved in the chemistry of living cells are organic, that is, they are based on a skeleton of covalently linked carbon atoms. Biological molecules have one or more *functional groups* attached to this skeleton; these are groupings of atoms with distinctive reactive properties, and are responsible for many of the chemical properties of the organic molecule. The possession of a functional group(s) frequently makes an organic molecule more polar and therefore more soluble in water.

Some of the most common functional groups are shown in Table 2.4. It can be seen that the functional groups occur in simpler organic molecules as well as in the macromolecules we consider below.

Biomacromolecules

Many of the most important molecules in biological systems are *polymers*, that is, large molecules made up of smaller subunits joined together by covalent bonds, and in some cases in a specific order.

Carbohydrates

The suffix -ose always denotes a carbohydrate.

Carbohydrates are made up of just three different elements, carbon, hydrogen and oxygen. The simplest carbohydrates are *monosaccharides*, or simple sugars; these

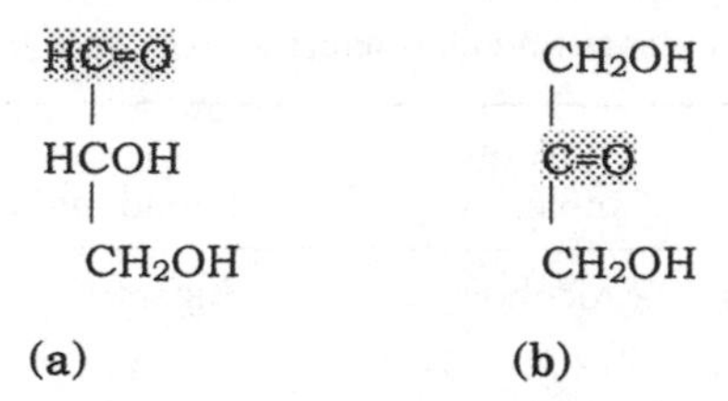

Figure 2.10 Monosaccharides may be aldoses or ketoses. The three carbon sugars (a) glyceraldehyde and (b) dihydroxyacetone share the same molecular formula, but have different functional groups. The two molecules are isomers (see Box 2.2)

have the general formula $(CH_2O)_n$. They are classed as either aldoses or ketoses, according to whether they contain an aldehyde group or a ketone group (Figure 2.10). Monosaccharides can further be classified on the basis of the number of carbon atoms they contain. The simplest are trioses (three carbons) and the most important biologically are hexoses (six carbons) (see Boxes 2.2 and 2.3).

Monosaccharides are generally crystalline solids which are soluble in water and have a sweet taste. They all reducing sugars, so called because they are able to reduce alkaline solutions of cupric ions (Cu^{2+}) to cuprous ions (Cu^{+}).

A *disaccharide* is formed when two monosaccharides (which may be of the same type or different), join together with a concomitant loss of a water molecule (Figure 2.11). Further monosaccharides can be added, giving chains of three, four, five

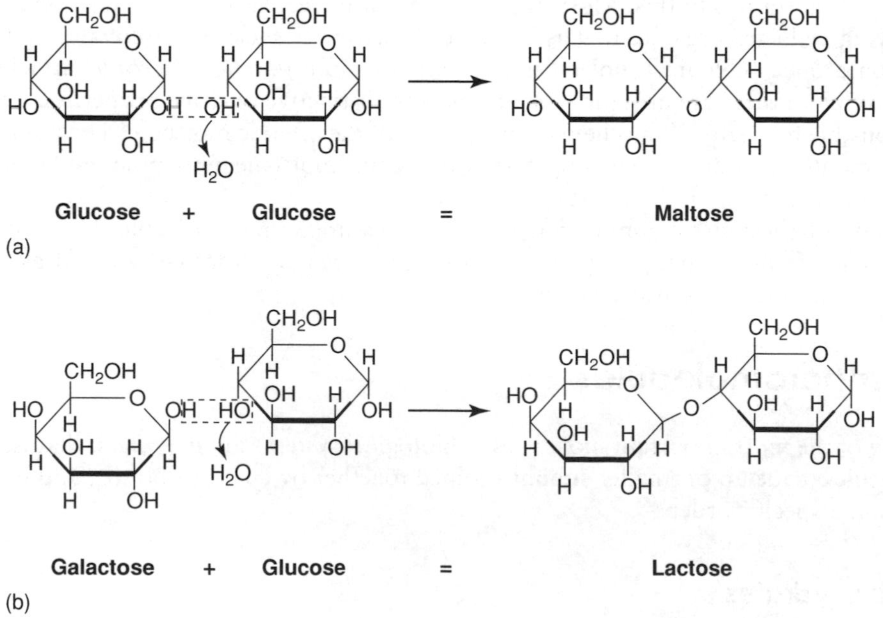

(a) Glucose + Glucose = Maltose

(b) Galactose + Glucose = Lactose

Figure 2.11 Monosaccharides such as two glucose molecules can be joined by a glycosidic linkage to form a disaccharide. The reaction is a condensation reaction, in which a molecule of water is lost. α- (a) and β-linkages (b) result in different orientations in space

Box 2.2 Isomers: same formula, different structure

The simplest monosaccharides are the trioses glyceraldehyde and dihydroxyacetone (Figure 2.10). Look carefully at the structures, and you will see that although they both share the same number of carbons (3), hydrogens(6) and oxygens (3), the way in which these atoms are arranged is different in the two sugars. Molecules such as these, which have the same chemical formula but different structural formulas, are said to be structural isomers. The different groupings of atoms lead to structural isomers having different chemical properties. When we come to look at the hexoses (six carbon sugars), we see that there are many structural possibilities for the general formula $C_6H_{12}O_6$; some of these are shown below.

CHO	CHO	CHO	CHO	CH_2OH
HO–C–H	H–C–OH	H–C–OH	HO–C–H	C=O
H–C–OH	HO–C–H	HO–C–H	HO–C–H	HO–C–H
HO–C–H	H–C–OH	HO–C–H	H–C–OH	H C–OH
HO–C–H	H–C–OH	H–C–OH	H–C–OH	H–C–OH
CH_2OH	CH_2OH	CH_2OH	CH_2OH	CH_2OH
D-glucose	L-glucose	D-galactose	D-mannose	D–fructose

Note that some of these structures are identical apart from the orientation of groups around the central axis; D-glucose and L-glucose for example differ only in the way H atoms and -OH groups are arranged to the right or left. They are said to be *stereoisomers* or optical isomers, and are mirror images of each other, just like your right and left hands. (D- and L- are short for *dextro*- and *laevo*-rotatory, meaning that the plane of polarised light is turned to the right and left respectively when passed through a solution of these substances). Generally, living cells will only synthesise one or other stereoisomer, and not both.

or more units. These are termed *oligosaccharides* (*oligo*, a few), and chains with many units are *polysaccharides*. The chemical bond joining the monosaccharide units together is called a *glycosidic linkage*. The bond between the two glucose molecules that make up maltose is called an α-glycosidic linkage; in lactose, formed from one glucose and one galactose, we have a β-glycosidic linkage. The two bonds are formed in the same way, with the elimination of water, but they have a different orientation in space. Thus disaccharides bound together by α- and β-glycosidic linkages have a different overall shape and as a result the molecules behave differently in cellular metabolism.

Biologically important molecules such as starch, cellulose and glycogen are all polysaccharides. Another is dextran, a sticky substance produced by some bacteria to aid their adhesion. They differ from monosaccharides in being generally insoluble in water, not tasting sweet and not being able to reduce cupric ions. Most polysaccharides

Box 2.3 Sugars are more accurately shown as ring structures

When dissolved in water, the aldehyde or ketone group reacts with a hydroxyl group on the fifth carbon to give a cyclic form. D-Glucose is shown in both forms below. The cyclic form of the molecule is shown below as a *Haworth projection*. The idea is that the ring is orientated at 90° to the page, with the edge which is shown thicker towards you, and the top edge away from you. Notice that there are even two forms of D-Glucose! Depending on whether the -OH on carbon-1 is below or above the plane of the ring, we have α- or β-D-Glucose.

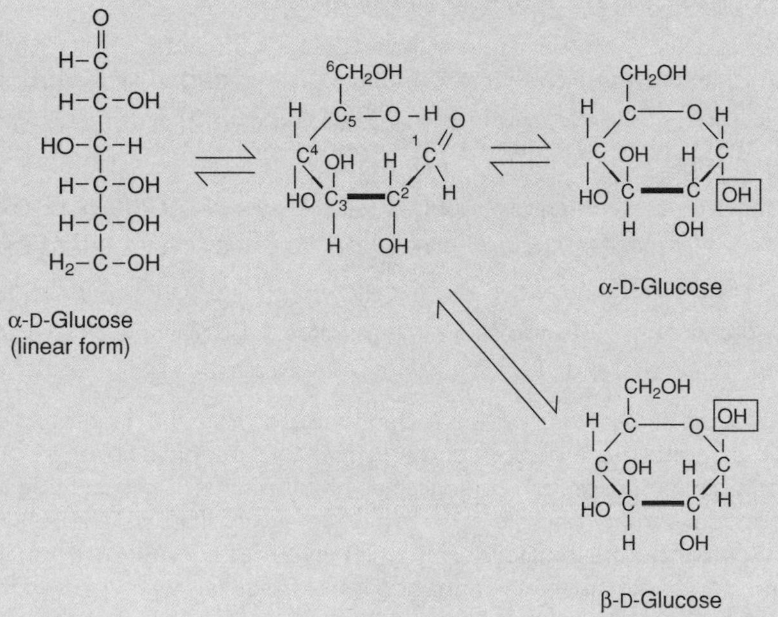

α-D-Glucose
(linear form)

α-D-Glucose

β-D-Glucose

are made up from either pentose or hexose sugars, and, like di- and oligosaccharides, can be broken down into their constituent subunits by hydrolysis reactions.

Proteins

Of the macromolecules commonly found in living systems, proteins are the most versatile, having a wide range of biological functions and this fact is reflected in their structural diversity.

The five elements found in most naturally occurring proteins are carbon, hydrogen, oxygen, nitrogen and sulphur. In addition, other elements may be essential components of certain specialised proteins such as haemoglobin (iron) and casein (phosphorus).

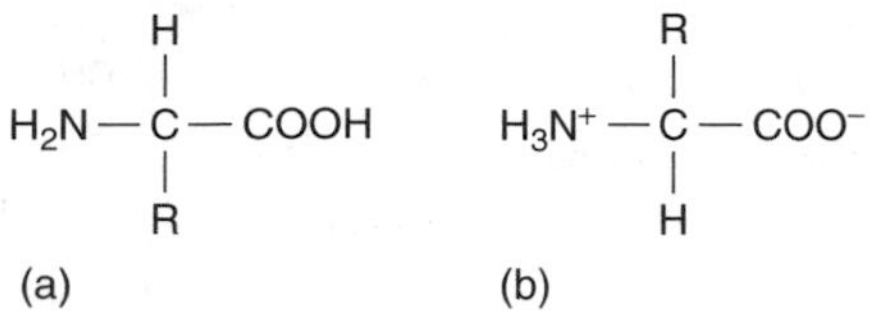

(a) $H_2N-C(H)(R)-COOH$ (b) $H_3N^+-C(R)(H)-COO^-$

Figure 2.12 Amino acid structure. (a) The basic structure of an amino acid. (b) In solution, the amino and carboxyl groups become ionised, giving rise to a *zwitterion* (a molecule with spatially separated positive and negative charges). All the 20 amino acids commonly found in proteins are based on a common structure, differing only in the nature of their 'R' group (see Figure 2.13)

Proteins can be very large molecules, with molecular weights of tens or hundreds of thousands. Whatever their size, and in spite of the diversity referred to above, all proteins are made up of a collection of 'building bricks' called *amino acids* joined together. Amino acids are thought to have been among the first organic molecules formed in the early history of the Earth, and many different types exist in nature. All these, including the 20 commonly found occurring in proteins, are based on a common structure, shown in Figure 2.12. It comprises a central carbon atom (known as the α-carbon) covalently bonded to an amino (NH_2) group, a carboxyl (COOH) group and a hydrogen atom. It is the group attached to the final valency bond of the α-carbon which varies from one amino acid to another; this is known as the *'R'-group*.

The 20 amino acids found in proteins can be conveniently divided into five groups, on the basis of the chemical nature of their 'R'-group. These range from a single hydrogen atom to a variety of quite complex side chains (Figure 2.13). It is unlikely nowadays that you would need to memorise the precise structure of all 20, as the author was asked to do in days gone by, but it would be advisable to familiarise yourself with the groupings and examples from each of them. The groups differentiate on the basis of a polar/non-polar nature and on the presence or absence of an ionisable 'R'-group. Box 2.4 shows how we normally refer to proteins in shorthand.

Note that one amino acid, proline, falls outside the main groups. This differs from the others in that it has one of its N—H linkages replaced by an N—C, which forms part of a cyclic structure (Figure 2.13). This puts certain conformational constraints upon proteins containing proline residues.

As can be seen from Figure 2.13, the simplest amino acid is glycine, whose R-group is simply a hydrogen atom. This means that the glycine molecule is symmetrical, with a hydrogen atom on opposite valency bonds. All the other amino acids however, are asymmetrical. The α-carbon acts as what is known as a chiral centre, giving the molecule right or left 'handedness'. Thus two stereoisomers known as the D- and L-forms are possible for each of the amino acids except glycine. All the amino acids found in naturally occurring proteins have the L-form; the D-form also occurs in nature but only in certain specific, non protein contexts.

Proteins, as we've seen, are polymers of amino acids. Amino acids are joined together by means of a *peptide bond*. This involves the -NH_2 group of one amino acid and the -COOH group of another. The formation of a peptide bind is a form of condensation reaction in which water is lost (Figure 2.14). The resulting structure of two linked amino

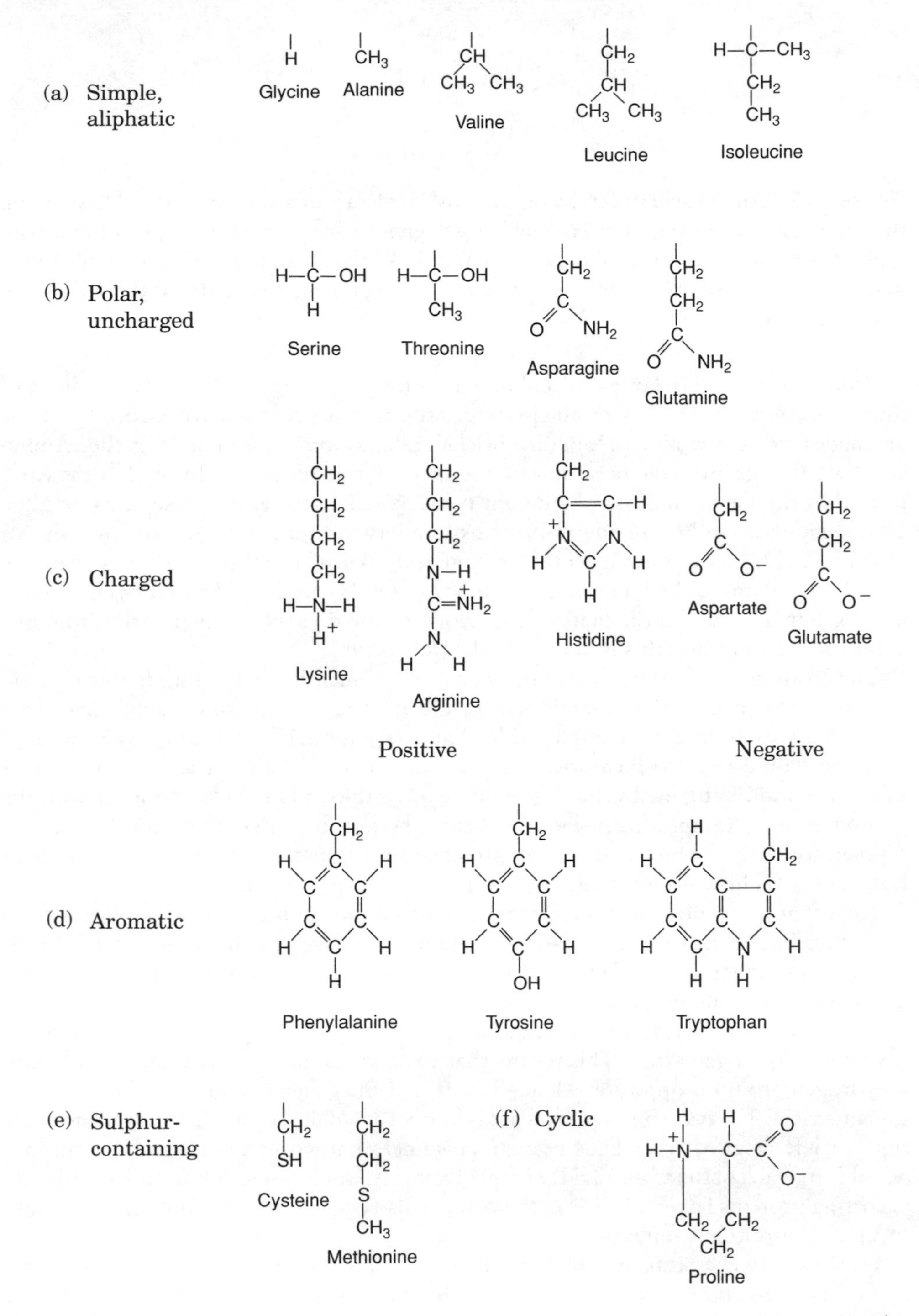

Figure 2.13 The 20 amino acids found in proteins. The 'R' group of each amino acid is shown. These range from the simplest, glycine, to more complex representatives such as tryptophan

Box 2.4 Amino acid shorthand

It is sometimes necessary to express in print the sequence of amino acids which make up the primary structure of a particular protein; clearly it would be desperately tedious to express a sequence of hundreds of bases in the form 'glycine, phenylalanine, tryptophan, methionine... etc', so a system of abbreviations for each amino acid has been agreed. Each amino acid can be reduced to a three letter code, thus you might see something like:

1	2	3	4	5	6	7	8	9	10	11
Gly	Phe	Try	Met	His	Lys	Gly	Ala	His	Val	Glu....and so on.

Note that each residue has a number; this *numbering always begins at the N-terminus*.

Each amino acid can also be represented by a single letter. The abbreviations using the two systems are shown below.

A	Ala	Alanine	M	Met	Methionine
B	Asx	Asparagine/aspartic acid	N	Asn	Asparagine
C	Cys	Cysteine	P	Pro	Proline
D	Asp	Aspartic acid	Q	Gln	Glutamine
E	Glu	Glutamic acid	R	Arg	Arginine
F	Phe	Phenylalanine	S	Ser	Serine
G	Gly	Glycine	T	Thr	Threonine
H	His	Histidine	V	Val	Valine
I	Ile	Isoleucine	W	Trp	Tryptophan
K	Lys	Lysine	Y	Tyr	Tyrosine
L	Leu	Leucine	Z	Glx	Glutamine/glutamic acid

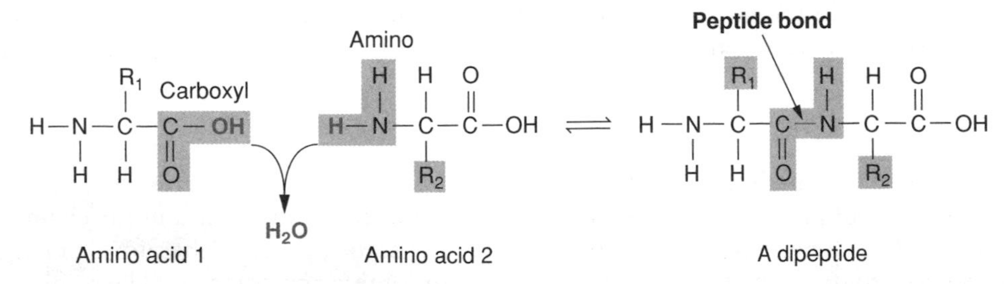

Figure 2.14 The carboxyl group of one amino acid is joined to the amino group of another. This is another example of a condensation reaction (c.f. Figure 2.11). No matter how many amino acids are added, the resulting structure always has a free carboxyl group at one end and a free amino group at the other

acids is called a *dipeptide*; note that this structure still retains an -NH_2 at one end and a -COOH at the other. If we were to add on another amino acid to form a tripeptide, this would still be so, and if we kept on adding them until we had a *polypeptide*, we would still have the same two groupings at the extremities of the molecule. These are referred to as the N-terminus and the C-terminus of the polypeptide. Since a water molecule has been removed at the formation of each peptide bond, we refer to the chain so formed as being composed of amino acid residues, rather than amino acids. The actual distinction between a protein and a polypeptide based on the number of amino acid residues is not clear-cut; generally, with over 100, we refer to proteins, but some naturally occurring proteins are a lot smaller than this.

So far, we can think of proteins as long chains of many amino acid residues, rather like a string of beads. This is called the *primary structure* of the protein; it is determined by the relative proportions of each of the 20 amino acids, and the order in which they are joined together. It is the basis of all the remaining levels of structural complexity, and it ultimately determines the properties of a particular protein. It is also what makes one protein different from another. Since the 20 types of amino acid can be linked together in any order, the number of possible sequences is astronomical, and it is this great variety of structural possibilities that gives proteins such diverse structures and functions.

> In theory, there are 20^{100} or some 10^{130} different ways in which 20 different amino acids could combine to give a protein 100 amino acid residues in length!

Some parts of the primary sequence are more important than others. If we took a protein of, say, 200 amino acid residues in length, took it apart and reassembled the amino acids in a different order, we would almost certainly alter (and probably lose completely) the properties of that protein. If we look at the primary sequence of a protein molecule which serves essentially the same function in several species, we find that nature has allowed slight alterations to occur during evolution, but these are often conservative substitutions, where an amino acid has been replaced by a similar one (one from the same group in Figure 2.13), and thus have little effect on the protein's properties. In certain parts of the primary sequence, such substitutions are less well tolerated, for example the few residues that make up the active site of an enzyme (see Chapter 6). In cases such as the one above, alterations have not been allowed at these points in the primary sequence, and the sequence is the same, or almost so, in all species possessing that protein. The sequence in question is said to have been conserved.

Higher levels of protein structure

The structure of proteins is a good deal more complicated than a just a linear chain of amino acids. A long thin chain is unlikely to be very stable; proteins therefore undergo a process of folding which makes the molecule more stable and compact. The results of this folding are the secondary and tertiary structures of a protein.

The *secondary structure* is due to hydrogen bonding between a carbonyl (-CO) group and an amido (-NH) group of amino acid residues on the peptide backbone (Figure 2.15). The 'R' group plays no part in secondary protein structure. Two regular patterns of folding result from this; the *α-helix* and the *β-pleated sheet*.

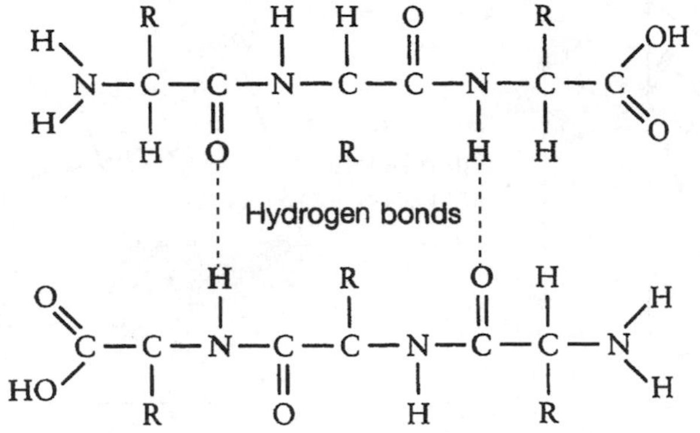

Hydrogen bonds

Figure 2.15 Secondary structure in proteins. Hydrogen bonding occurs between the -CO and -NH groups of amino acids on the backbone of a polypeptide chain. The two amino acids may be on the same or different chains

> Very small distances within molecules are measured in Angstrom units (A). One Angstrom unit is equal to one tenbillionth (10^{-10}) of a metre.

The α-helix occurs when hydrogen bonding takes place between amino acids close together in the primary structure. A stable helix is formed by the -NH group of an amino acid bonding to the -CO group of the amino acid four residues further along the chain (Figure 2.16a). This causes the chain to twist into the characteristic helical shape. One turn of the helix occurs every 3.6 amino acid residues, and results in a rise of 5.4 Å (0.54 nm); this is called the pitch height of the helix. The ability to form a helix like this is dependent on the component amino acids; if there are too many with large R-groups, or R-groups carrying the same charge, a stable helix will not be formed. Because of its rigid structure, proline (Figure 2.13) cannot be accommodated in an α-helix. Naturally occurring α-helices are always right-handed, that is, the chain of amino acids coils round the central axis in a clockwise direction. This is a much more stable configuration than a left-handed helix, due to the fact that there is less steric hindrance (overlapping of electron clouds) between the R-groups and the C=O group on the peptide backbone. Note that if proteins were made up of the D-form of amino acids, we would have the reverse situation, with a left-handed form favoured. In the β-pleated sheet, the hydrogen bonding occurs between amino acids either on separate polypeptide chains or on residues far apart in the primary structure (Figure 2.16b). The chains in a β-pleated sheet are fully extended, with 3.5 Å (0.35 nm) between adjacent amino acid residues (c.f. α-helix, 1.5 Å). When two or more of these chains lie next to each other, extensive hydrogen bonding occurs between the chains. Adjacent strands in a β-pleated sheet can either run in the same direction (e.g. N→C), giving rise to a parallel β-pleated sheet, or in opposite directions (antiparallel β-pleated sheet, as shown in Figure 2.16b).

A common structural element in the secondary structure of proteins is the β-turn. This occurs when a chain doubles back on itself, such as in an antiparallel β-pleated sheet. The -CO group of one amino acid is hydrogen bonded to the -NH group of the

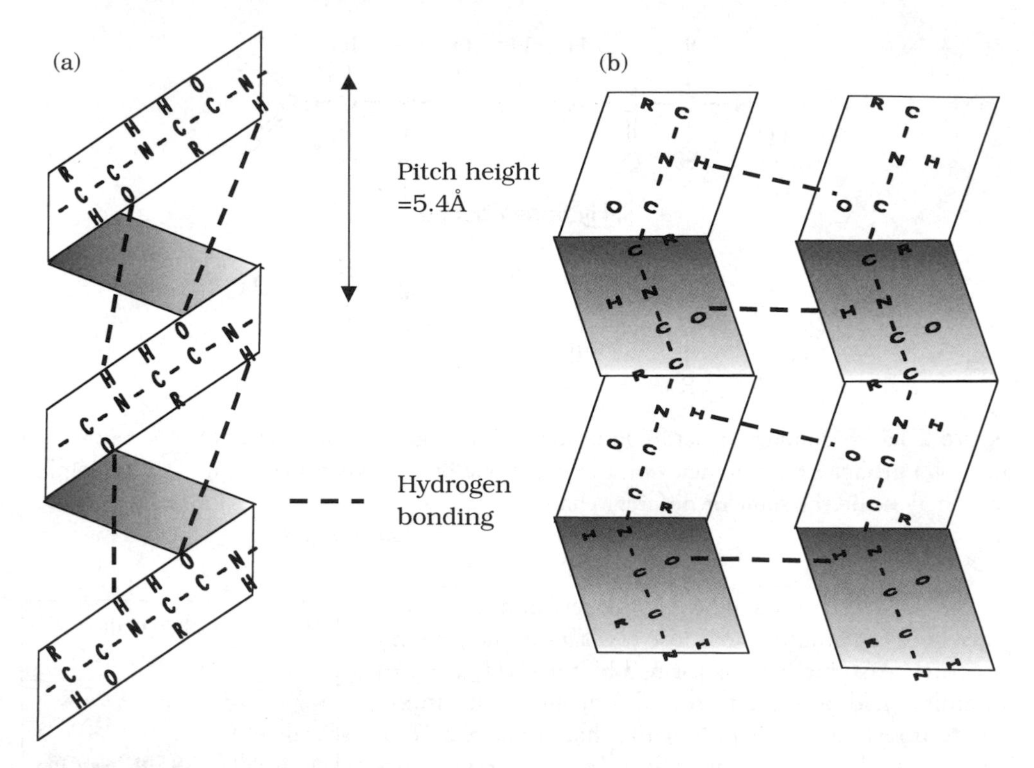

Figure 2.16 Secondary structure in proteins: the α-helix and β-pleated sheet. (a) Hydrogen bonding between amino acids four residues apart in the primary sequence results in the formation of an α-helix. (b) In the β-pleated sheet hydrogen bonding joins adjacent chains. Note how each chain is more fully extended than in the α-helix. In the example shown, the chains run in the same direction (parallel)

residue three further along the chain. Frequently, it is called a hairpin turn, for obvious reasons (Figure 2.17). Numerous changes in direction of the polypeptide chains result in a compact, globular shape to the molecule.

Typically about 50 per cent of a protein's secondary structure will have an irregular form. Although this is often referred to as *random coiling*, it is only random in the sense that there is no regular pattern; it still contributes towards the stability of the molecule. The proportions and combinations in which α-helix, β-pleated sheet and random coiling occur varies from one protein to another. Keratin, a structural protein found in skin, horn and feathers, is an example of a protein entirely made up of α-helix, whilst the lectin (sugar-binding protein) concanavalin A is mostly made up of β-pleated sheets.

The *tertiary structure* of a protein is due to interactions between side chains, that is, R-groups of amino acid residues, resulting in the folding of the molecule to produce a thermodynamically more favourable structure. The structure is formed by a variety of weak, non-covalent forces; these include hydrogen bonding, ionic bonds, hydrophobic interactions, and Van der Waals forces. The strength of these forces diminishes with distance, therefore the formation of a compact structure is encouraged. In addition, the -SH groups on separate cysteine residues can form a covalent -S—S- linkage. This is

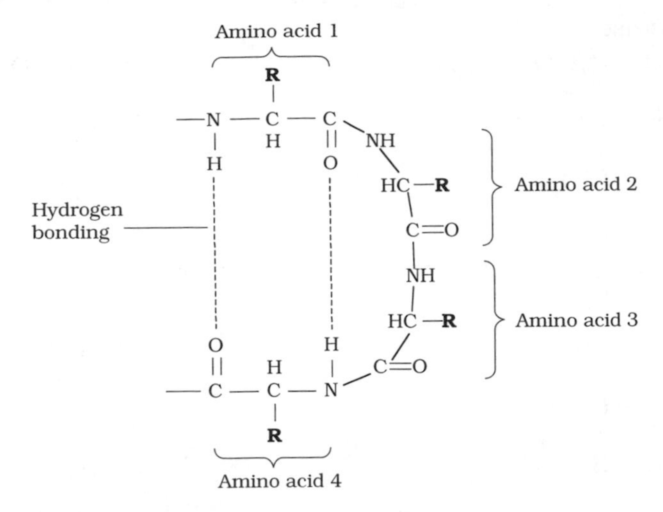

Figure 2.17 The β-turn. The compact folding of many globular proteins is achieved by the polypeptide chain reversing its direction in one or more places. A common way of doing this is with the β-turn. Hydrogen bonding between amino acid residues on the same polypeptide stabilizes the structure

known as a *disulphide bridge* and may have the effect of bringing together two cysteine residues that were far apart in the primary sequence (Figure 2.18).

In globular proteins, the R-groups are distributed according to their polarities; non-polar residues such as valine and leucine nearly always occur on the inside, away from the aqueous phase, while charged, polar residues including glutamic acid and histidine generally occur at the surface, in contact with the water.

The protein can be *denatured* by heating or treatment with certain chemicals; this causes the tertiary structure to break down and the molecule to unfold, resulting in a loss of the protein's biological properties. Cooling, or removal of the chemical agents, will lead to a restoration of both the tertiary structure and biological activity, showing that both are entirely dependent on the primary sequence of amino acids.

> Complex molecules such as globular proteins become denatured when their three-dimensional structure is disrupted, leading to a loss of biological function.

Even the tertiary structure is not always the last level of organisation of a protein, because some are made up of two or more polypeptide chains, each with its own secondary and tertiary structure, combined together to give the *quaternary structure* (Figure 2.19). These chains may be identical or different, depending on the protein. Like the tertiary structure, non-covalent forces between R-groups are responsible, the difference being that this time they link amino acid residues on separate chains rather than on the same one.

Such proteins lose their functional properties if dissociated into their constituent units; the quaternary joining is essential for their activity. Phosphorylase A, an enzyme involved in carbohydrate metabolism, is an example of a protein with a quaternary structure. It has four subunits, which have no catalytic activity unless joined together as a tetramer.

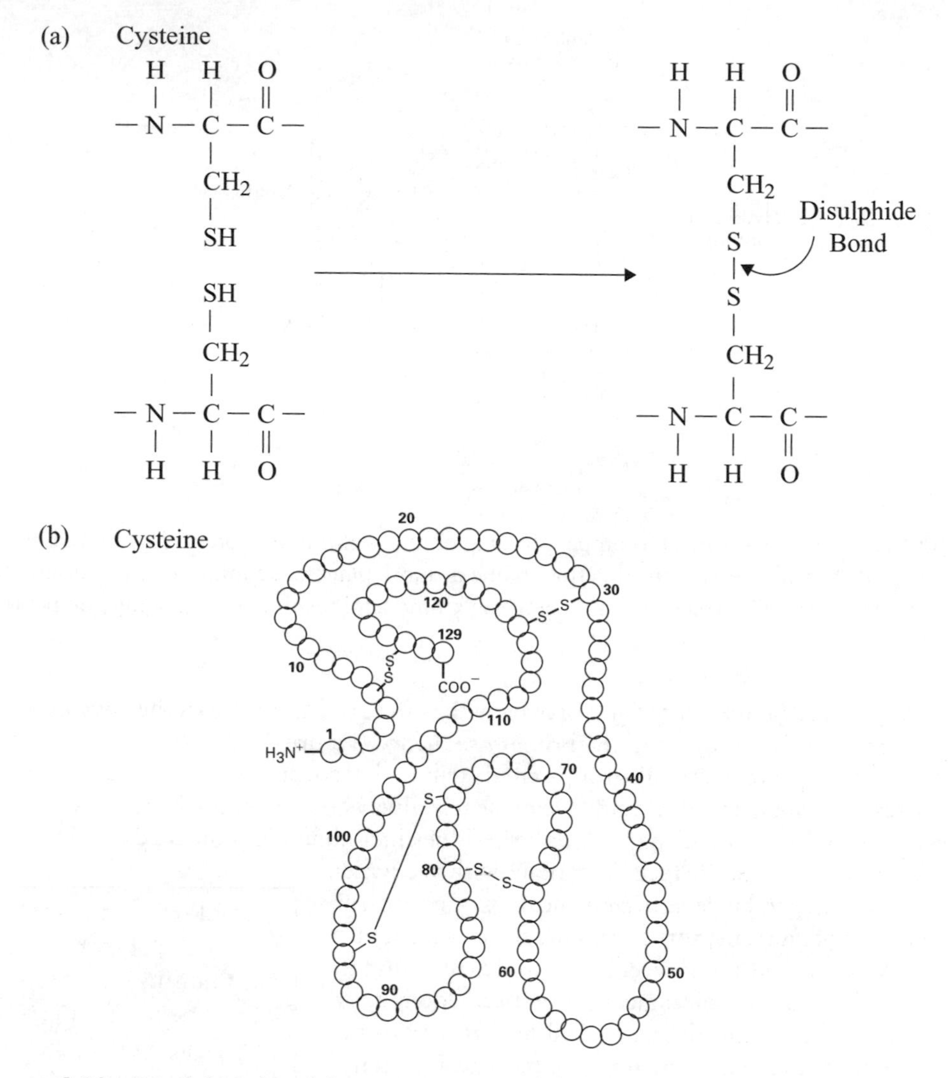

Figure 2.18 Disulphide bond formation. (a) Disulphide bonds formed by the oxidation of cysteine residues result in cross-linking of a polypeptide chain. (b) This can have the effect of bringing together residues that lie far apart in the primary amino acid sequence. Disulphide bonds are often found in proteins that are exported from the cell, but rarely in intracellular proteins

Although all proteins are polymers of amino acids existing in various levels of structural complexity as we have seen above, some have additional, non-amino acid components. They may be organic, such as sugars (glycoproteins) or lipids (lipoproteins) or inorganic, including metals (metalloproteins) or phosphate groups (phosphoproteins). These components, which form an integral part of the protein's structure, are called *prosthetic groups.*

> A prosthetic group is a non-polypeptide component of a protein, such as a metal ion or a carbohydrate

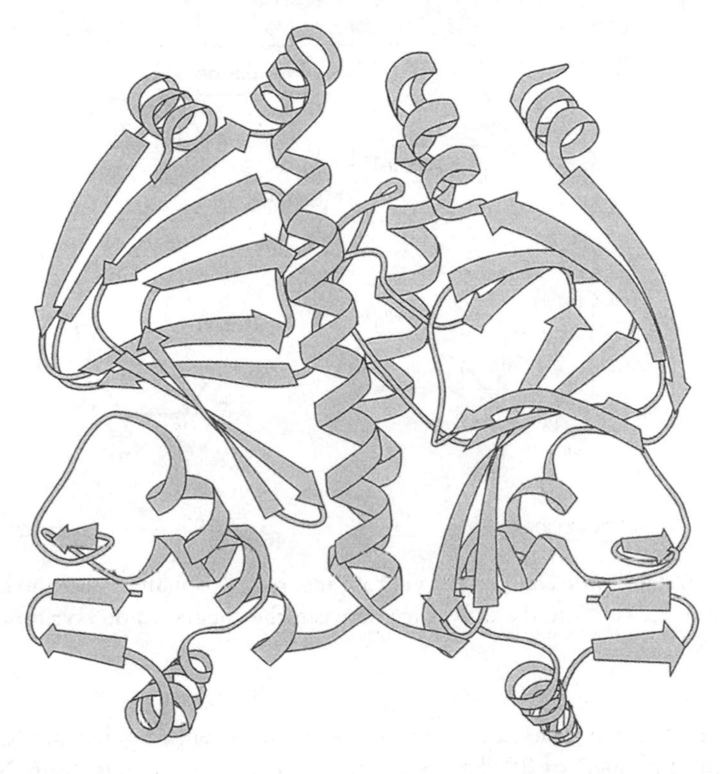

Figure 2.19 Polypeptide chains may join to form quaternary structure. The example shown comprises two identical polypeptide subunits. Coils indicate α-helical sequences, arrows are β-pleated sheets. From Bolsover, SR , Hyams, JS, Jones, S, Shepherd, EA & White, HA: From Genes to Cells, John Wiley & Sons, 1997. Reproduced by permission of the publishers

Nucleic acids

The third class of polymeric macromolecules are the nucleic acids. These are deoxyribonucleic acid (DNA) and ribonucleic acid (RNA), and both are polymers of smaller molecules called nucleotides. As we shall see, there are important differences both in the overall structures of RNA and DNA and in the nucleotides they contain, so we shall consider each of them in turn.

The structure of DNA

The composition of a DNA nucleotide is shown in Figure 2.20(a). It has three parts, a five-carbon sugar called deoxyribose, a phosphate group and a base. This base can be any one of four molecules; as can be seen in Figure 2.21, these are all based on a cyclic structure containing nitrogen. Two of the bases, cytosine and thymine, have a single ring and are called pyrimidines. The other two, guanine and adenine, have a double ring structure; these are the purines. The four bases are often referred to by their initial letter only, thus we have A, C, G and T.

One nucleotide differs from another by the identity of the base it contains; the rest of the molecule (sugar and phosphate) is identical. You will recall from the previous section that the properties of a protein depend on the order in which its constituent amino acids

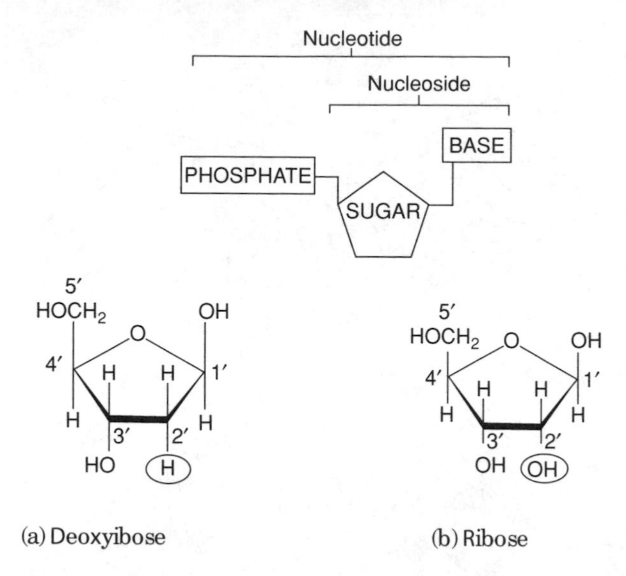

Figure 2.20 A nucleotide comprises a pentose sugar, a phosphate group and a nitrogenous base (see Figure 2.21). Note the difference between the sugars (a) deoxyribose (DNA) and (b) ribose (RNA)

are linked together; we have exactly the same situation with nucleic acids, except that instead of an 'alphabet' of 20 'letters', here we have one of only four. Nevertheless, because nucleic acid molecules are extremely long, and the bases can occur in almost any order, an astronomically large number of different sequences is possible.

The nucleotides join together by means of a *phosphodiester bond*. This links the phosphate group of one base to an -OH group on the 3-carbon of the deoxyribose sugar of another (Figure 2.22). The chain of nucleotides therefore has a free -OH group attached to the 3-carbon (the 3′ end) and a free phosphate group attached to the 5-carbon (the 5′ end). This remains the case however long the chain becomes.

The structure of DNA however is not just a single chain of linked nucleotides, but *two* chains wound around each other to give the *double helix* form made famous by the model of James Watson and Francis Crick in 1953 (Figure 2.23, see also Chapter 11). If we compare this to an open spiral staircase, alternate sugar and phosphate groups make up the 'skeleton' of the staircase, while the inward-facing bases pair up by hydrogen bonding to form the steps. Notice that each nucleotide pair always comprises three rings, resulting from a combination of one purine and one pyrimidine base. This means that the two strands of the helix are always evenly spaced. The way in which the bases pair is further governed by the phenomenon of *complementary base pairing*. A nucleotide containing thymine will only pair with one containing adenine, and likewise guanine always pairs with cytosine (Figure 2.24). Thus, the sequence of

> Erwin Chargaff measured the proportions of the different nucleotides in a range of DNA samples. He found that T always = A and C always = G. Watson and Crick interpreted this as meaning that the bases always paired up in this way.

(a)

PURINES

Adenine

Guanine

PYRIMIDINES

Thymine

Cytosine

(b)

Uracil

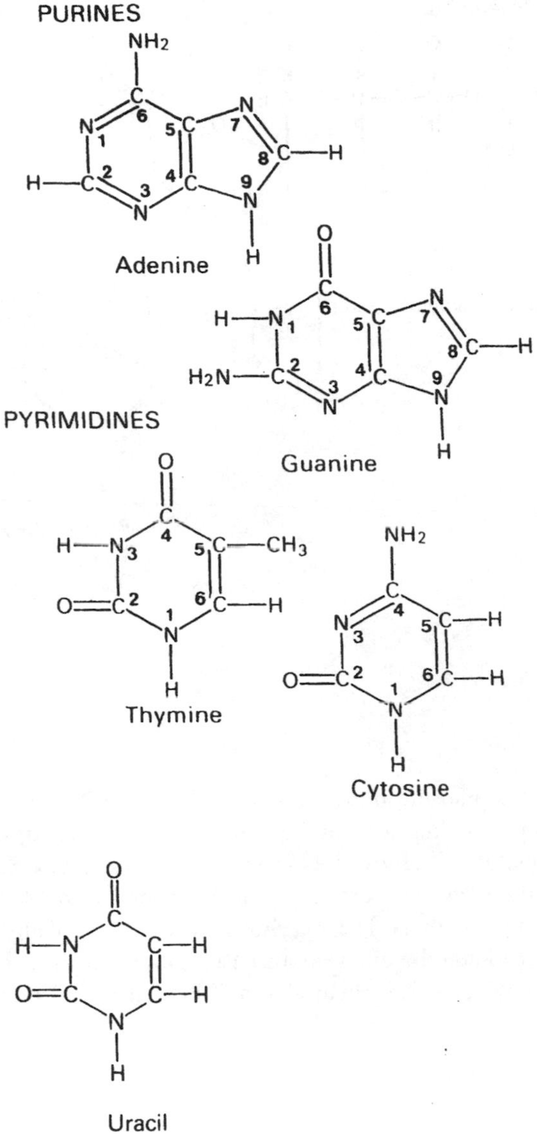

Figure 2.21 Bases belong to two classes. Nucleotides differ from each other in the identity of the nitrogenous base. (a) In DNA these are adenine (A), cytosine (C), guanine (G) or thymine (T). The purines (A and G) have a two-ring structure, while the pyrimidines (C and T) have only one ring. (b) In RNA, thymine is replaced by a similar molecule, uracil (U)

nucleotides on one strand of the double helix determines that of the other, as it has a complementary structure. Figure 2.23 shows how the two strands of the double helix are *antiparallel*, that is they run in opposite directions, one $5' \rightarrow 3'$ and the other $3' \rightarrow 5'$. In Chapter 12 we shall look at how this structure was used to propose a mechanism for the way in which DNA replicates and genetic material is copied.

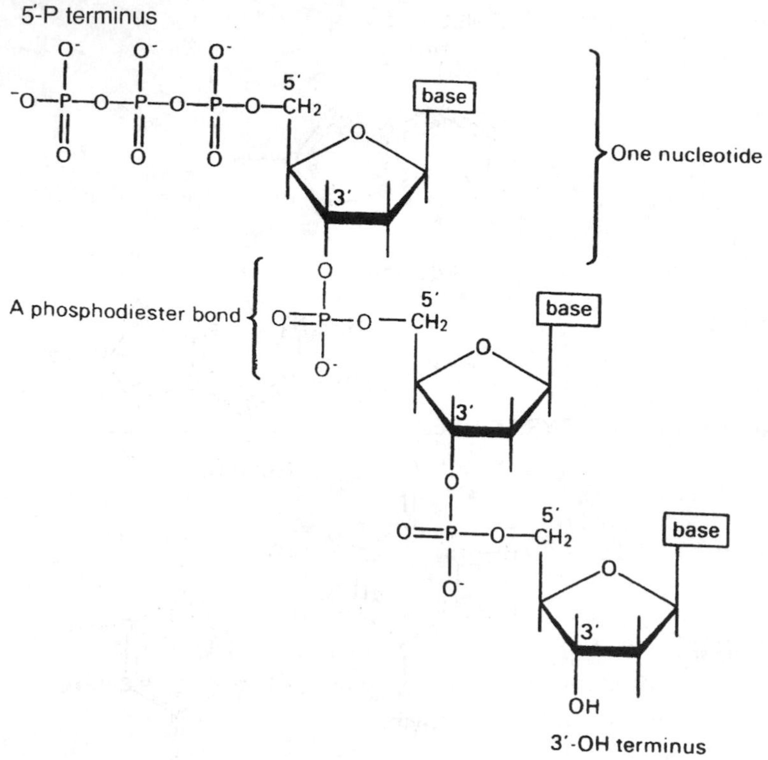

5'-P terminus

One nucleotide

A phosphodiester bond

3'-OH terminus

Figure 2.22 The phosphodiester bond. A chain of DNA is made longer by the addition of nucleotides containing not one but three phosphate groups; on joining the chain, two of these phosphates are removed. Nucleotides are joined to each other by a phosphodiester bond, linking the phosphate group on the 5-carbon of one deoxyribose to the -OH group on the 3-carbon of another. (These carbons are known as 5′ and 3′ to distinguish them from the 5- and 3-carbon on the nitrogenous base). Note that the resulting chain, however many nucleotides it may comprise, always has a 5′(PO_4) group at one end and a 3′(OH) group at the other

The structure of RNA

In view of the similarities in the structure of DNA and RNA, we shall confine ourselves here to a consideration of the major differences. There are two important differences in the composition of nucleotides of RNA and DNA. The central sugar molecule is not deoxyribose, but ribose; as shown in Figure 2.20, these differ only in the possession of an -H atom or an -OH group attached to carbon-2. Second, although RNA shares three of DNA's nitrogenous bases (A, C and G), instead of thymine it has uracil. Like thymine, this pairs specifically with adenine.

The final main difference between RNA and DNA is the fact that RNA generally comprises only a single polynucleotide chain, although this may be subject to secondary and tertiary folding as a result of complementary base pairing within the same strand. The roles of the three different forms of RNA will be discussed in Chapter 11.

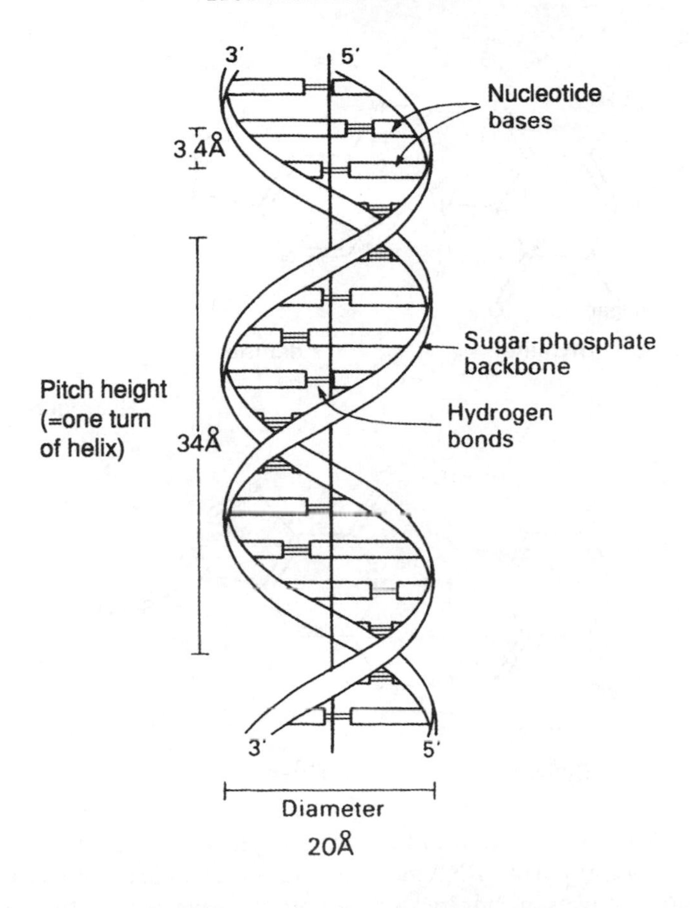

Figure 2.23 The model of DNA proposed by Watson and Crick has two chains of nucleotides joined together by hydrogen-bonded base pairs pointing inwards towards the centre of the helix. The rules of complementary base pairing means that the sequence of one chain can be predicted from the sequence of the other. Note how the chains run in opposite directions (antiparallel)

Lipids

Although lipids can be large molecules, they are not regarded as macromolecules because unlike proteins, polysaccharides and nucleic acids, they are not polymers of a basic subunit. Moreover, lipids do not share any single structural characteristic; they are a diverse group structurally, but have in common the fact that they are *insoluble in water*, but soluble in a range of organic solvents. This non-polar nature is due to the predominance of covalent bonding, mainly between atoms of carbon and hydrogen.

Fats are simple lipids, whose structure is based on *fatty acids* (see Box 2.5). Fatty acids are long hydrocarbon chains ending in a carboxyl (-COOH) group. They have the

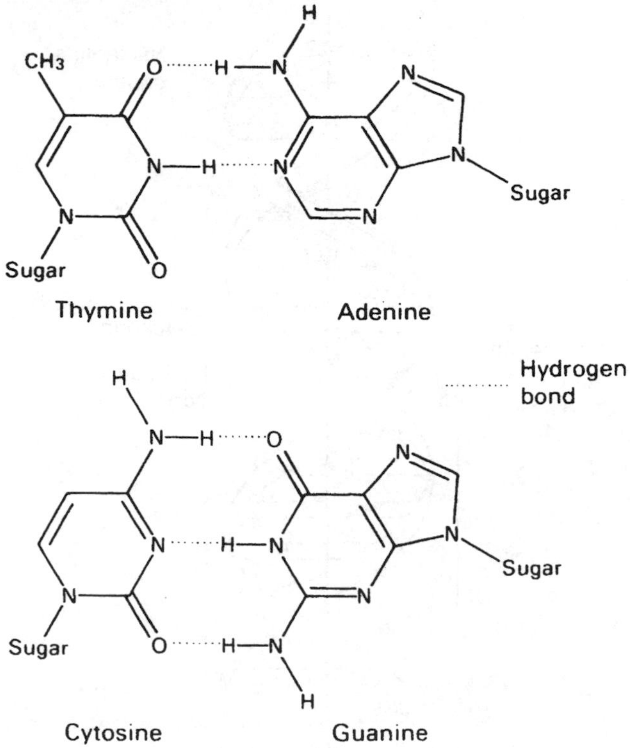

Figure 2.24 Adenine pairs only with thymine, and guanine with cytosine, thus if the sequence of bases in one strand of a DNA molecule is known, that of the other can be predicted. This critical feature of Watson and Crick's model offers an explanation for how DNA is able to replicate itself. Note that GC pairs are held together by three hydrogen bonds, while AT pairs only have two

general formula:

$$CH_3\text{—}(CH_2)_n\text{—}COOH$$

where n is usually an even number. They combine with glycerol according to the basic reaction:

$$\text{Alcohol} + \text{Acid} \rightarrow \text{Ester}$$

The bond so formed is called an ester linkage, and the result is an acylglycerol (Figure 2.25). One, two or all three of the -OH groups may be esterified with a fatty acid, to give respectively mono-, di- and *triacylglycerols*. Natural fats generally contain a mixture of two or three different fatty acids substituted at the three positions; consequently, a considerable diversity is possible among fats. Fats serve as energy stores; a higher proportion of C—C and C—H bonds in comparison with proteins or carbohydrates results in a higher energy-storing capacity.

Box 2.5 Saturated or unsaturated?

You may well have heard of saturated and unsaturated fats in the context of the sorts of foods we should and shouldn't be eating. This terminology derives from the type of fatty acids which make up the different types of fat.

Each carbon atom in the hydrocarbon chain of a *saturated* fatty acid such as stearic acid is bonded to the maximum possible number of hydrogen atoms (i.e. it is saturated with them).

Fatty acids containing one or more double bonds have fewer hydrogen atoms and are said to be *unsaturated*.

Compare the structures of stearic acid and oleic acid below. Both have identical structures except that oleic acid has two fewer hydrogen atoms and in their place a C=C double bond. A kink or bend is introduced into the chain at the point of the double bond; this means that adjacent fatty acids do not pack together so neatly, leading to a drop in the melting point. The presence of unsaturated fatty acids in membrane phospholipids makes the membrane more fluid.

Stearic acid (18.0)
(saturated)

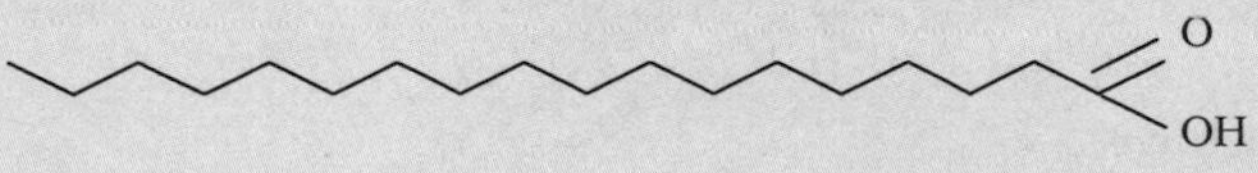

Oleic acid (18.1)
(monounsturated)

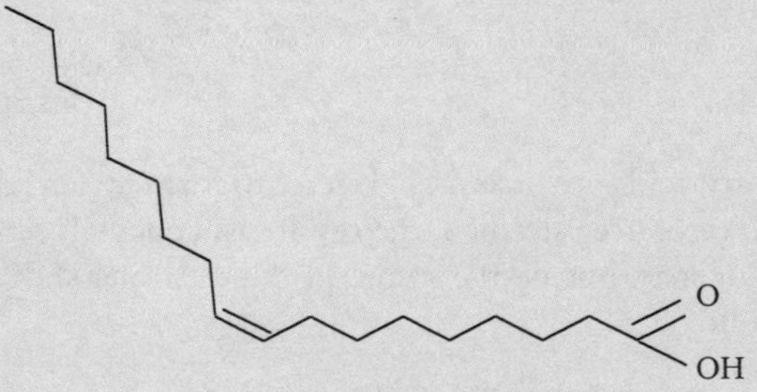

Linoleic acid (18.2)
(polyunsaturated)

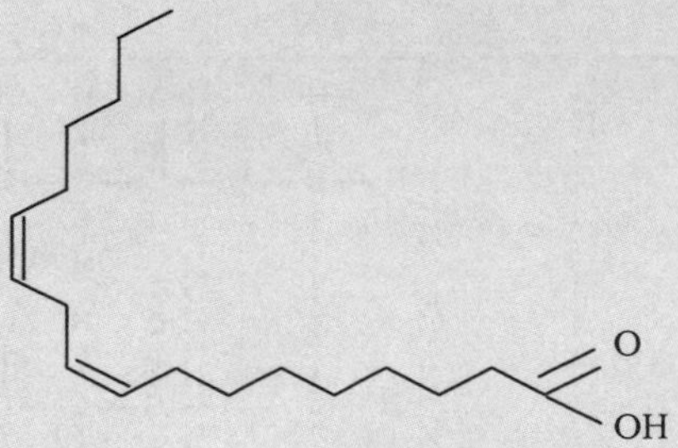

The second main group of lipids to be found in living cells are *phospholipids*. These have a similar structure to triacylglycerols, except that instead of a third fatty acid chain, they have a phosphate group joined to the glycerol (Figure 2.26), introducing a hydrophilic element to an otherwise hydrophobic molecule. Thus, phospholipids are an example of an *amphipathic* molecule, with a polar region at one end of the

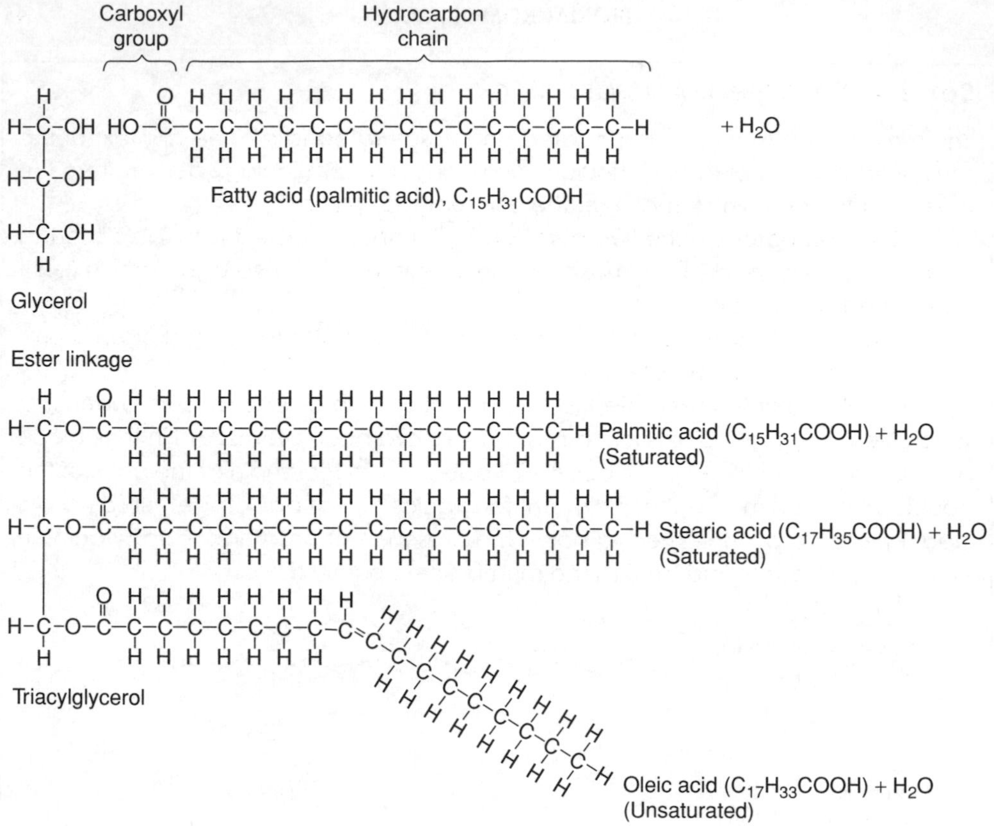

Carboxyl group

Hydrocarbon chain

$+ H_2O$

Fatty acid (palmitic acid), $C_{15}H_{31}COOH$

Glycerol

Ester linkage

Palmitic acid ($C_{15}H_{31}COOH$) + H_2O (Saturated)

Stearic acid ($C_{17}H_{35}COOH$) + H_2O (Saturated)

Triacylglycerol

Oleic acid ($C_{17}H_{33}COOH$) + H_2O (Unsaturated)

Figure 2.25 Fatty acids are linked to glycerol to form an acylglycerol. When all three -OH groups on the glycerol are esterified, the result is a triacylglycerol or triglyceride. The three fatty acids may or may not be the same. In the example shown, one of the fatty acids is unsaturated (see Box 2.5)

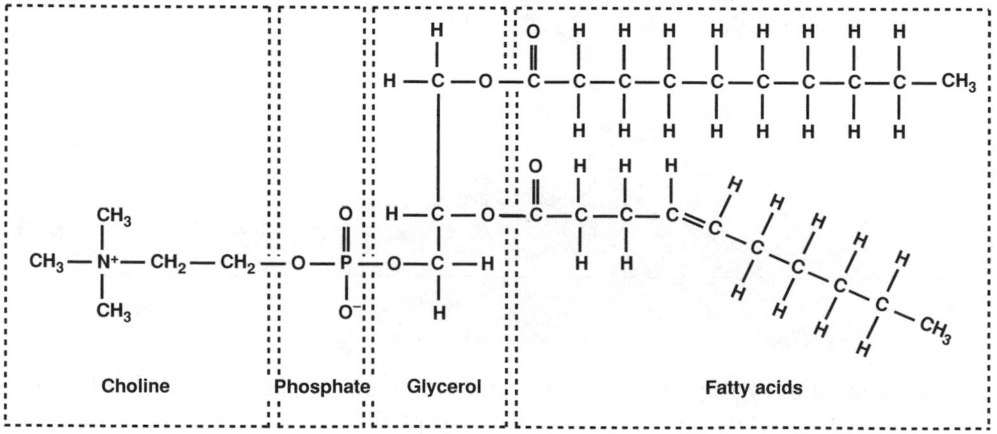

Choline

Phosphate

Glycerol

Fatty acids

Figure 2.26 Phospholipids introduce a polar element to acylglycerols by substituting a phosphate at one of the glycerol -OH groups. A second charged group may attach to the phosphate group; the phospholipid shown is phosphatidylcholine

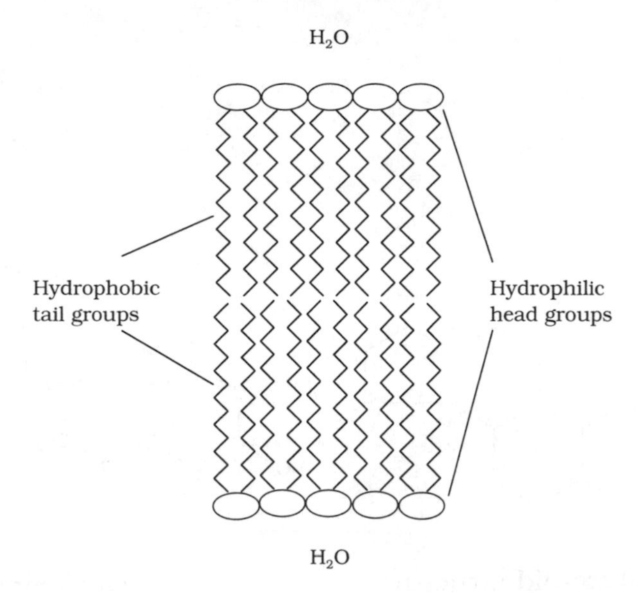

Figure 2.27 Phospholipids can form a bilayer in aqueous surroundings. A 'sandwich' arrangement is achieved by the polar phosphate groups facing outwards and burying the fatty acid chains within. Water is thus excluded from the hydrophobic region, a key property of biological membranes (see Figure 3.5)

molecule and a non-polar region at the other. This fact is essential for the formation of a bilayer when the phospholipid is introduced into an aqueous environment; the hydrophilic phosphate groups point outwards towards the water, while the hydrophobic hydrocarbon chains 'hide' inside (Figure 2.27, and c.f. Figure 2.8, micelle formation).

This bilayer structure forms the basis of all biological membranes (see Chapter 3), forming a barrier around cells and certain organelles. Phospholipids generally have another polar group attached to the phosphate; Figure 2.25 shows the effect of substituting serine.

The structural diversity of lipids can be illustrated by comparing fats and phospholipids with the final group of lipids we need to consider, the steroids. As can be seen from Figure 2.28, these have a completely different form, but still share in common the property of hydrophobicity. The four ring planar structure is common to all steroids, with the substitution of different side groups producing great differences in function. Cholesterol is an important component of many membranes.

It would be wrong to gain the impression that living cells contain only molecules of the four groups outlined above. Smaller organic molecules play important roles as precursors or intermediates in metabolic pathways (see Chapter 6), and several inorganic ions such as potassium, sodium and chloride play essential roles in maintaining the living cell. Finally, some macromolecules comprise elements of more than one group, for example, lipopolysaccharides (carbohydrate and lipid) and glycoproteins (protein and carbohydrate).

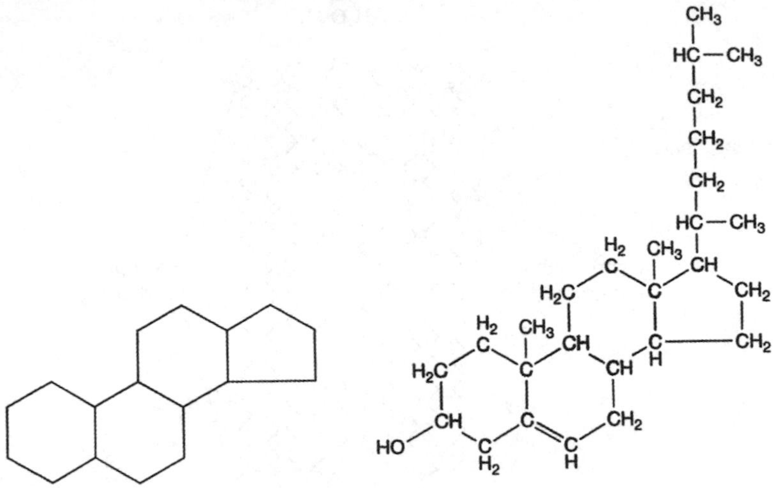

General steroid structure **Cholesterol**

Figure 2.28 All steroids are based on a four-ringed structure. The presence of an –OH group on the lower left ring makes the molecule a *sterol*. Cholesterol plays an important role in the fluidity of animal membranes by interposing itself among the fatty acid tails of phospholipids. The only bacterial group to contain sterols are the mycoplasma; however some other groups contain *hopanoids*, which have a similar structure and are thought to play a comparable role in membrane stability

Test yourself

1 The number of protons in an atom of an element is called the ________ ________ of that element.

2 The sum of the protons and neutrons in an atom is the ________ ________ of the element.

3 The transfer of an electron from one atom to another so that both achieve a full outer electron shell is called ________ bonding.

4 ________ bonding involves the sharing of one or more pairs of electrons.

5 A solution with a pH of 3.0 is ________ times more acidic than one with a pH of 6.0.

6 Possession of a functional group such as phosphate or aldehyde makes a molecule more ________ and therefore more readily ________ in water.

7 Simple carbohydrates may be classed on the basis of whether they have a ________ or ________ group, or according to how many ________ atoms they possess.

8 Sugars are joined together by ________ linkages.

9 No matter how long a peptide chain grows, it always has an ________ group at one end and a ________ group at the other.

10 An example of a negatively charged amino acid is ________.

11 Disulphide bonds are formed between residues of the amino acid ________.

12 Secondary protein structure is brought about by the formation of ________ bonds between a ________ group and a ________ group.

13 Naturally occurring amino acids are all the ________-isomer. This results in an α-helix taking on a ________ configuration for maximum stability.

14 Heating or treatment with certain chemicals cause proteins to lose their three-dimensional structure and become ________.

15 A short stretch of double stranded DNA has 52 adenine residues and 61 guanine residues. There are therefore ________ cytosine residues and a total of ________ hydrogen bonds joining the two strands.

16 The arrangement of the two strands of a DNA molecule is described as ________.

17 The nucleotides of RNA contain ________ instead of thymine.

18 Lipids are a diverse group of molecules, sharing the common property of ________ in ________.

19 Phospholipids are described as ________ because their molecules have both polar and non-polar regions.

20 In fats, fatty acids are joined to glycerol via ________ linkages.

3

Cell Structure and Organisation

The basic unit of all living things is the cell. The *cell theory* is one of the fundamental concepts of biology; it states that:

- all organisms are made up of cells,

  and that

- all cells derive from other, pre-existing cells.

As we shall see in this chapter, there may exist within a cell many smaller, subcellular structures, each with its own characteristics and function, but these are not capable of independent life.

An organism may comprise just a single cell (*unicellular*), a collection of cells that are not morphologically or functionally differentiated (*colonial*), or several distinct cell types with specialised functions (*multicellular*). Among microorganisms, all bacteria and protozoans are unicellular; fungi may be unicellular or multicellular, while algae may exist in all three forms. There is, however, one way that organisms can be differentiated from each other that is even more fundamental than whether they are uni- or multicellular. It is a difference that is greater than that between a lion and a mushroom or between an earthworm and an oak tree, and it exists at the level of the individual cell. All organisms are made up of one or other (definitely not both!) of two very distinct cell types, which we call *procaryotic* and *eucaryotic* cells, both of which exist in the microbial world. These differ from each other in many ways, including size, structural complexity and organisation of genetic material (Table 3.1).

> The names given to the two cell types derive from Greek words:
> *Procaryotic* = 'before nucleus'
> *Eucaryotic* = 'true nucleus'

The most fundamental difference between procaryotic and eucaryotic cells is reflected in their names; eucaryotic cells possess a true nucleus, and several other distinct subcellular organelles that are bounded by a membrane. Procaryotes have no such organelles. Most of these differences only became apparent after the development of electron microscopy techniques.

As can be seen from Table 3.2, the procaryotes comprise the simpler and more primitive types of microorganisms; they are generally single celled, and arose much earlier in evolutionary history than the eucaryotes. Indeed, as discussed later in this chapter, it

Table 3.1 Similarities and differences between procaryotic and eucaryotic cell structure

Similarities	
Cell contents bounded by a plasma membrane	
Genetic information encoded on DNA	
Ribosomes act as site of protein synthesis	
Differences	
Procaryotic	**Eucaryotic**
Size	
Typically 1–5 μm	Typically 10–100 μm
Genetic material	
Free in cytoplasm	Contained within a membrane-bound nucleus
Single circular chromosome or nucleoid	Multiple chromosomes, generally in pairs
Histones absent.	DNA complexed with histone proteins
Internal features	
Membrane-bound organelles absent	Several membrane-bound organelles present, including mitochondria, Golgi body, endoplasmic reticulum and (in plants & algae) chloroplasts
Ribosomes smaller (70S), free in cytoplasm	Ribosomes larger (80S), free in cytoplasm or attached to membranes
Respiratory enzymes bound to plasma membrane	Respiratory enzymes located in mitochondria
Cell wall	
Usually based on peptidoglycan (not Archaea)	When present, based on cellulose or chitin
External features	
Cilia absent	Cilia may be present
Flagella, if present, composed of flagellin. Provide rotating motility	Flagella, if present, have complex (9 + 2) structure. Provide 'whiplash' motility
Pili may be present	Pili absent
Outside layer (slime layer, capsule, glycocalyx) present in some types	Pellicle or test present in some types

is widely accepted that eucaryotic cells actually arose from their more primitive counterparts. Note that the viruses do not appear in Table 3.2, because they do not have a cellular structure at all, and are not therefore considered to be living organisms. (See Chapter 10 for further discussion of the viruses).

The use of DNA sequencing methods to determine *phylogenetic* relationships between organisms has revealed that within the procaryotes there is another fundamental division. One group of bacteria were shown to differ greatly from all the others; we now call these the *Archaea*, to differentiate them from the true *Bacteria*.

Phylogenetic: pertaining to the evolutionary relationship between organisms.

Table 3.2 Principal groups of procaryotic and eucaryotic organisms

Procaryotes	Eucaryotes
Bacteria	Fungi
Blue-green 'algae'*	Algae
	Protozoa
	Plants
	Animals

*An old-fashioned term: this group are in fact a specialized form of bacteria, and are known more correctly as the Cyanobacteria, or simply the blue-greens. They are discussed in more detail in Chapter 7. Animals and plants fall outside the scope of this book.

These two groups, together with the eucaryotes, are thought to have evolved from a common ancestor, and represent the three *domains* of life (Figure 3.1). The Archaea comprise a wide range of mostly anaerobic bacteria, including many of those that inhabit extreme environments such as hot springs. In this book we shall largely confine our discussions to the Bacteria, however in Chapter 7 there is a discussion of the principal features of the Archaea and their main taxonomic groupings.

Despite their differences, *Archaea* and *Bacteria* are both procaryotes.

Taxonomy is the science of classifying living (and once-living) organisms.

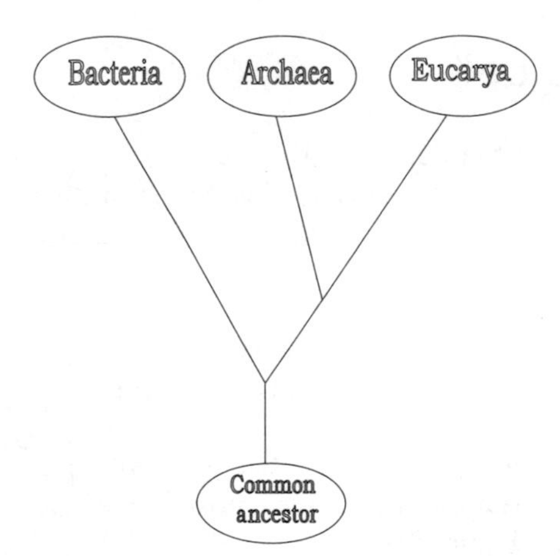

Figure 3.1 The three domains of life. All life forms can be assigned to one of three domains on the basis of their ribosomal RNA sequences. The Archaea are quite distinct from the true bacteria and are thought to have diverged from a common ancestral line at a very early stage, before the evolution of eucaryotic organisms. The scheme above is the one most widely accepted by microbiologists, but alternative models have been proposed

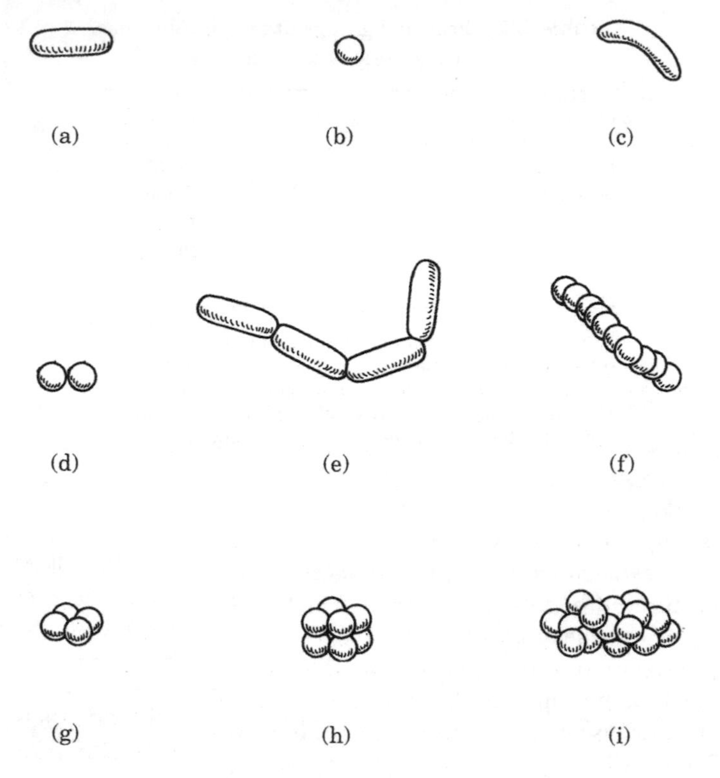

Figure 3.2 Bacterial shapes. Most bacteria are (a) rod shaped, (b) spherical or (c) curved. These basic shapes may join to form (d) pairs, (e and f) chains, (g) sheets, (h) packets or (i) irregular aggregates

The procaryotic cell

Bacteria are much smaller than eucaryotic cells; most fall into a size range of about 1–5μm, although some may be larger than this. Some of the smallest bacteria, such as the mycoplasma measure less than 1μm, and are too small to be resolved clearly by an ordinary light microscope.

Because of their extremely small size, it was only with the advent of the electron microscope that we were able to learn about the detailed structure of bacterial cells. Using the light microscope however, it is possible to recognise differences in the shape and arrangement of bacteria. Although a good deal of variation is possible, most have one of three basic shapes (Figure 3.2):

> In recent years, square, triangular and star-shaped bacteria have all been discovered!

- rod shaped (*bacillus*)
- spherical (*coccus*)
- curved: these range from comma-shaped (*vibrio*) to corkscrew-shaped (*spirochaete*)

All these shapes confer certain advantages to their owners; rods, with a large surface area are better able to take up nutrients from the environment, while the cocci are less prone to drying out. The spiral forms are usually motile; their shape aids their movement through an aqueous medium.

As well as these characteristic cell shapes, bacteria may also be found grouped together in particular formations. When they divide, they may remain attached to one another, and the shape the groups of cells assume reflects the way the cell divides. Cocci, for example, are frequently found as chains of cells, a reflection of repeated division in one plane (Figure 3.2(f)). Other cocci may form regular sheets or packets of cells, as a result of division in two or three planes. Yet others, such as the staphylococci, divide in several planes, producing the irregular and characteristic 'bunch of grapes' appearance. Rod-shaped bacteria only divide in a single plane and may therefore be found in chains, while spiral forms also divide in one plane, but tend not to stick together. Blue–greens form filaments; these are regarded as truly multicellular rather than as a loose association of individuals.

Procaryotic cell structure

When compared with the profusion of elaborate organelles encountered inside a typical eucaryotic cell, the interior of a typical bacterium looks rather empty. The only internal structural features are:

- a bacterial chromosome or *nucleoid*, comprising a closed loop of double stranded, supercoiled DNA. In addition, there may be additional DNA in the form of a *plasmid*
- thousands of granular *ribosomes*
- a variety of granular *inclusions* associated with nutrient storage.

All of these are contained in a thick aqueous soup of carbohydrates, proteins, lipids and inorganic salts known as the *cytoplasm*, which is surrounded by a *plasma membrane*. This in turn is wrapped in a *cell wall*, whose rigidity gives the bacterial cell its characteristic shape. Depending on the type of bacterium, there may be a further surrounding layer such as a *capsule* or *slime layer* and/or structures external to the cell associated with motility (*flagella*) or attachment (*pili/fimbriae*). Figure 3.3 shows these features in a generalised bacterial cell. In the following pages we shall examine these features in a little more detail, noting how each has a crucial role to play in the survival or reproduction of the cell.

Genetic material
Although it occupies a well defined area within the cell, the genetic material of procaryotes is not present as a true nucleus, as it lacks a surrounding nuclear membrane (c.f. the eucaryotic nucleus, Figure 3.12). The nucleoid or bacterial chromosome comprises a closed circle of double stranded DNA, many times the length of the cell and highly folded and compacted. (The common laboratory

> Not all bacteria conform to the model of a single circular chromosome; some have been shown to possess two with genes shared between them, while examples of linear chromosomes are also known.

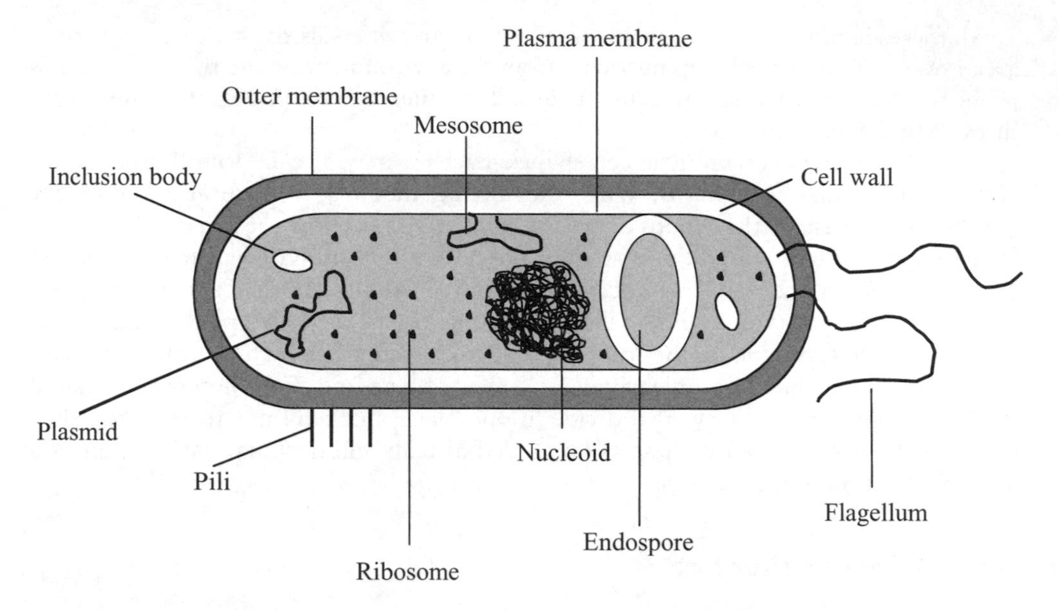

Figure 3.3 Structure of a generalised procaryotic cell. Note the lack of complex internal organelles (c.f. Figure 3.12). Gram-positive and Gram-negative bacteria differ in the details of their cell wall structure (see Figures 3.7 & 3.8)

bacterium *Escherichia coli* is around 3–4 μm in length, but contains a DNA molecule some 1400 μm in length!) The DNA may be associated with certain bacterial proteins, but these are not the same as the histones found in eucaryotic chromosomes. Some bacteria contain additional DNA in the form of small, self-replicating extrachromosomal elements called plasmids. These do not carry any genes essential for growth and reproduction, and thus the cell may survive without them. They can be very important however, as they may include genes encoding toxins or resistance to antibiotics, and can be passed from cell to cell (see Chapter 12).

> Plasmids are small loops of DNA independent of the chromosome. They are capable of directing their own replication.

Ribosomes

Apart from the nucleoid, the principal internal structures of procaryotic cells are the ribosomes. These are the site of protein synthesis, and there may be many thousands of these in an active cell, lending a speckled appearance to the cytoplasm. Ribosomes are composed of a complex of protein and RNA, and are the site of protein synthesis in the cell.

Although they carry out a similar function, the ribosomes of procaryotic cells are smaller and lighter than their eucaryotic counterparts. Ribosomes are measured in *Svedberg units* (*S*), a function of their size and shape, and determined by their rate of sedimentation in a centrifuge; procaryotic ribosomes are 70S, while those of eucaryotes are 80S. Some types of antibiotic exploit this difference by

Table 3.3 Comparison of procaryotic and eucaryotic ribosomes

	Procaryotic	Eucaryotic
Overall size	70S	80S
Large subunit size	50S	60S
Large subunit RNA	23S & 5S	28S, 5.8S & 5S
Small subunit size	30S	40S
Small subunit RNA	16S	18S

targeting the procaryotic form and selectively disrupting bacterial protein synthesis (see Chapter 14).

All ribosomes comprise two unequal subunits (in procaryotes, these are 50S and 30S, in eucaryotes 60S and 40S: Table 3.3)). Each subunit contains its own RNA and a number of proteins (Figure 3.4). Many ribosomes may simultaneously be attached to a single mRNA molecule, forming a threadlike *polysome*. The role of ribosomes in bacterial protein synthesis is discussed in Chapter 11.

A *polyribosome* (polysome) is a chain of ribosomes attached to the same molecule of mRNA.

Inclusion bodies

Within the cytoplasm of certain bacteria may be found granular structures known as inclusion bodies. These act as food reserves, and may contain organic compounds such as starch, glycogen or lipid. In addition, sulphur and polyphosphate can be stored as inclusion bodies, the latter being known as volutin or metachromatic granules. Two special types of inclusion body are worthy of mention. Magnetosomes, which contain a form of iron oxide, help some types of bacteria to orientate themselves downwards into favourable conditions, whilst gas vacuoles maintain bouyancy of the cell in blue greens and some halobacteria.

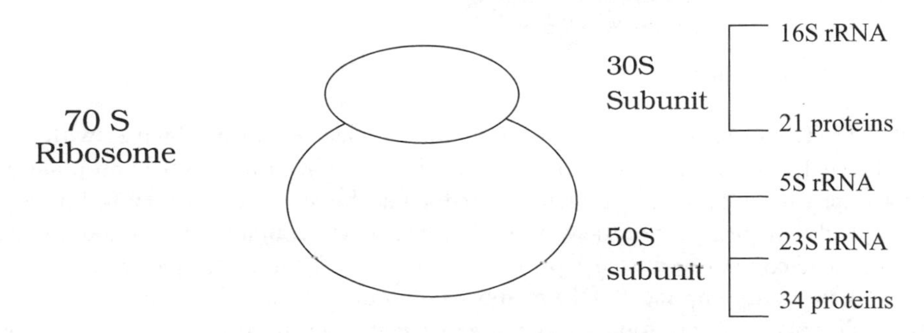

Figure 3.4 The bacterial ribosome. Each subunit comprises rRNA and proteins. The nucleotide sequence of small subunit (16S) rRNA is widely used in determining the phylogenetic (evolutionary) relationship between bacteria (see Chapter 7)

Endospores

Certain bacteria such as *Bacillus* and *Clostridium* produce endospores. They are dormant forms of the cell that are highly resistant to extremes of temperature, pH and other environmental factors, and germinate into new bacterial cells when conditions become more favourable. The spore's resistance is due to the thick coat that surrounds it.

> Endospores of pathogens such as *Clostridium botulinum* can resist boiling for several hours. It is this resistance that makes it necessary to autoclave at 121°C in order to ensure complete sterility.

The plasma membrane

The cytoplasm and its contents are surrounded by a plasma membrane, which can be thought of as a bilayer of phospholipid arranged like a sandwich, together with associated proteins (Figure 3.5). The function of the plasma membrane is to keep the contents in, while at the same time allowing the selective passage of certain substances in and out of the cell (it is a semipermeable membrane).

Phospholipids comprise a compact, hydrophilic (= water-loving) head and a long hydrophobic tail region (Figure 2.27); this results in a highly ordered structure when the membrane is surrounded by water. The tails 'hide' from the water to form the inside of the membrane, while the heads project outwards. Also included in the membrane are a variety of proteins; these may pass right through the bilayer or be associated with the inner (cytoplasmic) or outer surface only. These proteins may play structural or functional roles in the life of the cell. Many enzymes associated with the metabolism of nutrients and the production of energy are associated with the plasma membrane in procaryotes. As we will see later in this chapter, this is fundamentally different from

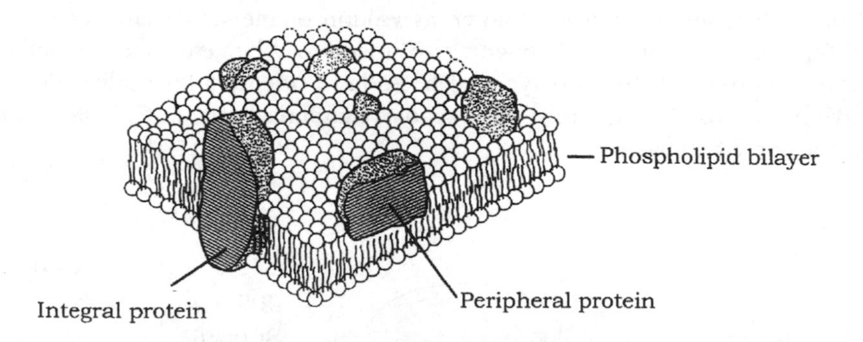

Figure 3.5 The plasma membrane. Phospholipid molecules form a bilayer, with the hydrophobic hydrocarbon chains pointing in towards each other, leaving the hydrophilic phosphate groups to face outwards. Proteins embedded in the membrane are known as integral proteins, and may pass part of the way or all of the way through the phospholipid bilayer. The amino acid composition of such proteins reflects their location; the part actually embedded among the lipid component of the membrane comprises non-polar (hydrophobic) amino acids, while polar ones are found in the aqueous environment at either side. Singleton, P: Bacteria in Biology, Biotechnology and Medicine, 5th edn, John Wiley & Sons, 1999. Reproduced by permission of the publishers

eucaryotic cells, where these reactions are carried out on specialised internal organelles. Proteins involved in the active transport of nutrients (see Chapter 4) are also to be found associated with the plasma membrane. The model of membrane structure as depicted in Figure 3.5 must not be thought of as static; in the widely accepted *fluid mosaic model*, the lipid is seen as a fluid state, in which proteins float around, rather like icebergs in an ocean.

The majority of bacterial membranes do not contain sterols (c.f. eucaryotes: see below), however many do contain molecules called hopanoids that are derive from the same precursors. Like sterols, they are thought to assist in maintaining membrane stability. A comparison of the lipid components of plasma membranes reveals a distinct difference between members of the Archaea and the Bacteria.

The bacterial cell wall

Bacteria have a thick, rigid cell wall, which maintains the integrity of the cell, and determines its characteristic shape. Since the cytoplasm of bacteria contains high concentrations of dissolved substances, they generally live in a hypotonic environment (i.e. one that is more dilute than their own cytoplasm). There is therefore a natural tendency for water to flow into the cell, and without the cell wall the cell would fill and burst (you can demonstrate this by using enzymes to strip off the cell wall, leaving the naked *protoplast*).

A protoplast is a cell that has had its cell wall removed.

The major component of the cell wall, which is responsible for its rigidity, is a substance unique to bacteria, called *peptidoglycan* (murein). This is a high molecular weight polymer whose basic subunit is made up of three parts: *N*-acetylglucosamine, *N*-acetylmuramic acid and a short peptide chain (Figure 3.6). The latter comprises the amino acids L-alanine, D-alanine, D-glutamic acid and either L-lysine or diaminopimelic acid (DAP). DAP is a rare amino acid, only found in the cell walls of procaryotes. Note that some of the amino acids of peptidoglycan are found in the D-configuration. This is contrary to the situation in proteins, as you may recall from Chapter 2, and confers protection against proteases specifically directed against L-amino acids.

Proteases are enzymes that digest proteins.

Precursor molecules for peptidoglycan are synthesised inside the cell, and transported across the plasma membrane by a carrier called bactoprenol phosphate before being incorporated into the cell wall structure. Enzymes called *transpeptidases* then covalently bond the tetrapeptide chains to one another, giving rise to a complex network (Figure 3.7); it is this cross-linking that gives the wall its mechanical strength. A number of antimicrobial agents exert their effect by inhibiting cell wall synthesis; β-lactam antibiotics such as penicillin inhibit the transpeptidases, thereby weakening the cell wall, whilst bacitracin prevents transport of peptidoglycan precursors out of the cell. The action of antibiotics will be discussed further in Chapter 14. Although all bacteria (with a few exceptions) have a cell wall containing peptidoglycan, there are two distinct structural types. These are known as *Gram-positive* and *Gram-negative*. The names derive from the Danish scientist Christian Gram, who, in the 1880s developed a rapid staining technique that could differentiate bacteria as belonging to one of two basic types (see Box 1.2). Although the usefulness of the Gram stain was recognised for many years, it

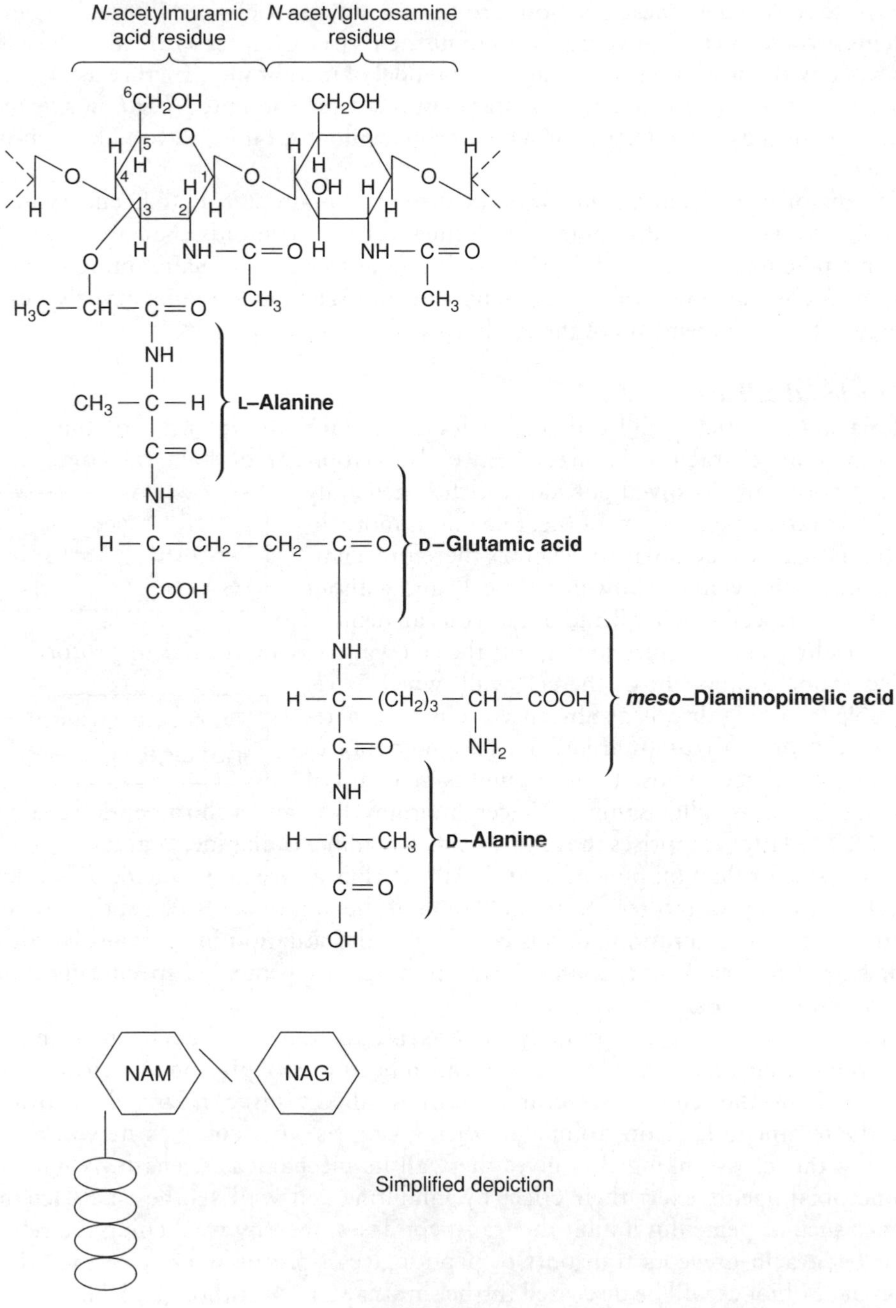

N-acetylmuramic acid residue
N-acetylglucosamine residue
L-Alanine
D-Glutamic acid
meso-Diaminopimelic acid
D-Alanine
NAM
NAG
Simplified depiction

Figure 3.6 Peptidoglycan structure. Peptidoglycan is a polymer made up of alternating molecules of *N*-acetylglucosamine (NAG) and *N*-acetylmuramic acid (NAM). A short peptide chain is linked to the NAM residues (see text for details). This is important in the cross-linking of the straight chain polymers to form a rigid network (Figure 3.6). The composition of *E. coli* peptidoglycan is shown; the peptide chain may contain different amino acids in other bacteria. Partly from Hardy, SP: Human Microbiology, Taylor and Francis, 2002. Reproduced by permission of Thomson Publishing Services

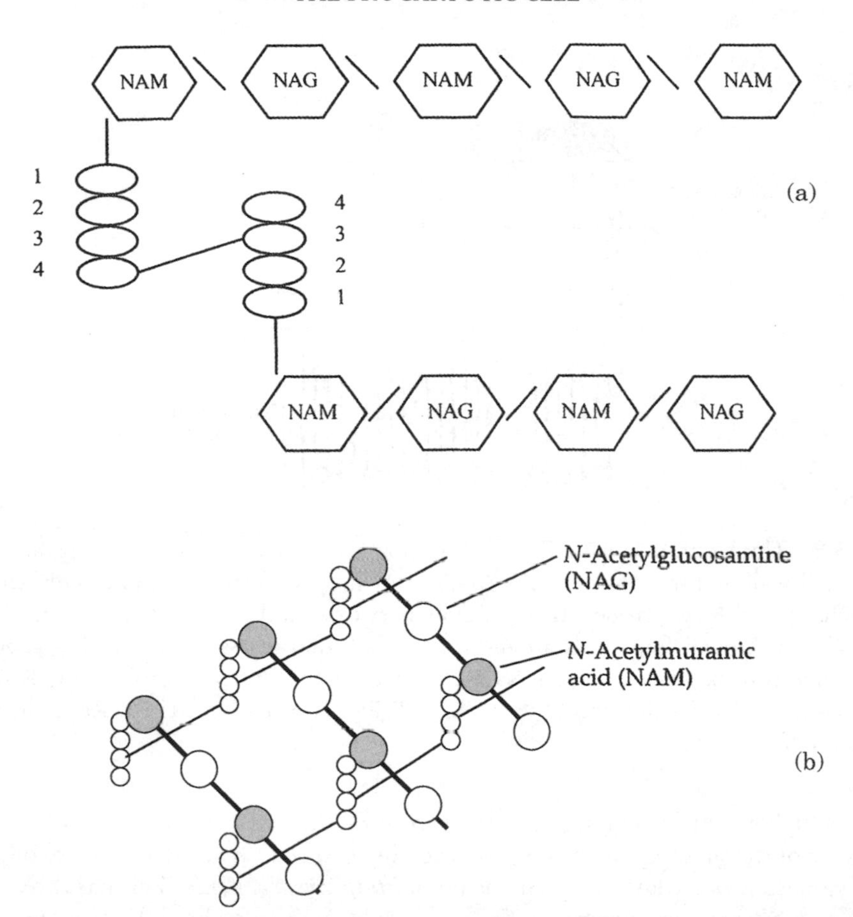

Figure 3.7 Cross-linking of peptidoglycan chains in *E. coli*. (a) The D-alanine on the short peptide chain attached to the *N*-acetylmuramic acid cross-links to a diaminopimelic acid residue on another chain. In other bacteria, the precise nature of the cross-linking may differ. From Hardy, SP: Human Microbiology, Taylor and Francis, 2002. Reproduced by permission of Thomson Publishing Services. (b) Further cross-linking produces a rigid network of peptidoglycan. The antibiotic penicillin acts by inhibiting the transpeptidase enzymes responsible for the cross-linking reaction (see Chapter 15)

was only with the age of electron microscopy that the underlying molecular basis of the test could be explained, in terms of cell wall structure.

Gram-positive cell walls are relatively simple in structure, comprising several layers of peptidoglycan connected to each other by cross-linkages to form a strong, rigid scaffolding. In addition, they contain acidic polysaccharides called *teichoic acids*; these contain phosphate groups that impart an overall negative charge to the cell surface. A diagram of the gram-positive cell wall is shown in Figure 3.8.

Gram-negative cells have a much thinner layer of peptidoglycan, making the wall less sturdy, however the structure is made more complex by the presence of a layer of lipoprotein, polysaccharide and phospholipid known as the *outer membrane*

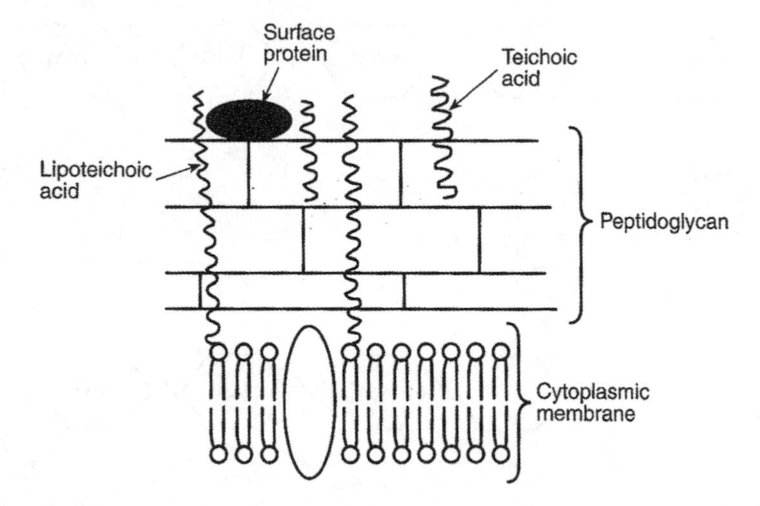

Figure 3.8 The Gram-positive cell wall. Peptidoglycan is many layers thick in the Gram-positive cell wall and may account for 30–70% of its dry weight. Teichoic acids are negatively charged polysaccharides; they are polymers of ribitol phosphate and cross-link to peptidoglycan. Lipoteichoic acids are teichoic acids found in association with glycolipids. From Henderson, B, Wilson, M, McNab, R & Lax, AJ: Cellular Microbiology: Bacteria-Host Interactions in Health and Disease, John Wiley & Sons Inc., 1999. Reproduced by permission of the publishers

(Figure 3.9). This misleading name derives from the fact that it superficially resembles the bilayer of the plasma membrane; however, instead of two layers of phospholipid, it has only one, the outer layer being made up of *lipopolysaccharide*. This has three parts: lipid A, core polysaccharide and an O-specific side chain. The lipid A component may act as an *endotoxin*, which, if released into the bloodstream, can lead to serious conditions such as fever and toxic shock. The O-specific antigens are carbohydrate chains whose composition often varies between strains of the same species. Serological methods can distinguish between these, a valuable tool in the investigation, for example, of the origin of an outbreak of an infectious disease. Proteins incorporated into the outer membrane and penetrating its entire thickness form channels that allow the passage of water and small molecules to enter the cell. Unlike the plasma membrane, the outer membrane plays no part in cellular respiration.

Box 3.1 Mesosomes – the structures that never were?

When looked at under the electron microscope, Gram-positive bacteria often contained localised in-foldings of the plasma membrane. These were given the name *mesosomes*, and were thought by some to act as attachment points for DNA during cell division, or to play a role in the formation of cross-walls. Others thought they were nothing more than artefacts produced by the rather elaborate sample preparation procedures necessary for electron microscopy. Nowadays, most microbiologists support the latter view.

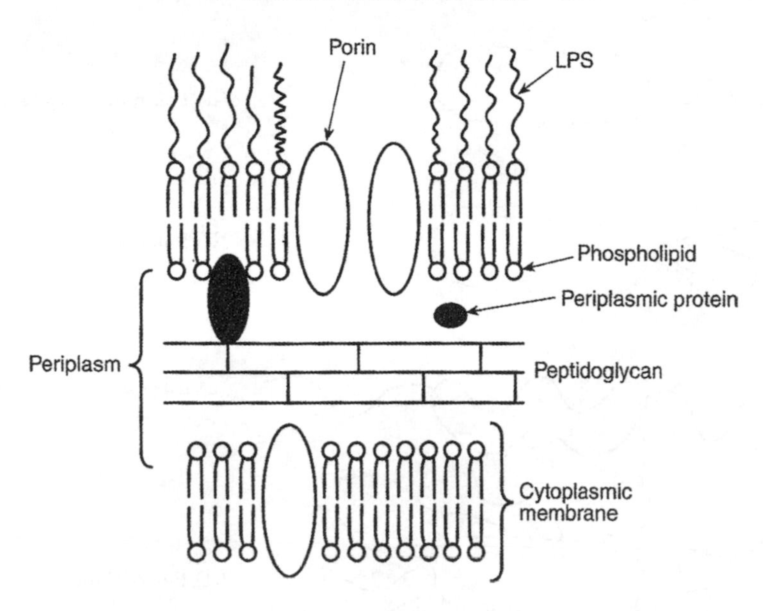

Figure 3.9 The Gram-negative cell wall. Note the thinner layer of peptidoglycan compared to the gram-positive cell wall (Figure 3.8). It accounts for <10% of the dry weight. Beyond this lies the outer membrane, with its high lipopolysaccharide content. Channels made of *porins* allow the passage of certain solutes into the cell. From Henderson, B, Wilson, M, McNab, R & Lax, AJ: Cellular Microbiology: Bacteria-Host Interactions in Health and Disease, John Wiley & Sons Inc., 1999. Reproduced by permission of the publishers

Members of the Archaea have a cell wall chemistry quite different to that described above (see Chapter 7). Instead of being based on peptidoglycan, they have other complex polysaccharides, although a distinction between gram-positive and gram-negative types still occurs.

Beyond the cell wall

A number of structural features are to be found on the outer surface of the cell wall; these are mainly involved either with locomotion of the cell or its attachment to a suitable surface.

Perhaps the most obvious extracellular structures are *flagella* (sing: flagellum), thin hair-like structures often much longer than the cell itself, and used for locomotion in many bacteria. There may be a single flagellum, one at each end, or many, depending on the bacteria concerned (Figure 3.10). Each flagellum is a hollow but rigid cylindrical filament made of the protein flagellin, attached via a hook to a basal body, which secures it to the cell wall and plasma membrane (Figure 3.11). The basal body comprises a series of rings, and is more complex in Gram-negative than Gram-positive bacteria. Rotation of the flagellum is an energy-dependent process driven by the basal body, and the direction of rotation determines the nature of the resulting cellular movement. Clockwise rotation of a single flagellum results in a directionless 'tumbling',

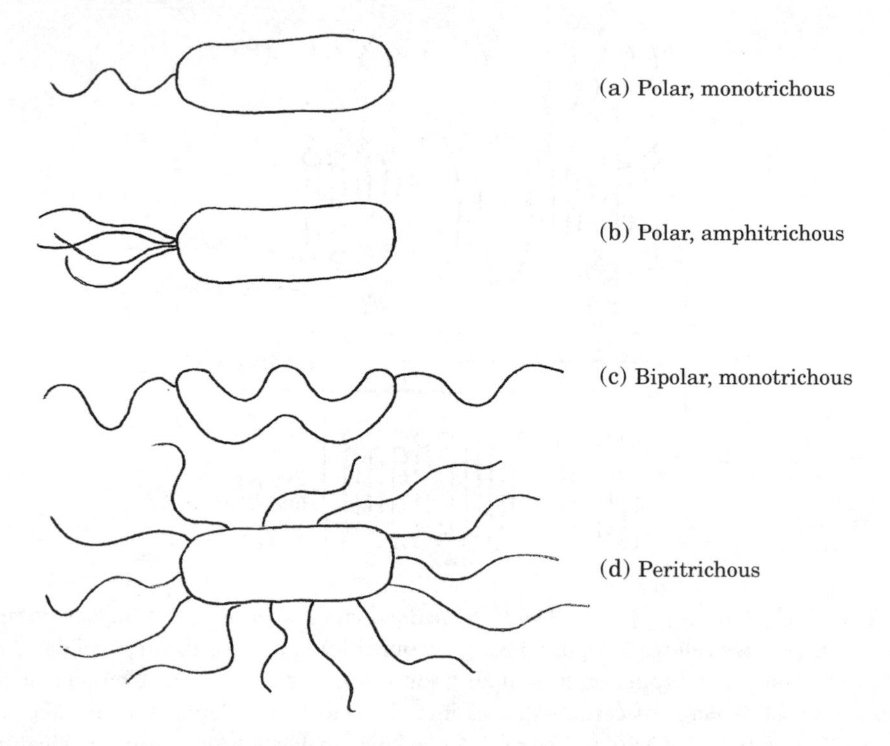

Figure 3.10 Flagella may be situated at one end (a & b), at both ends (c) or all over the cell surface (d)

but if it rotates anticlockwise, the bacterium will 'run' in a straight line (Figure 3.12). Likewise, anticlockwise rotation causes bunched flagella to 'run' by winding around each other and acting as a single structure, whilst spinning in the opposite direction gives rise to multiple independent rotations and tumbling results once more.

Pili (sing: pilus) are structures that superficially resemble short flagella. They differ from flagella, however, in that they do not penetrate to the plasma membrane, and they are not associated with motility. Their function, rather, is to anchor the bacterium to an appropriate surface. Pathogenic (disease-causing) bacteria have proteins called adhesins on their pili, which adhere to specific receptors on host tissues. Attachment pili are sometimes called *fimbriae*, to distinguish them from another distinct type of pilus, the *sex pilus*, which as its name suggests, is involved in the transfer of genetic information by conjugation. This is discussed in more detail in Chapter 12.

Outside the cell wall, most bacteria have a polysaccharide layer called a glycocalyx. This may be a diffuse and loosely bound slime layer or a better defined, and generally thicker capsule. The slime layer helps protect against desiccation, and is instrumental in the attachment of certain bacteria to a substratum (the bacteria that stick to your teeth are a good example of this). Capsules offer protection to certain pathogenic bacteria against the phagocytic cells of the immune system. Both capsules and slime layers are key components of biofilms, which form at liquid/solid interfaces, and can be highly significant in such varied settings as wastewater treatment systems, indwelling catheters and the inside of your mouth!

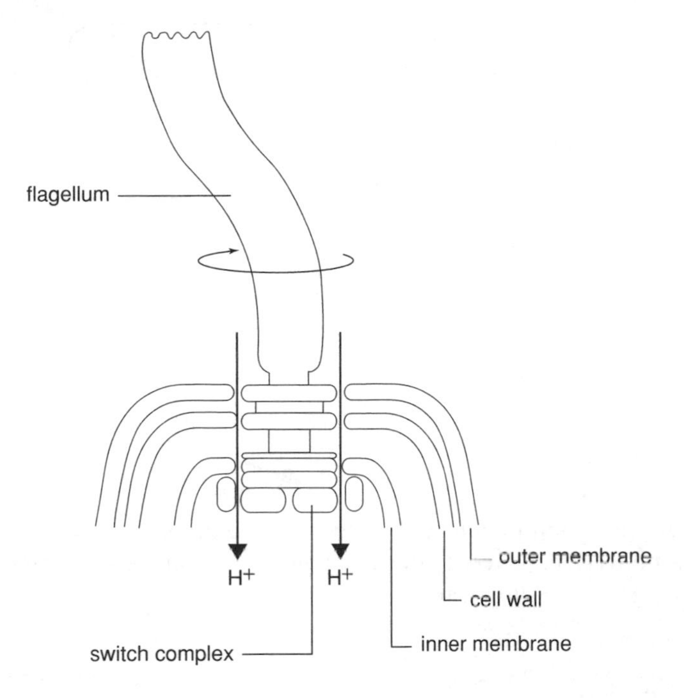

Figure 3.11 Bacterial flagella are anchored in the cell wall and plasma membrane. The filament of the flagellum is anchored by a basal body. In Gram-positive organisms, this comprises two rings inserted in the plasma membrane. In Gram-negative organisms (as shown), there are additional rings associated with the outer membrane and the peptidoglycan layer. Some modern interpretations of flagellar structure view the M and S rings as a single structure. Energy for rotation of the flagellum is derived from the proton motive force generated by the movement of protons across a membrane (see Chapter 6). From Bolsover, SR, Hyams, JS, Jones, S, Shepherd, EA & White, HA: From Genes to Cells, John Wiley & Sons, 1997. Reproduced by permission of the publishers

The eucaryotic cell

We have already seen that eucaryotic cells are, for the most part, larger and much more complex than procaryotes, containing a range of specialised subcellular organelles (Figure 3.13). Within the microbial world, the major groups of eucaryotes are the fungi and the protists (protozoans and algae); all of these groups have single-celled representatives, and there are multicellular forms in the algae and fungi.

Our survey of eucaryotic cell structure begins once more with the genetic material, and works outwards. However, since many internal structures in eucaryotes are enclosed in a membrane, it is appropriate to preface our description by briefly considering eucaryotic membranes. These are, in fact, very similar to the fluid mosaic structure we described earlier in this chapter, as depicted in Figure 3.5. The main difference is that eucaryotic membranes contain lipids called sterols, which enhances their rigidity. We shall consider the significance of this when we discuss the plasma membrane of eucaryotes below. Cholesterol, which we usually hear about in a very negative context, is a very important sterol found in the membranes of many eucaryotes.

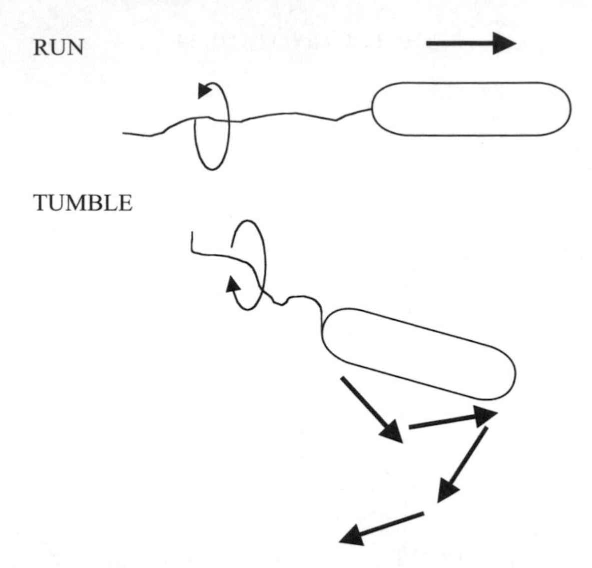

Figure 3.12 Running and tumbling. Anticlockwise rotation of the flagellum gives rise to 'running' in a set direction. Reversing the direction of rotation causes 'tumbling', and allows the bacterial cell to change direction

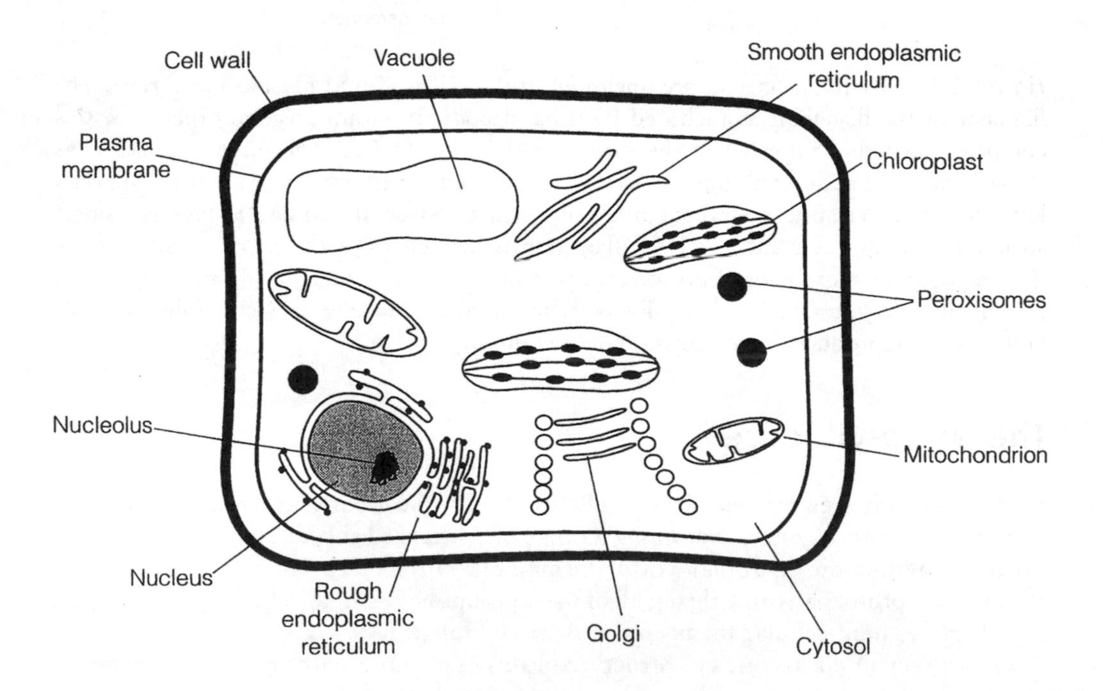

Figure 3.13 This example of eucaryotic cell structure shows a plant cell. Other eucaryotic cells may differ with respect to the cell wall and the possession of choroplasts. Note the much more elaborate internal structure compared to a typical procaryotic cell (Figure 3.3), in particular the presence of membrane-bounded organelles such as mitochondria, chloroplasts, endoplasmic reticulum and a true nucleus. From Nicklin, J, Graeme-Cook, K & Killington, R: Instant Notes in Microbiology, 2nd edn, Bios Scientific Publishers, 2002. Reproduced by permission of Thomson Publishing Services

The nucleus

The principal difference between procaryotic and eucaryotic cells, and the one that gives the two forms their names, lies in the accommodation of their genetic material. Eucaryotic cells have a true nucleus, surrounded by a nuclear membrane. This is in fact a double membrane; it contains pores, through which messenger RNA leaves the nucleus on its way to the ribosomes during protein synthesis (see Chapter 11).

The organisation of genetic material in eucaryotes is very different from that in procaryotes. Instead of existing as a single closed loop, the DNA of eucaryotes is organised into one or more pairs of *chromosomes*. The fact that they occur in pairs highlights another important difference from procaryotes: eucaryotes are genetically *diploid* in at least some part of their life cycle, while procaryotes are *haploid*. The DNA of eucaryotic chromosomes is linear in the sense that it has free ends; however, because there is so much of it, it is highly condensed and wound around proteins called *histones*. These carry a strong positive charge and associate with the negatively charged phosphate groups on the DNA.

As well as the chromosomes, the nucleus also contains the *nucleolus*, a discrete structure rich in RNA, where ribosomes are assembled. The ribosomes themselves have the same function as their procaryotic counterparts; the differences in size have already been discussed (see Table 3.3). They may be found free in the cytoplasm or associated with the endoplasmic reticulum (see below), depending on the type of protein they synthesise.

A cell containing only one copy of each chromosome is said to be haploid. The term is also applied to organisms made up of such cells The haploid state is often denoted as N. (c.f. diploid (2N): containing two copies of each chromosome)

A histone is a basic protein found associated with DNA in eucaryotic chromosomes.

Endoplasmic reticulum

Running throughout the cell and taking up much of its volume, the endoplasmic reticulum (ER) is a complex membrane system of tubes and flattened sacs. The presence of numerous ribosomes on their surface gives those parts of the ER involved in protein synthesis a granular appearance when seen under the electron microscope, giving rise to the name *rough ER*. Areas of the ER not associated with ribosomes are known as *smooth ER*; this is where the synthesis of membrane lipids takes place. The ER also serves as a communications network, allowing the transport of materials between different parts of the cell.

Golgi apparatus

The Golgi apparatus is another membranous organelle, comprising a set of flattened vesicles, usually arranged in a stack called a *dictyosome*. The function of the Golgi apparatus is to package newly synthesised substances such as proteins and assist in their transport away from the cell. The substances are contained in vesicles that are released from the main part of the complex, and fuse with the cytoplasmic membrane. The Golgi apparatus is poorly defined in certain fungi and protozoans.

Lysosomes

Another function of the Golgi apparatus is to package certain hydrolytic (digestive) enzymes into membrane-bound packets called *lysosomes*. The enzymes are needed to digest nutrient molecules that enter the cell by *endocytosis* (Figure 3.14), and would break down the fabric of the cell itself if they were not contained within the lysosomes. *Peroxisomes* are similar to lysosomes, but smaller, and also contain degradative enzymes. They contain the enzyme catalase, which breaks down the potentially toxic hydrogen peroxide generated by other breakdown reactions within the peroxisome.

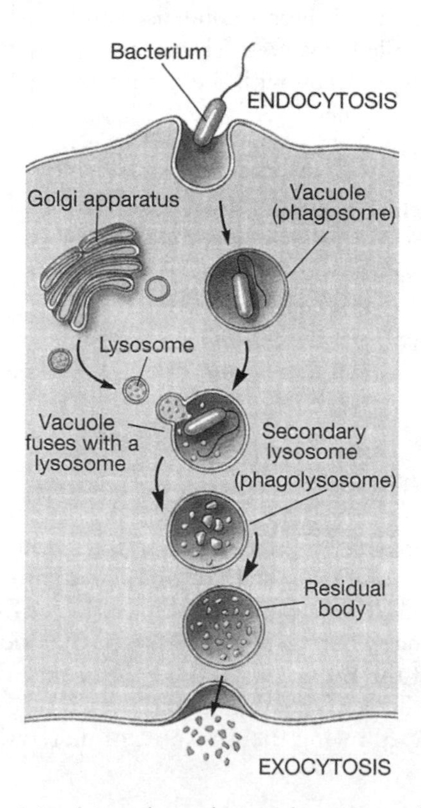

Figure 3.14 Endocytosis. Membrane-bound vacuoles surround a food particle and internalise it in the form of a *phagosome*. This fuses with a lysosome, which releases digestive enzymes, resulting in the breakdown of the contents. The process of endocytosis is unique to eucaryotic cells. From Black, JG: Microbiology: Principles and Explorations, 4th edn, John Wiley & Sons Inc., 1999. Reproduced by permission of the publishers

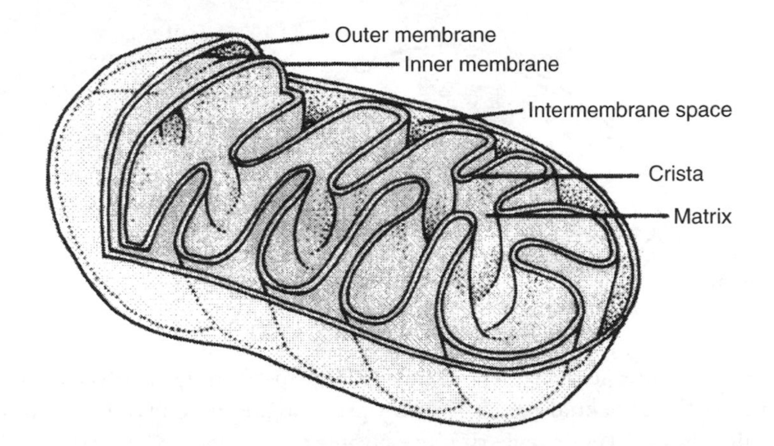

Figure 3.15 Mitochondrial structure. The inner membrane, the location of the electron transport chain in aerobic respiration, is formed by the invagination of the more permeable outer membrane. Mitochondria have similar dimensions to many bacteria (approx. 1–3 μm), but may vary in shape due to the plasticity of their membranes

Mitochondria

Whereas in procaryotes the enzymes involved in adenosine triphosphate generation (see Chapter 6) are associated with the plasma membrane, in eucaryotes they are found in specialised organelles called mitochondria. These are generally rod-shaped and may be present in large numbers. They are enclosed by a double membrane, the inner surface of which is folded into finger-like projections called *cristae*. Respiratory enzymes are located on the increased surface area this provides, while other metabolic reactions take place in the semi-fluid matrix (Figure 3.15) (see also Chapter 6).

The mitochondrial cristae of algae, fungi and protozoans each have their own characteristic shapes. Until very recently, a few primitive protozoans, such as *Giardia*, appeared to lack mitochondria completely, and were thought to represent an intermediate stage in the evolution of the eucaryotic condition. Recent research, however, has shown them to possess highly reduced remnants of mitochondria, which have been given the name *mitosomes*. It seems that such organisms did, after all, once possess mitochondria, but have subsequently lost much of their function – an example of so-called reductive evolution.

Chloroplasts

Chloroplasts are specialised organelles involved in the process of *photosynthesis*, the conversion of light into cellular energy. As such, they are characteristic of green plants and algae. Like mitochondria, chloroplasts are surrounded by a double membrane, and serve as the location for energy-generating reactions. Inside the chloroplast are

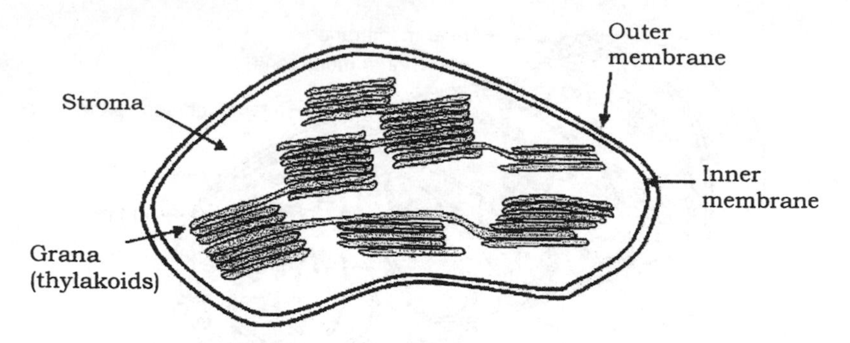

Figure 3.16 Chloroplast structure. Adenosine triphosphate generation from photosynthesis occurs on the thylakoid membranes. In green algae these take the form of discrete structures called grana. The enzyme ribulose bisphosphate carboxylase, responsible for fixing carbon dioxide via the Calvin cycle (see Chapter 6) is located in the stroma. The outer membrane of chloroplasts is relatively permeable, allowing the diffusion of the products of photosynthesis into the surrounding cytoplasm. Reproduced by permission of Dr Lance Gibson, Iowa State University

flattened membranous sacs known as *thylakoids*, which contain the photosynthetic pigment *chlorophyll*. Thylakoids are arranged in stacks called *grana* (Figure 3.16).

Mitochondria and chloroplasts both contain 70S ribosomes (similar to those found in procaryotes), a limited amount of circular DNA and the means to replicate themselves. This is seen as key evidence for the *endosymbiotic theory* of eucaryotic evolution, which envisages that specialised organelles within eucaryotic cells arose from the ingestion of small procaryotes, which over a long period of time lost their independent existence.

Vacuoles

Vacuoles are membrane-covered spaces within cells, and derive from the Golgi apparatus. They act as stores for various nutrients, and also for waste products. Some types of vacuole are important in regulating the water content of the cell.

Plasma membrane

Many eucaryotes do not have cell walls, so the plasma membrane represents the outermost layer of the cell. The sterols mentioned earlier are important in helping these cells to resist the effects of osmotic pressure. The only procaryotes to contain sterols are the mycoplasma, which are unusual in not possessing the typical bacterial cell wall. Although the eucaryotic plasma membrane does not have the role in cellular respiration associated with its procaryotic counterpart, it does have additional functions. The process of endocytosis (and its reverse, *exocytosis*), by which particles or large soluble molecules are enveloped and brought into the cell, is carried out at the plasma membrane. Also, carbohydrate residues in the membrane act as receptors for cell-to-cell recognition, and may be involved in cell adhesion.

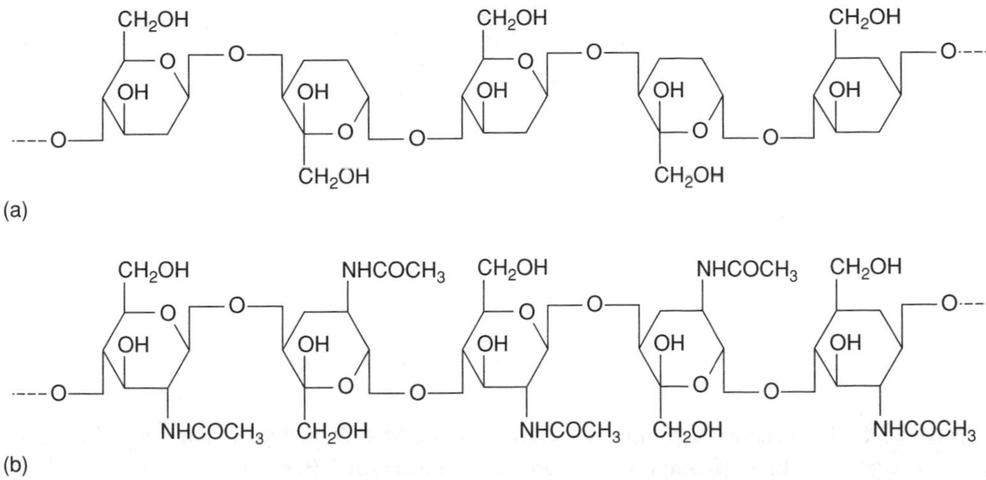

Figure 3.17 The structures of (a) cellulose and (b) chitin. Cellulose is composed of repeating glucose units joined by β-1,4 linkages, and chitin is a polymer of *N*-acetylglucosamine

Cell wall

As we have just noted, not all eucaryotes possess a cell wall; among those that do are fungi, algae and plants. Whilst the function, like that of procaryotes, is to give strength to the cell, the chemical composition is very different, generally being a good deal simpler. The cell walls of plants, algae and lower members of the fungi are based on *cellulose* (Figure 3.17a), a repeating chain of glucose molecules joined by β-1,4 linkages, and may also include pectin and hemicellulose, both also polymers of simple sugars. Most fungi such as yeasts and mushrooms contain *chitin*, a polymer of *N*-acetylglucosamine (Figure 3.17b: we have encountered *N*-acetylglucosamine before, as a component of peptidoglycan in bacterial walls.) Chitin is also to be found as the major component of insect and crustacean exoskeletons, where the function is also to provide strength and rigidity. As in procaryotes, the cell wall plays little part in the exchange of materials between the cell and its environment, a role fulfilled by the plasma membrane.

Some protozoans and unicellular algae are surrounded by a flexible *pellicle* made of protein.

Flagella and cilia

Motility in eucaryotic cells may be achieved by means of flagella or *cilia*; cilia can be thought of as, essentially, short flagella. Both are enclosed within the plasma membrane and anchored by means of a basal body. Flagellated cells generally have a single flagellum, whereas cilia are often present in very large numbers on each cell. In the microbial world, flagella are found in protozoans and motile algal forms, whilst cilia are mostly found in a class of protozoans called the Ciliophora. Flagella and cilia are not found in members of the Fungi. Although they share the same thread-like gross morphology,

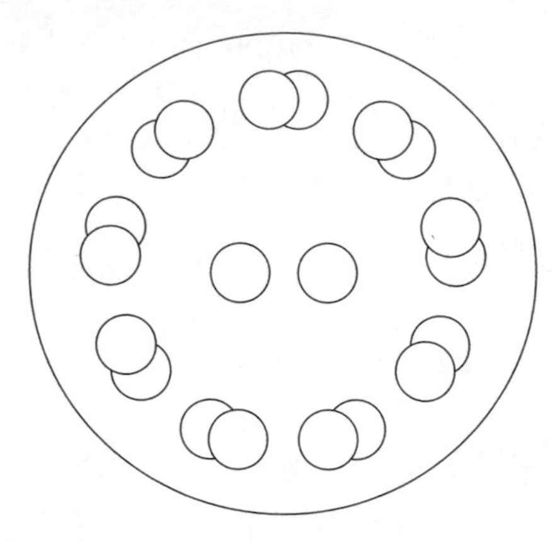

Figure 3.18 Eucaryotic flagella have a characteristic '9 + 2' structure. Although functionally analogous to their procaryotic counterparts, eucaryotic flagella differ appreciably in their fine structure. A membrane surrounds an arrangement of proteinaceous microtubules, in which nine pairs surround a single central pair. Movement of eucaryotic flagella is by means of an adenosine triphosphate-driven whiplike motion

eucaryotic flagella differ dramatically in their ultrastructure from those of procaryotes. Seen in cross-section, they have a very characteristic appearance, made up of two central *microtubules*, surrounded by a further nine pairs arranged in a circle (Figure 3.18). The microtubules are made of a protein called *tubulin*. Flagella in eucaryotes beat in waves, rather than rotating; cilia, present in large numbers, beat in a coordinated fashion so that some are in forward motion while others are in the recovery stroke (rather like a 'Mexican wave'!). In animals, ciliary motion has been adapted to move particulate matter across a tissue surface; ciliated cells of the respiratory tract, for example, act as a first line of defence in the removal of inhaled particles, such as bacteria from the airways.

Cell division in procaryotes and eucaryotes

In, unicellular procaryotes, cell division by *binary fission* leads to the creation of a new individual. Growth occurs in individual cells until a maximum size is achieved and a cross-wall forms. Before cell division takes place, the genetic material must replicate itself (see Chapter 11), and one copy pass to each new daughter cell (Figure 3.19).

Cell division in eucaryotes also results in two identical daughter cells. In the case of unicellular eucaryotes, this results in two individual organisms (asexual reproduction), while in multicellular forms there is an increase in overall size. Cell division is preceded by a process of nuclear division called *mitosis*, which ensures that both daughter cells receive a full complement of chromosomes. The principal phases of mitosis are summarised in Figure 3.20(a). In *interphase*, the chromosomes are not clearly visible under the microscope; DNA replication takes place during this period. The duplicated chromosomes, held together as sister *chromatids* by the centromere, move towards the centre of the cell during *prophase*. A series of microtubules form a spindle between

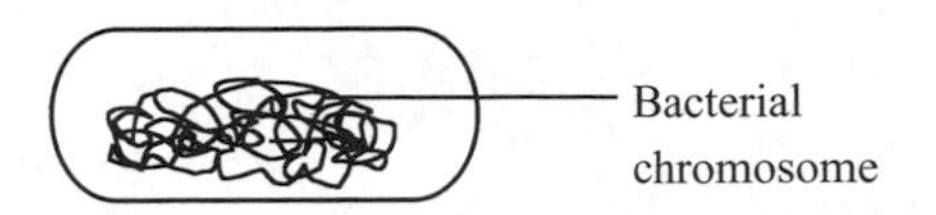

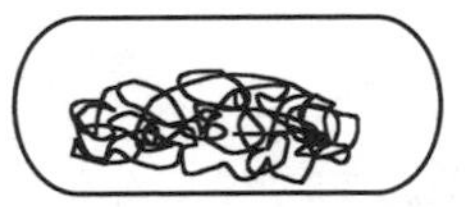

DNA replicates, making a
second copy of the chromosome.
Origins of replication migrate to
ends of cell.

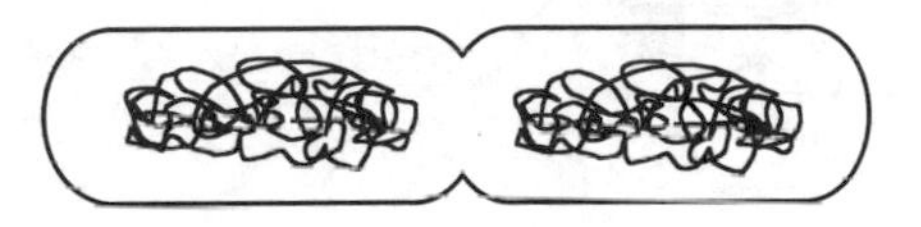

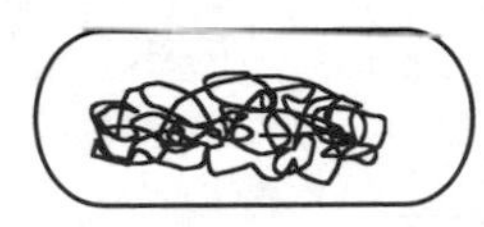
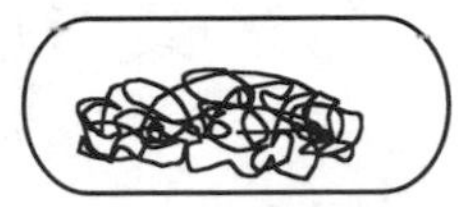

Cell lengthens and new
cell wall is laid down

starts to grow inwards

Septum formation is
complete and daughter

Figure 3.19 Binary fission in *E. coli*. Replication of the single circular chromosome is accompanied by an increase in cell size. The plasma membrane invaginates, and a new cross-wall is synthesised, resulting in two new daughter cells

the centrioles, and the chromosomes line up along this during *metaphase*. Also, during this phase the nuclear membrane breaks down, and each centromere duplicates. One chromosome from each pair then migrates away from the centre to opposite ends of the spindle. This stage is called *anaphase*. Finally, in *telophase*, new nuclear membranes surround the two sets of chromosomes, to form two nuclei. Mitosis is followed by cell division. Overall, the process of mitosis results in two identical nuclei containing the original (diploid) chromosome number.

At various stages of eucaryotic life cycles, a process of *meiosis* may occur, which halves the total number of chromosomes, so that each nucleus only contains one copy of each. In sexual reproduction, the haploid gametes are formed in this way, and the diploid condition is restored when two different gametes fuse. In some eucaryotes, not just the gametes but a substantial part of the life cycle may occur in the haploid form (see Chapters 8 & 9). Meiosis (Figure 3.20b) comprises two nuclear divisions, the second of which is very similar to the process of mitosis just described. In the first meiotic division, homologous chromosomes (i.e. the two members of a pair) line up on the spindle together and eventually migrate to opposite poles. While they are together, it

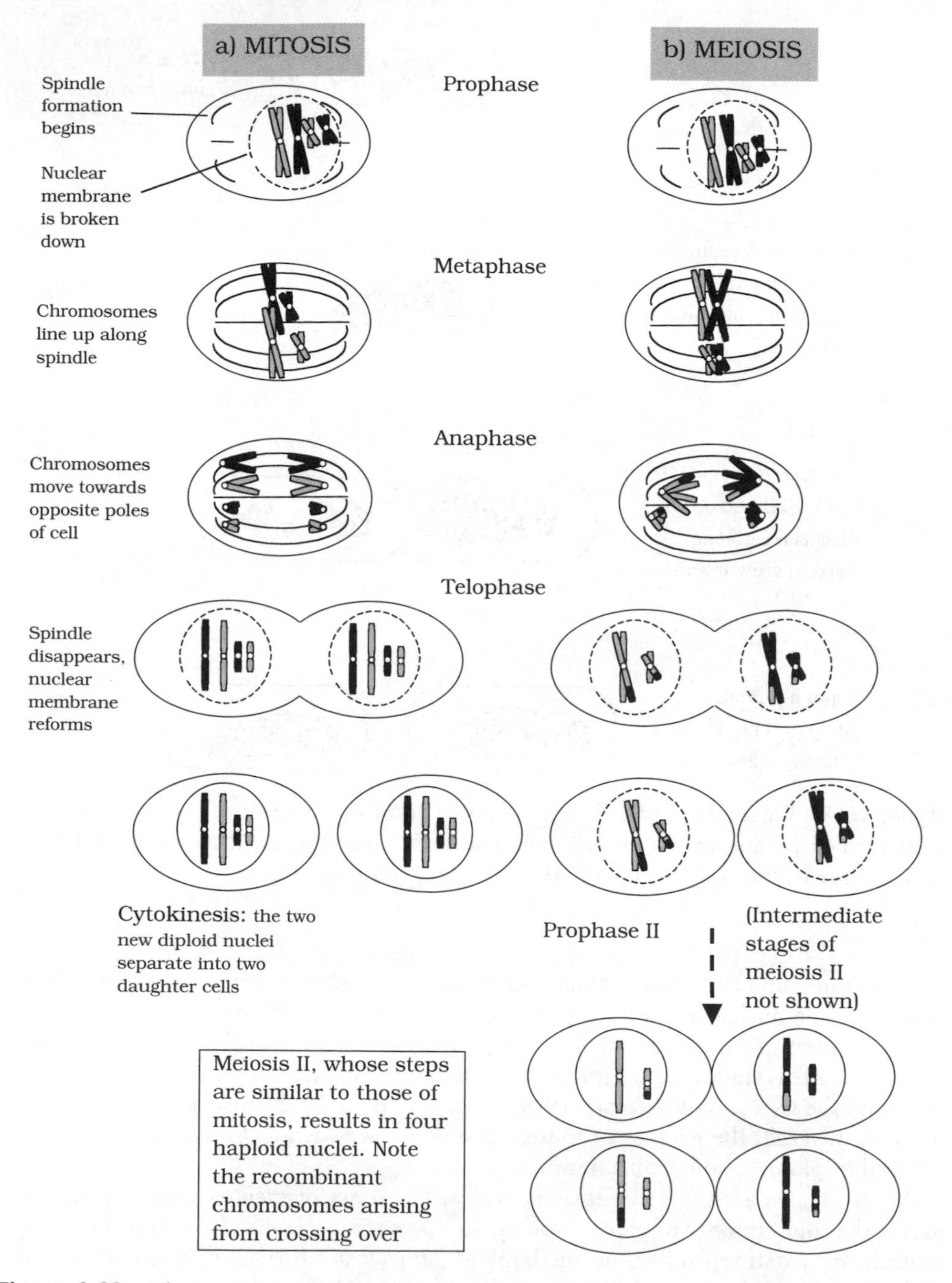

Figure 3.20 The main steps of (a) mitosis and (b) meiosis in an organism whose diploid number (2n) =4. Mitosis results in two cells identical to the parent. Meiosis results in a reduction in the chromosome number and introduces genetic variation by means of crossing over. For details see the text

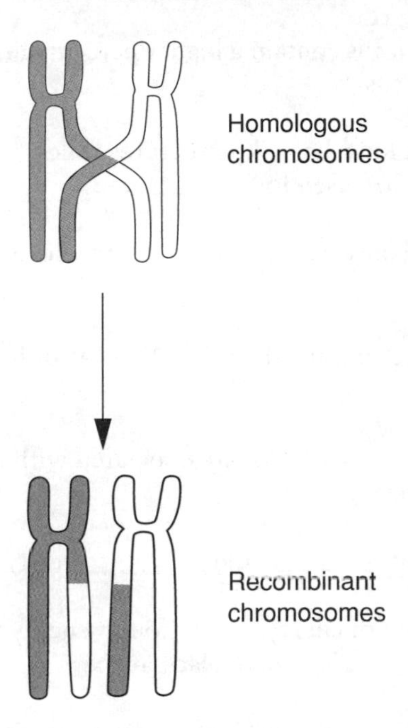

Figure 3.21 Crossing over leads to recombination of genetic material. During crossing over, portions of homologous chromosomes are exchanged. This forms the basis of genetic recombination in eucaryotes, and ensures that offspring contain new combinations of genetic material

is possible for *crossing over* to occur, a process by which the two chromosomes swap homologous stretches of DNA (Figure 3.21). Since these may not be identical, crossing over serves to introduce genetic variation into the daughter nuclei. In the second meiotic division, sister chromatids separate as before, resulting in four haploid nuclei.

Test yourself

1 Procaryotic cells have a much simpler structure than eucaryotes, lacking internal ________ and a true ________.

2 Most bacterial cells are rod-shaped (________), spherical (________) or curved (________).

3 Many bacteria commonly carry extrachromosomal pieces of DNA called ________, which are able to ________ independently of the bacterial chromosome.

4 Protein synthesis takes place at ________.

5 The main components of cell membranes are ________ and ________.

6 Gram-positive cell walls contain a higher percentage of ________ than those of Gram-negative cells.

7 Many bacteria have long, hair-like structures called ________ projecting from the cell wall. These are used for ________.

8 The DNA of eucaryotes is organized into chromosomes and associated with proteins called ________.

9 In eucaryotic cells, extranuclear DNA is also found in ________ and ________.

10 Eucaryotic ribosomes may be found associated with the ________ ________ or free in the cytoplasm.

11 The Golgi apparatus ________ and ________ newly synthesised substances.

12 ________ are the site of energy generation in eucaryotic cells. In procaryotic cells, some of these reactions take place at the ________ ________.

13 The photosynthetic membranes of chloroplasts are called ________.

14 The cell walls of algae are mostly made up of ________.

15 The structure of eucaryotic flagella is more complex than that of procaryotes, comprising an arrangement of ________ made of ________.

Part II

Microbial Nutrition, Growth and Metabolism

4

Microbial Nutrition and Cultivation

In Chapter 2 we introduced the major groups of macromolecules found in living cells; the raw materials from which these are synthesised are ultimately derived from the organism's environment in the form of nutrients (Table 4.1). These can be conveniently divided into those required in large quantities* (macronutrients) and those which are needed only in trace amounts (micronutrients or trace elements).

You will recall that carbon forms the central component of proteins, carbohydrates, nucleic acids and lipids; indeed, the living world is based on carbon, so it should come as no surprise that this is the most abundant element in all living cells, microbial or otherwise. Of the other macronutrients, nitrogen, oxygen, hydrogen, sulphur and phosphorus are also constituents of biological macromolecules, while the remainder (magnesium, potassium, sodium, calcium and iron in their ionised forms) are required in lesser quantities for a range of functions that will be described in due course. Micronutrients are all metal ions, and frequently serve as cofactors for enzymes.

All microorganisms must have a supply of the nutrients described above, but they show great versatility in the means they use to satisfy these requirements.

The metabolic processes by which microorganisms assimilate nutrients to make cellular material and derive energy will be reviewed in Chapter 6. In the following section we briefly describe the role of each element, and the form in which it may be acquired.

Carbon is the central component of the biological macromolecules we discussed in Chapter 2. Carbon incorporated into biosynthetic pathways may be derived from organic or inorganic sources (see below); some organisms can derive it from CO_2, while others require their carbon in 'ready-made', organic form.

Hydrogen is also a key component of macromolecules, and participates in energy generation processes in most microorganisms. In autotrophs (see 'Nutritional categories' below), hydrogen is required to reduce carbon dioxide in the synthesis of macromolecules.

Oxygen is of central importance to the respiration of many microorganisms, but in its molecular form (O_2), it can be toxic to some forms (see Chapter 5). These obtain the oxygen they need for the synthesis of macromolecules from water.

* Everything is relative in the microbial world; a typical bacterial cell weighs around three tenmillion millionths (3×10^{-13}) of a gram!

Table 4.1 Elements found in living organisms

Element	Form in which usually supplied	Occurrence in biological systems
Macronutrients		
Carbon (C)	CO_2, organic compounds	Component of all organic molecules, CO_2
Hydrogen (H)	H_2O, organic compounds	Component of biological molecules, H^+ released by acids
Oxygen (O)	O_2, H_2O, organic compounds	Component of biological molecules; required for aerobic metabolism
Nitrogen (N)	NH_3, NO_3^-, N_2, organic N compounds	Component of proteins, nucleic acids
Sulphur (S)	H_2S, SO_4^{2-}, organic S compounds	Component of proteins; energy source for some bacteria
Phosphorus (P)	PO_4^{3-}	Found in nucleic acids, ATP, phospholipids
Potassium (K)	In solution as K^+	Important intracellular ion
Sodium (Na)	In solution as Na^+	Important extracellular ion
Chlorine (Cl)	In solution as Cl^-	Important extracellular ion
Calcium (Ca)	In solution as Ca^{2+}	Regulator of cellular processes
Magnesium (Mg)	In solution as Mg^{2+}	Coenzyme for many enzymes
Iron (Fe)	In solution as Fe^{2+} or Fe^{3+} or as FeS, $Fe(OH_3)$ etc	Carries oxygen; energy source for some bacteria
Micronutrients	Present as contaminants at very low concentrations	
Copper (Cu)	In solution as Cu^+, Cu^{2+}	Coenzyme; microbial growth inhibitor
Manganese (Mn)	In solution as Mn^{2+}	Coenzyme
Cobalt (Co)	In solution as Co^{2+}	Vitamin B_{12}
Zinc (Zn)	In solution as Zn^{2+}	Coenzyme; microbial growth inhibitor
Molybdenum (Mo)	In solution as Mo^{2+}	Coenzyme
Nickel (Ni)	In solution as Ni^{2+}	Coenzyme

Nitrogen is needed for the synthesis of proteins and nucleic acids, as well as for important molecules such as ATP (you will learn more about ATP and its role in the cell's energy relations in Chapter 6). Microorganisms range in their demands for nitrogen from those that are able to assimilate ('fix') gaseous nitrogen (N_2) to those that require all 20 amino acids to be provided preformed. Between these two extremes come species that are able to assimilate nitrogen from an inorganic source such as nitrate, and those that utilise ammonium salts or urea as a nitrogen source.

Table 4.2 Selected microbial growth factors

Growth factor	Function
Amino acids	Components of proteins
p-Aminobenzoic acid	Precursor of folic acid, involved in nucleic acid synthesis
Niacin (nicotinic acid)	Precursor of NAD^+ and $NADP^+$
Purines & pyrimidines	Components of nucleic acids
Pyridoxine (vitamin B_6)	Amino acid synthesis
Riboflavin (vitamin B_2)	Precursor of FAD

Sulphur is required for the synthesis of proteins and vitamins, and in some types is involved in cellular respiration and photosynthesis. It may be derived from sulphur-containing amino acids (methionine, cysteine), sulphates and sulphides.

Phosphorus is taken up as inorganic phosphate, and is incorporated in this form into nucleic acids and phospholipids, as well as other molecules such as ATP.

Metals such as copper, iron and magnesium are required as *cofactors* in enzyme reactions.

A cofactor is a non-protein component of an enzyme (often a metal ion) essential for its normal functioning.

Many microorganisms are unable to synthesise certain organic compounds necessary for growth and must therefore be provided with them in their growth medium. These are termed *growth factors* (Table 4.2), of which three main groups can be identified: amino acids, purines and pyrimidines (required for nucleic acid synthesis) and vitamins. You will already have read about the first two of these groups in Chapter 2. Vitamins are complex organic compounds required in very small amounts for the cell's normal functioning. They are often either *coenzymes* or their precursors (see Chapter 6). Microorganisms vary greatly in their vitamin requirements. Many bacteria are completely self-sufficient, while protozoans, for example, generally need to be supplied with a wide range of these dietary supplements. A vitamin requirement may be absolute or partial; an organism may be able, for example, to synthesise enough of a vitamin to survive, but grow more vigorously if an additional supply is made available to it.

Nutritional categories

Microorganisms can be categorised according to how they obtain their carbon and energy. As we have seen, carbon is the most abundant component of the microbial cell, and most microorganisms obtain their carbon in the form of organic molecules, derived directly or indirectly from other organisms. This mode of nutrition is the one that is familiar to us as humans (and all other animals); all the food we eat is derived as complex organic molecules from plants and other animals (and even some representatives of the microbial world such as mushrooms!). Microorganisms which obtain their carbon in this way are described as *heterotrophs*, and include all the fungi and protozoans as well as most types of bacteria. Microorganisms as a group are able to incorporate the carbon from an incredibly wide range of organic compounds into cellular material. In fact there is hardly any such compound occurring in nature that cannot be metabolised by some microorganism or other, explaining in part why microbial life is to be found thriving in the most unlikely habitats. Many synthetic materials can also serve as carbon sources for some microorganisms, which can have considerable economic significance.

A heterotroph must use one or more organic compounds as its source of carbon.

A significant number of bacteria and all of the algae do not, however, take up their carbon preformed as organic molecules in this way, but derive it instead from carbon dioxide. These organisms are called *autotrophs*, and again we can draw a parallel with higher organisms, where all members of the plant kingdom obtain their carbon in a similar fashion.

An autotroph can derive its carbon from carbon dioxide.

We can also categorise microorganisms nutritionally by the way they derive the energy they require to carry out essential cellular reactions. Autotrophs thus fall into two categories. *Chemoautotrophs* obtain their energy as well as their carbon from inorganic sources; they do this by the oxidation of inorganic molecules such as sulphur or nitrite. *Photoautotrophs* have photosynthetic pigments enabling them to convert light energy into chemical energy. The mechanisms by which this is achieved will be discussed in Chapter 6.

A chemotroph obtains its energy from chemical compounds. A phototroph uses light as its source of energy.

The great majority of heterotrophs obtain energy as well as carbon from the same organic source. Such organisms release energy by the chemical oxidation of organic nutrient molecules, and are therefore termed *chemoheterotrophs*. Those few heterotrophs which do not follow this mode of nutrition include the green and purple non-sulphur bacteria. These are able to carry out photosynthesis and are known as *photoheterotrophs*.

There is one final subdivision of nutritional categories in microorganisms! Whether organisms are chemotrophs or phototrophs, they need a molecule to act as a source of electrons (reducing power) to drive their energy-generating systems (see Chapter 6). Those able to use an inorganic electron donor such as H_2O, H_2S or ammonia are called *lithotrophs*, while those requiring an organic molecule to fulfil the role are *organotrophs*. Most (but not all) microorganisms are either photolithotrophic autotrophs (algae, blue-greens) or chemo-organotrophic heterotrophs (most bacteria). For the latter category, a single organic compound can often act as the provider of carbon, energy and reducing power. The substance used by chemotrophs as an energy source may be organic (chemoorganotrophs) or inorganic (chemolithotrophs).

A lithotroph is an organism that uses inorganic molecules as a source of electrons. An organotroph uses organic molecules for the same purpose.

How do nutrients get into the microbial cell?

Having found a source of a given nutrient, a microorganism must:

- have some means of taking it up from the environment
- possess the appropriate enzyme systems to utilise it.

The plasma membrane represents a selective barrier, allowing into the cell only those substances it is able to utilise. This selectivity is due in large part to the hydrophobic nature of the lipid bilayer. A substance can be transported across the cell membrane in one of three ways, known as simple diffusion, facilitated diffusion and active transport.

In *simple* diffusion, small molecules move across the membrane in response to a concentration gradient (from high to low), until concentrations on either side of the membrane are in equilibrium. The ability to do this depends on being small (H_2O, Na^+, Cl^-) or soluble in the lipid component of the membrane (non-polar gases such as O_2 and CO_2).

Larger polar molecules such as glucose and amino acids are unable to enter the cell unless assisted by membrane-spanning *transport proteins* by the process of *facilitated* diffusion (Figure 4.1). Like enzymes, these proteins are specific for a single/small number of related solutes; another parallel is that they too can become saturated by too much 'substrate'. As with simple diffusion, there is no expenditure of cellular energy, and an inward concentration gradient is required. The transported substance tends to be metabolised rapidly once inside the cell, thus maintaining the concentration gradient from outside to inside.

Diffusion is only an effective method of internalising substances when their concentrations are greater outside the cell than inside. Generally, however, microorganisms find themselves in very dilute environments; hence the concentration gradient runs in the other direction, and diffusion into the cell is not possible. *Active transport* enables the cell to overcome this unfavourable gradient. Here, regardless of the direction of the gradient, transport takes place in one direction only, *into* the cell. Energy, derived from

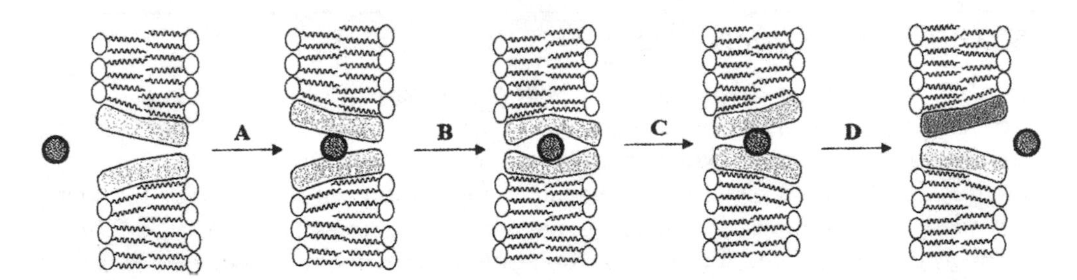

Figure 4.1 In facilitated diffusion, substances can move across the plasma membrane by binding to an embedded transport protein (shown shaded). No energy input is required, but the diffusion can only occur from an area of high concentration to one of low concentration. From Thomas, G: Medicinal Chemistry, an Introduction, John Wiley & Sons Inc., 2000. Reproduced by permission of the publishers

hydrolysis of adenosine triphosphate (see Chapter 6) is required to achieve this, and again specific transmembrane proteins are involved. They bind the solute molecule with high affinity outside of the cell, and then undergo a conformational change that causes them to be released into the interior. Procaryotic cells can carry out a specialised form of active transport called *group translocation*, whereby the solute is chemically modified as it crosses the membrane, preventing its escape. A well-studied example of this is the phosphorylation of glucose in *E. coli* by the phosphotransferase system. Glucose present in very low concentrations outside the cell can be concentrated within it by this mechanism. Glucose is unable to pass back across the membrane in its phosphorylated form (glucose-6-phosphate), however it can be utilised in metabolic pathways in this form.

Often it may be necessary to employ extracellular enzymes to break down large molecules before any of these mechanisms can be used to transport nutrients into the cell.

Laboratory cultivation of microorganisms

Critical to the development of microbiology during its 'golden age' was the advance in culturing techniques, enabling the isolation and pure culture of specific microorganisms. The study of pure cultures made it possible to determine the properties of a specific organism such as its metabolic characteristics or its ability to cause a particular disease. It also opened up the possibility of classifying microorganisms, on the basis of the characteristics they display in pure culture.

The artificial culture of any organism requires a supply of the necessary nutrients, together with the provision of appropriate conditions such as temperature, pH and oxygen concentration. The nutrients and conditions provided in the laboratory are usually a reflection of those found in the organism's natural habitat. It is also essential that appropriate steps are taken to avoid contamination (Box 4.1). In the next section we shall describe the techniques used to isolate and propagate microorganisms in the laboratory. The section refers specifically to the culture of bacteria; laboratory propagation of algae, fungi and viruses will be referred to in the chapters devoted to those groups.

Box 4.1 Aseptic technique

Most commonly used culture media will support the growth of a number of different bacteria. It is therefore essential when working in the microbiology laboratory that suitable precautions are taken to prevent the growth of unwanted contaminants in our cultures. These simple practical measures are termed *aseptic technique*, and it is essential to master them if reliable experimental results are to be obtained. Any glassware and equipment used is sterilised before work begins. Containers such as tubes, flasks and plates are kept open for the minimum amount of time, and the necks of bottles and tubes are passed through a flame to maintain their sterility. The wire loops and needles used to transfer small volumes of microbial cultures are sterilised by heating them to redness in a flame. Your instructor will normally demonstrate aseptic technique to you in an early practical session.

Obtaining a pure culture

Microorganisms in the natural world do not live in pure cultures; they exist as part of complex ecosystems comprising numerous other organisms. The first step in the cultivation of microorganisms is therefore the creation of a pure culture. A key development for the production of pure cultures was the ability to grow microorganisms on a solid medium. Koch had noticed that when a nutrient surface such as cut potato was exposed to air, individual microbial *colonies* grew up, and he inferred from this that these had each arisen from the numerous divisions of single cells.

Bacteria may be cultured using either liquid or solid media. Solid media are particularly useful in the isolation of bacteria; they are also used for their long-term storage. Liquid (broth) cultures are used for rapid and large-scale production of bacteria.

It soon became apparent that a number of organisms would not grow on potatoes, so Koch and his colleagues turned to gelatin as a means of solidifying a synthetic nutrient growth medium. Horizontal slabs were cut, and covered to help keep them free from atmospheric contaminants. Gelatin was a convenient means of solidifying media, as it could be boiled and then allowed to set in the desired vessel. There were two main drawbacks to its use, however; many organisms needed to be incubated at around body temperature (37 °C), and gelatin melted before this temperature was reached. Also, it was found that a number of bacteria were capable of utilising gelatin as a nutrient source, resulting in the liquefaction of the gel.

A culture consisting entirely of one strain of organism is called a *pure* or *axenic* culture. In theory, such a culture represents the descendants of a single cell.

A more suitable alternative was soon found in the form of *agar*. This is a complex polysaccharide derived from seaweeds, and was suggested by the wife of one of Koch's colleagues, who had used it as a setting agent in jam making. Agar does not melt until near boiling point; this means that cultures can be incubated at 37 °C or above without the medium melting. Moreover, when it cools, agar remains molten until just over 40 °C, allowing heat-sensitive media components such as blood to be added. In addition, most bacteria can tolerate a short exposure to temperatures in this range, so they too can be inoculated into molten agar (see pour plate method below). Crucially, agar is *more or less inert nutritionally*; only a very few organisms are known that are able to use agar as a food source; consequently, it is the near ideal setting agent, resisting both thermal and microbial breakdown. Agar soon became the setting agent of choice, and has remained so ever since; shortly afterwards, Richard Petri developed the two-part culture dish that was named after him, and which could be sterilised separately from the medium and provided protection from contamination by means of its lid,. This again is still standard equipment today, although the original glass has been largely replaced by presterilised, disposable plastic.

A petri dish is the standard vessel for short-term growth of solid medium cultures in the laboratory. It comprises a circular dish with an overlapping lid.

The standard method of obtaining a pure bacterial culture is the creation of a *streak plate* (Figure 4.2). A wire inoculating loop is used to spread out a drop of bacterial suspension on an agar plate in such a way that it becomes progressively more dilute;

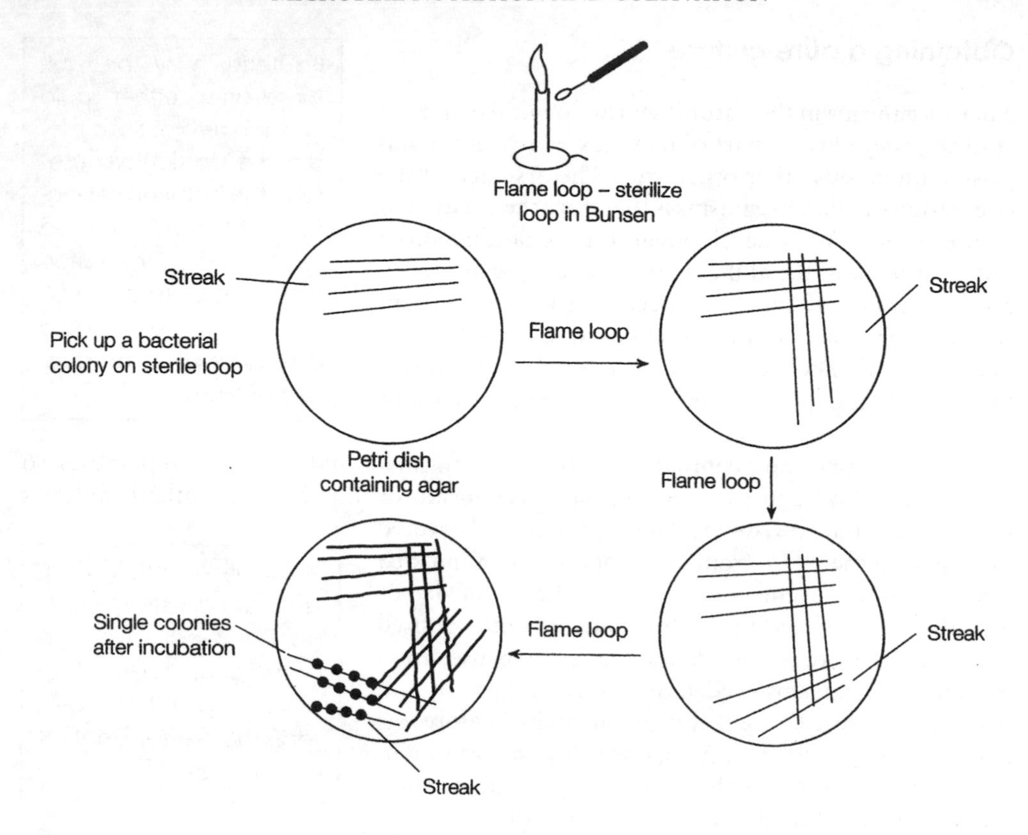

Figure 4.2 The streak plate. Streaking the sample across the agar surface eventually results in individual cells being deposited. Repeated cycles of cell division lead to the production of visible, isolated colonies. From Nicklin, J, Graeme-Cook, K & Killington, R: Instant Notes in Microbiology, 2nd edn, Bios Scientific Publishers, 2002. Reproduced by permission of Thomson Publishing Services

eventually, individual cells will be deposited on the agar surface. Following incubation at an appropriate temperature, a succession of cell divisions occurs, resulting in the formation of a bacterial *colony*, visible to the naked eye. Colonies arise because movement is not possible on the solid surface and all the progeny stay in the same place. A colony represents, in theory at least, the offspring of a single cell and its members are therefore genetically identical. (In reality, a clump of cells may be deposited together and give rise to a colony; this problem can be overcome by repeated isolation and restreaking of single colonies.)

An alternative method for the isolation of pure cultures is the *pour plate* (Figure 4.3). In this method, a dilute suspension of bacteria is mixed with warm molten agar, and poured into an empty petri plate. As the agar sets, cells are immobilised, and once again their progeny are all kept together, often within, as well as on, the agar. This method is especially useful for the isolation of bacteria that are unable to tolerate atmospheric levels of oxygen.

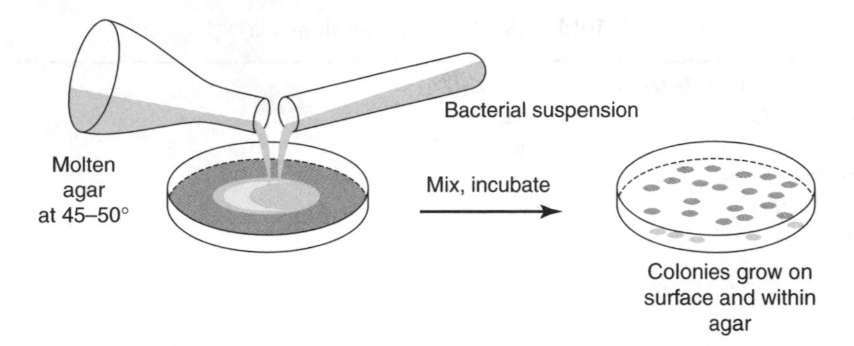

Figure 4.3 The pour plate. A sample of diluted bacterial suspension is mixed with molten agar and poured into a petri dish. Most bacteria can tolerate a short exposure to the agar, which is held at a temperature just above its setting point

Growth media for the cultivation of bacteria

A synthetic growth medium may be *defined*, that is, its exact chemical composition is known, or *undefined*. A defined growth medium may have few or many constituents, depending on the nutritional requirements of the organism in question. Examples of each are given in Table 4.3. An undefined or *complex* medium may have a variable composition due to the inclusion of a component such as blood, yeast extract or tap water (Table 4.4). Peptones are also commonly found in complex media; these are the products of partially digesting protein sources such as beef or casein. The exact composition of a complex medium is neither known nor critically important. A medium of this type would generally be chosen for the cultivation of *fastidious* bacteria such as *Neisseria gonorrhoeae* (the causative agent of gonorrhoea); it is easier and less expensive to supply the many nutrients required by such an organism in this form rather than supplying them all individually. Bacteria whose specific nutrient requirements are not known are also grown on complex media.

> A defined medium is one whose precise chemical composition is known.

> An undefined or complex medium is one whose precise chemical composition is not known.

> A fastidious organism is unable to synthesise a range of nutrients and therefore has complex requirements in culture.

Whilst media such as nutrient agar are used to support the growth of a wide range of organisms, others are specifically designed for the isolation and identification of particular types. *Selective* media such as bismuth sulphite medium preferentially support the growth of particular bacteria. The bismuth ion inhibits the growth of Gram-positive organisms as well as many Gram-negative types; this medium is used for the isolation of the

> A selective medium is one that favours the growth of a particular organism or group of organisms, often by suppressing the growth of others.

Table 4.3 Defined growth media

(a) Medium for *Acidithiobacillus ferrooxidans*

$FeSO_4.7H_2O$	40 g
$(NH_4)_2SO_4$	2 g
KH_2PO_4	0.5 g
$MgSO_4.7H_2O$	0.5 g
KCl	0.1 g
$Ca(NO_3)_2$	0.01 g
Distilled H_2O (pH 3.0) to	1 litre

(b) Medium for *Leuconostoc mesenteroides*

Component	Amount	Component	Amount
Glucose	25 g	Phenylalanine	100 mg
Sodium acetate	20 g	Proline	100 mg
NH_4Cl	3 g	Serine	50 mg
KH_2PO_4	0.6 g	Threonine	200 mg
K_2HPO_4	0.6 g	Tryptophan	40 mg
NaCl	3 g	Tyrosine	100 mg
$MgSO_4 \cdot 7H_2O$	0.2 g	Valine	250 mg
$MnSO_4 \cdot 4H_2O$	20 mg	Adenine	10 mg
$FeSO_4.7H_2O$	10 mg	Cytosine	10 mg
Alanine	200 mg	Guanine	10 mg
Arginine	242 mg	Uracil	10 mg
Aspartic acid	100 mg	Nicotinic acid	1 mg
Asparagine	400 mg	Pyridoxine	1 mg
Cysteine	50 mg	Riboflavin	0.5 mg
Glutamic acid	300 mg	Thiamine	0.5 mg
Glycine	100 mg	Ca pantothenate	0.5 mg
Histidine	62 mg	Pyridoxamine	0.3 mg
Isoleucine	250 mg	Pyridoxal	0.3 mg
Leucine	250 mg	*p*-Aminobenzoic acid	0.1 mg
Lysine	250 mg	Biotin	1 μg
Methionine	100 mg	Folic acid	10 μg
Distilled H_2O to	1 litre		

Examples of defined (synthetic) media for (a) the iron-oxidising bacterium *Acidithiobacillus ferrooxidans* and (b) the lactic acid bacterium *Leuconostoc mesenteroides*. Note how *L. mesenteroides* must be provided with numerous amino acids, nucleotides and vitamins as well as glucose as a carbon source, whereas *A. ferrooxidans* requires only mineral salts, including reduced iron to act as an energy source.

Table 4.4 Composition of an undefined growth medium

Calf brain infusion	200 g
Beef heart infusion	250 g
Proteose peptone	10 g
Glucose	2 g
NaCl	5 g
Na_2HPO_4	2.5 g
H_2O (pH 7.4)	To 1 litre

Brain heart infusion broth contains three undefined components. It is used for the culture of a wide variety of fastidious species, both bacterial and fungal.

pathogenic bacterium *Salmonella typhi*, one of the few organisms that can tolerate the bismuth. Specific media called *differential* media can be used to distinguish between organisms whose growth they support, usually by means of a coloured indicator. MacConkey agar contains lactose and a pH indicator, allowing the differentiation between lactose fermenters (red colonies) and non-lactose fermenters (white/pale pink colonies). Many media act both selectively and differentially; MacConkey agar, for example, also contains bile salts and the dye crystal violet, both of which serve to inhibit the growth of unwanted Gram-positive bacteria. Mannitol salt agar is also both selective and differential. The high (7.5 per cent) salt content suppresses growth of most bacteria, whilst a combination of mannitol and an indicator permits the detection of mannitol fermenters in a similar fashion to that just described. Sometimes, it is desirable to isolate an organism that is present in small numbers in a large mixed population (e.g. faeces or soil). *Enrichment* media provide conditions that selectively encourage the growth of these organisms; the use of blood agar in the isolation of streptococci provides an example of such a medium. Blood agar can act as a differential medium, in allowing the user to distinguish between haemolytic and non-haemolytic bacteria (see Chapter 7).

A differential medium allows colonies of a particular organism to be differentiated from others growing in the same culture.

An enrichment culture uses a selective medium to encourage the growth of an organism present in low numbers.

If we are to culture microorganisms successfully in the laboratory, we must provide appropriate physical conditions as well as providing an appropriate nutrient medium. In the next chapter, we shall examine how physical factors such as pH and temperature influence the growth of microorganisms, and describe how these conditions are provided in the laboratory.

Preservation of microbial cultures

Microbial cultures are preserved by storage at low temperatures, in order to suspend growth processes. For short periods, most organisms can be kept at refrigerator temperature (around 4 °C), but for longer-term storage, more specialised treatment is necessary. Using deep freezing or freeze-drying, cultures can be kept for many years, and then resurrected and re-cultured. Deep freezing requires rapid freezing to −70 °C to −95 °C, while freeze-drying (lyophilisation) involves freezing at slightly less extreme temperatures and removing the water content under vacuum. Long-term storage may be desirable to avoid the development of mutations or loss of cell viability.

Test yourself

1 Heterotrophic organisms acquire their carbon in an ________ form, whilst autotrophic organisms acquire theirs in an ________ form.

2 Some autotrophs can derive energy from the Sun; these are termed ________.

3 The passage of solutes into a cell across a concentration gradient is known as ________.

4 Agar has a high ________ point and a relatively low ________ point.

5 The use of agar as a setting agent was a crucial step in the development of ________ ________ techniques.

6 Successive divisions of a single cell lead to the formation of a ________ on a solid medium.

7 The chemical composition of an undefined medium is ________.

8 ________ organisms must have a wide range of organic nutrients supplied.

9 ________ media encourage the growth of chosen species, whilst ________ media prevent the growth of unwanted forms.

10 ________ (removal of the water content at low temperatures) is used in the preservation of microbial cultures.

5

Microbial Growth

When we consider growth as applied to a multicellular organism such as a tree, a fish or a human being, we think in terms of an ordered increase in the size of an individual. Growth in unicellular microorganisms such as bacteria, yeasts and protozoans, however, is more properly defined in terms of an increase in the size of a given *population*. This may be expressed as an increase in either the number of individuals or the total amount of *biomass*. Methods employed in the measurement of growth of unicellular microorganisms may be based on either of these. In this chapter we shall describe some of these methods, before considering the dynamics of microbial growth and some of the factors that affect it.

Biomass is the total amount of cellular material in a system.

Estimation of microbial numbers

Several methods exist for the measurement of bacterial numbers, most of which are also applicable to the enumeration of other unicellular forms such as yeasts. Such methods fall into two main categories: those that count total cell numbers, and those that count viable cells only.

Total cell counts are generally done by direct microscopic examination. A specialised glass slide is employed, which carries an etched grid of known area (Figure 5.1). The depth of the liquid sample is also known, so by counting the number of cells visible in the field of view, the number of cells per unit volume can be determined. The method may be made more accurate by the use of a fluorescent dye such as acridine orange, which binds to DNA, and hence avoids confusion with non-cellular debris. However, such methods cannot differentiate between living and non-living cells. Their usefulness is further limited by the fact that the smallest bacteria are difficult to resolve as individual cells by light microscopy. Other total cell count methods use cell-sorting devices, originally developed for separating blood cells in medical research. These pass the cell suspension through an extremely fine nozzle, and a detector registers the conductivity change each time a particle passes it. Again, no distinction can be made between viable and non-viable cells.

A *viable cell count*, on the other hand, is a measure of the number of *living* cells in a sample, or more specifically those capable of multiplying and producing a visible colony of cells. It is most commonly estimated by spreading a known volume of cell suspension onto an agar plate, and counting the number of colonies that arise after a period of

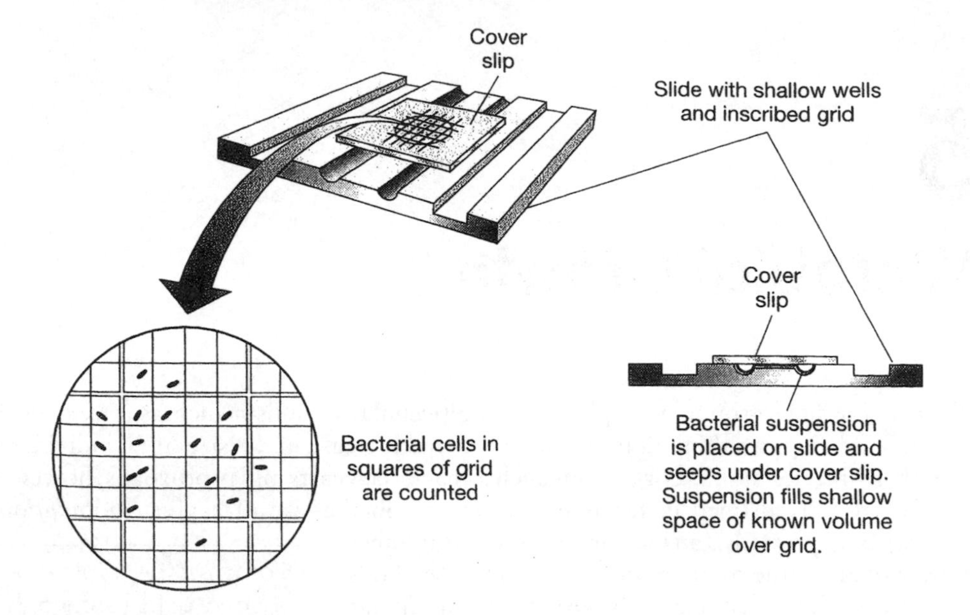

Figure 5.1 Estimation of total cell numbers by direct microscopic measurement. The Petroff-Hauser counting chamber is a specialised glass slide with an etched grid of known area. A droplet of cell suspension is placed on top, followed by a coverslip. Since the depth of liquid trapped is known, the volume covering the grid can be calculated. The number of cells present in several random squares is counted, and an average value obtained. The method does not distinguish between living and dead cells. From Black, JG: Microbiology: Principles and Explorations, 4th edn, John Wiley & Sons Inc., 1999. Reproduced by permission of the publishers

incubation (Box 5.1). The method is based on the premise that each visible colony has derived from the repeated divisions of a single cell. In reality, it is accepted that this is not always the case, and so viable counts are expressed in *colony-forming units (cfu)*, rather than cells, per unit volume. It is generally necessary to dilute the suspension before plating out, otherwise the resulting colonies will be too numerous to count. In order to improve statistical reliability, plates are inoculated in duplicate or triplicate, and the mean value is taken.

Viable cell counts can also be made using liquid media, in the *most probable number (MPN)* technique (Box 5.2). Here, a series of tubes containing a broth are inoculated with a sample of a progressively more dilute cell suspension, incubated, and examined for growth. The method is based on the statistical probability of each sample containing viable cells. It is well suited to the testing of drinking water, where low bacterial densities are to be expected.

Another method employed for the enumeration of bacteria in water is the *membrane filter test*. Here, a large volume of water is passed through a membrane filter with a pore size (0.45 μm) suitable for trapping bacteria (Figure 5.2). The filter is placed on an appropriate solid growth medium and colonies allowed to develop.

None of the methods described above provides a particularly rapid result, yet sometimes it is desirable to have an estimate of bacterial numbers immediately. A useful

Box 5.1 Estimation of viable cell numbers

In order that the sample to be plated out contains an appropriate number of cells, the original sample is subjected to *serial dilution*. In the example below, it is diluted by a factor of ten at each stage to give a final dilution of 10^{-5} (one in a hundred thousand).

Samples of 0.1 ml of each dilution are plated out on a suitable solid medium and colonies allowed to develop.

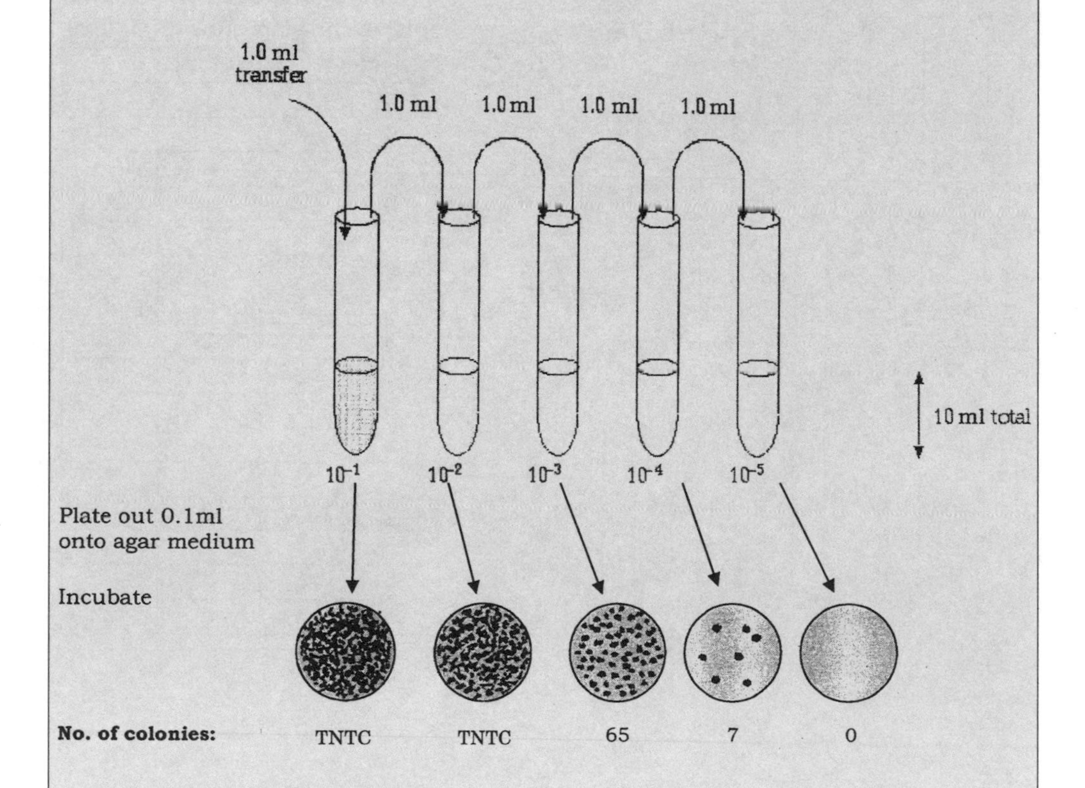

TNTC = too numerous to count

In the earlier tubes, the suspension of cells is too concentrated, resulting in too many colonies to count. In the final tube, the suspension is so dilute that there are no cells in the sample taken. The 10^{-3} dilution is used to calculate the concentration of cells in the original culture, as it falls within the range of 30–300 colonies regarded as being statistically reliable.

Colony count (10^{-3} dilution) = 65

$\therefore$ Number of cfu per ml diluted suspension $= 65 \times 10 = 650$

$\therefore$ Number of cfu per ml original susp'n $= 650 \times 10^3 = 6.5 \times 10^5$

Box 5.2 The most probable number (MPN) method

In the example below, three sets of five tubes of broth were inoculated with 10 ml, 1 ml and 0.1 ml of a water sample. The tubes were incubated to allow any bacteria present to multiply in number, and were scored as 'growth' (dark shading) or 'no growth' (no shading). The cell density statistically most likely to give rise to the result obtained (5-3-1) is then looked up on a set of MPN tables. The table (only part shown) indicates that there is a 95% probability that the sample fell within the range 40–300 cells/ml, with 110 cells/ml being the most likely value.

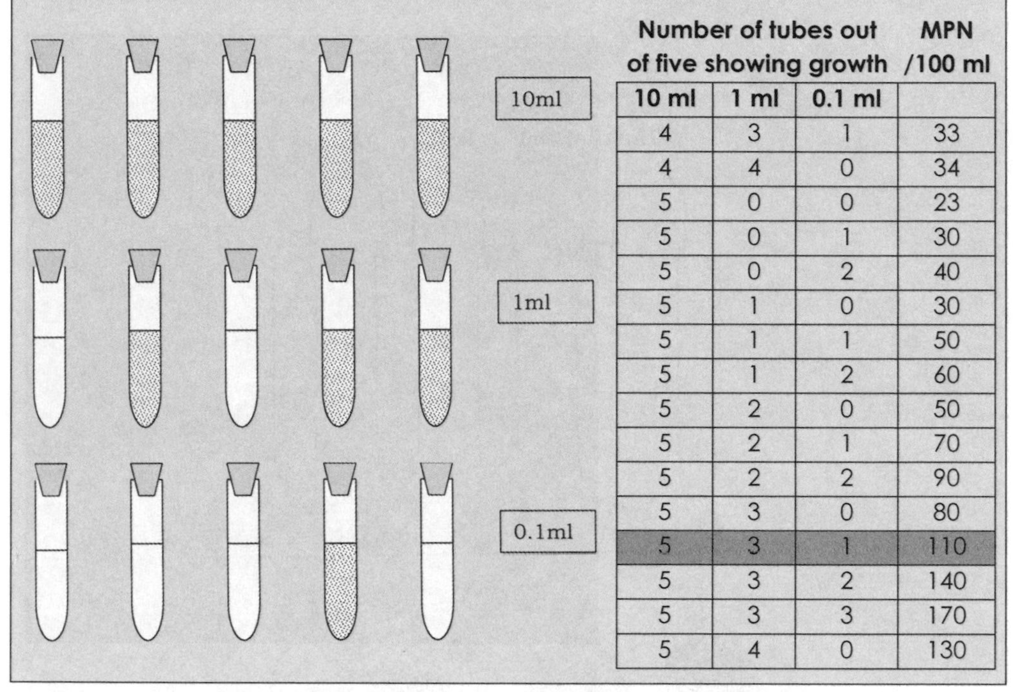

Number of tubes out of five showing growth			MPN /100 ml
10 ml	1 ml	0.1 ml	
4	3	1	33
4	4	0	34
5	0	0	23
5	0	1	30
5	0	2	40
5	1	0	30
5	1	1	50
5	1	2	60
5	2	0	50
5	2	1	70
5	2	2	90
5	3	0	80
5	3	1	110
5	3	2	140
5	3	3	170
5	4	0	130

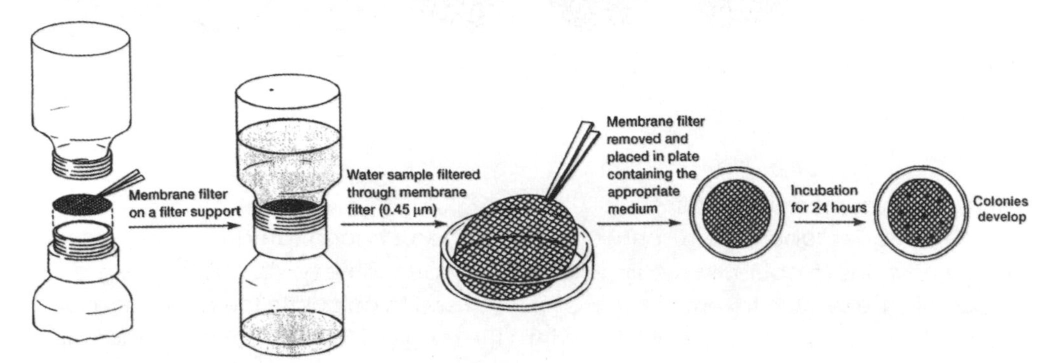

Figure 5.2 Estimation of cell numbers by membrane filtration. Microbial cells present in a water sample are trapped on a membrane, which is then placed on an agar medium to allow colony development. The technique is used to concentrate the cells present at low densities in a large volume. From Prescott, LM, Harley, JP & Klein, DA: Microbiology 5th edn, McGraw Hill, 2002. Reproduced by permission of the publishers

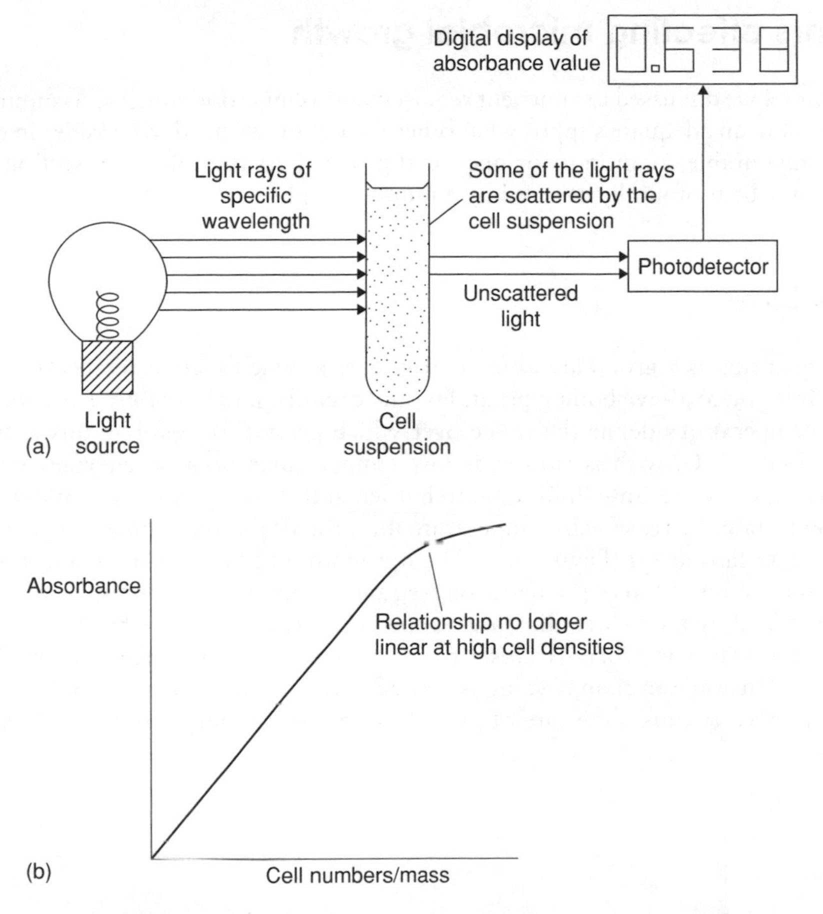

Figure 5.3 Indirect measurement of cell numbers by turbidimetric measurements. (a) Turbidimetry offers an immediate estimate of bacterial density by measuring the degree to which a culture scatters light shone through it in a spectrophotometer. Absorbance is a measure of the amount of light scattered by the cell suspension. (b) Within certain limits, there is a linear relationship between cell numbers and optical density. By determining cell numbers or cell mass for samples of known optical density, a calibration graph can be produced

method for doing this is based on how cloudy or turbid the liquid growth medium becomes due to bacterial growth. *Turbidimetric* methods measure the change in optical density or absorbance of the medium, that is, how much a beam of light is scattered by the suspended particulate matter (Figure 5.3). They can be carried out very quickly by placing a sample in a spectrophotometer. Values of optical density can be directly related to bacterial numbers or mass by reference to a standard calibration curve. Thus, an estimate of bacterial numbers, albeit a fairly approximate one, can be obtained almost instantaneously during an experimental procedure. Other indirect methods of measuring cell density include wet and dry weight estimations, and the measurement of cell components such as total nitrogen, protein or nucleic acid.

Factors affecting microbial growth

In Chapter 4 we discussed the nutrient requirements of microorganisms. Assuming these are present in an adequate supply, what other factors do we need to consider in order to provide favourable conditions for microbial growth? As the following section shows, growth may be profoundly affected by a number of physical factors.

Temperature

Microorganisms as a group are able to grow over a wide range of temperatures, from around freezing to above boiling point. For any organism, the *minimum* and *maximum* growth temperatures define the range over which growth is possible; this is typically about 25–30 °C. Growth is slower at low temperatures because enzymes work less efficiently and also because lipids tend to harden and there is a loss of membrane fluidity. Growth rates increase with temperature until the *optimum* temperature is reached, then the rate falls again (Figure 5.4). The optimum and limiting temperatures for an organism are a reflection of the temperature range of its enzyme systems, which in turn are determined by their three-dimensional protein structures (see Chapter 6). The optimum temperature is generally closer to the maximum growth temperature than the minimum. Once the optimum value is passed, the loss of activity caused by denaturation of enzymes causes the rate of growth to fall away sharply (see also Figures 6.3 and 6.4).

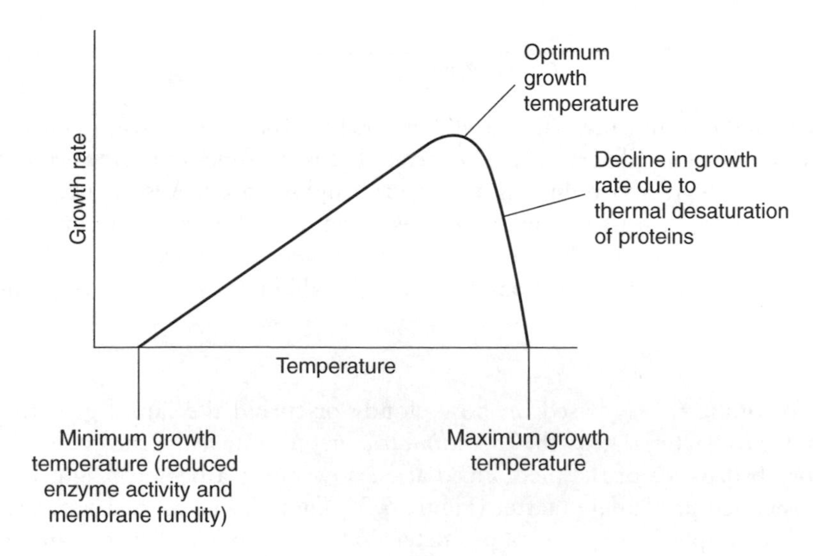

Figure 5.4 Effect of temperature on microbial growth rate. The factors governing the minimum, optimum and maximum temperatures for a particular organism are indicated. The curve is asymmetrical, with the optimum temperature being closer to the maximum than the minimum

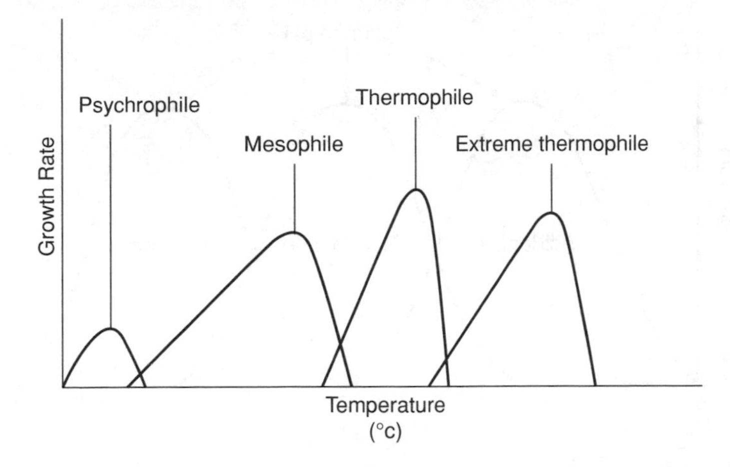

Figure 5.5 Microorganisms can be categorised according to the temperature range at which they grow

The majority of microorganisms achieve optimal growth at 'middling' temperatures of around 20–45 °C; these are called *mesophiles* (Figure 5.5). Contrast these with *thermophiles*, which have become adapted to not only surviving, but thriving at much higher temperatures. Typically, these would be capable of growth within a range of about 40–80 °C, with an optimum around 50–65 °C. *Extreme thermophiles* have optimum values in excess of this, and can tolerate temperatures in excess of 100 °C. In 2003, a member of the primitive bacterial group called the Archaea (see Chapter 7) was reported as growing at a temperature of 121 °C, a new world record! *Psychrophiles* occupy the other extreme of the temperature range; they can grow at 0°C, with optimal growth occurring at 15 °C or below. Such organisms are not able to grow at temperatures above 25 °C or so. *Psychrotrophs*, on the other hand, although they can also grow at 0 °C, have much higher temperature optima (20–30 °C). Members of this group are often economically significant due to their ability to grow on refrigerated foodstuffs.

In the laboratory, appropriate temperatures for growth are provided by culturing in an appropriate incubator. These come in a variety of shapes and sizes, but all are thermostatically controlled and generally hold the temperature within a degree or two of the desired value.

pH

Microorganisms are strongly influenced by the prevailing pH of their surroundings. As with temperature, we can define minimum, optimum and maximum values for growth of a particular type (Figure 5.6). The pH range (between minimum and maximum values) is greater in fungi than it is in bacteria. Most microorganisms grow best around neutrality (pH 7). Many bacteria prefer slightly alkaline conditions but relatively few

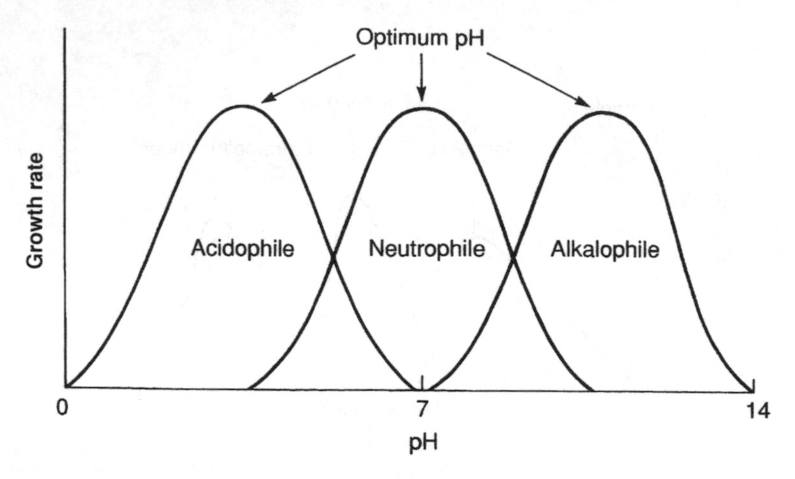

Figure 5.6 Effect of pH on microbial growth rate. Individual species of microorganism occupy a relatively narrow range of pH. Although for most species this is around neutrality, both acidophilic and alkalophilic forms exist. The shape of the curve reflects the properties of a particular organism's enzymes and other proteins

are tolerant of acid conditions, and fewer still are *acidophilic*. Fungi, on the other hand, generally prefer slightly acid conditions and therefore tend to dominate bacteria when these prevail. The reason for the growth rate falling away either side of the optimum value is again due to alterations in three-dimensional protein structure.

> Acidophilic = 'acid-loving'; a term applied to organisms that show optimal growth in acid conditions (pH < 5.5).

The pH value of growth media is adjusted to the desired value by the addition of acid or alkali during its preparation. The metabolic activities of microorganisms often means that they change the pH of their environment as growth proceeds, so it is important in a laboratory growth medium that a desirable pH is not only set but maintained. This is achieved by the use of an appropriate *buffer* system. Phosphate buffers are widely used in the microbiology laboratory; they enable media to minimise changes in their pH when acid or alkali is produced (see Box 5.3).

Box 5.3 Buffers

The pH of a batch culture may change from its intended value as a result of an organism's metabolic activity. Buffers are added to a medium in order to minimise the effect of these changes. Such buffers must be non-toxic to the organism in question, and phosphate buffers are commonly used in microbial systems. Phosphate combines with hydrogen ions or hydroxyl ions to form respectively a weak acid or a weak base:

$$H^+ + HPO_4^{2-} \longrightarrow H_2PO_4^-$$

$$OH^- + H_2PO_4^- \longrightarrow HPO_4^{2-} + H_2O$$

This reduces the impact on the pH of the medium.

Oxygen

Oxygen is present as a major constituent (20 per cent) of our atmosphere, and most life forms are dependent upon it for survival and growth. Such organisms are termed *aerobes*. Not all organisms are aerobes however; some *anaerobes* are able to survive in the absence of oxygen, and for some this is actually a necessity.

> An aerobe is an organism that grows in the presence of molecular oxygen, which it uses as a terminal electron acceptor in aerobic respiration.
>
> An anaerobe is an organism that grows in the absence of molecular oxygen

Aerobic organisms require oxygen to act as a terminal electron acceptor in their respiratory chains (see Chapter 6). Such organisms, when grown in laboratory culture, must therefore be provided with enough oxygen to satisfy their requirements. For a shallow layer of medium such as that in a petri dish, sufficient oxygen is available dissolved in surface moisture. In a deeper culture such as a flask of broth however, aerobes will only grow in the surface layers unless additional oxygen is provided (oxygen is poorly soluble in water). This is usually done by shaking or mechanical stirring.

Obligate anaerobes cannot tolerate oxygen at all (see Box 5.4). They are cultured in special anaerobic chambers, and oxygen excluded from all liquid and solid media. *Facultative anaerobes* are able to act like aerobes in the presence of oxygen, but have the added facility of being able to survive when conditions become anaerobic. *Aerotolerant anaerobes* are organisms that are basically anaerobic; although they are not inhibited by atmospheric oxygen, they do not utilise it. *Microaerophiles* require oxygen, but are only able to tolerate low concentrations of it (2–10 per cent), finding higher concentrations harmful. Organisms inoculated into a static culture medium will grow at positions that reflect their oxygen preferences (Figure 5.7).

Carbon dioxide

In Chapter 4 we saw that autotrophic organisms are able to use carbon dioxide as a carbon source; when grown in culture, these are provided with bicarbonate in their growth medium or incubated in a CO_2-enriched atmosphere. However, heterotrophic bacteria also require small amounts of carbon dioxide, which is incorporated into various metabolic intermediates. This dependency can be demonstrated by the failure of these organisms to grow if carbon dioxide is deliberately removed from the atmosphere.

> **Box 5.4** How can oxygen be toxic?
>
> It seems strange to us to think of oxygen as a toxic substance, however it can be converted by metabolic enzymes into highly reactive derivatives such as the superoxide free radical (O_2^-), which are very damaging to cells. Aerobes and most facultative anaerobes convert this to hydrogen peroxide, by means of the enzyme *superoxide dismutase*. This is further broken down by *catalase*. Obligate anaerobes do not possess either enzyme, and so cannot tolerate oxygen.

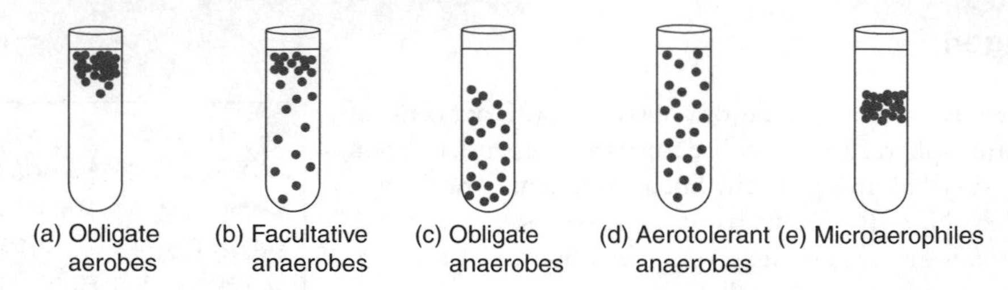

Figure 5.7 Microorganisms have different oxygen requirements. In a static culture, microorganisms occupy different regions of the medium, reflecting their pattern of oxygen usage. (a) Obligate aerobes must grow at or near the surface, where oxygen is able to diffuse. (b) Facultative anaerobes are able to adjust their metabolism to the prevailing oxygen conditions. (c) Obligate anaerobes, in contrast, occupy those zones where no oxygen is present at all. (d) Aerotolerant anaerobes do not use oxygen, but neither are they inhibited by it. (e) Microaerophiles have specific oxygen requirements, and can only grow within a narrow range of oxygen tensions

Osmotic pressure

Osmosis is the diffusion of water across a semipermeable membrane from a less concentrated solution to a more concentrated one, equalising concentrations. The pressure required to make this happen is called the *osmotic pressure*. If a cell were placed in a hypertonic solution (one whose solute concentration is higher), osmosis would lead to a loss of water from the cell (*plasmolysis*). This is the basis of using high concentrations of salt or other solutes in preserving foods against microbial attack. In the opposite situation, water would pass from a dilute (hypotonic) solution into the cell, causing it to swell and burst. The rigid cell walls of bacteria prevent them from bursting; this, together with their minute size, makes them less sensitive to variations in osmotic pressure than other types of cell. They are generally able to tolerate NaCl concentrations of between 0.5 and 3.0 per cent. *Haloduric* ('salt-tolerant') bacteria are able to tolerate concentrations ten times as high, but prefer lower concentrations, whereas *halophilic* ('salt-loving') forms are adapted to grow best in conditions of high salinity such as those that prevail in the Dead Sea in the Middle East. In order to do this without plasmolysis occurring, they must build up a higher internal solute concentration, which they do by actively concentrating potassium ions inside the cell.

> Plasmolysis is the shrinkage of the plasma membrane away from the cell wall, due to osmotic loss of water from the cell.

Light

Phototrophic organisms require light in order to carry out photosynthesis. In the laboratory, care must be taken that light of the correct wavelength is used, and that the source used does not also act as a heat source. Fluorescent light produces little heat, but does not provide the wavelengths in excess of 750 nm needed by purple and green photosynthetic bacteria.

The kinetics of microbial growth

Unicellular organisms divide by *binary fission*; each cell grows to full size, replicates its genetic material then divides into two identical daughter cells. By identical means, two cells divide into four, four into eight and so on, leading to an exponential increase in cell numbers:

$$1 \longrightarrow 2 \longrightarrow 4 \longrightarrow 8 \longrightarrow 2^n$$

If we were to plot the number of cells in a population against time, we would get an exponential curve (Figure 5.8a). It is more convenient when plotting a growth curve

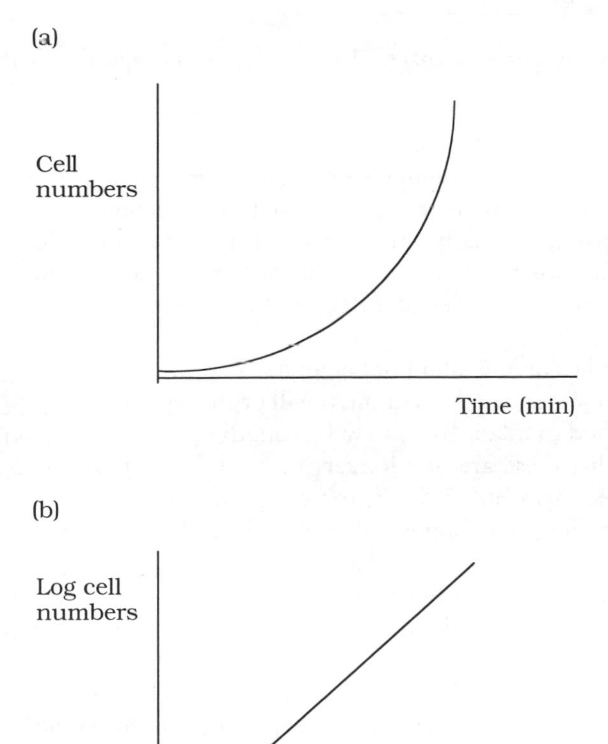

Figure 5.8 (a) Under ideal physicochemical conditions, the number of cells in a population of a unicellular organism increase exponentially. (b) A plot of the log of cell numbers against time during exponential growth gives a straight line

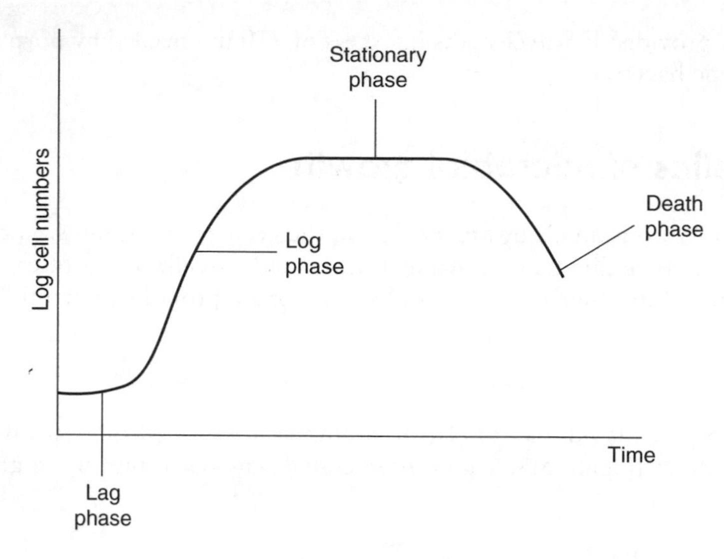

Figure 5.9 A microbial growth curve. The four phases of a typical growth curve are shown. See text for details

to plot the logarithm of cell numbers of against time, giving us a straight line (Figure 5.8b). Such exponential growth cannot continue indefinitely, however, and growth usually slows down due to either the supply of nutrients becoming exhausted, or because metabolism leads to an accumulation of harmful waste substances. Unicellular growth usually occurs in a series of different phases (Figure 5.9):

> *Inoculum* is the term given to the cells used to 'seed' a new culture.

1 *Lag phase*. When an inoculum of bacteria is first introduced into some growth medium, it will probably require a period to adapt to its new surroundings – the less familiar these are, the longer the period of adaptation (see also Box 5.5). If, for example, the carbon source in the medium is unfamiliar, the cells will need time to synthesise the necessary enzymes for its metabolism. The length of the lag phase will also depend on the age and general health of the cells in the inoculum. During this period, there is no net increase in bacterial numbers, however the cells are metabolically active.

2 *Log (exponential) phase*. When the bacteria have acclimatised to their new environment and synthesised the enzymes needed to utilise the available substrates, they are able to start regular division by binary fission. This leads to the exponential increase in numbers referred to above. Under optimal conditions, the population of cells will double in a constant and predictable length of time, known as the *generation (doubling) time*. The value for the widely used laboratory bacterium *E. coli* is 20 min, and for most organisms it is less than an hour. There are some bacteria, however, whose generation time is many hours. Thus, during exponential growth, the number of cells can be expressed as:

$$N_T = N_0 \times 2^n$$

Box 5.5 Diauxic growth

When *E. coli* grows in a medium containing glucose and lactose, it preferentially metabolises the former since it is more energy efficient to do so. The cell has a regulatory mechanism that suppresses the synthesis of lactose-metabolising enzymes until all the glucose has been used up (see Chapter 11). At this point a second lag phase is entered, while the lactose metabolising enzymes are synthesised. Such growth is termed *diauxic*, and the resulting growth curve is characteristically biphasic.

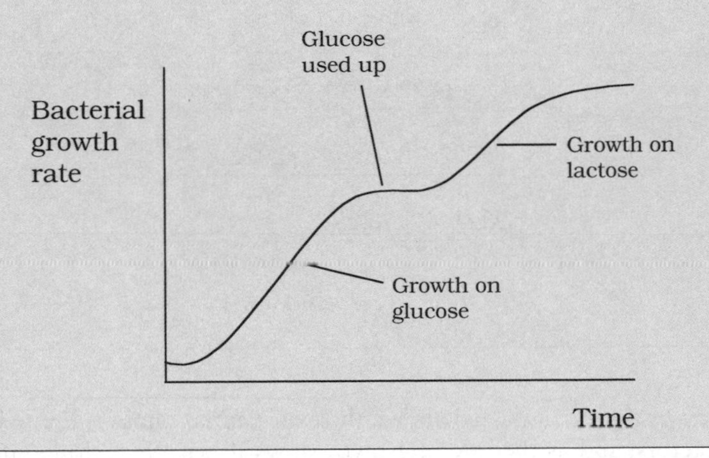

where N_0 is the number of cells at the start of exponential growth, N_T is the number of cells after time T, L is the length of the lag phase and n is the number of doubling times that have elapsed. n is therefore equal to T/T_d, where T_d is the doubling time. Substituting into the first equation:

$$N_T = N_0 \times 2^{T/T_d}$$

This can be expressed more conveniently by using logarithms to the base 2 (don't worry too much about how this is done!):

$$\log_2 N_T = \log_2 N_0 + \frac{T - L}{T_d}$$

Therefore,

$$\frac{\log_2 N_T - \log_2 N_0}{T - L} = \frac{1}{T_d}$$

Thus, if we know the number of cells at the start and end of a period of exponential growth, we can calculate the doubling time. See Box 5.6 for a worked example. We can also determine the *mean growth rate constant* (K); this is a measure of the number of doublings of the population per unit time, and is equal to $1/T_d$ *.

Many antibiotics such as penicillin (Chapter 14) are only effective when cells are actively dividing, since they depend on disrupting new cell wall synthesis.

* You may also come across the *instantaneous growth rate* (μ), which equals $\frac{0.693}{T_d}$

Box 5.6 Calculating an increase in microbial numbers
Example:

An inoculum of 10^7 bacterial cells was introduced into a flask of culture medium and growth monitored. No change was seen for 18 minutes (the lag phase), then growth occurred rapidly. After a further 76 minutes, the population had increased to 4.32×10^8 cells. What is the doubling time (T_d) of the culture?

(To obtain values of $\log_2$, multiply $\log_{10}$ values by 3.322)

$$\frac{1}{T_d} = \frac{\log_2 N_T - \log_2 N_o}{T - L}$$

$$\frac{1}{T_d} = \frac{\log_2 (4.32 \times 10^8) - \log_2 10^7}{76}$$

$$\frac{1}{T_d} = \frac{(3.322 \times 8.6355) - (3.322 \times 7)}{76}$$

$$\frac{28.6871 - 23.254}{76} = \frac{1}{T_d}$$

$$T_d = \frac{76}{5.4331} = 14\,\text{minutes}$$

3 *Stationary phase.* As discussed above, the exponential phase is limited by environmental factors, and as the rate of growth slows down, the culture enters the next phase. The levelling out of the growth curve does not mean that cell division has ceased completely, but rather that the increase due to newly formed cells is cancelled out by a similar number of cell deaths. Eventually, however, as the death rate increases, the overall numbers fall and we enter the final phase of growth.

4 *Death phase.* As cells die off and the culture is unable to replace them, the total population of viable cells falls. This is the *death* (or *decline*) *phase.*

Batch culture and continuous culture

The phases of growth described above apply to a *batch culture*. In this form of culture, appropriate nutrients and other conditions are provided for growth, then an inoculum is added and the culture incubated. No further nutrients are added and no waste products are removed, thus conditions in the culture are continually changing. This results in active growth being of limited duration for the reasons outlined above.

Sometimes it is desirable to keep the culture in the logarithmic phase, for example if the cells are being used to produce alcohol or antibiotics. In a *continuous culture*, nutrient concentrations and other conditions are held constant, and the cells are held in a state of exponential growth. This is achieved by continuously adding fresh culture medium and removing equal volumes of the old. Parameters such as pH can also be monitored and adjusted. The equipment used to do this is called a chemostat (Figure 5.10); it

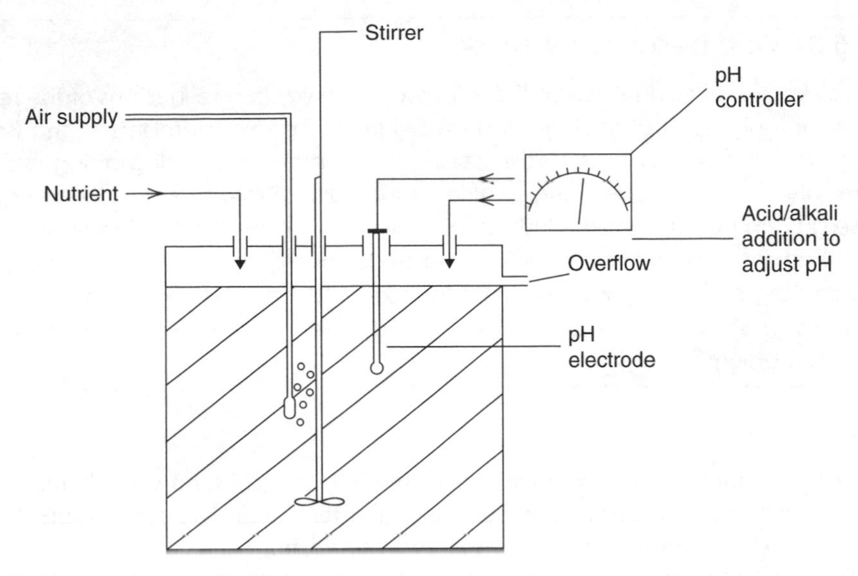

Figure 5.10 Continuous culture of microorganisms: the chemostat. By the continuous addition of new nutrients and removal of waste, the culture is kept in a steady state, with both cell numbers and nutrient composition remaining constant

produces a steady-state culture whose population size is kept constant by careful control of flow rates and nutrient concentrations.

Growth in multicellular microorganisms

If uninterrupted, growth in fungi proceeds radially outwards from the initiating spore (Box 5.7), allowing the fungal colony to colonise new regions potentially rich in nutrients (Box 5.8). Actual growth occurs solely at the hyphal tip; as this happens, the terminal cell grows longer, until eventually a new cross wall or septum is formed. Cells away

Box 5.7 See for yourself!

You can observe growth at a hyphal tip for yourself with a petri dish culture of a fungus such as *Mucor* and a light microscope fitted with an eyepiece graticule. Focus on the outer visible edge of a colony, then identify the tip of one of the ouwardly radiating hyphae. Align it with the graticule scale and note the position of the tip. Return after half an hour and note the new value, being careful not to disturb the microscope in the meantime. You may be surprised at how far the hypha has travelled in such a short time!

Box 5.8 Who believes in fairies?

The well-known phenomenon of 'fairy rings' can be explained in terms of the radial growth of fungi. As the underground mycelium of certain members of the Basidiomycota extends outwards, it releases enzymes into the soil, degrading organic matter ahead of it and releasing nutrients such as soluble nitrogen for the grass, whose growth becomes more lush at this point, and forms the familiar ring. Further back, the branching mycelium outcompetes the overlying grass and deprives it of minerals. Fairy rings are more likely to be found on cultivated land such as lawns and golf courses, because in order to spread uniformly they require a relatively homogeneous medium

from the tip do not become any longer during hyphal extension, however hyphae in this region may develop into aerial reproductive structures. Older hyphae at some distance from the tip may become completely empty of cytoplasm.

Cell counts and turbidometric measurements are not appropriate to estimate growth of fungi; however total mycelial mass can be measured and its change plotted against time. A fungal growth cycle shows roughly the same phases of growth as described above for bacteria.

Test yourself

1 Total cell counts include both ________ and ________ cells.

2 ________ cell counts only include those that are capable of forming ________.

3 ________ methods measure the degree to which suspended cells scatter a beam of light.

4 Microorganisms which show optimal growth at low and high temperatures are termed, respectively, ________ and ________.

5 Loss of growth activity at high temperatures is due to ________ of enzymes.

6 Fungi are generally more tolerant than bacteria of ________ pH conditions.

7 ________ are often required in laboratory growth media to prevent significant changes in pH.

8 ________ can only tolerate oxygen at low concentrations.

9 ________ ________ cannot tolerate oxygen at all, and live only in its absence.

10 ________ bacteria have evolved to thrive in conditions of extremely high salinity.

11 Bacteria produce two identical daughter cells by a process of ________ ________.

12 Given ideal growth conditions, the size of a population of bacteria would increase in a ________ fashion.

13 The four phases of a typical growth curve are ________, ________, ________ and ________.

14 The small sample of cells used to start off a culture is called an ________.

15 A ________ maintains a population of cells constantly in ________ phase, by carefully controlling nutrients and flow rates.

6

Microbial Metabolism

You may have wondered why it was necessary to learn all that biochemistry in Chapter 2; well, you are about to find out! In the following pages, you will learn about the processes by which microorganisms obtain and use *energy*.

Why is energy needed?

Like all other living things, microorganisms need to acquire energy in order to survive. Energy is required:

- to maintain the structural integrity of the cell by repairing any damage to its constituents
- to synthesise new cellular components such as nucleic acids, polysaccharides and enzymes
- to transport certain substances into the cell from its surroundings
- for the cell to grow and multiply
- for cellular movement.

Metabolism is the term used to describe all the biochemical reactions that take place inside a cell; it includes those reactions that release energy into the cell, and those that make use of that energy. Figure 6.1 summarises these processes.

Catabolism is the term used to describe reactions that break down large molecules, usually coupled to a release of energy.

Anabolism is the term used to describe reactions involved in the synthesis of macromolecules, usually requiring an input of energy

As we saw in Chapter 4, most microorganisms obtain their energy from the nutrients they take into the cell; these may come from an organic or an inorganic source. Once inside the cell, these nutrients must then be biochemically processed by reactions that trap some of their chemical energy, at the same time breaking them down into smaller molecules. These then serve as building blocks for the synthesis of new cellular components. Chemical compounds contain potential energy within their molecular structure, and some of this can be

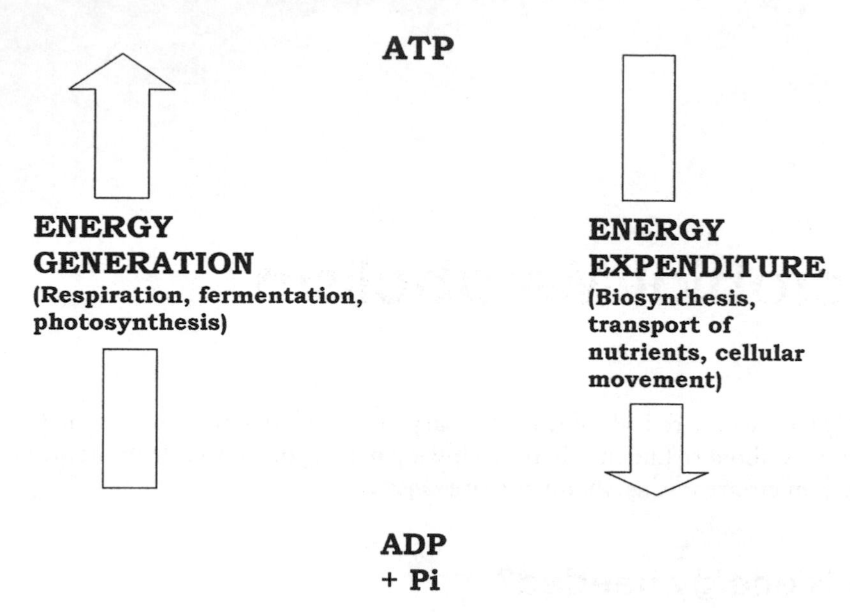

Figure 6.1 Microorganisms use a variety of processes to generate biochemical energy in the form of the compound adenosine triphosphate (ATP). As ATP is broken down to ADP and inorganic phosphate, the energy released is used for the maintenance, reproduction and survival of the cell

released when they are broken down. In other metabolic types, energy is obtained from the sun by means of *photosynthesis*; once again, however, the energy is used for synthetic purposes.

Central to the metabolic processes of any cell are *enzymes*. Without them, the many biochemical reactions referred to above simply wouldn't take place at a fast enough rate for living cells to maintain themselves. We shall start our consideration of metabolism by taking a look at enzymes: what they are, and how they work. In the later sections of the chapter, we shall consider in more detail those processes by which energy is acquired and spent.

> An *enzyme* is a cellular catalyst (usually protein), specific to a particular reaction or group of reactions.

Enzymes

An enzyme is a cellular catalyst; it makes biochemical reactions proceed many times more rapidly than they would if uncatalysed. The participation of an enzyme can increase the rate of a reaction by a factor of millions, or even billions.

Traditionally, all enzymes have been thought of as globular proteins, but around twenty years ago it was demonstrated (surprisingly) that certain RNA molecules also have catalytic properties. These *ribozymes* however, are very much in the minority, carrying out specific cut-and-splice reactions on RNA molecules, and in the present

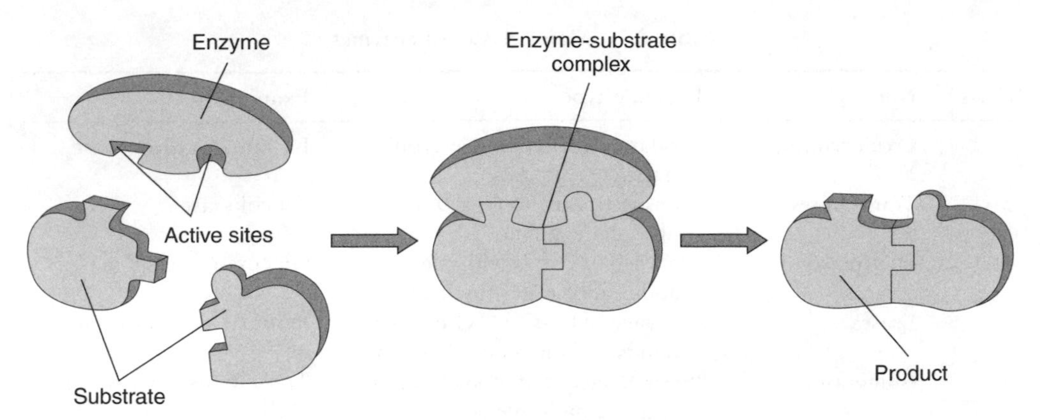

Figure 6.2 Enzyme–substrate interaction. An enzyme interacts with its substrate(s) to form an enzyme–substrate complex, leading to the formation of a product. In the example shown, two substrate molecules are held in position by the enzyme and joined together. From Black, JG: Microbiology: Principles and Explorations, 4th edn, John Wiley & Sons Inc., 1999. Reproduced by permission of the publishers

context can be ignored. In this book we shall confine our discussion of enzymes to the protein type.

Like any other catalyst, an enzyme remains unchanged at the end of a reaction. It must, however, at some point during the reaction bind to its *substrate* (the substance upon which it acts) to form an enzyme–substrate complex (Figure 6.2) by multiple weak forces such as electrostatic forces and hydrogen bonding. Only a small part of the enzyme's three-dimensional structure is involved in this binding; these few amino acids make up the *active site*, which forms a groove or dent in the enzyme's surface, into which the appropriate part of the substrate molecule fits (Figure 6.3). The amino

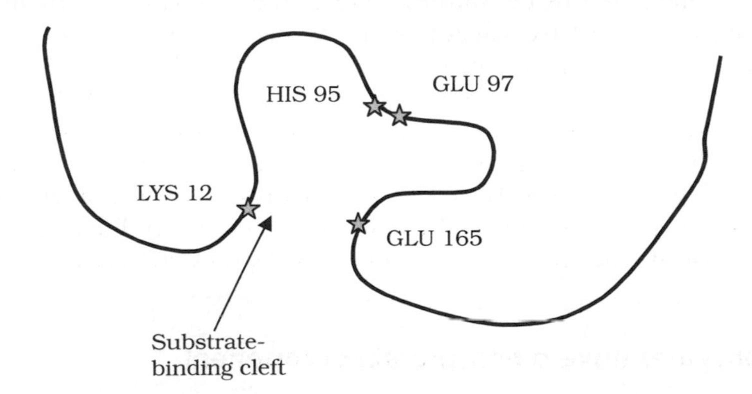

Figure 6.3 Catalytic activity occurs at the active site of an enzyme. Four amino acids play an important role in the active site of triose phosphate isomerase; note how they are far apart in the primary sequence, but are brought together by subsequent protein folding

Table 6.1 Major classes of enzymes

Class	Name	Reaction type	Example
1	Oxidoreductases	Oxidation/reduction (electron transfer) reactions	Lactate dehydrogenase
2	Transferases	Transfer of functional groups e.g. phosphate, amino	Glucokinase
3	Hydrolases	Cleavage of bonds with the addition of water (hydrolysis)	Glucose-6-phosphatase
4	Lyases	Cleavage of C–C, C–O or C–N bonds to form a double bond	Pyruvate decarboxylase
5	Isomerases	Rearrangement of atoms/groups within a molecule	Triose-phosphate isomerase
6	Ligases	Joining reactions, using energy from ATP	DNA ligase

acid residues that go to make up the active site may be widely separated in the enzyme's primary structure, but by means of the secondary and tertiary folding of the molecule, they are brought together to give a specific conformation, complementary to that of the substrate. It is this precise formation of the active site that accounts for one of the major characteristics of enzymes, their *specificity*. You should not think, however, that these few residues making up the active site are the only ones that matter; the enzyme can only fold in this way because the order and arrangement of the other amino acids allows it.

Enzyme classification

Most enzymes have names that end in the suffix *–ase*. The first part of the name often gives an indication of the substrate; for example, urease and pyruvate decarboxylase. Other enzymes have names that are less helpful, such as trypsin, and others have several alternative names to confuse the issue further. To resolve such problems, an internationally agreed system of nomenclature has been devised. All enzymes are assigned initially to one of six broad groups according to the type of reaction they carry out, as shown in Table 6.1. Each enzyme is then placed into successively more specific groupings, each with a number. Thus regardless of any colloquial or alternative names, each enzyme has its own unique and unambiguous four-figure Enzyme Commission 'signature' (pyruvate decarboxylase, mentioned above, is EC 4.1.1.1).

Certain enzymes have a non-protein component

Many enzymes require the involvement of an additional, non-protein component in order to carry out their catalytic action. These 'extra' parts are called *cofactors*; they

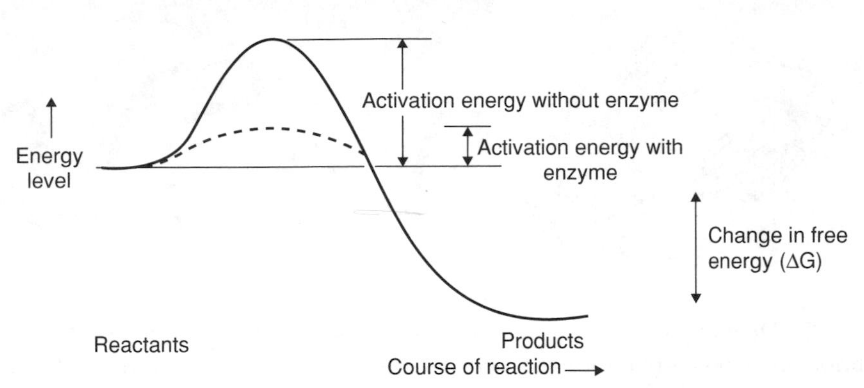

Figure 6.4 An enzyme lowers the activation energy of a reaction. By lowering the amount of energy that must be expended in order for a reaction to commence, enzymes enable it to proceed much more quickly. Note that ΔG, the change in free energy, remains the same whether the reaction is catalysed or not

are usually either metal ions (e.g. Mg^{2+}, Zn^{2+}) or complex organic molecules called *coenzymes*. Some of the most important coenzymes act by transferring electrons between substrate and product in redox reactions (see below).

The purely protein component of an enzyme is known as the *apoenzyme*. The complex of apoenzyme and cofactor is called the *holoenzyme*. The apoenzyme on its own does not have biological activity.

How do enzymes speed up a reaction?

For any chemical reaction to take place there must be a small input of energy. This is called the *activation energy*, and is often likened to the small push that is needed to loosen a boulder and allow it to roll down a hill. It is the energy needed to convert the molecules at the start of a reaction into intermediate forms known as *transition states*, by the rearrangement of chemical bonds. The great gift of enzymes is that they can greatly *lower the activation energy* of a reaction, so that it requires a smaller energy input, and may therefore occur more readily (Figure 6.4).

Environmental factors affect enzyme activity

The rate at which an enzyme converts its substrate into product is called its *velocity (v)*, and is affected by a variety of factors.

Temperature

The rate of any chemical reaction increases with an increase in temperature due to the more rapid movement of molecules, and so it is with enzyme-catalysed reactions, until a peak is reached (the optimum temperature) after which the rate rapidly falls away. What

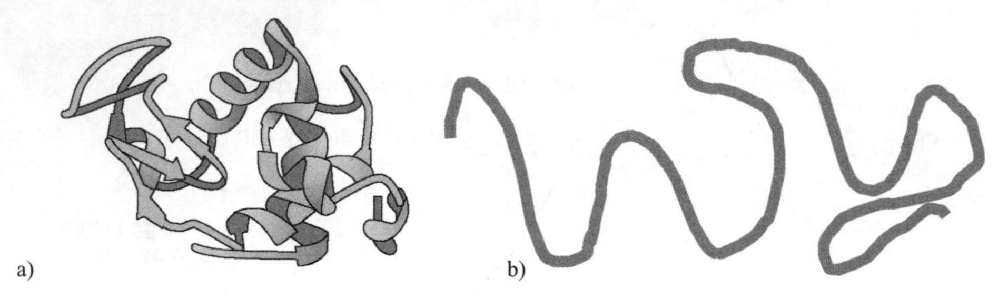

Figure 6.5 Disruption of an enzyme's three-dimensional structure causes denaturation. Disruption of the bonds that form the secondary and tertiary protein structure of an enzyme lead to a loss of catalytic activity, as the amino acids forming the active site are pulled apart. a) From Bolsover, SR , Hyams, JS, Jones, S, Shepherd, EA & White, HA: From Genes to Cells, John Wiley & Sons, 1997. Reproduced by permission of the publishers

causes this drop in the velocity? Recall from Chapter 2 that the very ordered secondary and tertiary structure of a protein molecule is due to the existence of numerous weak molecular bonds, such as hydrogen bonds. Disruption of these by excessive heat results in *denaturation* (Figure 6.5), that is, an unfolding of the three-dimensional structure. In the case of an enzyme, this leads to changes in the configuration of the active site, and a loss of catalytic properties. The effect of temperature on enzyme activity is shown in Figure 6.6. The graph can be thought of as a composite of two lines, one increasing with temperature due to the rise in thermal energy of the substrate molecules, and one falling due to denaturation of the protein structure. Before the optimum temperature it is the former which dominates, then the effect of the latter becomes more pronounced, and takes over completely.

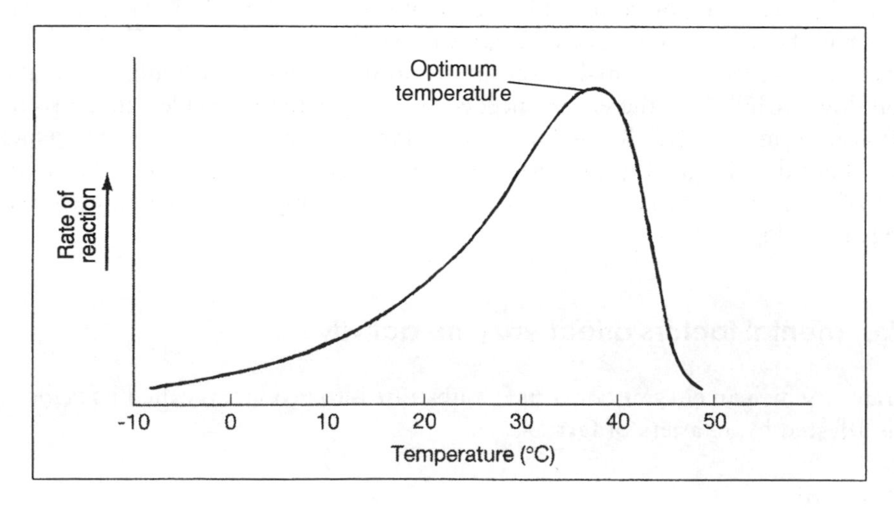

Figure 6.6 Effect of temperature on enzyme activity. Below the optimum temperature, the rate of reaction increases as the temperature rises. Above the optimum, there is a sharp falling off of reaction rate due to thermal denaturation of the enzyme's three-dimensional structure

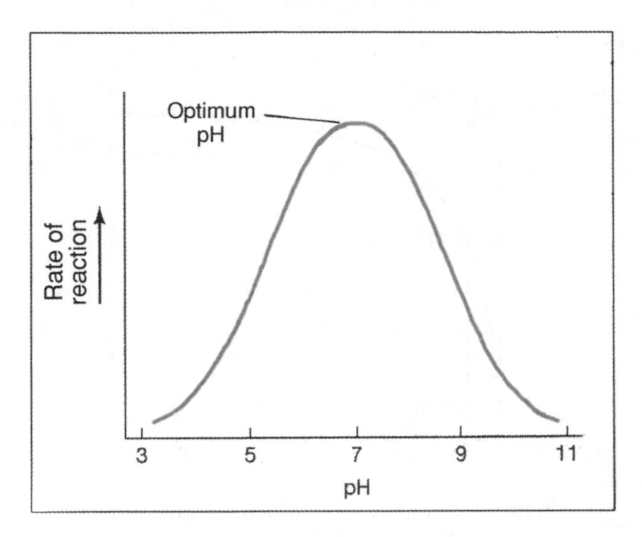

Figure 6.7 Effect of pH on enzyme activity. Either side of the optimum pH value, changes in ionisation of amino acid side chains lead to protein denaturation and a loss of enzyme activity

pH

Enzyme velocity is similarly affected by the prevailing pH. Once again, this is due to alterations in three-dimensional protein structure. Changes in the pH affect the ionisation of charged 'R'-groups on amino acids at the active site and elsewhere, causing changes in the enzyme's precise shape, and a reduction in catalytic properties. As with temperature, enzymes have an optimum value at which they operate most effectively; when the pH deviates appreciably from this in either direction, denaturation occurs, leading to a reduction of enzyme activity (Figure 6.7).

Microorganisms are able to operate in a variety of physicochemical environments, a fact reflected in the diversity of optimum values of temperature and pH encountered in their enzymes.

Substrate concentration

Under conditions where the active sites of an enzyme population are not saturated, an increase in substrate concentration will be reflected in a proportional rise in the rate of reaction. A point is reached, however, when the addition of further substrate has no effect on the rate (Figure 6.8). This is because all the active sites have been occupied and the enzymes are working flat out; this is called the *maximum velocity (V_{max})*. A measure of the affinity an enzyme has for its substrate (i.e. how tightly it binds to it) is given by its *Michaelis constant (K_m)*. This is the substrate concentration at which the rate of reaction is half of the V_{max} value. Values of V_{max} and K_m are more easily determined experimentally by plotting the reciprocals of [S] and V to obtain a straight line (Figure 6.9).

> The Michaelis–Menten equation relates the rate of a reaction to substrate concentration, [S]:
>
> $$v = \frac{V_{max}[S]}{[S] + K_m}$$

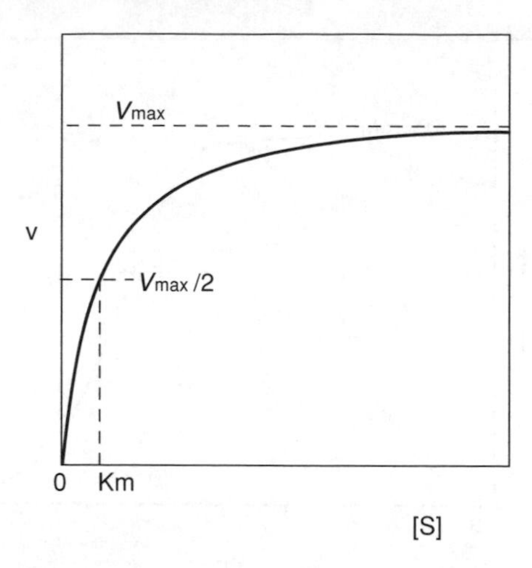

Figure 6.8 Enzyme activity is influenced by substrate concentration. The initial rate of reaction (v_o) is proportional to substrate concentration at low values of [S]. However, when the active sites of the enzyme molecules become saturated with substrate, a maximum rate of reaction (V_{max}) is reached. This cannot be exceeded, no matter how much the value of [S] increases. The curve of the graph fits the Michaelis–Menten equation. K_m is the value of [S] where $v = \frac{V_{max}}{2}$.

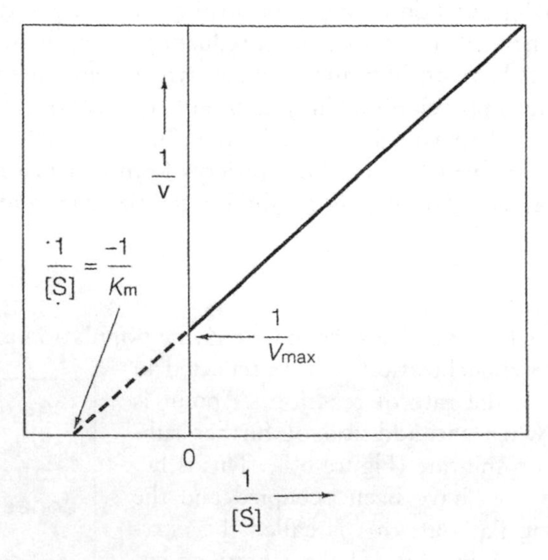

Figure 6.9 The Lineweaver–Burk plot. Plotting the reciprocal values of V_o and [S] enables values of K_m and V_{max} to be derived from the intercepts on a straight-line graph

Some enzymes do not obey Michaelis–Menten kinetics. The activity of *allosteric enzymes* is regulated by effector molecules which bind at a position separate from the active site. By doing so, they induce a conformational change in the active site that results in activation or inhibition of the enzyme. Thus effector molecules may be of two types, activators or inhibitors.

Enzyme inhibitors

Many substances are able to interfere with an enzyme's ability to catalyse a reaction. As we shall see in Chapter 14, enzyme inhibition forms the basis of several methods of microbial control, so a consideration of the main types of inhibitor is appropriate here.

Perhaps the easiest form of enzyme inhibition to understand is *competitive inhibition*. Here, the inhibitory substance competes with the normal substrate for access to the enzyme's active site; if the active site is occupied by a molecule of inhibitor, it can't bind a molecule of substrate, thus the reaction will proceed less quickly (Figure 6.10a).

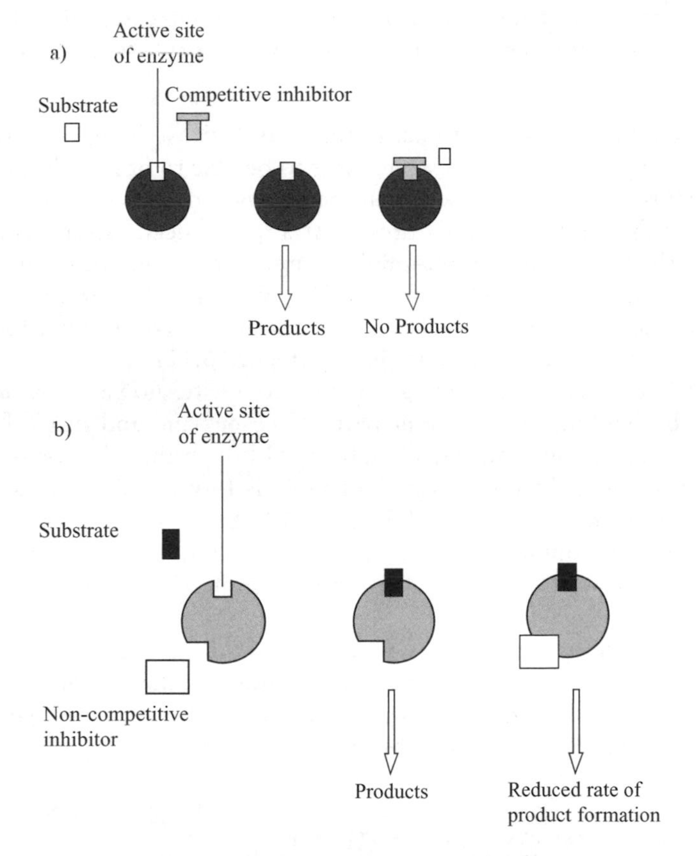

Figure 6.10 Enzyme inhibition. (a) A competitive inhibitor mimics the structure of the normal substrate molecule, enabling it to fit into the active site of the enzyme. Although it is not acted on by the enzyme and no products are formed, such an inhibitor prevents the normal substrate gaining access to the active site. (b) A non-competitive substance binds to a second site on the enzyme and thus does not affect substrate binding; however distortion of the enzyme molecule makes catalysis less efficient

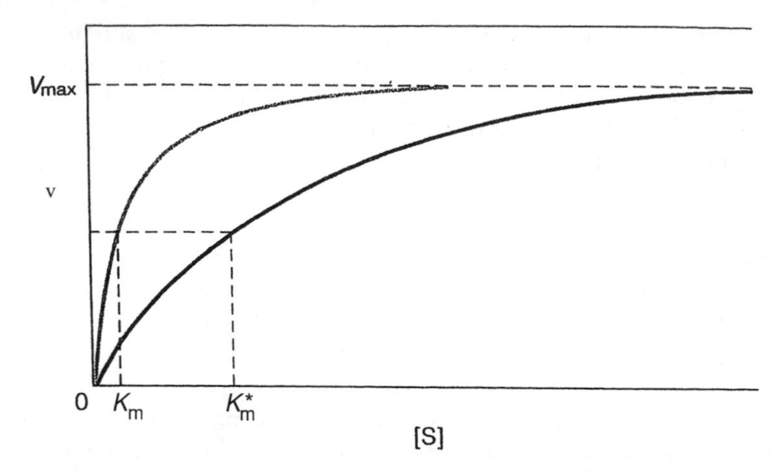

Figure 6.11 Competitive inhibition. In the presence of a competitive inhibitor (lower curve) V_{max} is reached eventually, but the apparent value of K_m is increased. Since there are fewer enzyme molecules in circulation, the apparent affinity for its substrate is diminished

The competitive inhibitor is able to act in this way because its molecular structure is sufficiently similar to that of the substrate for it to be able to fit into the active site. The effect is competitive because it depends on the relative concentrations of the substrate and inhibitor. If the inhibitor is only present at a low concentration, its effect will be minimal, since the number of enzyme–inhibitor interactions will be greatly outweighed by reactions with the 'proper' substrate. The V_{max} value for the enzyme is not reduced, but it is only reached more gradually. The apparent affinity of the enzyme for its substrate is decreased, reflected by an increase in the K_m (Figure 6.11).

Not all inhibitors act by competing for the active site, however. *Non-competitive inhibitors* act by binding to a different part of the enzyme and in so doing alter its three-dimensional configuration (Figure 6.10b). Although they do not affect substrate binding, they do reduce the rate at which product is formed. V_{max} cannot be reached; however, the value of K_m, is unchanged (Figure 6.12). Such inhibitors may bind to either the enzyme–substrate complex or to free enzyme. Both competitive and non-competitive forms of inhibition are *reversible*, since the inhibitor molecule is relatively weakly bound and can be displaced.

Irreversible inhibition, on the other hand, is due to the formation of a strong covalent linkage between the inhibitor and an amino acid residue on the enzyme. As a result of its binding, the inhibitor effectively makes a certain percentage of the enzyme population permanently unavailable to catalyse substrate conversion.

Principles of energy generation

In this section, we shall consider how enzyme-catalysed reactions are involved in the cellular capture and utilisation of energy.

Energy taken up by the cell, be it in the form of nutrients or sunlight, must be converted into a usable form. A simple analogy is selling goods for cash, which you can then use to buy exactly what you want. The 'cash' of cellular metabolism is a

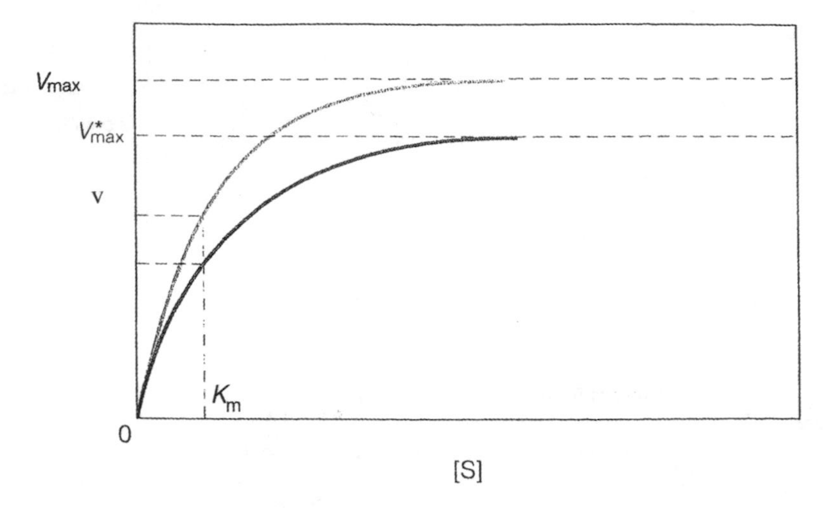

Figure 6.12 Non-competitive inhibition. Since product is not formed as efficiently in the presence of a non-competitive inhibitor (lower curve), V_{max} is not reached. K_m is not altered

compound called *adenosine triphosphate (ATP)*. ATP is by far the most important of a class of compounds known as high-energy transfer compounds, which store the energy* from the breakdown of nutrients (or trapped by photosynthetic pigments) and release it when required by the cell. In *catabolic* reactions, in which molecules such as glucose are broken down, energy is released in the form of ATP, which can then be utilised in *anabolic* (synthetic) reactions.

ATP has a structure very similar to the nucleotides found in RNA, except it has two additional phosphate groups (Figure 6.13). The bond that links the third phosphate group requires a lot of energy for its formation, and is often referred to as a *'high-energy'* phosphate bond. Importantly for the cell, when this bond is broken, the same large amount of energy is released, so when ATP is broken down to ADP and a free phosphate group, energy is made available to the cell. It should be noted that 'high-energy' refers to the amount of energy needed to make or break the bond, and not to any intrinsic property of the bond itself. The process of adding or removing a phosphate group is called *phosphorylation* or *dephosphorylation*, respectively.

Oxidation–reduction reactions

Many metabolic reactions involve the transfer of electrons from one molecule to another; these are called *oxidation-reduction* or *redox* reactions. When a molecule (or atom or ion) loses an electron, it is said to be *oxidized*. (Note, that despite the terminology, oxygen does not necessarily take part in the reaction.) Conversely, when an electron is gained, the recipient is *reduced* (Figure 6.14).

Many metabolic reactions involve the loss of a hydrogen atom; since this contains one proton and one electron, the reaction is regarded as an oxidation, because an electron

*Not all of the energy is converted into ATP. A proportion of it is lost as heat, some of which allows the enzyme-mediated reactions to proceed at a faster rate.

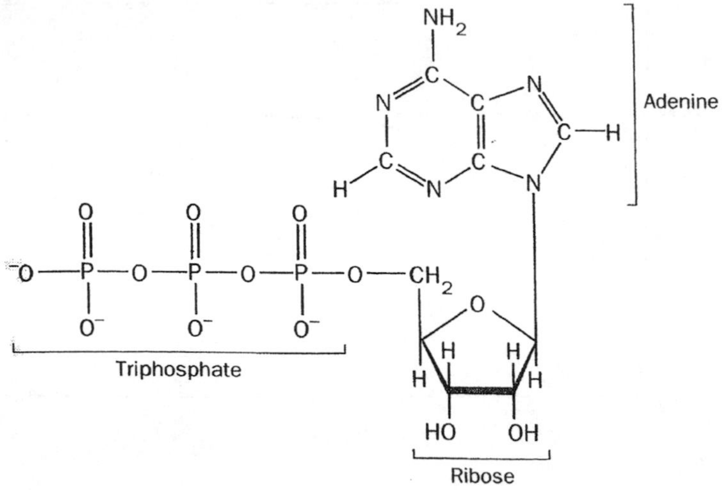

Figure 6.13 Adenosine triphosphate (ATP). ATP is a nucleotide such as similar to those depicted in Figures 2.20 and 2.21. Note the extra phosphate groups

has been lost:

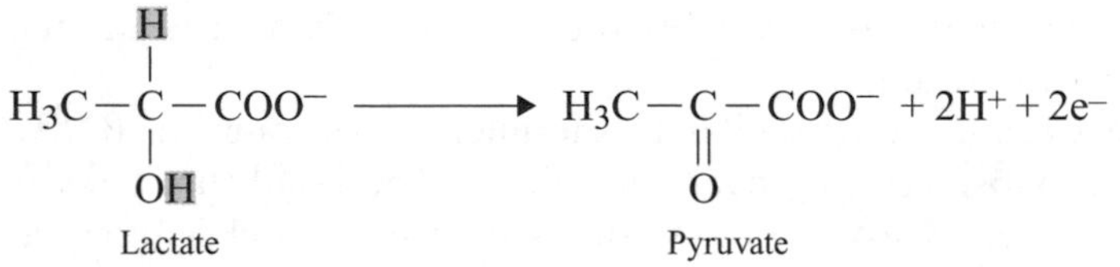

$$H_3C-CH(OH)-COO^- \longrightarrow H_3C-C(=O)-COO^- + 2H^+ + 2e^-$$

Lactate → Pyruvate

The lactate in the example above, by losing two hydrogen atoms, has automatically lost two electrons and thus become *oxidised* to pyruvate. Oxidation reactions are

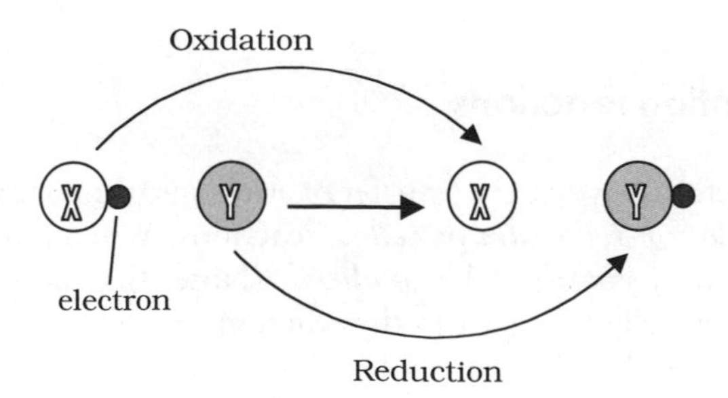

Figure 6.14 Oxidation–reduction reactions. When one molecule is oxidised, another is simultaneously reduced. In the example shown, 'X' loses an electron and thus becomes oxidised. By receiving the electron, 'Y' becomes reduced

Box 6.1 Coupled reactions

Many reactions in metabolic pathways, including glycolysis, can only take place if they are coupled to a secondary reaction. The oxidising power necessary for the conversion of glyceraldehyde-3-phosphate into 1,3-diphosphoglycerate in step 6 of glycolysis for example, is provided by the coenzyme NAD^+. In the coupled reaction, this becomes oxidised to NADH:

$$\text{Glyceraldehyde-3-phosphate} \xrightarrow{NAD^+ \rightarrow NADH} \text{1, 3-diphosphoglycerate}$$

always associated with the transfer of energy from the oxidised substance to the reduced substance.

Two important molecules that we shall encounter a number of times in the following pages are the coenzymes nicotinamide adenine dinucleotide and nicotinamide adenine dinucleotide phosphate; you will be relieved to learn that they are nearly always referred to by their abbreviations, *NAD*$^+$ and *NADP*$^+$, respectively! Both are derivatives of the B vitamin niacin, and each can exist in an oxidised and a reduced form:

$$NAD^+ + H^+ + 2e^- \longleftrightarrow NADH$$
$$NADP^+ + H^+ + 2e^- \longleftrightarrow NADPH$$

Oxidised — Reduced

They are to be found associated with redox reactions (see Box 6.1), acting as carrier molecules for the transfer of electrons. In the oxidation of lactate shown above, the oxidising power is provided by the reduction of NAD^+, so the full story would be:

$$\text{Lactate} \xrightarrow{NAD^+ \rightarrow NADH + H^+} \text{Pyruvate}$$

As the lactate is oxidised, so the NAD^+ in the coupled reaction is reduced. It is said to act as the *electron acceptor*. NAD^+/NADH is generally involved in catabolic reactions, and $NADP^+$/NADPH in anabolic ones.

As the equation above shows, there can be no oxidation without reduction, and vice versa; the two are irrevocably linked. The tendency of an compound to lose or gain electrons is termed its *redox potential* (E_o) (Box 6.2).

In the following section, we shall examine chemoheterotrophic metabolism, used by the majority of microorganisms to derive cellular energy from the oxidation of carbohydrates. Other groups have evolved their own systems of energy capture, and these will be considered later in the chapter.

Figure 6.15 provides a summary of the catabolic (breakdown) pathways used by heterotrophs. Complex nutrients such as proteins and polysaccharides must be

Box 6.2 Redox potentials

Substances vary in the affinity they have for binding electrons; this can be measured as their oxidation--reduction potential or redox potential, relative to that hydrogen. The flow of electrons in the electron transport chain (see Fig. 6.21) occurs because the carriers are arranged in order of their redox potentials, with each having a greater electron affinity (more positive redox potential) than its predecessor. Thus electrons are donated to carriers with a more positive redox potential.

Reduction Potential	
−	Strongly negative redox potential (Good electron donors)
0	
+	Strongly positive redox potential (Good electron acceptors)

enzymatically broken down and converted to substances that can then enter one of the degradative pathways that lead to energy production.

Glucose is the carbohydrate most widely used as an energy source by cells, and the processes by which it is broken down in the presence of oxygen to give carbon dioxide and water are common to many organisms. These have been very thoroughly studied and can be summarised:

$$C_6H_{12}O_6 + 6O_2 \longrightarrow 6CO_2 + 6H_2O$$

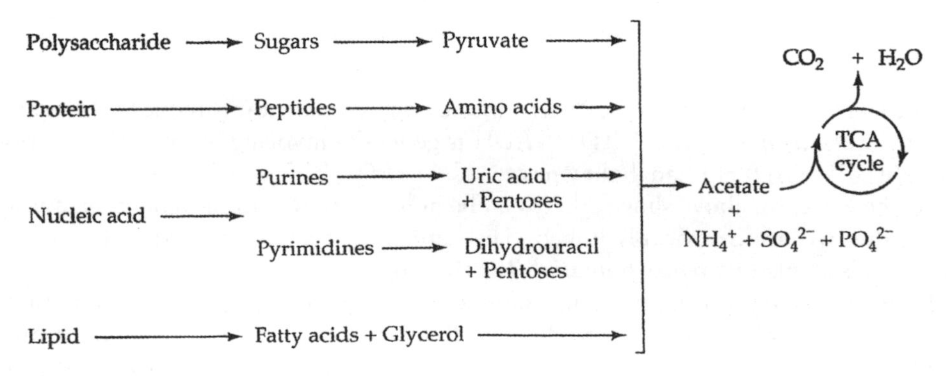

Figure 6.15 Catabolic pathways in heterotrophs. Pathways for the catabolism of proteins, nucleic acids and lipids as well as carbohydrates can all feed into the tricarboxylic acid cycle

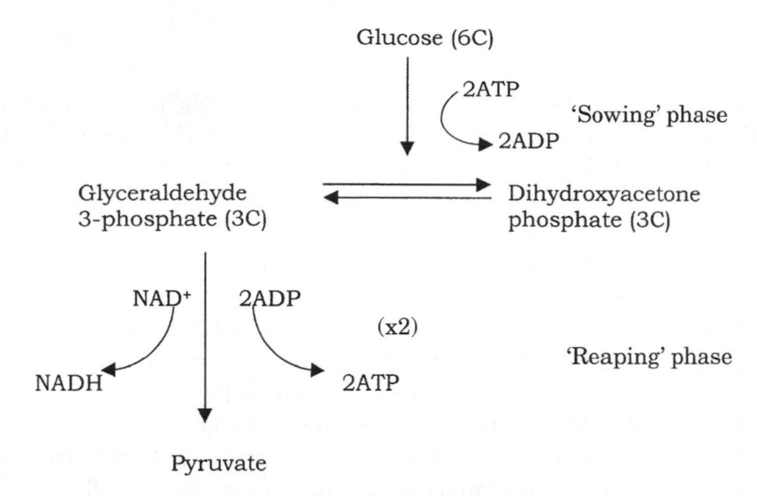

Figure 6.16 Glycolysis. Two molecules of ATP are 'spent' in the first stage of glycolysis, in which glucose is converted into the 3 carbon compounds glyceraldehyde 3-phosphate and dihydroxyacetone phosphate. During the second stage, four ATPs are produced per molecule of glucose, so there is a net gain of two ATPs. In addition, reducing power is generated in the form of NADH (two molecules per molecule of glucose). See Figure 6.17 for a more detailed depiction of the reactions involved in glycolysis

What this equation, crucially, does *not* show is that as a result of this process, energy is released, and stored in the form of 38 molecules of ATP, so for completeness, we need to add to the respective sides:

$$38\text{ADP} + 38\text{Pi} \longrightarrow 38\text{ATP}$$
$$(\text{Pi} = \text{inorganic phosphate group})$$

The release of the energy contained within a molecule of glucose does not occur in a single reaction, but happens gradually, as the result of numerous reactions linked together in biochemical pathways, the first of which is *glycolysis* (Figure 6.16). Glycolysis can occur with or without oxygen, and is common to both aerobic and anaerobic organisms. Oxygen is essential, however, for *aerobic respiration*, by which ATP is generated from the products of glycolysis. Anaerobes proceed down their own pathways following glycolysis as we shall see, but these lack the ATP-generating power of the aerobic process.

Why glucose?

By concentrating on glucose catabolism in this way, you may think we are ignoring the fate of other nutrient molecules. If you take another look at Figure 6.15, however, you will notice that the breakdown products of lipids, proteins and nucleic acids also find their way into our pathway sooner or later, having undergone transformations of their own.

Glycolysis

The initial sequence of reactions, in which a molecule of glucose is converted to two molecules of *pyruvate*[*], is called *glycolysis* (Figures 6.16 and 6.17). In the first phase of glycolysis, glucose is phosphorylated and its six-carbon ring structure rearranged, before being cleaved into two three-carbon molecules. In the second phase, each of these undergoes oxidation, resulting in pyruvate.

Also known as the *Embden–Meyerhof pathway*, glycolysis is used for the metabolism of simple sugars not just by microorganisms, but by most living cells. The pathway, which takes place in the cytoplasm, comprises a series of 10 linked reactions, in which each molecule of the six-carbon glucose is converted to two molecules of the three-carbon pyruvate, with a net gain of two molecules of ATP.

The full pathway of reactions is shown in Figure 6.17. Note how, in terms of energy, glycolysis can be divided conveniently into a 'sowing' phase, in which two molecules of ATP are *expended* per molecule of glucose, followed by a 'reaping' phase which *yields* four molecules of ATP. The overall energy balance is therefore a *gain of two molecules of ATP* for each molecule of glucose oxidised to pyruvate. In addition, the second phase features the conversion of two molecules of NAD^+ to NADH, which, as we will see, act as an important source of reducing power in subsequent pathways.

The reactions by which ATP is generated from ADP in the second phase of glycolysis are examples of *substrate-level phosphorylation*, so-called because the phosphate group is transferred directly from a donor molecule.

What happens next to the pyruvate produced by glycolysis depends on the organism concerned, and on whether the environment is aerobic or anaerobic; we shall look at these possibilities in due course.

Glycolysis is not the only way to metabolise glucose

Although glycolysis is widespread in both the microbial and nonmicrobial worlds, several bacterial types use alternative pathways to oxidise glucose. For certain Gram-negative groups, notably the pseudomonads (see Chapter 7), the main route used is the *Entner–Doudoroff pathway*, producing a mixture of pyruvate and glyceraldehyde-3-phosphate (Figure 6.18). The former, like that produced in glycolysis, can enter a number of pathways, while the latter can feed into the later stages of glycolysis. The net result of catabolism by the Entner–Doudoroff pathway is the production of one molecule each of ATP, NADH and NADPH per molecule of glucose degraded.

A secondary pathway, which may operate in tandem with glycolysis or the Entner–Doudoroff pathway, is the *pentose phosphate pathway*, sometimes known as the *hexose monophosphate shunt* (Figure 6.19). Like glycolysis, the pathway can operate in the presence or absence of oxygen. Although glyceraldehyde-3-phosphate can once again enter the glycolytic pathway and lead to ATP generation, for most organisms the pathway has a mainly anabolic (biosynthetic) function, acting as a source of precursor molecules for other metabolic pathways. The pentose phosphate pathway is a useful

[*] At physiological pH, carboxylic acids such as pyruvic acid and citric acid are found in their ionised form (pyruvate, citrate); however long-established traditions persist, and you may well find reference elsewhere to the '-ic acid' forms.

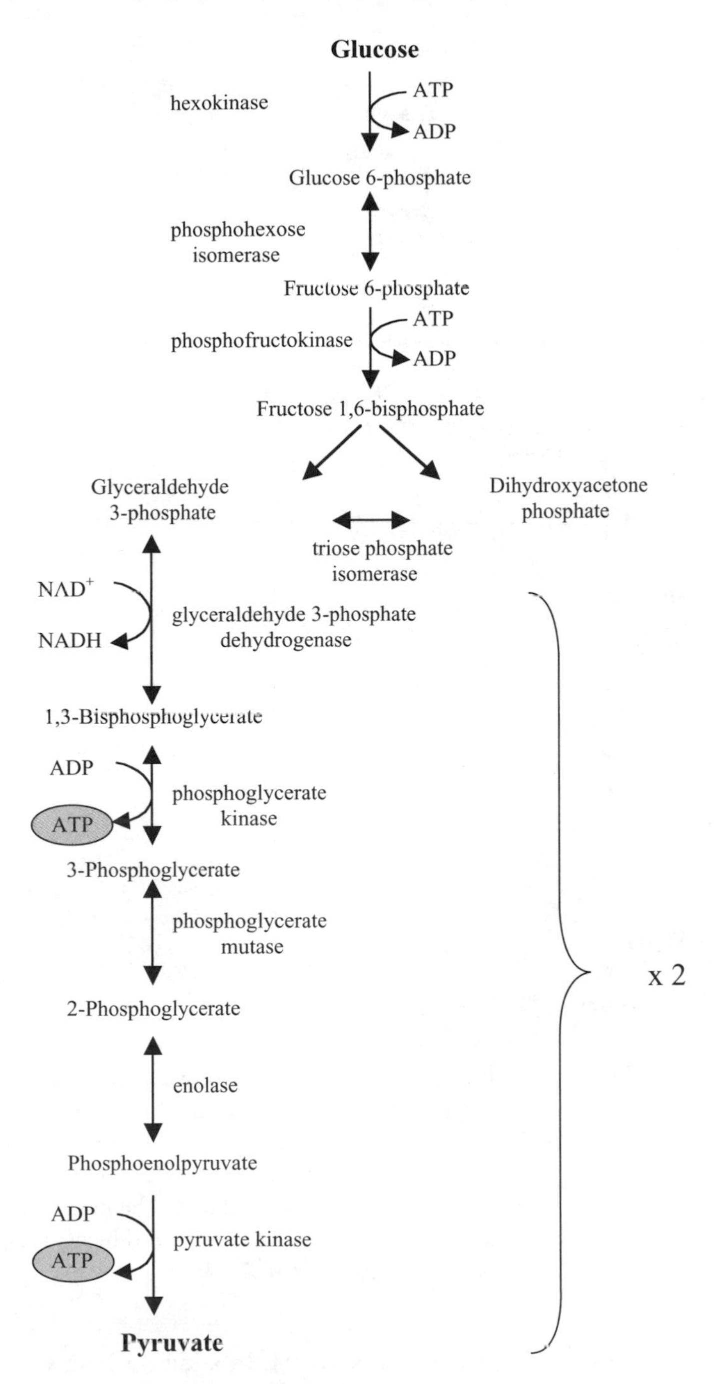

Figure 6.17 Glycolysis: a more detailed look. Glycolysis comprises 10 separate enzyme-catalysed reactions. Of the two 3-carbon compounds formed in the first stage, dihydroxyacetone phosphate cannot directly enter the later part of the pathway, but must first be converted to its isomer glyceraldehyde-3-phosphate. For each molecule of glucose, two molecules of each compound are therefore produced from this point onwards, and the yield of ATP and NADH is likewise doubled

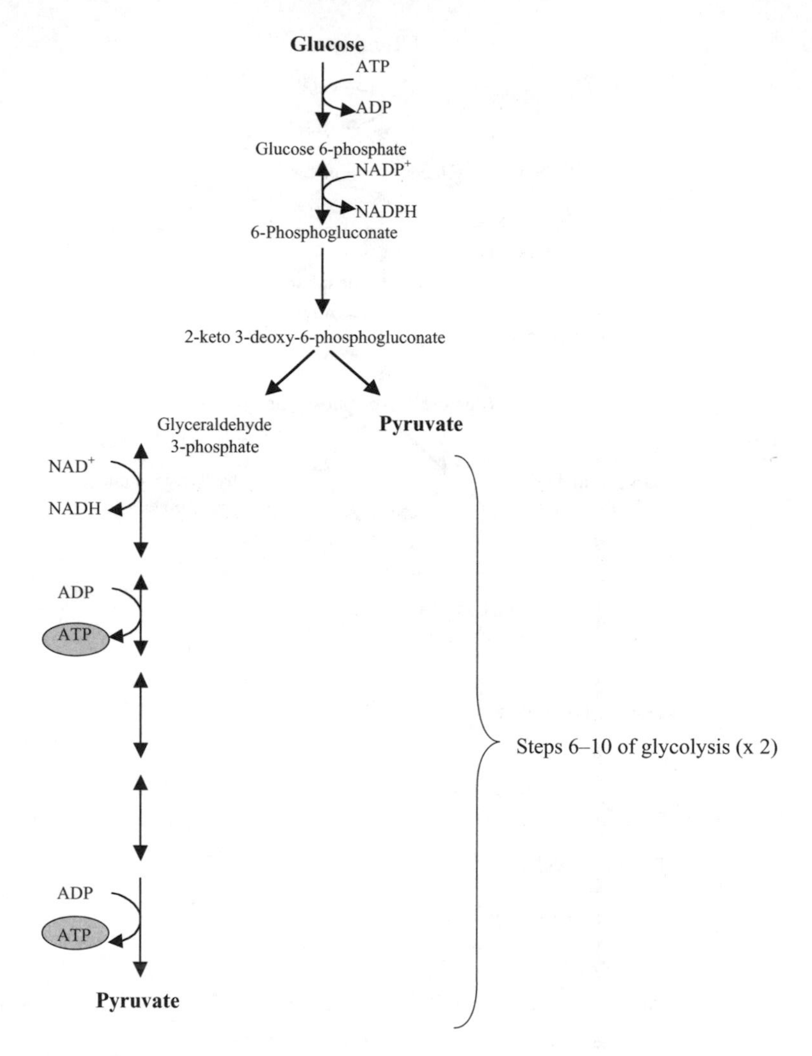

Figure 6.18 The Entner–Doudoroff pathway: an alternative way to metabolise glucose. The products of the pathway are pyruvate and glyceraldeyde-3-phosphate (G3-P). There is a loss of one molecule of ATP in the opening reaction, however when the G3-P joins the later stages of glycolysis, two ATPs are generated, giving a net balance of +1. In addition, the pathway yields one molecule each of NADH and NADPH. The intermediate compound 6-phosphogluconate can enter the pentose phosphate pathway and be decarboxylated to the 5-carbon compound ribulose-5-phosphate (see Figure 6.19)

source of reducing power in the form of NADPH. In addition it acts as an important source of precursors in the synthesis of essential molecules; ribose-5-phosphate is an important precursor in the synthesis of nucleotides, while the four-carbon erythrose 4-phosphate is required for the synthesis of certain amino acids and ribulose-5-phosphate is an intermediate in the Calvin cycle of carbon fixation (see later in this chapter).

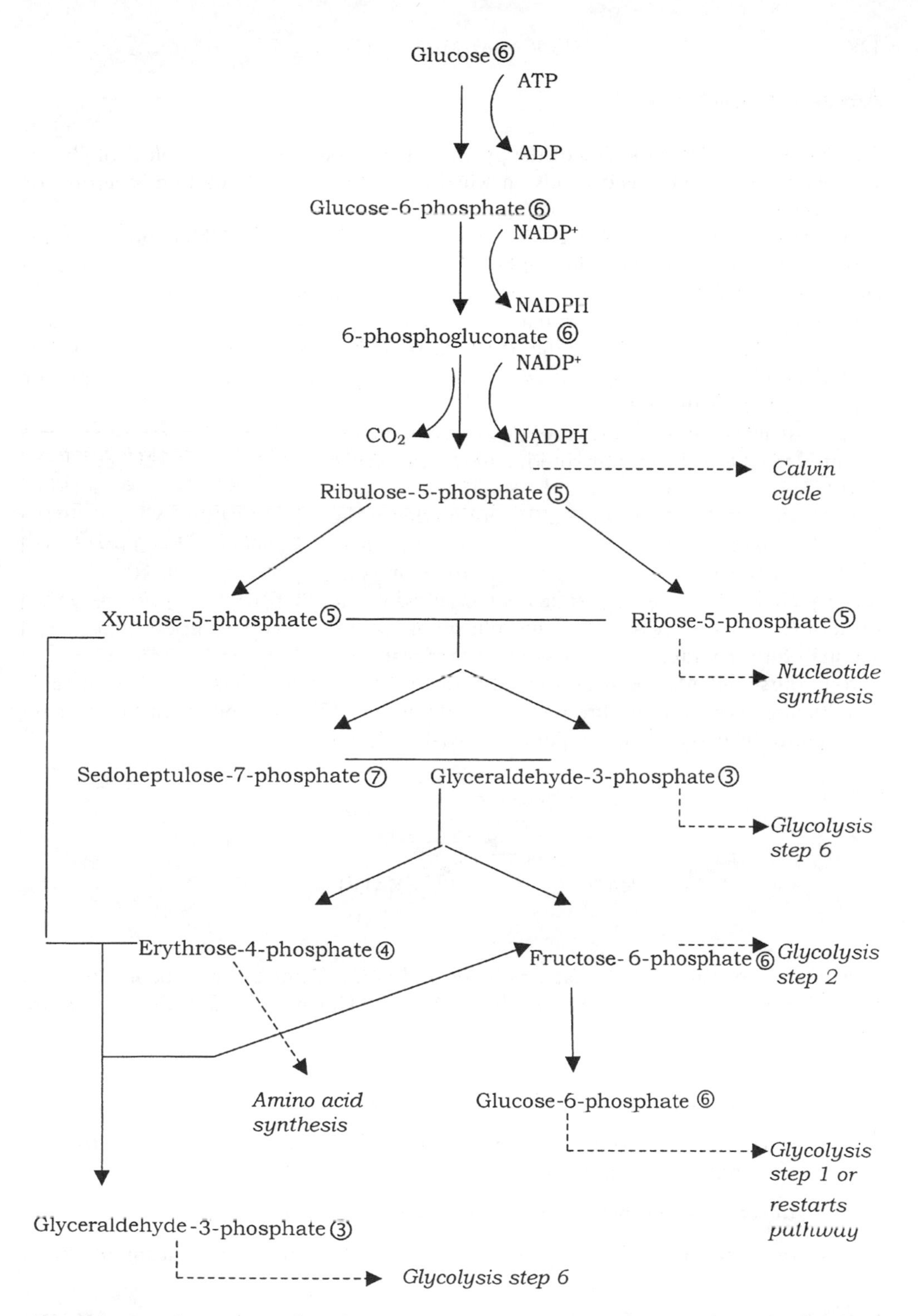

Figure 6.19 The pentose phosphate pathway. Operating simultaneously with glycolysis, the pathway serves as a source of precursors for other metabolic pathways. The metabolic fate of intermediates is indicated in italics. Circled numbers next to each molecule denote the number of carbons

Aerobic respiration

We shall now examine the fate of the pyruvate produced as the end-product of glycolysis. As we have seen, this depends on whether the organism in question is aerobic or anaerobic.

You will recall that during glycolysis, NAD^+ was reduced to NADH. In order for glucose metabolism to continue, this supply of NAD^+ must be replenished; this is achieved either by *respiration* or *fermentation*. Respiration is the term used to describe those ATP-generating processes, aerobic or anaerobic, by which oxidation of a substrate occurs, with an inorganic substance acting as the final *electron acceptor*. In *aerobic* respiration, that substance is oxygen; in *anaerobic* respiration, a substance such as nitrate or sulphate can fulfil the role.

In most aerobic organisms, the pyruvate is completely oxidised to CO_2 and water by entering the *tricarboxylic acid (TCA) cycle*, also known as the *Krebs cycle* or simply the *citric acid cycle* (Figure 6.20). During this cycle, a series of redox reactions result in the gradual transfer of the energy contained in the pyruvate to coenzymes (mostly NADH). This energy is finally conserved in the form of ATP by a process of oxidative phosphorylation. We shall turn our attention to these important reactions in due course, but first let us examine the role of the TCA cycle in a little more detail.

> The TCA cycle is a series of reactions that oxidize acetate to CO_2, generating reducing power in the form of NADH and $FADH_2$ for use in the electron transport chain.

Pyruvate does not itself directly participate in the TCA cycle, but must first be converted into the two-carbon compound *acetyl-Coenzyme A*:

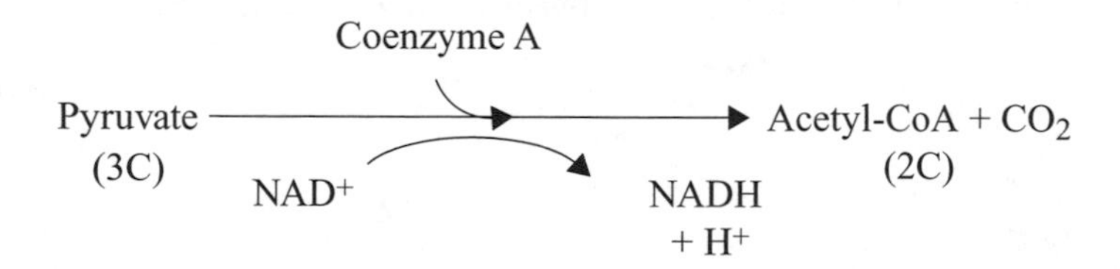

This is an important intermediate, as lipids and amino acids can also be metabolised into this form, and thereby feed into the TCA cycle. The main features of the cycle are as follows:

- each reaction is catalysed by a separate enzyme
- four of the reactions involve substrate oxidation, with energy, in the form of electrons, passing to form NADH (mainly) and $FADH_2$
- the two carbons present in acetyl-CoA are removed as CO_2
- one reaction involves the generation of ATP by substrate-level phosphorylation.

For each 'turn' of the citric acid cycle, *one molecule of ATP, three molecules of NADH* and *one molecule of $FADH_2$* are produced ($FADH_2$ is the reduced form of another coenzyme, FAD). Since these derive from oxidation of a single acetyl-CoA molecule,

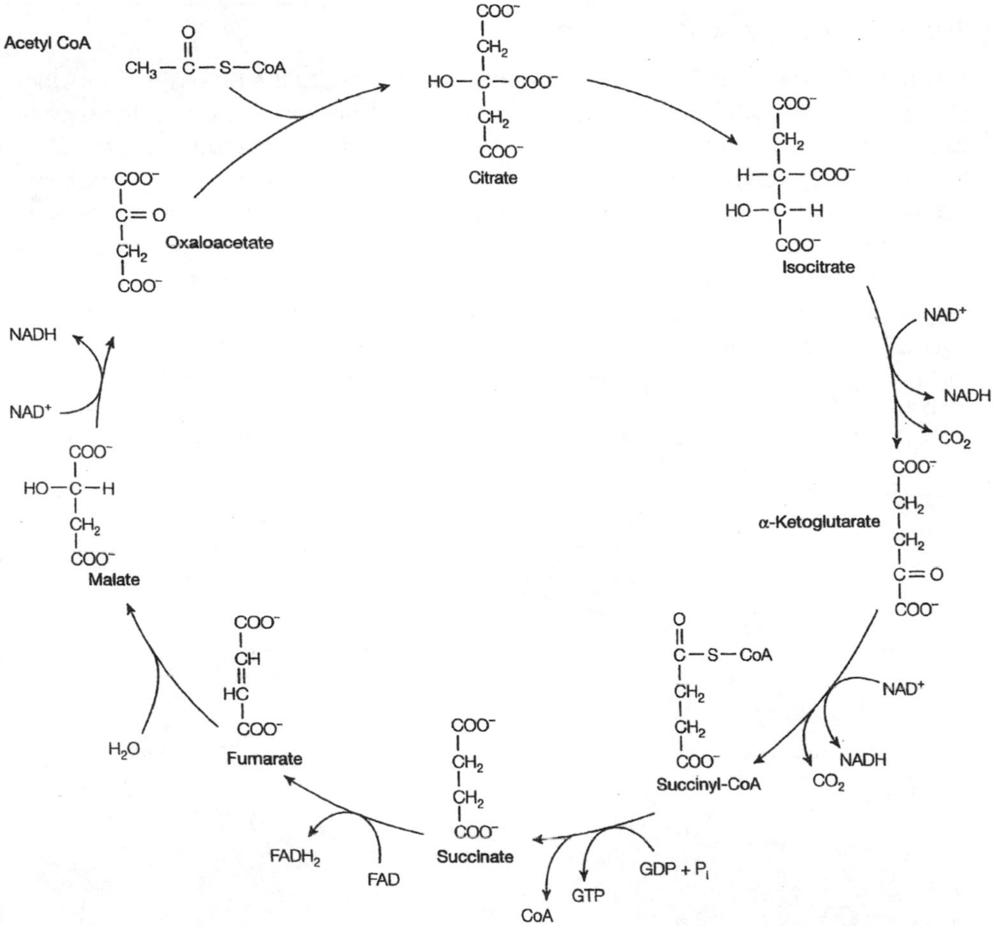

Figure 6.20 The TCA cycle. Acetyl-CoA may derive from the pyruvate of glycolysis or from lipid or amino acid metabolism. It joins with the four-carbon oxaloacetate to form the six-carbon citric acid. Two decarboxylation steps reduce the carbon number back to four and oxaloacetate re-enters the cycle once more. Although no ATP results directly from the cycle, the third phosphate on GTP can be easily transferred to ADP (GTP + ADP = GDP + ATP), thus generating one molecule of ATP per cycle. In addition, substantial reducing power is generated in the form of NADH and $FADH_2$. These carry electrons to the electron transport chain, where further ATPs are generated

we need to double these values per molecule of glucose originally entering glycolysis. Several of the intermediate molecules in the TCA cycle also act as precursors in other pathways, such as the synthesis of amino acids, fatty acids or purines and pyrimidines (see Anabolic metabolism, below). Other pathways regenerate such intermediates for continued use in the TCA cycle (see Box 6.3).

So far, we are a long way short of the 38 molecules of ATP per molecule of glucose mentioned earlier; we have only managed two ATPs from glycolysis and a further two

Box 6.3 The glyoxylate cycle

The components of the TCA cycle may act as precursors for the biosynthesis of other molecules (e.g., both α-ketoglutarate and oxaloacetate can be used for the synthesis of amino acids). For the TCA cycle to continue, it must replace these compounds. Many microorganisms are able to do this by converting pyruvate to oxaloacetate via a carboxylation reaction. A pathway that replenishes intermediate compounds of another in this way is termed *anaplerotic*. Organisms that use acetate (or molecules that give rise to it e.g. fatty acids) as sole carbon source regenerate TCA intermediates by means of the *glyoxylate cycle* (sometimes known as the glyoxylate shunt or bypass). This resembles the TCA cycle, but the two decarboxylation reactions (i.e. those where CO_2 is removed) are missed out (compare with Figure 6.20).

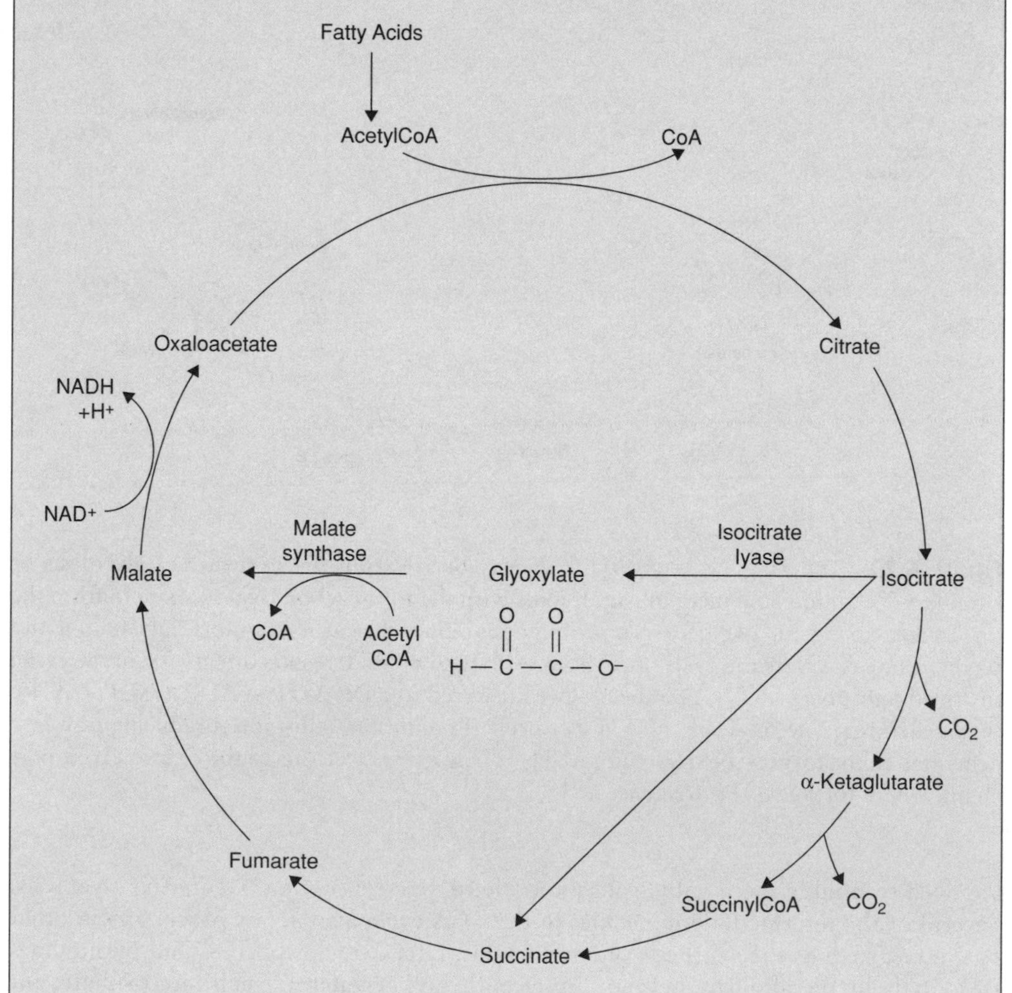

Thus, isocitrate is converted directly to succinate and glyoxylate, and in another unique reaction, the glyoxylate is joined by acetyl-coA to form malate. The result of this is that succinate can be removed to participate in a biosynthetic pathway, but oxaloacetate is still renewed via glyoxylate and malate.

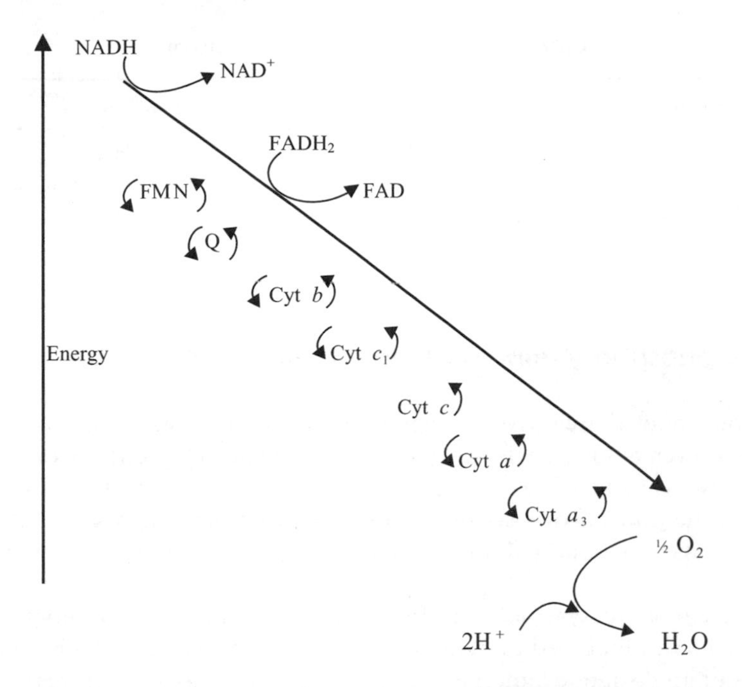

Figure 6.21 The electron transport chain. Electrons from NADH and $FADH_2$ pass from one electron carrier to another, with a gradual release of energy as ATP by chemiosmosis (see Figure 6.22). The electron carriers are arranged in order of their reduction potential (tendency to gain electrons) and oscillate between the oxidised and the reduced state. FMN = flavin mononucleotide, Q = coenzyne Q.

from the TCA cycle. Where do all the rest come from? Most of the energy originally stored in the glucose molecule is now held in the form of the reduced coenzymes (NADH and $FADH_2$) produced during glycolysis and the TCA cycle. This is now converted to no less than 34 molecules of ATP per glucose molecule by oxidative phosphorylation in the remaining steps in aerobic respiration (three from each molecule of NADH and two from each of $FADH_2$).

In the final phase of aerobic respiration, electrons are transferred from NADH and $FADH_2$, via a series of carrier molecules known collectively as the *electron transport* (or *respiratory*) *chain* to oxygen, the terminal electron acceptor (Figure 6.21). This in turn is reduced to the molecules of water you will remember from our overall equation on page 122. In procaryotes, this electron transfer occurs at the plasma membrane, while in eucaryotes it takes place on the inner membrane of mitochondria. Table 6.2 summarises the locations of the reactions in the different phases of carbohydrate metabolism.

The electron transport chain is a series of donor/acceptor molecules that transfer electrons from donors (e.g. NADH) to a terminal electron acceptor (e.g. O2).

Table 6.2 Location of respiratory enzymes

Reaction	Procaryotes	Eucaryotes
Glycolysis	Cytoplasm	Cytoplasm
TCA cycle	Cytoplasm	Mitochondrial matrix
Electron Transport	Plasma membrane	Mitochondrial inner membrane

Oxidative phosphorylation and the electron transport chain

The components of the electron transport chain differ between procaryotes and eucaryotes, and even among bacterial systems, thus details may differ from the example outlined below. The purpose of the electron transport is the same for all systems, however, that is, the transfer of electrons from NADH/$FADH_2$ via a series of carriers to, ultimately, oxygen. Around half of the energy released during this process is conserved as ATP.

The carrier molecules, which act alternately as acceptors and donors of electrons, are mostly complex modified proteins such as flavoproteins and cytochromes, together with a class of lipid-soluble molecules called ubiquinones (also called coenzyme Q). The carriers are arranged in the chain such that each one has a more positive redox potential than the previous one. In the first step in the chain, NADH passes electrons to flavin mononucleotide (FMN), and in so doing becomes converted back to NAD^+, thereby ensuring a ready supply of the latter for the continuation of glycolysis (Figure 6.21). From FMN, the electrons are transferred to coenzyme Q, and thence to a series of cytochromes; at each transfer of electrons the donor reverts back to its oxidised form, ready to pick up more electrons. You may recall that $FADH_2$ yields only two, rather than three molecules of ATP per molecule; this is because it enters the electron transport chain at a later point than NADH, thereby missing one of the points where export of protons occurs. The final cytochrome in the chain transfers its electrons to molecular oxygen, which, as we have seen, acts as the terminal oxygen acceptor. The negatively charged oxygen combines with protons from its surroundings to form water. Four electrons and protons are required for the formation of each water molecule:

$$O_2 + 4e^- + 4H^+ \longrightarrow 2H_2O$$

Since two electrons are released by the oxidation of each NADH, it follows that two NADH are needed for the oxidation of each oxygen.

How does this transfer of electrons lead to the formation of ATP? The *chemiosmotic theory* proposed by Peter Mitchell in 1961 offers an explanation. Although it was not immediately accepted, the validity of the chemiosmotic model is now widely recognised, and in 1978 Mitchell received a Nobel Prize for his work. As envisaged by Mitchell, sufficient energy is released at three points in the electron transport chain for the transfer of protons to the outside of the membrane, resulting in a gradient of both concentration and charge (*proton motive force*). The protons are able to return across the membrane and achieve an equilibrium through specific protein channels within the enzyme *ATP*

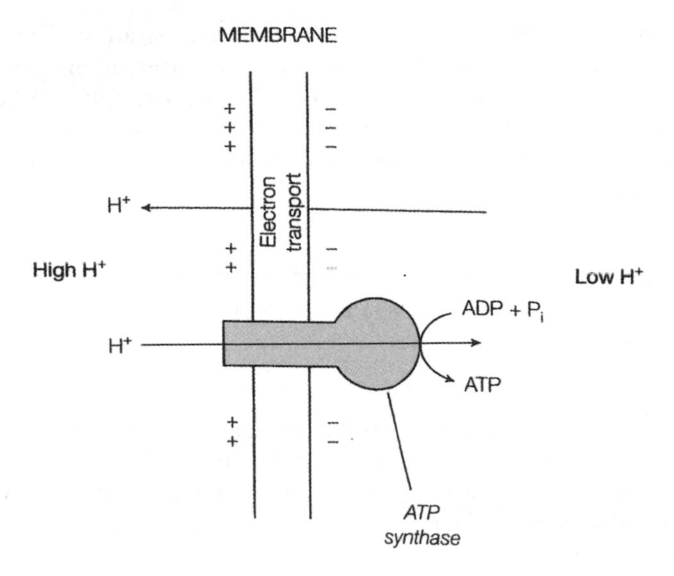

Figure 6.22 Chemiosmosis. The active transport of protons across the membrane creates a gradient of charge and concentration (proton motive force). Special channels containing ATP synthase allow the return of the protons; the energy released is captured as ATP. From Hames, BD, Hooper, NM & Houghton, JD: Instant Notes in Biochemistry, Bios Scientific Publishers, 1997. Reproduced by permission of Thomson Publishing Services

synthase. The energy released by the protons as they return through these channels enables the ATP synthase to convert ADP to ATP (Figure 6.22).

Aerobic respiration in eucaryotes is slightly less efficient than in procaryotes due to the fact that the three stages take place at separate locations (see Table 6.2). Thus the total number of ATPs generated is 36 rather than the 38 in procaryotes (Table 6.3).

Fermentation

We turn now to the fate of pyruvate when oxygen is not available for aerobic respiration to take place. Microorganisms are able, by means of fermentation, to oxidise the pyruvate incompletely to a variety of end products.

Table 6.3 Yield of ATP by aerobic respiration in procaryotes

Process	ATP Yield (per glucose molecule)
Glycolysis	2
TCA Cycle	2
Electron transport chain	34*

*Derived from the oxidative phosphorylation of 2 × NADH from glycolysis, 8 × NADH and 2 × $FADH_2$ from TCA cycle.

It is worth spending a moment defining the word fermentation. The term can cause some confusion to students, as it has come to have different meanings in different contexts. In everyday parlance, it is understood to mean simply alcohol production, while in an industrial context it generally means any large scale microbial process such as beer or antibiotic production, which may be aerobic or anaerobic. To the microbiologist, the meaning is more precise:

a microbial process by which an organic substrate (usually a carbohydrate) is broken down without the involvement of oxygen or an electron transport chain, generating energy by substrate-level phosphorylation

Two common fermentation pathways result in the production of respectively lactic acid and ethanol. Both are extremely important in the food and drink industries, and are discussed in more detail in Chapter 17. *Alcoholic fermentation*, which is more common in yeasts than in bacteria, results in pyruvate being oxidised via the intermediate compound acetaldehyde to ethanol.

$$\underset{\text{Pyruvate}}{CH_3COCOO^-} \xrightarrow[\text{pyruvate decarboxylase}]{\nearrow CO_2} \underset{\text{Acetaldehyde}}{CH_3CHO} \xrightarrow[\text{alcohol dehydrogenase}]{NADH \quad NAD^+} \underset{\text{Ethanol}}{CH_3CH_2OH}$$

Electrons pass from the reduced coenzyme NADH to acetaldehyde, which acts as an electron acceptor, and NAD^+ is thereby regenerated for use in the glycolytic pathway. No further ATP is generated during these reactions, so the only ATP generated in the fermentation of a molecule such as glucose is that produced by the glycolysis steps. Thus, in contrast to aerobic respiration, which generates 38 molecules of ATP per molecule of glucose, fermentation is a very inefficient process, producing just two. Note that all the ATP produced by any fermentation is due to substrate-level phosphorylation, and does not involve an electron transport chain.

A variety of microorganisms carry out *lactic acid fermentation*. Some, such as *Streptococcus* and *Lactobacillus*, produce lactic acid as the only end product; this is referred to as *homolactic fermentation*.

$$\underset{\text{Pyruvate}}{CH_3COCOO^-} \xrightarrow[\text{Lactate dehydrogenase}]{NADH \quad NAD^+} \underset{\text{Lactate}}{CH_3CHOHCOO^-}$$

Certain other microorganisms, such as *Leuconostoc*, generate additional products such as alcohols and acids in a process called *heterolactic* fermentation.

In both alcohol and lactic acid fermentation, the two NADH molecules produced per molecule of glucose have been reoxidised to NAD^+, ready to re-enter the glycolytic pathway.

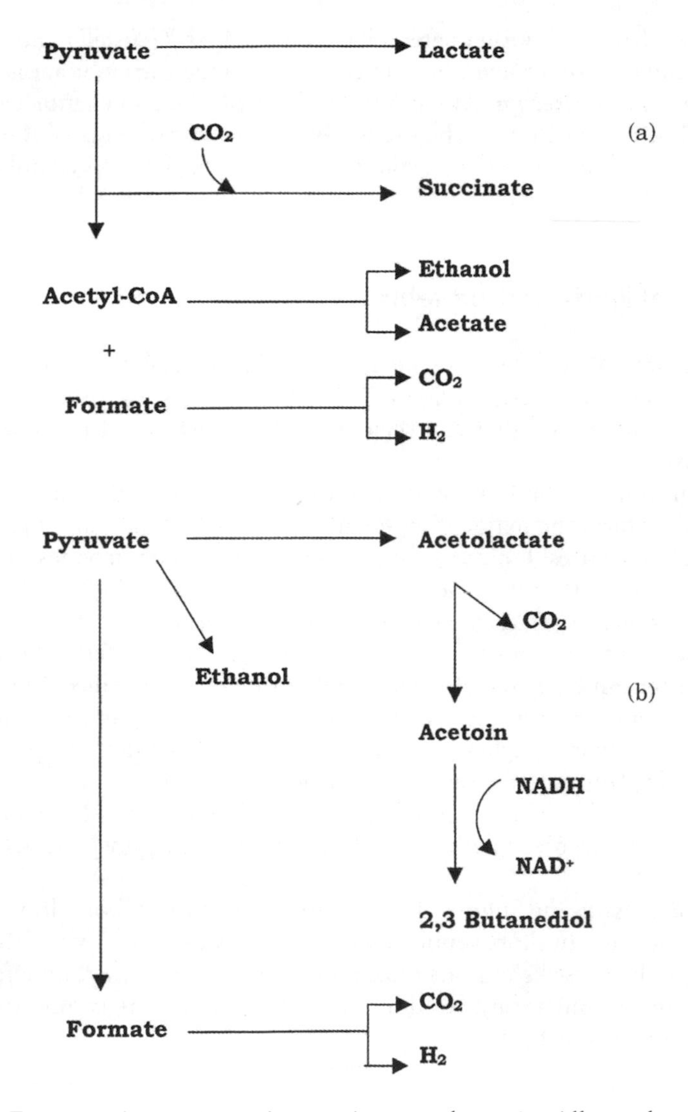

Figure 6.23 Fermentation patterns in enteric proteobacteria. All members of the group produce pyruvate via the Embden–Meyerhof pathway (glycolysis), but subsequent reactions fall into one of two main types. (a) Mixed acid fermentation results in ethanol and a mixture of acids, mainly acetic, lactic, succinic and formic, e.g. *Escherichia, Salmonella.* (b) Butanediol fermentation involves the conversion of pyruvate to acetoin, then to 2,3-butanediol, e.g. *Enterobacter, Klebsiella.* The ratio of CO_2 to H_2 production is much greater in butanediol fermentation

Other types of fermentation

Members of the enteric bacteria metabolise pyruvate to a variety of organic compounds; two principal pathways are involved, both of them involving formic acid. In *mixed acid fermentation*, pyruvate is reduced by the NADH to give succinic, formic and acetic acids, together with ethanol (Figure 6.23a). *Escherichia*, *Shigella* and *Salmonella* belong to this group (see Chapter 7). Other enteric bacteria, such as *Klebsiella* and *Enterobacter*, carry out *2,3-butanediol fermentation*. In this, the products are not acidic, and include an intermediate called *acetoin*. Much more CO_2 is produced per molecule of glucose than in mixed acid fermentation (Figure 6.23b). The end products of the two types of fermentation provide a useful diagnostic test for the identification of unknown enteric bacteria.

Metabolism of lipids and proteins

We have concentrated so far on the metabolism of carbohydrates, but both lipids and proteins may also act as energy sources. Both are converted by a series of reactions to an intermediate compound that can then enter the pathways of metabolism we have discussed above.

Lipids often form an important energy source for microorganisms; they are plentiful in nature, as they form the major component of cell membranes, and may also exist as cellular storage structures. Lipids are hydrolysed by a class of enzymes called *lipases* to their constituent parts; the fatty acids so produced enter the cyclic β-oxidation pathway. In this, fatty acids are joined to coenzyme A to form an *acyl-CoA*, and shortened by two carbons in a series of reactions (Figure 6.24). Molecules of NADH and $FADH_2$ derived from β-oxidation can enter the electron transport chain to produce ATP. Acetyl-CoA, you will recall from earlier in this chapter, serves as the entry point into the TCA cycle. When you consider that a single turn of the TCA cycle gives rise to the production of 14 molecules of ATP, you can appreciate what a rich source of energy a 16- or 18-carbon fatty acid represents. The glycerol component of a lipid requires only slight modification in order to enter the glycolytic pathway as dihydroxyacetone phosphate (see 'Glycolysis' above).

Proteins are a less useful source of energy than lipids or carbohydrates, but may be utilised when these are in short supply. Like lipids, they are initially hydrolysed to their constituent 'building blocks', in this case, amino acids. These then undergo the loss of an amino group (deamination), resulting in a compound that is able to enter, either directly or indirectly, the TCA cycle.

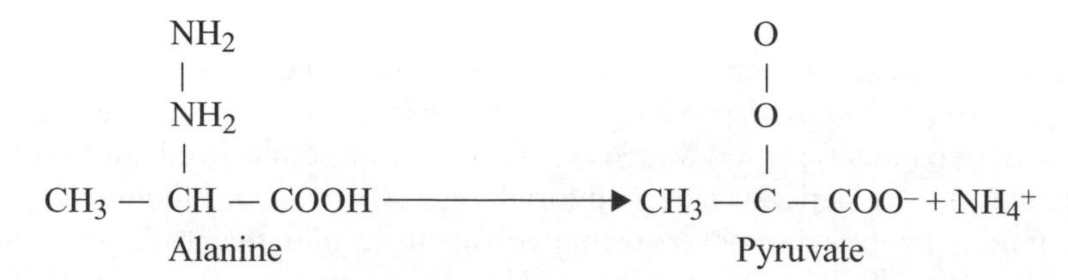

Figure 6.24 β-oxidation of lipids. β–Oxidation comprises a series of four reactions, repeated for the removal of each two-carbon unit. Formation of the acyl CoA ester requires the expenditure of ATP, but there is a net gain in reducing power (NADH + $FADH_2$), which can feed into the electron transport chain. The shortened acyl chain at the end of the process can re-enter the cycle and become further shortened, while acetyl CoA can enter directly into the TCA cycle (CoA-SH = coenzyme A)

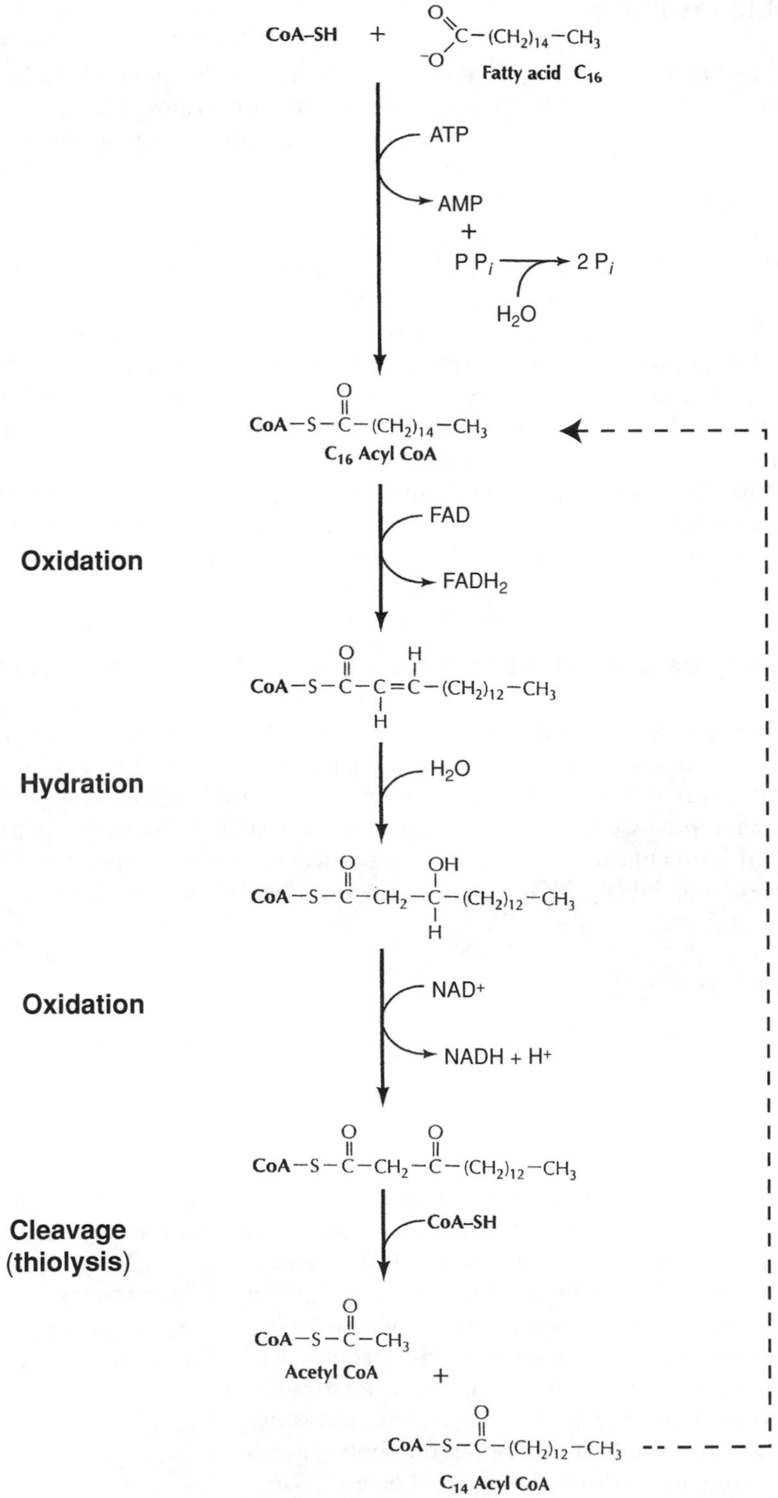

$$\text{CoA–SH} + {}^{-}\text{O}(\text{O=})\text{C}-(\text{CH}_2)_{14}-\text{CH}_3 \quad \text{(Fatty acid } C_{16})$$

ATP → AMP + PP_i; PP_i + H_2O → 2 P_i

$$\text{CoA}-\text{S}-\overset{\text{O}}{\overset{\|}{\text{C}}}-(\text{CH}_2)_{14}-\text{CH}_3$$

C_{16} Acyl CoA

Oxidation — FAD → $FADH_2$

$$\text{CoA}-\text{S}-\overset{\text{O}}{\overset{\|}{\text{C}}}-\underset{\text{H}}{\text{C}}=\overset{\text{H}}{\text{C}}-(\text{CH}_2)_{12}-\text{CH}_3$$

Hydration — H_2O

$$\text{CoA}-\text{S}-\overset{\text{O}}{\overset{\|}{\text{C}}}-\text{CH}_2-\underset{\text{H}}{\overset{\text{OH}}{\text{C}}}-(\text{CH}_2)_{12}-\text{CH}_3$$

Oxidation — NAD^+ → NADH + H^+

$$\text{CoA}-\text{S}-\overset{\text{O}}{\overset{\|}{\text{C}}}-\text{CH}_2-\overset{\text{O}}{\overset{\|}{\text{C}}}-(\text{CH}_2)_{12}-\text{CH}_3$$

Cleavage (thiolysis) — CoA–SH

$$\text{CoA}-\text{S}-\overset{\text{O}}{\overset{\|}{\text{C}}}-\text{CH}_3$$

Acetyl CoA +

$$\text{CoA}-\text{S}-\overset{\text{O}}{\overset{\|}{\text{C}}}-(\text{CH}_2)_{12}-\text{CH}_3$$

C_{14} Acyl CoA

Anaerobic respiration

In the process of anaerobic respiration, carbohydrate can be metabolised by a process that utilises oxidative phosphorylation via an electron transport chain, but instead of oxygen serving as the terminal electron acceptor a (usually) inorganic molecule such as nitrate or sulphate is used. These processes are referred to, respectively, as dissimilatory nitrate or sulphate reduction. *Obligate* anaerobes carry out this process, as they are unable to utilise oxygen; in addition, other organisms may turn to this form of respiration if oxygen is unavailable (*facultative* anaerobes). Other examples of inorganic electron acceptors for anaerobic respiration include Fe^{3+}, CO_2 and Mn^{4+}. In certain circumstances, an organic molecule such as fumarate may be used instead.

Anaerobic respiration is not as productive as its aerobic counterpart in terms of ATP production, because electron acceptors such as nitrate or sulphate have less positive redox potentials than oxygen. Anaerobic respiration tends to occur in oxygen-depleted environments such as waterlogged soils.

It must be stressed that anaerobic respiration is *not* the same as fermentation. The latter process does not involve the components of the electron transport chain (i.e. there is no oxidative phosphorylation), and much smaller amounts of energy are generated.

Energy may be generated by the oxidation of inorganic molecules

In the previous pages, we have seen how electrons derived from a variety of organic sources can be channelled into the glycolytic pathway (or one of its alternatives), and how energy is generated by either oxygen or an organic/inorganic molecule acting as an electron acceptor. Certain bacteria, however, are able to derive their energy from the oxidation of inorganic substrates; these are termed *chemolithotrophs* (see Chapter 4). Molecules such as NH_4^+, NO_2^-, Fe^{2+} and S^0 can be oxidised, with the concomitant generation of ATP.

$$NH_4^+ \longrightarrow NO_2^-$$

$$NO_2^- \longrightarrow NO_3^-$$

$$Fe^{2+} \longrightarrow Fe^{3+}$$

$$S^0 \longrightarrow SO_4^{2-}$$

The ΔG (change in free energy) for each of these reactions is much smaller than that for aerobic respiration. The value of ΔG is a measure of how much energy is released by a reaction. Thus bacteria using this form of metabolism need to oxidise a larger amount of their substrate in order to synthesise the same amount of cellular material. In most cases, such bacteria are autotrophs, fixing carbon from carbon dioxide via the *Calvin cycle*. This is also used by phototrophic organisms, and is described at greater length in the section on photosynthesis below. If organic carbon is available, however, some

The Calvin cycle is a pathway for the fixation of carbon dioxide used by photosynthetic organisms and some chemolithotrophs.

organisms are able to act as heterotrophs, deriving their carbon, but not energy, from such molecules. The overall energy yield from inorganic oxidation is much lower than that from aerobic respiration.

Photosynthesis

Photosynthetic organisms are differentiated from all other forms of life by their ability to derive their cellular energy not from chemical nutrients, but from the energy of the Sun itself. A number of different microbial types are able to carry out photosynthesis, which we can regard as having two distinct forms:

- oxygenic photosynthesis, in which oxygen is produced; found in algae, cyanobacteria (blue-greens) and also green plants
- anoxygenic photosynthesis, in which oxygen is not generated; found in the purple and green photosynthetic bacteria.

Both forms of photosynthesis are dependent on a form of the pigment *chlorophyll*, similar, but not identical, to the chlorophyll found in green plants. We might at this point mention that there is a third method of phototrophic growth, quite distinct from the other two in that chlorophyll plays no role. This form, which is found in halophilic members of the Archaea (see Chapter 7), uses instead a pigment called *bacteriorhodopsin*, similar in structure and function to the rhodopsin found in the retina of animals. In view of the fact that carbon dioxide is not fixed by this mechanism, and no form of chlorophyll is involved, it does not qualify to be described as photosynthesis by some definitions.

The reactions that make up photosynthesis fall into two distinct phases: in the *'light' reactions*, light energy is trapped and some of it conserved as ATP, and in the *'dark' reactions* the energy in the ATP is used to drive the synthesis of carbohydrate by the reduction of carbon dioxide.

In the description of photosynthesis that follows, we shall discuss first the reactions of oxygenic photosynthesis, and then consider how these are modified in the anoxygenic form.

Oxygenic photosynthesis

The overall process of oxygenic photosynthesis can be summarised by the equation:

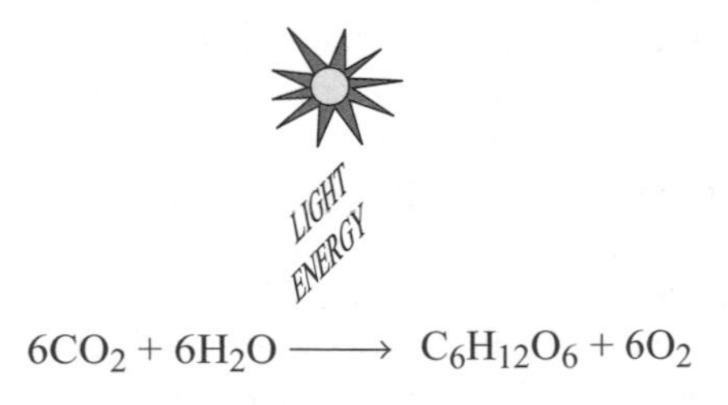

$$6CO_2 + 6H_2O \longrightarrow C_6H_{12}O_6 + 6O_2$$

You will perhaps have noticed that the equation is in essence the reverse of the one we saw earlier when discussing aerobic respiration. In fact, the equation above is not strictly accurate, since it is known that all the oxygen is derived from water. It is therefore necessary to rebalance the equation thus:

$$6CO_2 + 12H_2O \longrightarrow C_6H_{12}O_6 + 6O_2 + 6H_2O$$

Unlike the metabolic reactions we have encountered so far, this reaction requires an input of energy, so the value of ΔG is positive rather than negative.

Where does photosynthesis take place?

In photosynthetic eucaryotes, photosynthesis takes place in specialised organelles, the *chloroplasts*. The light-gathering pigments are located on the stacks of flattened *thylakoid* membranes, while the dark reactions occur within the *stroma* (see Figure 3.15). The light reactions of cyanobacteria also take place on structures called thylakoids; however, since procaryotic cells do not possess chloroplasts, these exist free in the cytoplasm. Their surface is studded with knob-like *phycobilisomes*, which contain unique accessory pigments called phycobilins.

Thylakoids are photosynthetic membranes found in chloroplasts or free in the cytoplasm (in cyanobacteria). They contain photosynthetic pigments and components of the electron transport chain.

'Light' reactions

The light reactions result in:

- splitting of water to release O_2 (photolysis)
- reduction of $NADP^+$ to NADPH
- synthesis of ATP.

This first stage of photosynthesis is dependent on the ability of chlorophyll to absorb photons of light. Absorption of light causes a rearrangement of the electrons in the chlorophyll, so that the molecule attains an 'excited' state (see Box 6.4). Chlorophyll belongs to a class of organic compounds called tetrapyrroles, centred on a magnesium atom (Figure 6.25); a hydrophobic side chain allows the chlorophyll to embed itself in the thylakoid membrane where photosynthetic reactions take place. Several variants of the chlorophyll molecule exist, which differ slightly in their structure and the wavelengths of light they absorb. In organisms carrying out oxygenic photosynthesis, chlorophyll *a* and *b* predominate, while various *bacteriochlorophylls* operate in the anoxygenic phototrophs. Chlorophyll *a* absorbs light in the red and blue parts of the spectrum and reflects or transmits the green part (Figure 6.26) thus cells containing chlorophyll *a* appear green (unless masked by another pigment). Although other pigments are capable of absorbing light, only chlorophyll is able to pass the excited electrons via a series of electron acceptors/donors to convert $NADP^+$ to its reduced form, NADPH. Associated with chlorophyll are several *accessory pigments* such as *carotenoids* or *phycobilins* (in cyanobacteria) with their own absorbance characteristics. They absorb light and transfer the energy to chlorophyll, enabling the organism to utilise light from a broader range of wavelengths.

Over in a flash! A molecule of chlorophyll may remain in its excited state for less than 1 picosecond (10^{-12} s).

Box 6.4 How does a molecule become excited?

When a photon of light strikes a pigment molecule, it is *absorbed*, instead of being reflected or going straight through, and the pigment molecule is raised to an excited (= higher energy) state. An electron in the pigment molecule moves to a higher electron shell (i.e. one further out); this is an unstable state, and the energy it has absorbed is either lost as light (fluorescence) or heat, or is transferred to an acceptor molecule. The pigment molecule then returns to its ground state. Chlorophyll in a test tube will re-emit light as fluorescence, but when present in intact chloroplasts or blue-green cells, it is able to pass on the energy to the next component of one or other photosystem.

The light-gathering process occurs in collections of hundreds of pigment molecules known as *photosynthetic units*. Most of these act as antenna molecules that absorb photons of radiant light. A large number of antenna molecules 'funnel' light towards a single reaction centre, allowing an organism to operate efficiently even at reduced light intensities. Following excitation of the chlorophyll molecule by a photon of light, an

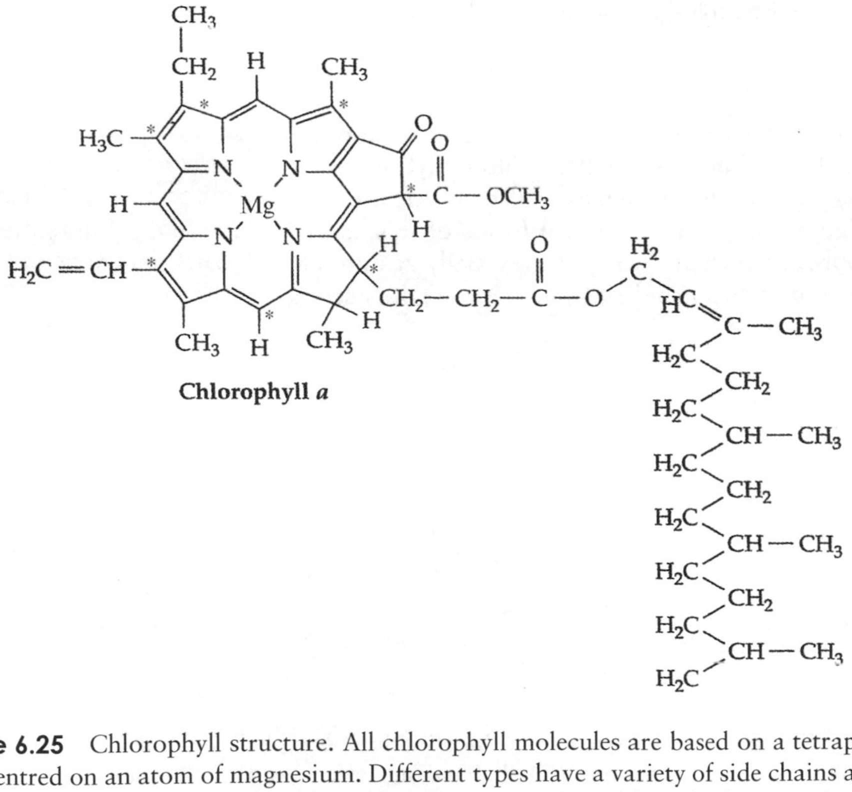

Chlorophyll *a*

Figure 6.25 Chlorophyll structure. All chlorophyll molecules are based on a tetrapyrrole ring centred on an atom of magnesium. Different types have a variety of side chains around the molecule, resulting in different light absorbance properties. The side chains at the points marked * vary between different types of chlorophyll.

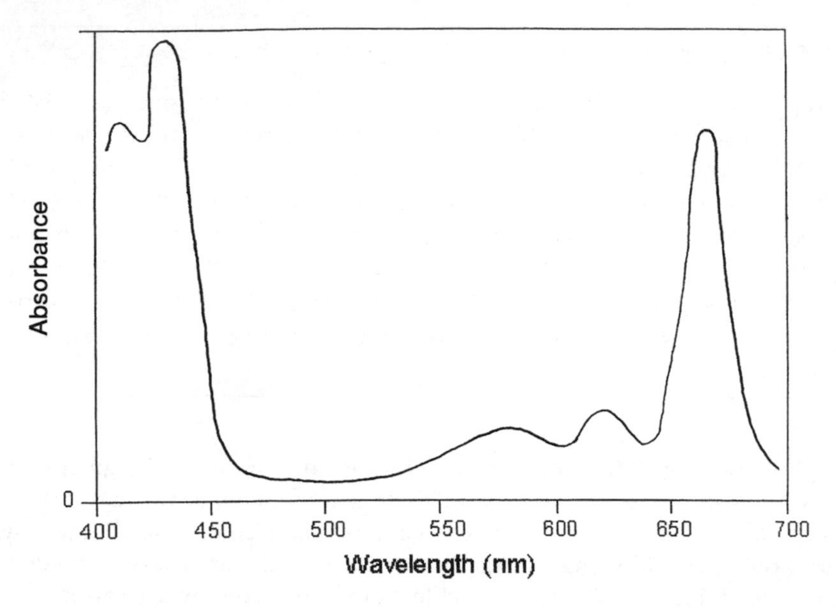

Figure 6.26 Absorbance spectrum of chlorophyll *a*. Chlorophyll *a* absorbs light maximally in the blue and red regions of the spectrum. The green and yellow regions are reflected, giving the characteristic coloration to plants and algae

electron is transferred from one antenna to another, eventually reaching a specialised chlorophyll molecule in a pigment/protein complex called a *reaction centre* (Figure 6.27). There are two forms of this special chlorophyll, which absorb light maximally at different wavelengths, 680 nm and 700 nm. These are the starting

> Wavelengths of light are measured in nanometres (nm). A nanometre is a millionth of a millimetre (10^{-9} M).

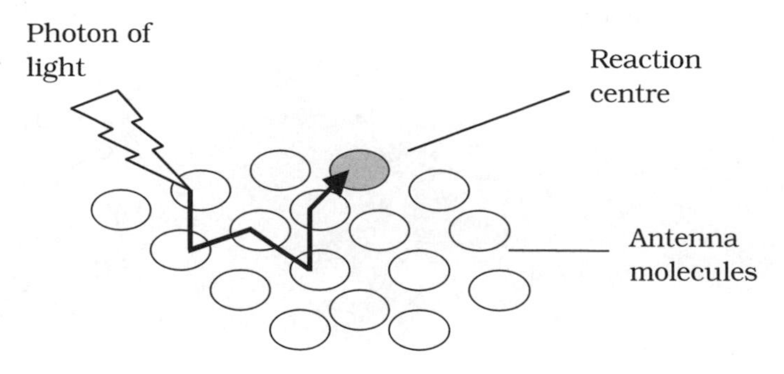

Figure 6.27 The photosynthetic unit. Most of the chlorophyll molecules do not take part in converting light energy into ATP, but transfer electrons to specialised reaction centre chlorophylls

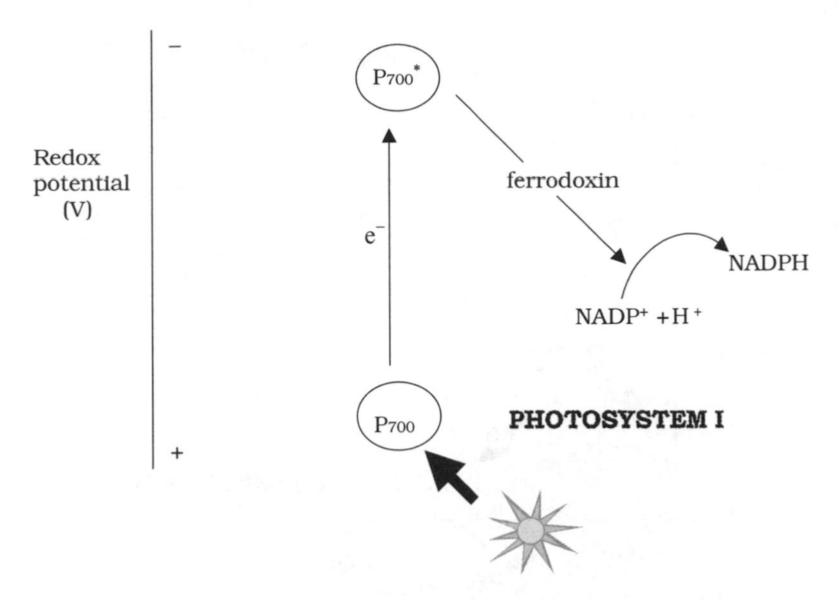

Figure 6.28 Photosystem I. Electrons pass from excited chlorophyll P_{700} through a series of electron acceptors, eventually reducing $NADP^+$. Hydrogen ions are provided by the photolysis of water. The asterisk indicates the excited form of reaction centre chlorophyll

points of two different pathways, known as *photosystem I* and *photosystem II*, which act in sequence together (as we shall see later in this chapter, anoxygenic photosynthesisers only have one of these systems).

The reactions of photosystem I are responsible for the reduction of $NADP^+$ (Figure 6.28). In its excited state, the chlorophyll molecule has a negative redox potential and sufficient energy to donate an excited electron to an acceptor molecule called *ferredoxin*. This is an electron carrier capable of existing in oxidised and reduced forms, like those we encountered in the electron transport chain of aerobic respiration. From here, the electron is passed through a series of successively less electronegative electron carriers until $NADP^+$ acts as the terminal electron acceptor, becoming reduced to NADPH. The hydrogen ions necessary for this are derived from the splitting of water (see below). In order for $NADP^+$ to become reduced as described above, the chlorophyll P_{700} must receive a constant supply of electrons. There are two ways in which this may be done: the first involves photosystem II. It is here that water is cleaved to produce oxygen in a process involving a light-sensitive enzyme:

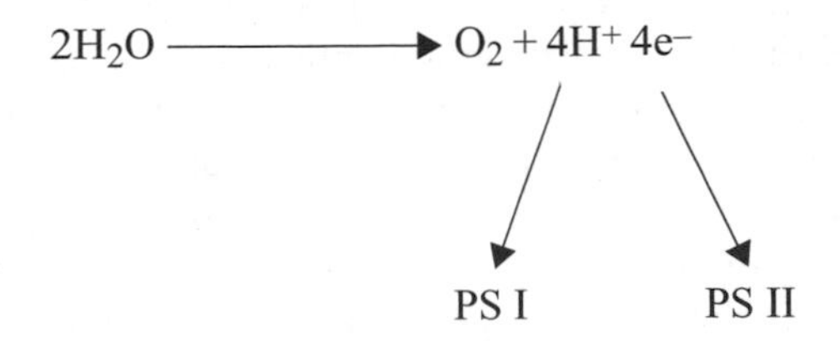

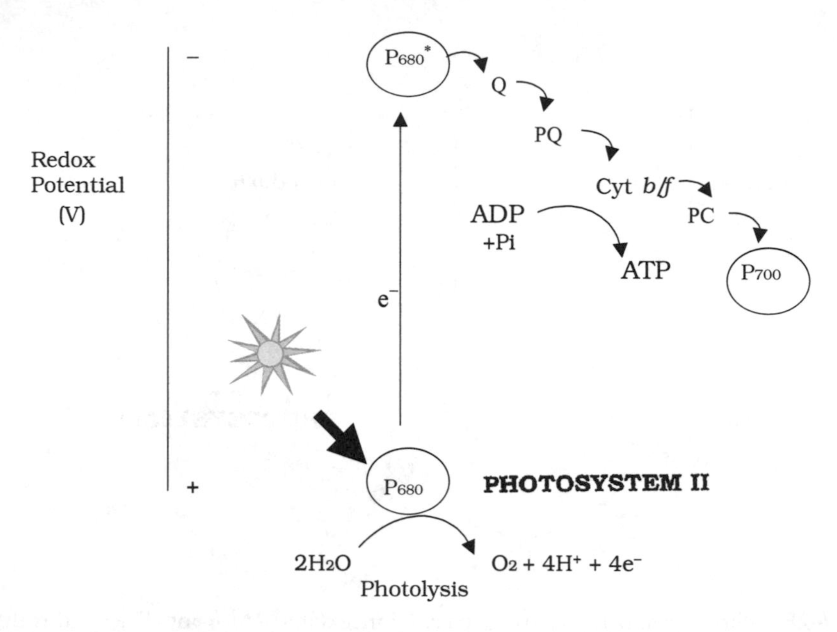

Figure 6.29 Photosystem II. Electrons released from excited molecules of chlorophyll P_{680} pass down an electrochemical gradient and replenish the supply for the chlorophyll P_{700} of photosystem I. The electron flow causes protons to be pumped across the photosynthetic membrane and drive the chemiosmotic synthesis of ATP. Electrons lost by chlorophyll P_{680} are replaced by electrons from the photolysis of water

> Photophosphorylation is the synthesis of ATP using light energy.

Chlorophyll P_{680} of photosystem II absorbs a photon of light, and is raised to an excited state, causing an electron to be released down an electrochemical gradient via carrier molecules as described above (Figure 6.29). The terminal electron acceptor for photosystem II is the chlorophyll P_{700} of photosystem I. The electron flow during this process releases sufficient energy to drive the synthesis of ATP. This process of *photophosphorylation* occurs by means of a chemiosmotic mechanism similar to that involved in the electron transport chain of aerobic respiration.

When we look at the way the two photosystems combine to produce ATP and NADPH from light energy and water, it is easy to see why this is sometimes known as the 'Z' scheme (Figure 6.30). Only by having the two photosystems operating in series and at different energy levels can sufficient energy be generated to extract an electron from water on the one hand and generate ATP and NADPH on the other.

As indicated above, there is an alternative pathway by which the electron supply of the chlorophyll P_{700} can be replenished. You will recall that in photosystem I, having reached ferredoxin and reduced it, electrons pass through further carriers before reaching $NADP^+$. Sometimes, however, they may take a different route, joining the electron transport chain at *plastoquinone (PQ)*, and generating further ATP (Figure 6.31). The electron ends up back at the P_{700}, so the pathway is termed *cyclic photophosphorylation*. Note that because no splitting of water is involved in the cyclic pathway, no

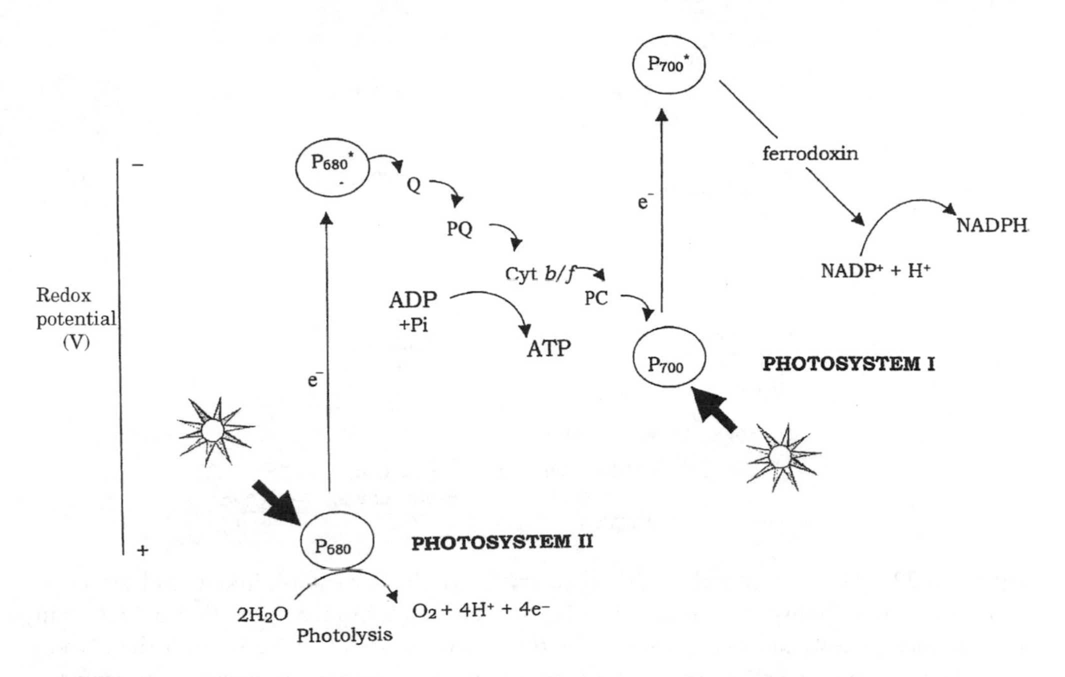

Figure 6.30 Non-cyclic photophosphorylation: the 'Z' scheme. Photosystems I and II interact to convert water and light energy into ATP and NADPH. P_{680} and P_{700} indicate the wavelength of light maximally absorbed by the two photosystems

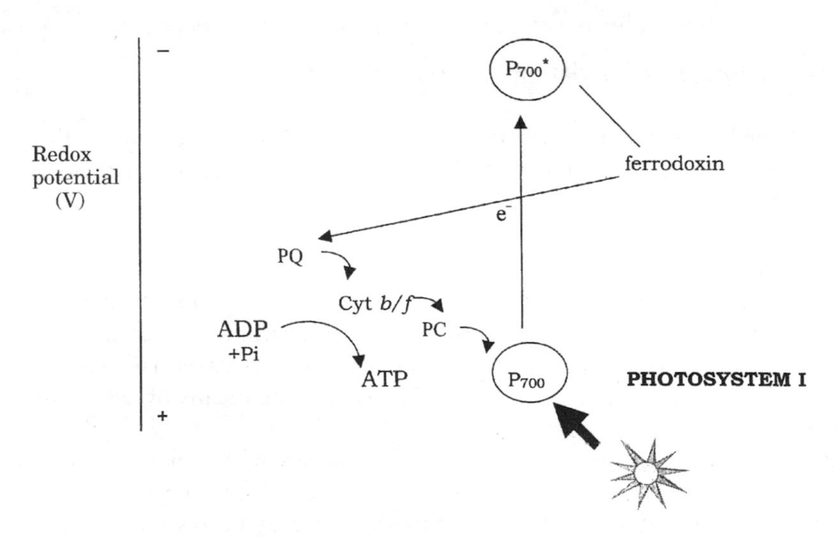

Figure 6.31 Cyclic photophosphorylation. Photosystem I electrons may enter the electron transport chain at plastoquinone and return to reaction centre chlorophyll P_{700}, generating a proton motive force for use in ATP production

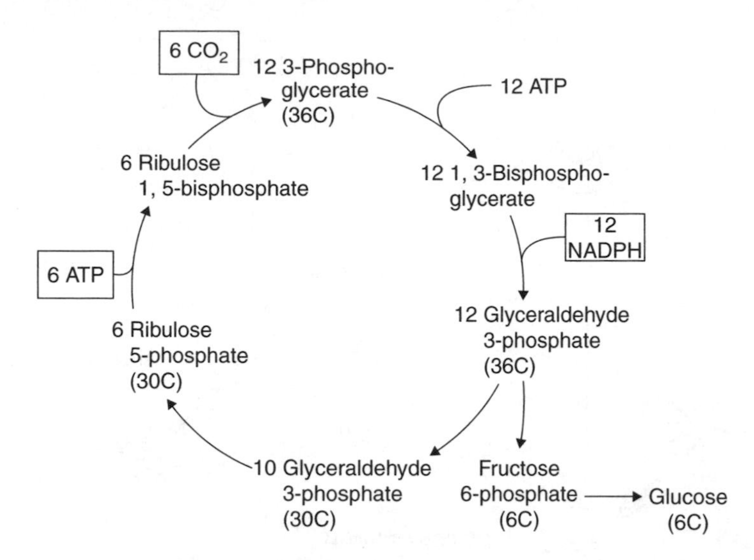

Figure 6.32 The Calvin cycle. Carbon enters the cycle as carbon dioxide and leaves as glyceraldehyde-3-phosphate, from which hexose sugars are formed. Note that only a small part of each hexose molecule produced by this pathway derives directly from the carbon dioxide (only one carbon out of six), thus six complete turns of the cycle would be required to generate a single glucose molecule. (The numbers in the figure relate to six turns.) Although used by most autotrophs, the Calvin cycle is not found in members of the Archaea

oxygen is generated. Also, because the electrons follow a different pathway, NADPH is not produced.

Non-cyclic photophosphorylation: ATP + NADPH + O_2 produced

Cyclic photophosphorylation: only ATP produced

As we shall see below, the biosynthetic reactions of the Calvin cycle that follow the light reactions require more ATP than NADPH. This additional ATP is provided by cyclic photophosphorylation.

'Dark' reactions

The term 'dark reactions' is somewhat misleading; while they *can* take place in the dark, they are dependent upon the ATP and NADPH produced by the light reaction.

The most widely used mechanism for the incorporation of carbon dioxide into cellular material is the Calvin cycle (Figure 6.32). This is also the means by which many other, non-photosynthetic, autotrophs fix CO_2.

Carbon dioxide entering the Calvin cycle combines with a five-carbon compound called ribulose-1,5-bisphosphate to form two molecules of 3-phosphoglycerate via a transient six-carbon intermediate (not shown). The enzyme responsible for this fixation of CO_2 is called ribulose bisphosphate carboxylase (*rubisco*), and it is the most abundant protein in the natural world. The 3-phosphoglycerate is then reduced to give glyceraldehyde 3-phosphate (G3-P), in a process that uses both ATP and NADPH from the light reaction. By a series of reactions that are essentially the reverse of the opening steps of glycolysis, some of this G3-P is converted to glucose. The remaining steps of the Calvin cycle are concerned with regenerating the five-carbon ribulose-5-phosphate. We can summarise the incorporation of CO_2 into glucose as:

$$6CO_2 + 18ATP + 12NADPH + 2H^+ + 12H_2O \rightarrow C_6H_{12}O_6 + 18ADP + 18Pi + 12NADP^+$$

Anoxygenic photosynthesis

Photosynthesis as carried out in the purple and green bacteria is different to the process just described in a number of respects. Some of the main differences are summarised below:

- no oxygen is generated during this type of photosynthesis, and the bacteria involved are growing anaerobically
- bacteriochlophylls absorb light maximally at longer wavelengths than chlorophyll *a* and *b*, allowing them more effectively to utilise the light available in their own particular habitat
- purple and green bacteria are not able to utilise water as a donor of electrons, and must instead use a compound that is oxidised more easily, such as hydrogen sulphide or succinate
- only a single photosystem is involved in the light reactions of anoxygenic photosynthesis. In the green bacteria this is similar to photosystem I, while in the purples it more closely resembles photosystem II
- thylakoid membranes are not found in the green and purple bacteria; light reactions take place in lamellar invaginations of the cytoplasmic membrane in the purples and in vesicles called *chlorosomes* in the greens.

In the generation of ATP, a form of cyclic photophosporylation is employed; the bacteriochlorophyll acts as both donor and acceptor of electrons (Figure 6.33). In order to generate reducing power in the form of reduced coenzymes, an external electron source is necessary, since this process is non-cyclic. In the green and purple sulphur bacteria, this role is served by sulphur or reduced sulphur compounds such as sulphide or thiosulphate. The non-sulphur bacteria utilise an organic molecule such as succinate as an electron donor. In some cases, the electron donor has a more positive redox potential than the $NADP^+$, so in order to reduce the coenzyme, electrons would have to flow against the electrochemical gradient. An input of energy in the form of ATP is needed to make this possible, in a process known as *reverse electron flow*. This also happens in many chemolithotrophs such as *Acidithiobacillus* and *Nitrobacter*. Table 6.4 compares the different groups of anoxygenic photosynthesisers. A special form of anoxygenic photosynthesis occurs in the Heliobacteria (see Chapter 7); these carry out photoheterotrophy using a unique form of bacteriochlorophyll called bacteriochlorophyll *b*. They are unable to use carbon dioxide as a carbon source.

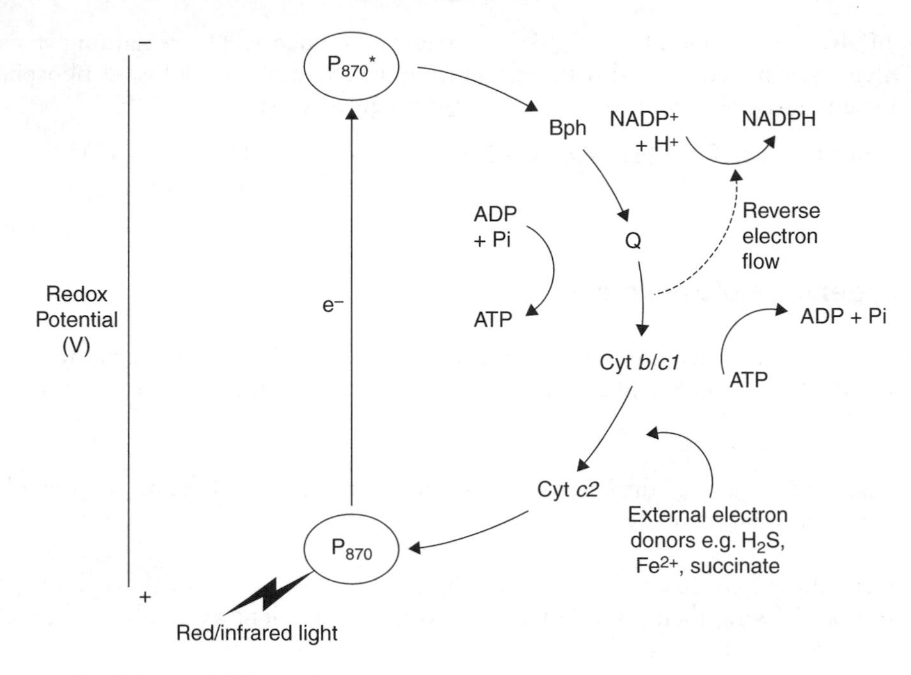

Figure 6.33 Electron flow in the anoxygenic photosynthesis of a purple bacterium. ATP is generated by the passage of electrons down an electron transport chain back to the reaction centre bacteriochlorophyll. Anoxygenic photosynthetic bacteria use molecules such as sulphur and hydrogen sulphide instead of water as external electron donors, hence no oxygen is generated. NADPH for use in CO_2 fixation must be generated by reverse electron flow. Bph = bacteriophaeophytin (bacteriochlorophyll *a* minus its magnesium atom)

Anabolic reactions

So far, in describing respiration and photosynthesis, we have considered those reactions whereby a microorganism may generate cellular energy from its environment. As we saw at the beginning of this chapter, one of the uses to which this may be put is the synthesis of new cellular materials. In all of the pathways described in the following

Table 6.4 Comparison of anoxygenic photosynthetic bacteria

	Green sulphur	Green non-sulphur	Purple sulphur	Purple non-sulphur
Bacteriochlorophyll	*c, d, e*	*a, c*	*a, b*	*a, b*
Electron donor	H_2S/reduced S/H_2	Organic compounds	H_2S/reduced S/H_2	Organic compounds
CO_2 fixation	Reverse TCA cycle	Calvin cycle	Calvin cycle	Calvin cycle
Photoheterotrophy	No	Yes	Some	Mainly

section, the conversion of ATP to ADP is required at some point. The term *biosynthesis* is used to describe those reactions by which nutrients are incorporated first into small molecules such as amino acids and sugars and subsequently into biomacromolecules such as proteins and polysaccharides.

Biosynthesis of carbohydrates

We have already seen in the preceding pages that autotrophic organisms (not necessarily phototrophic ones) are able to incorporate inorganic carbon as CO_2 or HCO_3^- into hexose sugars, most commonly via the Calvin cycle. Heterotrophic organisms are unable to do this, and must convert a range of organic compounds into glucose by a series of reactions called *gluconeogenesis* (Figure 6.34). Many compounds such as lactate or certain amino acids can be converted to pyruvate directly or via other members of the TCA cycle, and thence to glucose. To all intents and purposes, gluconeogenesis reverses the steps of glycolysis (see above), although not all the enzymes involved are exactly the same. This is because three of the reactions are essentially irreversible, so different enzymes must be used to overcome this. These reactions are highlighted in Figure 6.34.

Once glucose or fructose has been produced, it can be converted to other hexose sugars by simple rearrangement reactions. Building up these sugars into bigger carbohydrates (polysaccharides) requires them to be in an energised form: this usually takes the form of either an ADP- or UDP-sugar, and necessitates an input of energy. Pentose sugars such as ribose are important in the synthesis of nucleotides for nucleic acids and coenzymes (see below).

Biosynthesis of lipids

Fatty acids are synthesised by a stepwise process that involves the addition of two-carbon units to form a chain, most commonly of 16–18 carbons. The starting point of fatty acid metabolism is the two-carbon compound

The basic building blocks in the synthesis of fatty acids are acetyl-CoA (two-carbon) and malonyl-CoA (three-carbon). We have encountered acetyl-CoA before, when discussing the TCA cycle; malonyl-CoA is formed by the carboxylation of acetyl-CoA.

$$\begin{array}{ccc} \text{Acetyl–CoA} + CO_2 & \rightarrow & \text{Malonyl–CoA} \\ 2 - \text{carbon} & & 3 - \text{carbon} \end{array}$$

Carbon dioxide is essential for this step, but is not incorporated into the fatty acid as it is removed in a subsequent decarboxylation step. In order to take part in the biosynthesis of fatty acids, both molecules have their coenzyme A element replaced by an *acyl carrier protein (ACP)*. In a condensation reaction, one carbon is lost as CO_2 and one of the ACPs is released. The resulting four-carbon molecule is reduced, with the involvement of two NADPH molecules, to *butyryl-ACP*. This is then extended two carbon atoms at a time by a series of further condensations with malonyl-ACP (Figure 6.35).

Thus, extending a fatty acid chain by two carbons involves the expenditure of one ATP and two NADPH molecules. The overall equation for the synthesis of a 16-carbon

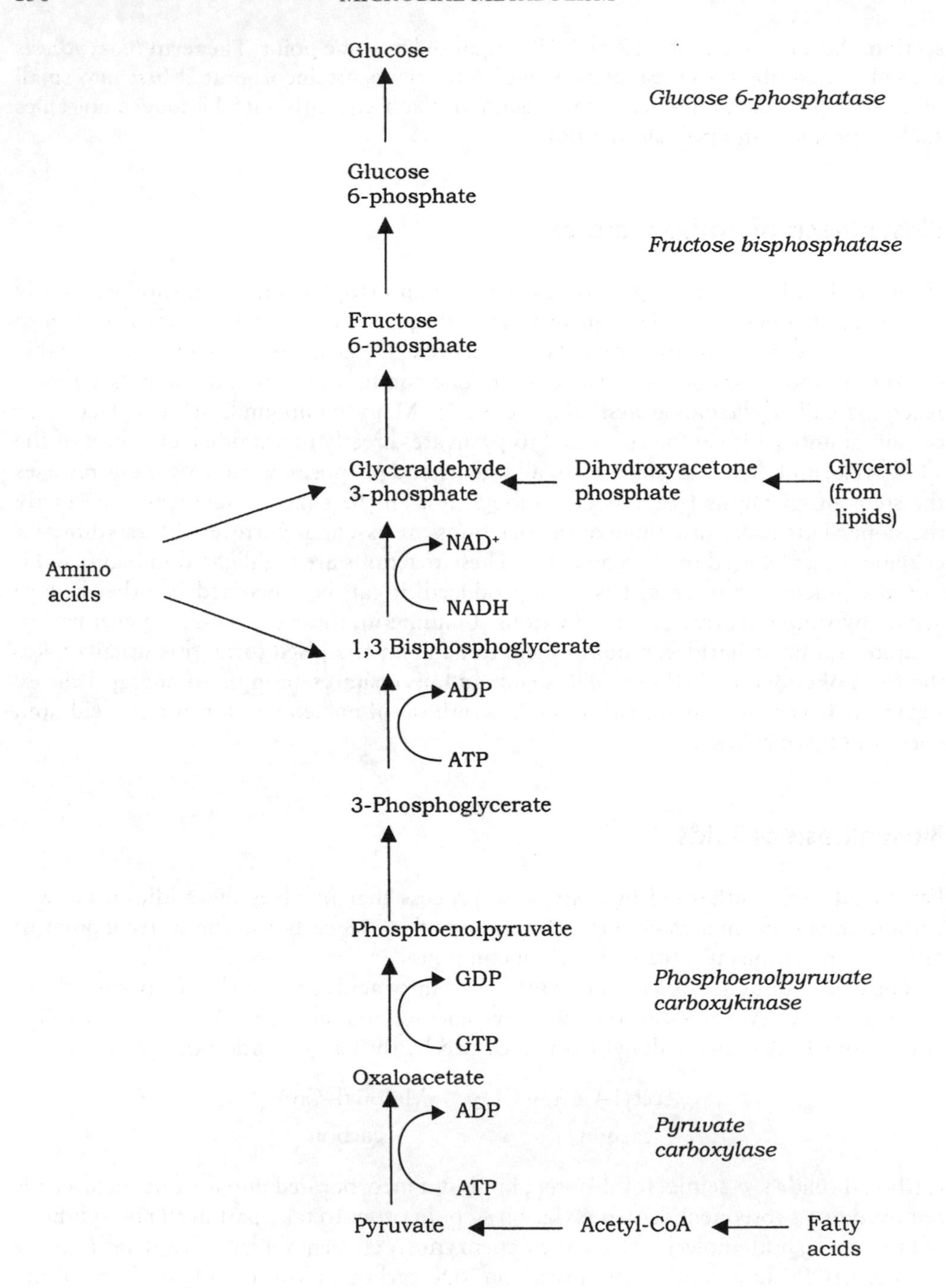

Figure 6.34 Gluconeogenesis. Non-carbohydrate precursors can feed into a pathway that converts pyruvate to glucose in a series of reactions that are mostly the reverse of glycolysis. Enzymes not found in glycolysis are shown in italics

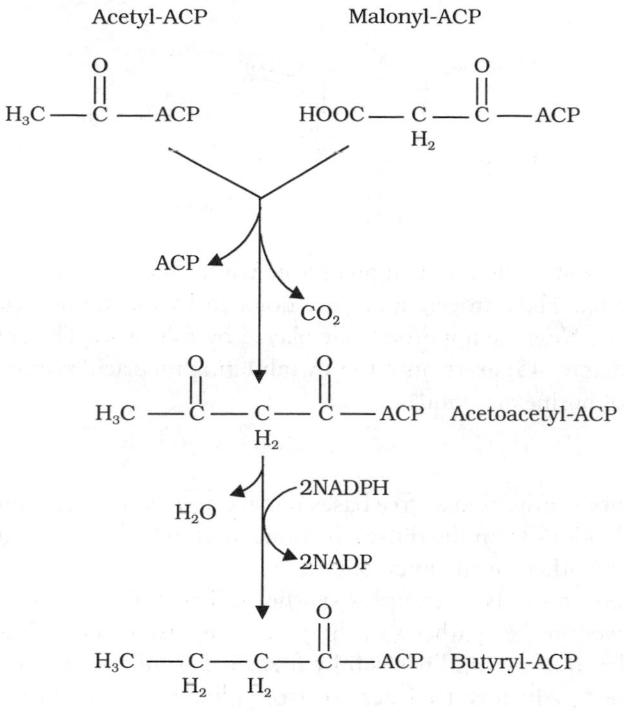

Figure 6.35 Fatty acid biosynthesis. Acetyl- and malonyl-ACPs condense with the loss of CO_2 to give a four-carbon molecule butyryl-ACP. The addition of further two-carbon acetyl groups is achieved by re-entering the pathway and reacting with further molecules of malonyl-ACP

fatty acid such as palmitic acid can be represented:

$$8\ \text{Acetyl-CoA} + 7\ \text{ATP} + 14\ \text{NADPH} + 6\text{H}^+ \longrightarrow \text{Palmitate} + 14\ \text{NADP}^+ + 8\ \text{CoA} + 6\ \text{H}_2\text{O} + 7\ \text{ADP} + 7\ \text{Pi}$$

Once formed, fatty acids may be incorporated into phospholipids, the major form of lipid found in microbial cells. Recall from Chapter 2 that a phospholipid molecule has three parts: fatty acid, glycerol and phosphate. These last two are provided in the form of glycerol phosphate, which derives from the dihydroxyacetone phosphate of glycolysis. Glycerol phosphate replaces the ACP of two fatty acid-ACP conjugates to yield phosphatidic acid, an important precursor for a variety of membrane lipids. The energy for this reaction is provided, unusually, not by ATP but by CTP (cytidine triphosphate).

Biosynthesis of nucleic acids

Most microorganisms are able to synthesise the purines and pyrimidines that comprise the nitrogenous bases of DNA and RNA. These compounds are synthesised from a number of sources in reactions that require an input of ATP. The contribution of different compounds towards the purine skeleton of guanine or adenine is shown in Figure 6.36.

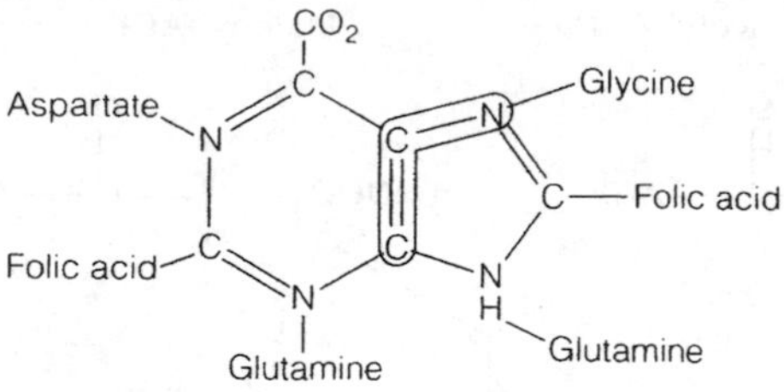

Figure 6.36 Purine structure. Several precursors contribute towards the formation of the purine base skeleton. The nitrogen atoms are donated by the amino acids glutamine, aspartate and glycine. Note the important role played by folic acid. The antimicrobial agent sulphonamide (Chapter 15) exerts its effect by inhibiting folic acid synthesis, which in turn affects synthesis of purine nucleotides

The purines are not synthesised as free bases but are associated with ribose-5-phosphate as complete nucleotides from the outset. Inosinic acid, which is formed initially, acts as a precursor for the other purine nucleotides.

Pyrimidines have a similarly complex synthesis. The amino acids aspartate and glutamine are involved in the synthesis of the precursor orotic acid. Note that unlike the purines, the skeleton of pyrimidines is fully formed *before* association with the ribose-5-phosphate moiety, which is itself derived from glucose (see biosynthesis of carbohydrates, above).

Ribonucleotides (as contained in RNA) are converted to deoxyribonucleotides (as contained in DNA) by a reduction reaction that may involve vitamin B_{12} acting as a cofactor.

Biosynthesis of amino acids

A very limited number of microorganisms are able to utilise molecular nitrogen from the atmosphere by incorporating it initially into ammonia and subsequently into organic compounds (see Chapter 7). Most organisms, however, need to have their nitrogen supplied as nitrate, nitrite, or ammonia itself. Ammonia can be incorporated into organic nitrogen compounds in several ways, including glutamate formation from α-ketoglutarate (see TCA cycle, above):

$$\alpha\text{-ketoglutarate} + NH_4^+ + NADPH^* + H^+ \longrightarrow \text{glutamate} + NADP^+ + H_2O$$

(*Some species utilise NADH as their electron donor)

The amino group can subsequently be transferred from the glutamate to make other amino acids by *transamination* reactions involving other keto acids:

$$\text{e.g. glutamate} + \text{oxaloacetate} \longrightarrow \text{aspartate} + \alpha\text{-ketoglutarate}$$

Glutamate plays a central role in the biosynthesis of other amino acids, as it usually donates the primary amino group of each:

$$\begin{array}{c} COO^- \\ | \\ CH_2 \\ | \\ CH_2 \\ | \\ H\text{-}C\text{-}\mathbf{NH_2} \\ | \\ COOH \\ \text{Glutamate} \end{array} + \begin{array}{c} CH_3 \\ | \\ C{=}O \\ | \\ COOH \\ \text{Pyruvic acid} \end{array} \longrightarrow \begin{array}{c} COO^- \\ | \\ CH_2 \\ | \\ CH_2 \\ | \\ C{=}O \\ | \\ COOH \\ \alpha\text{-Ketoglutarate} \end{array} + \begin{array}{c} CH_3 \\ | \\ H\text{-}C\text{-}\mathbf{NH_2} \\ | \\ COOH \\ \text{Alanine} \end{array}$$

According to the precursor molecule from which they derive, amino acids can be placed into six 'families' (Figure 6.37). The precursors are intermediates in metabolic pathways we have already encountered in this chapter, such as glycolysis or the TCA cycle. When

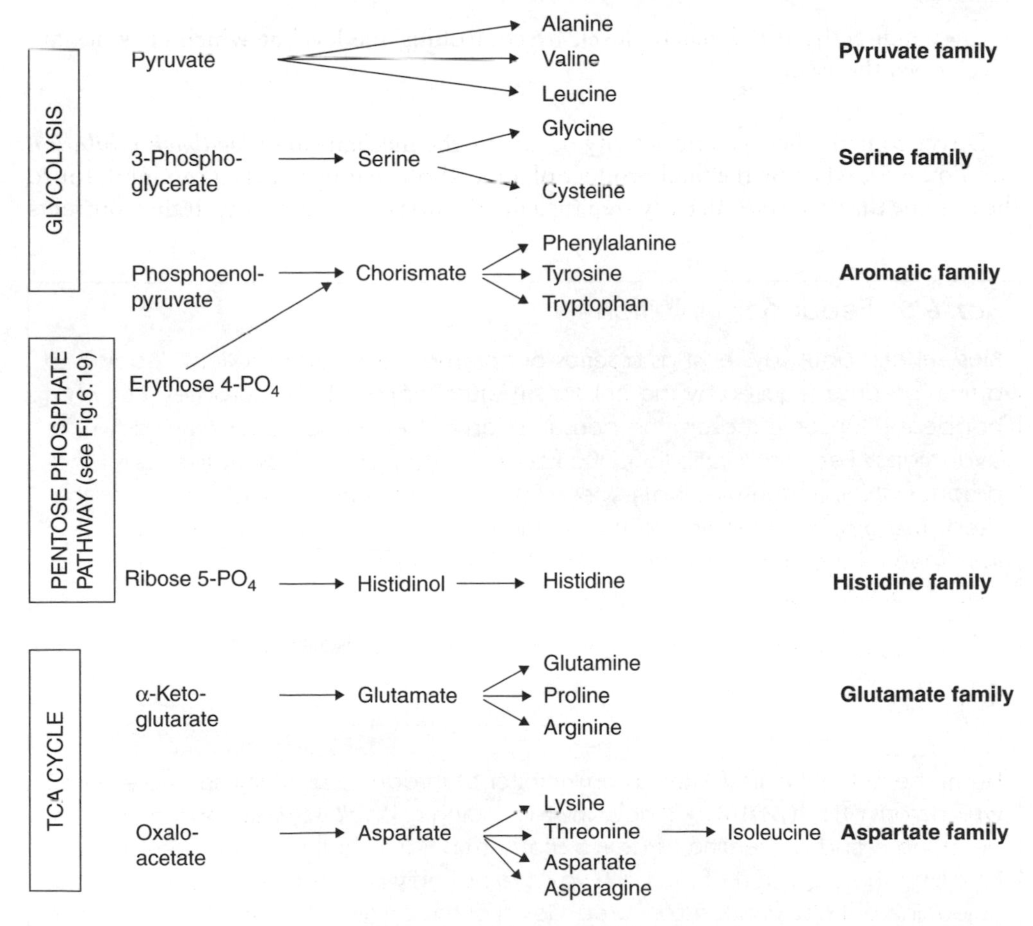

Figure 6.37 Amino acid biosynthesis. The carbon skeleton of amino acids is obtained from a limited number of precursor molecules, mainly intermediates in glycolysis or the TCA cycle. The amino group originally derives from inorganic sources, but can then be transferred from one organic molecule to another

amino acids are broken down, they are likewise broken down into a handful of metabolic intermediates, which then feed into the TCA cycle.

The regulation of metabolism

Microorganisms, like the rest of us, live in a changing world, and their needs do not always remain the same. It would be highly inefficient and (frequently wasteful) if all their metabolic reactions were going on with equal intensity all the time, regardless of whether they were needed. Over evolutionary time, regulation systems have developed, so that metabolism is tailored to the prevailing conditions.

Essentially, this regulation involves controlling the activity of enzymes which direct the many biochemical reactions occurring in each cell. This can be done by:

- directly affecting enzyme activity, or
- indirectly, at the genetic level, by controlling the level at which enzymes are synthesised.

Direct control of enzymatic activity occurs by the mechanism of *feedback inhibition* (see Box 6.5), whereby the final product of a metabolic pathway acts as an inhibitor to the enzyme that catalyses an early step (usually the first) in the pathway. It thus prevents

Box 6.5 Feedback inhibition

Biosynthetic pathways exist as a series of enzyme-mediated reactions, leading to a final product required by the cell for structural or metabolic purposes. But what happens if for some reason, the consumption of the final product slows down, or even stops? Feedback inhibition, also known, perhaps more descriptively, as '*end-product inhibition*', ensures that excess amounts of the end product are not synthesised. The pathways leading to the synthesis of many amino acids are regulated in this way, for example isoleucine, which is synthesised from another amino acid, threonine, via a series of intermediates:

Threonine → → → → → Isoleucine

Here, the isoleucine itself acts as an inhibitor of threonine deaminase, the enzyme which starts off the pathway. It does this by binding to an *allosteric* site on the enzyme, distorting it and preventing its active site from binding to threonine. Note, that by inhibiting the early part of the pathway, we not only prevent further production of isoleucine but also unnecessary breakdown of threonine. When levels of isoleucine starts to run low, less will be available to block the threonine deaminase, and thus the pathway starts to function again.

more of its own formation. When the concentration of the product subsequently falls below a certain level, it is no longer inhibitory, and biosynthesis resumes.

Regulation of metabolic pathways can also be achieved by controlling whether or not an enzyme is synthesised in the first place, and if so, the rate at which it is produced. This is done at the DNA level, by one of two mechanisms, *induction* and *repression*, which respectively 'switch on' and 'switch off' the machinery of protein synthesis discussed earlier in this chapter. These are discussed under the heading 'Regulation of gene expression' in Chapter 11

Test yourself

1 Reactions which involve the breakdown of compounds are collectively termed ________ and synthetic (building up) ones are termed ________.

2 Enzymes lower the ________ ________ of a reaction.

3 The ________ ________ of an enzyme is the part that is actually involved in catalysis.

4 At pH values away from the optimum, ________ of 'R' groups causes denaturation of proteins.

5 When a molecule ________ an electron, it is said to be oxidised.

6 Compounds which are good electron donors have a strongly ________ redox potential.

7 The energy 'currency' of cellular metabolism is ________ ________ ________.

8 The first step in glucose metabolism is ________, which can occur with or without ________.

9 ATP may be formed from ADP by ________ phosphorylation, ________ ________ phosphorylation or ________ phosphorylation.

10 Before entering the TCA cycle, pyruvate must be converted to the two-carbon compound ________. This type of reaction, in which a molecule of carbon dioxide is removed, is called a ________.

11 The main product of the TCA cycle is reducing power, in the form of ________ and ________.

12 The role of oxygen in aerobic respiration is to act as the ________ ________ in the electron transport chain.

13 As envisaged by the chemiosmotic theory of Peter Mitchell, protons are pumped out across a membrane, generating a ________ ________ ________.

14 ________ anaerobes are able to utilise anaerobic respiration if oxygen is unavailable.

15 The two commonest products of fermentation reactions are ________ and ________ ________.

16 Fermentation yields far less ATP than aerobic respiration because it does not involve an ________ ________ ________.

17 Fatty acids are broken down ________ carbons at a time by the process of ________.

18 ________ are organisms that can derive their energy from the oxidation of inorganic compounds such as ammonia or reduced iron.

19 Photosynthesis carried out by green and purple bacteria differs from that carried out by algae and cyanobacteria in that no ________ is generated.

20 In the light reactions of photosynthesis, light energy and an electron donor are used to generate ________ and ________.

21 In the dark reactions that make up the ________ cycle, carbon dioxide is assimilated into glyceraldehyde-3-phosphate, which is then converted to glucose in a series of steps that are the reverse of ________.

22 Anoxygenic photosynthetic bacteria are unable to use ________ as an electron donor, and therefore use compounds such as hydrogen sulphide or succinate.

23 Organisms unable to synthesise glucose from inorganic carbon must derive it from other organic compound by the process of ________.

24 In amino acid synthesis, -NH_2 groups may be transferred between compounds in ________ reactions.

25 The inhibition of a metabolic pathway by its final product is termed end-product or ________ inhibition.

Part III

Microbial Diversity

A few words about classification

In this section, we examine the wide diversity of microbial life. In each of the next four chapters, we shall discuss major structural and functional characteristics, and outline the main taxonomic divisions within each group. We shall also consider some specific examples, particularly with respect to their effect on humans. By way of introduction, however, we need to say something on the subject of the classification of microorganisms.

In any discussion on biological classification, it is impossible to avoid mentioning Linnaeus, the Swedish botanist who attempted to bring order to the naming of living things by giving each type a Latin name. He even gave himself one – his real name was Carl von Linné! It was Linnaeus who was responsible for introducing the *binomial* system of nomenclature, by which each organism was assigned a *genus* and a *species*. To give a few familiar examples, you and I are *Homo sapiens*, the fruit fly that has contributed so much to our understanding of genetics is *Drosophila melanogaster*, and, in the microbial world, the bacterium responsible for causing anthrax is *Bacillus anthracis*. Note the following conventions, which apply to the naming of all living things (the naming of viruses is something of a special case, which we'll consider in Chapter 10):

- the generic (genus) name is always given a *capital* letter
- the specific (species) name is given a *small* letter
- the generic and specific name are *italicised*, or, if this isn't possible, underlined

The science of *taxonomy* involves not just naming organisms, but grouping them with other organisms that share common properties. In the early days, classification appeared relatively straightforward, with all living things apparently fitting into one of two *kingdoms*. To oversimplify the matter, if it ran around, it was an animal, if it was green and didn't, it was a plant! As our awareness of the microbial world developed, however, it was clear that such a scheme was not satisfactory to accommodate all life forms, and in the mid-19th century, Ernst Haeckel proposed a third kingdom, the *Protista*, to include the bacteria, fungi, protozoans and algae.

A *taxon* is a collection of related organisms grouped together for purposes of classification. Thus, genus, family, etc. are taxons.

In the 20th century, an increased focus on the cellular and molecular similarities and dissimilarities between organisms led to proposals for further refinements to the three-kingdom system. One of the most widely accepted of these has been the *five-kingdom system* proposed by Robert Whittaker in 1969 (Figure A1). Like some of its predecessors, this took into account the fundamental difference in cell structure between procaryotes and eucaryotes (Chapter 3), and so placed procaryotes (bacteria) in their own kingdom, the *Monera*, separate from single-celled eucaryotes. Another feature of Whittaker's scheme was to assign the *Fungi* to their own kingdom, largely on account of their distinctive mode of nutrition. Table A1 shows some of the characteristic features of each kingdom.

Molecular studies in the 1970s revealed that the *Archaea* differed from all other bacteria in their 16S rRNA sequences, as well as in their cell wall structure, membrane lipids

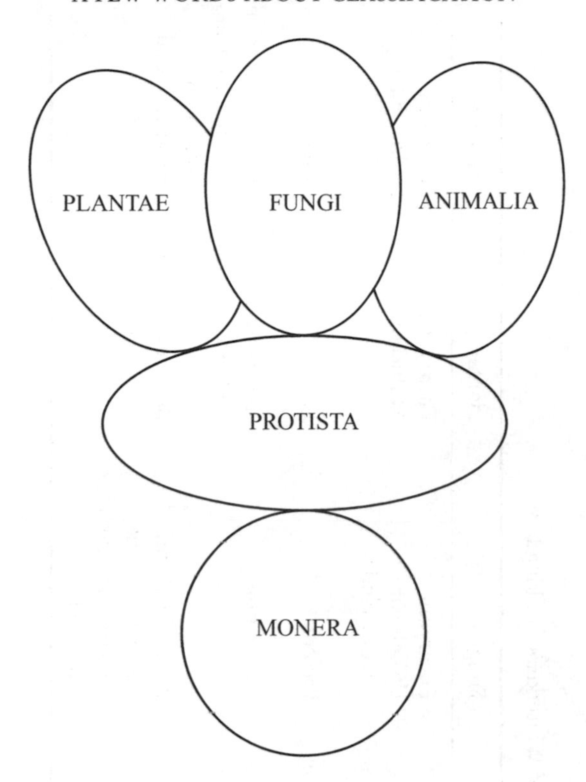

Figure A.1 Whittaker's five-kingdom system of classification

and aspects of protein synthesis. These differences were seen as sufficiently important for the recognition of a third basic cell type to add to the procaryotes and eucaryotes. This led to the proposal of a *three-domain* scheme of classification, in which procaryotes are divided into the Archaea and the Bacteria (Figure 3.1). The third domain, the Eucarya represents all eucaryotic organisms. The domains thus represent a level of classification that goes even higher than the kingdoms. Although the Archaea (the word means 'ancient') represent a more primitive bacterial form than the Bacteria, they are in certain respects more closely related to the eucaryotes, causing biologists to revise their ideas about the evolution of the eucaryotic state.

In *hierarchical* systems of classification, related species are grouped together in the same genus, genera sharing common features are placed in the same family, and so on. Table A2 shows a modern classification scheme for the gut bacterium *Escherichia coli*.

The boundaries between long-standing divisions such as algae and protozoa have become considerably blurred in recent years, and alternative classifications based on molecular data have been proposed. This is very much a developing field, and no definitive alternative classification has yet gained universal acceptance.

In the following chapters, we shall broadly follow the five-kingdom scheme, although of course the plant and animal kingdoms, having no microbial members, do not concern us. Thus, both the Archaea and Bacteria are considered in Chapter 7, and multicellular fungi form the focus of Chapter 8. Chapter 9 examines the Protists in the modern use

Table A.1 Characteristics of Whittaker's five kingdoms

	Monera (procaryotae)	Protista	Fungi	Plantae	Animalia
Cell type	Procaryotic	Eucaryotic	Eucaryotic	Eucaryotic	Eucaryotic
Cell organization	Unicellular; occasionally grouped	Unicellular; occasionally multicellular	Unicellular or multicellular	Multicellular	Multicellular
Cell wall	Present in most	Present in some, absent in others	Present	Present	Absent
Nutrition	Absorption, some photosynthetic, some chemosynthetic	Ingestion or absorption, some photosynthetic	Absorption	Absorptive, photosynthetic	Ingestion; occasionally in some parasites by absorption
Reproduction	Asexual, usually by binary fission	Mostly asexual, occasionally both sexual and asexual	Both sexual and asexual, often involving a complex life cycle	Both sexual and asexual	Primarily sexual

Table A.2 A modern hierarchical classification for *E. coli*; note that in this classification, there are no kingdoms

Domain	Bacteria
Phylum	Proteobacteria
Class	Zymobacteria
Order	Enterobacteriales
Family	Enterobacteriaceae
Genus	*Escherichia*
Species	*coli*

of the word, that is, unicellular eucaryotic forms. It retains the traditional distinction between protozoans, algae and other protists (water moulds and slime moulds), but also offers an alternative, 'molecular' scheme, showing the putative phylogenetic relationship between the various groups of organisms. Microbiology has traditionally embraced anomalies such as the giant seaweeds, as it has encompassed all organisms that fall outside of the plant and animal kingdoms. This book offers only a brief consideration of such macroscopic forms, and for the most part confines itself to the truly microbial world.

The viruses, it ought to be clear by now, are special cases, and are considered in isolation in Chapter 10. Because an understanding of viruses requires an appreciation of the basics of DNA replication and protein synthesis, it may help to jump ahead and read the relevant sections of Chapter 11 before embarking on Chapter 10.

7
Procaryote Diversity

Ever since bacteria were first identified, microbiologists have attempted to bring order to the way they are named and classified. The range of morphological features useful in the differentiation of bacteria is fairly limited (compared, say, to animals and plants), so other characteristics have also been employed. These include metabolic properties, pathogenicity, nutritional requirements, staining reactions and antigenic properties. The first edition of *Bergey's Manual of Systematic Bacteriology* (henceforth referred to as '*Bergey*'), published in the mid-1980s, mainly used phenotypic characteristics such as these to classify bacteria. The result placed bacteria into taxonomic groups that may or may not reflect their evolutionary relationship to one another. In the years since the first edition of *Bergey*, the remarkable advances made in molecular genetics have led to a radical reappraisal of the classification of bacteria. Comparison of nucleic acid sequences, notably those of 16S ribosomal RNA genes, has led to a new, phylogenetically based scheme of classification, that is, one based on how closely different groups of bacteria are thought to be related, rather than what morphological or physiological features they may share. Ribosomal RNA occurs in all organisms, and serves a similar function, thus to a large extent these sequences are *conserved* (remain similar) in all organisms. The nature and extent of any differences that have crept in during evolution will, therefore, be an indication of the relatedness of different organisms.

> The *phenotype* of an organism refers to its observable characteristics.
> The *genotype* of an organism refers to its genetic make-up.

The second edition of *Bergey* aims to reflect this change of approach and reassign many bacteria according to their phylogenetic relationship, as deduced from molecular evidence. Due to be issued in five volumes over a number of years, the first volume was published in 2001. As an example, the genus *Pseudomonas* previously contained some 70 species on the basis of phenotypic similarities, but in the second edition of *Bergey*, taking into account 16S rRNA information, many of these are assigned to newly created genera.

It must be stressed that *Bergey* (second edition) does not represent the definitive final word on the subject, and that the classification of bacteria is very much a developing science, in a constant process of evolution. Indeed, microbiologists are by no means unanimous in their acceptance of the 'molecular' interpretation of bacterial taxonomy. Some point to perceived inadequacies in the collection of data for the

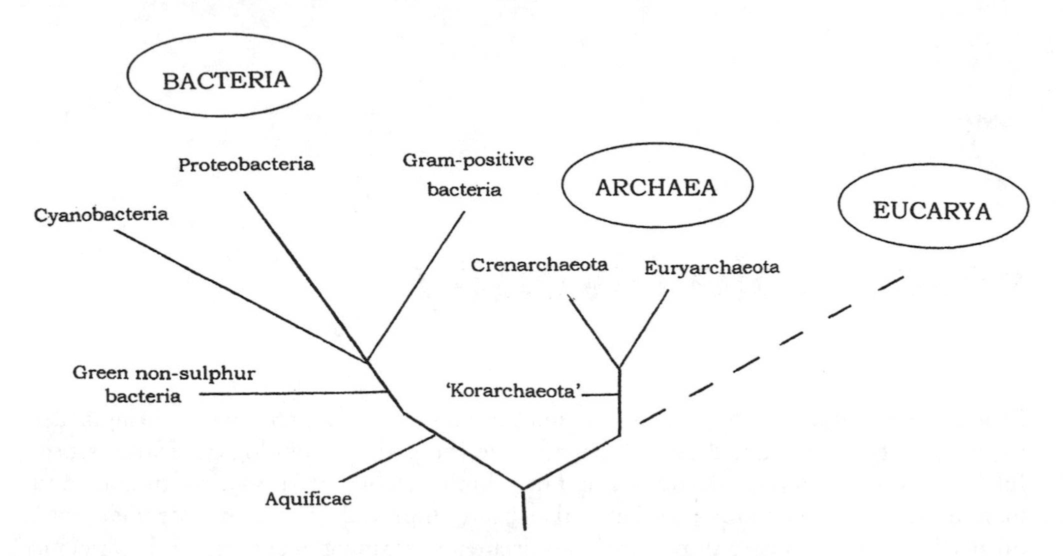

Figure 7.1 Phylogenetic relationships in the Procaryota. A phylogenetic tree based on 16S ribosomal RNA data, showing the relationships between members of the Archaea and the Bacteria. The position on the tree of the Eucarya (all eucaryotic organisms) is also indicated

scheme, as well as errors in the data arising from the sequencing and amplification techniques utilised. Other critics question the validity of a scheme based on 16S rRNA data when it seems increasingly likely that lateral gene transfer played an important role in bacterial evolution.

In the following pages, the major taxonomic groupings are discussed according to their arrangement in the second edition of *Bergey*. Figure 7.1 shows a phylogenetic tree, reflecting current ideas on the relationship between the major bacterial groups, as determined by 16S rRNA sequencing.

> Bacteria may acquire genes from other organisms by a variety of genetic transfer mechanisms (see Chapter 11). This is known as horizontal or *lateral gene transfer*, to distinguish it from vertical inheritance, in which the parental genotype is passed to the offspring.

Domain: Archaea

Studies on 16S ribosomal RNA sequences by Carl Woese and colleagues allowed the construction of phylogenetic trees for the procaryotes, showing their evolutionary relatedness. Figure 7.1 shows how the major procaryotic groups are thought to be related, based on 16S rRNA data. The work of Woese also revealed that one group of procaryotes differed from all the others. As described in Chapter 3, the Archaea are now regarded as being quite distinct from the Bacteria (sometimes called Eubacteria). Together with the Eucarya, these form the three domains of life (Figure 3.1). As can be seen in Table 7.1, archaea

> Archaea are procaryotes differing from true bacteria in cell wall and plasma membrane chemistry as well as 16S rRNA sequences.

Table 7.1 The three domains of life: Archaea share some features with true bacteria and others with eucaryotes

	Archaea	Bacteria	Eucarya
Main genetic material	Single closed circle of dsDNA	Single closed circle of dsDNA	True nucleus with multiple linear chromosomes
Histones	Present	Absent	Present
Gene structure	Introns absent	Introns absent	Introns present
Plasmids	Common	Common	Rare
Polycistronic mRNA	Present	Present	Absent
Ribosomes	70S	70S	80S
Protein synthesis	Not sensitive to streptomycin, chloramphenicol	Sensitive to streptomycin, chloramphenicol	Not sensitive to streptomycin, chloramphenicol
Initiator tRNA	Methionine	N-formyl methionine	Methionine
Membrane fatty acids	Ether-linked, branched	Ester-linked, straight chain	Ester-linked, straight chain
Internal organelles	Absent	Absent	Present
Site of energy generation	Cytoplasmic membrane	Cytoplasmic membrane	Mitochondria
Cell wall	Muramic acid absent	Muramic acid present	Muramic acid absent

share some features in common with other bacteria and some with eucaryotes. Extending nucleic acid analysis to other genes has shown that members of the Archaea possess many genes not found in any other type of bacteria.

The domain is the highest level of taxonomic grouping.

General features of the Archaea

Members of the Archaea show considerable diversity of both morphology and physiology. In view of the fact that the Archaea remained unidentified as a separate group for so many years, it should come as no surprise that they do not display any obvious morphological differences from true bacteria, and all the main cell shapes (see Chapter 2) are represented. More unusual shapes are also encountered in archaea; members of the genus *Haloarcula* have flattened square or triangular cells! Both Gram-positive and Gram-negative forms of archaea are found, but neither possesses true peptidoglycan. Some types have a so-called *pseudomurein*, composed of different substituted polysaccharides and L-amino acids (Figure 7.2). Most archaea, however, have cell walls composed of a layer of proteinaceous subunits known as an *S-layer*, directly associated with the cell membrane. This difference in cell wall chemistry means that members of the Archaea are not susceptible to antibacterial agents such as lysozyme and penicillin, whose action is directed specifically towards peptidoglycan. Differences are also found in the make-up

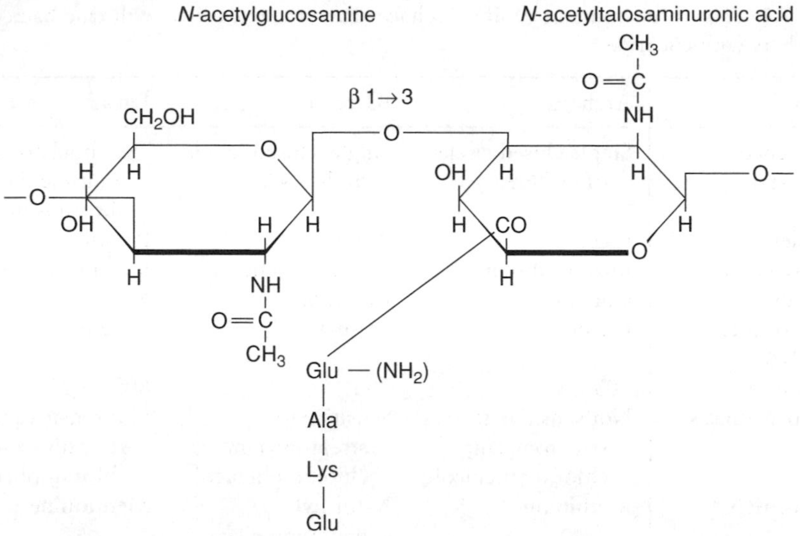

Figure 7.2 Pseudomurein, found in the cell walls of certain members of the Archaea, comprises subunits of *N*-acetylglucosamine and *N*-acetyltalosaminuronic acid. The latter replaces the *N*-acetylmuramic acid in true peptidoglycan (c.f. Figure 3.6). As with peptidoglycan, the amino acids in the peptide chain may vary, however in pseudomurein they are always of the L-form. The components in parentheses are not always present.

of archaean membranes, where the lipid component of membranes contains branched isoprenes instead of fatty acids, and these are joined to glycerol by ether-linkages, rather than the ester-linkages found in true bacteria (Figure 7.3). The diversity of archaea extends into their adopted means of nutrition and metabolism: aerobic/anaerobic and autotrophic/heterotrophic forms are known. Many members of the Archaea are found in extreme environments such as deep-sea thermal vents and salt ponds. Some extreme thermophiles are able to grow at temperatures well over 100 °C, while psychrophilic forms constitute a substantial proportion of the microbial population of Antarctica. Similarly, examples are to be found of archaea that are active at extremes of acidity, alkalinity or salinity. Initially it was felt that archaea were limited to such environments because there they faced little competition from true bacteria or eucaryotes. Recent studies have shown however that archaea are more widespread in their distribution, making up a significant proportion of the bacterial biomass found in the world's oceans, and also being found in terrestrial and semiterrestrial niches. The reason that this lay undetected for so long is that these organisms cannot as yet be cultured in the laboratory, and their presence can only be inferred by the use of modern DNA-based analysis.

Classification of the Archaea

According to the second edition of *Bergey*, the Archaea are divided into two phyla, the Euryarchaeota and Crenarchaeota. A third phylum, the Korarchaeota has been

(a)

Glycerol / Ester bond

$CH_2-O-C(=O)$

$CH-O-C(=O)$

CH_2-OH

(b)

Ether bond

CH_2-O-

CH_2-O-

CH_2-OH

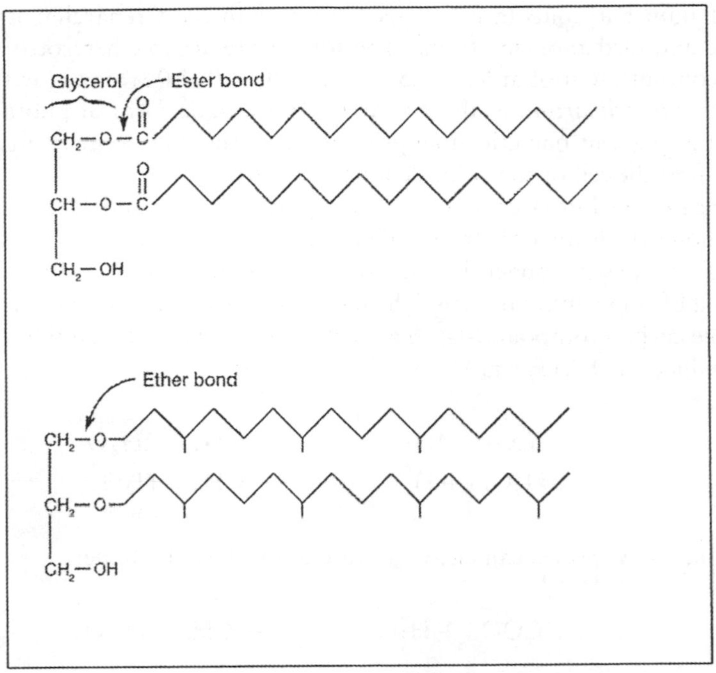

Figure 7.3 Membrane lipids in Archaea and Bacteria. Compositional differences in membrane lipids between (a) Bacteria and (b) Archaea. Note the ether linkages and branched fatty acids in (b)

proposed, whose members are known only from molecular studies, while the recent discovery of *Nanoarchaeum equitans* has led to the proposal of yet another archaean phylum (see Box 7.1). Countless more species of archaea are thought to exist, which like the Korarchaeota, have not yet been successfully cultured in the laboratory.

Box 7.1 A new phylum?

In 2002 a novel microorganism, *Nanoarchaeum equitans* was isolated from a hydrothermal vent deep below the sea off the Icelandic coast. Found at temperatures around boiling point, its tiny spherical cells were attached to the surface of another archaean, *Ignicoccus* sp., without which it cannot apparently be cultured. Ribosomal RNA studies showed *N. equitans* to be sufficiently unlike other members of the Archaea to justify the creation of a new archaean phylum, the Nanoarchaeota. Now fully sequenced, the genome of *N. equitans* is one of the smallest living things to date (490kb). It appears to belong to a very deeply rooted branch of the Archaea, leading to speculation that it may resemble the first living cells. At 0.4 μm in diameter, *N. equitans* is also among the smallest known living organisms. Although at present it is the only recognised species of the Nanarchaeota, it seems likely to be joined by isolates reported from other locations around the world.

The phylum Eurychaeota is a bigger group than the Crenarchaeota, and includes halophilic and methanogenic forms. The former are aerobic heterotrophs, requiring a chloride concentration of at least 1.5 M (generally 2.0–4.0 M) for growth. One species, *Halobacterium salinarum*, is able to carry out a unique form of photosynthesis using the bacterial pigment bacteriorhodopsin, and uses the ATP so generated for the active transport into the cell of the chloride ions it requires.

Members of the Euryarchaeota such as *Methanococcus* and *Methanobacterium* are unique among all life forms in their ability to generate methane from simple carbon compounds. They are strict anaerobes found in environments such as hot springs, marshes and the gut of ruminant mammals. The methane is derived from the metabolism of various simple carbon compounds such as carbon dioxide or methanol in reactions linked to the production of ATP. e.g.

$$CO_2 + 4H_2 \longrightarrow CH_4 + 2H_2O$$
$$CH_3OH + H_2 \longrightarrow CH_4 + H_2O$$

In addition, a few species can cleave acetate to produce methane:

$$CH_3COO^- + H_2O \longrightarrow CH_4 + HCO_3{}^-$$

This acetotrophic reaction is responsible for the much of the methane production in sewage sludges. Although sharing the unique facility to generate methane, some of the methanogenic genera are quite distantly related to one another.

Other representatives of the Eurychaeota include the Thermoplasmata and the Thermococci. Thermoplasmata are highly acidophilic and moderately thermophilic; they completely lack a cell wall, and are pleomorphic. A unique membrane lipid composition allows them to withstand temperatures of well over 50 °C. Thermococci are anaerobic extreme thermophiles found in anoxic thermal waters at temperatures as high as 95 °C. Enzymes isolated from thermococci have found a variety of applications. A thermostable DNA polymerase from *Pyrococcus furiosus* is used as an alternative to *Taq* polymerase (see Phylum 'Deinococcus-Thermus' later in this chapter) in the polymerase chain reaction (PCR).

Pleomorphic means lacking a regular shape.

Representative genera: *Methanobacterium, Halobacterium*

Members of the **Crenarchaeota** are nearly all extreme thermophiles, many of them capable of growth at temperatures in excess of 100 °C, including *Pyrolobus fumarii,* which has an optimum growth temperature of 106 °C, and can survive autoclaving at 121 °C.

Many utilise inorganic sulphur compounds as either a source or acceptor of electrons (respectively, oxidation to H_2SO_4 or reduction to H_2S).

Crenarchaeotes are mostly anaerobic, and are thought by many to resemble the common ancestors of all bacteria.

Representative genera: *Thermoproteus, Sulfolobus*

Domain: Bacteria

All the remaining bacterial groups belong to the domain Bacteria. This is divided into 23 phyla (Table 7.2), the more important of which are discussed in the following pages, according to their description in the second edition of *Bergey*. As with the Archaea, many other forms are known only through molecular analysis and it is estimated that these represent at least another 20 phyla.

Phylum: Proteobacteria

We start our survey of the Bacteria with the Proteobacteria. This is by far the biggest single phylum, and occupies the whole of volume 2 in the second edition of *Bergey*. The size of the group is matched by its diversity, both morphological and physiological; most forms of metabolism are represented, and the wide range of morphological forms gives rise to the group's name. (Proteus was a mythological Greek god who was able to assume many different forms.) The reason such a diverse range of organisms

Table 7.2 Phyla of domain Bacteria

Phylum	**Aquificae**
Phylum	**Thermotogae**
Phylum	Thermodesulfobacteria
Phylum	**'Deinococcus–Thermus'** *
Phylum	Chrysiogenetes
Phylum	**Chloroflexi**
Phylum	Thermomicrobia
Phylum	**Nitrospira**
Phylum	Deferribacteres
Phylum	**Cyanobacteria**
Phylum	**Chlorobi**
Phylum	**Proteobacteria**
Phylum	**Firmicutes**
Phylum	**Actinobacteria**
Phylum	**Planctomycetes**
Phylum	**Chlamydiae**
Phylum	**Spirochaetes**
Phylum	Fibrobacteres
Phylum	Acidobacteria
Phylum	**Bacteroidetes**
Phylum	Fusobacteria
Phylum	**Verrucomicrobia**
Phylum	Dictyoglomi

* This Phylum has not yet been assigned a formal name.
Those phyla discussed in the text are shown in bold print.

have been assigned to a single taxonomic grouping is that their 16S rRNA indicates a common ancestor (thought to be photosynthetic, though few members now retain this ability). At the time of writing more than 460 genera and 1600 species had been identified, all of them Gram-negative and representing almost half of all accepted bacterial genera. These include many of the best known Gram-negative bacteria of medical, industrial and agricultural importance. For taxonomic purposes, the Proteobacteria have been divided into five classes reflecting their presumed lines of descent and termed the Alphaproteobacteria, Betaproteobacteria, Gammaproteobacteria, Deltaproteobacteria and Epsilonproteobacteria (Figure 7.4). It should be stressed that because classification is based on molecular relatedness rather than shared phenotypic traits, few if any morphological or physiological properties can be said to be characteristic of all members of each class. Equally, organisms united by a particular feature may be found in more than one of the proteobacterial classes, for example nitrifying bacteria are to be found in the α, β and γ Protobacteria. For this reason, in the following paragraphs we describe the Proteobacteria in terms of their *phenotypic* characteristics rather than attempt to group them phylogenetically.

Photosynthetic Proteobacteria

The purple sulphur and purple non-sulphur bacteria are the only members of the Proteobacteria to have retained the photosynthetic ability of their presumed ancestor. The type of photosynthesis they carry out, however, is quite distinct from that carried out by plants, algae and cyanobacteria (see later in this chapter), differing in two important respects:

- it is anoxygenic – no oxygen is produced by the process
- it utilizes bacteriochlorophyll *a* and/or *b*, which have different absorbance properties to chlorophylls *a* and *b*.

Like organisms that carry out green photosynthesis, however, they incorporate CO_2 into carbohydrate by means of the Calvin cycle (see Chapter 6). All are at least facultatively

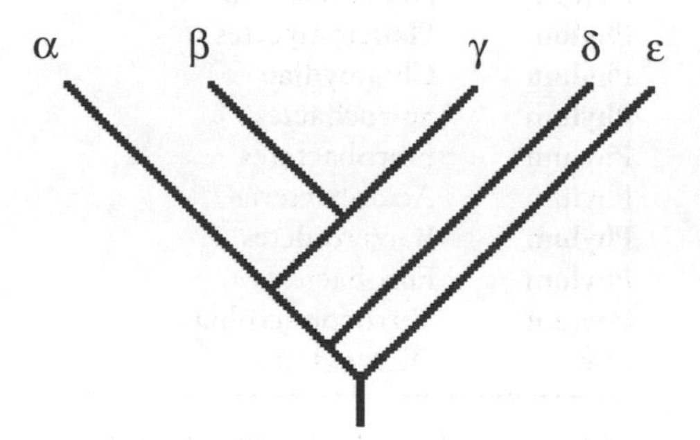

Figure 7.4 The phylogenetic relationships of the Proteobacteria, based on 16S rRNA sequences

anaerobic, and are typically found in sediments of stagnant lakes and salt marsh pools, where they may form extensive coloured blooms. Because the absorption spectrum of bacteriochlorophylls lies mostly in the infrared part of the spectrum, they are able to utilise light energy that penetrates beyond the surface layers of water.

The coloration, ranging from orange/brown to purple, is due to the presence of carotenoid pigments such as lycopene and spirillixanthin, which mask the blue/green colour of the bacteriochlorophylls. The photosynthetic pigments are located on highly folded extensions of the plasma membrane. Photosynthetic proteobacteria include rods, cocci and spiral forms.

Under anaerobic conditions, the purple sulphur bacteria typically utilise hydrogen sulphide or elemental sulphur as an electron donor for the reduction of CO_2.

$$H_2S + CO_2 \longrightarrow (CH_2O)_n + S^0$$
$$S^0 + CO_2 + H_2O \longrightarrow (CH_2O)_n + H_2SO_4$$

Many store sulphur in the form of intracellular granules. The purple sulphur bacteria all belong to the γ-Protobacteria. They are typically found in surface muds, and sulphur springs, habitats that provide the right combination of light and anaerobic conditions.

Representative genera: *Thiospirillum, Chromatium*

The purple non-sulphur bacteria were distinguished from the above group because of their apparent inability to use H_2S as an electron donor. It is now known, however, that the majority *can* do this, but are able to tolerate much lower concentrations in comparison with the purple sulphur bacteria. The purple non-sulphur bacteria are facultative anaerobes able to grow as photoheterotrophs, that is, with light as an energy source and a range of organic molecules such as carbohydrates and organic acids as sources of both carbon and electrons. In addition, many are able to grow aerobically as chemoheterotrophs in the absence of light. Under present classification systems, purple nonsulphur bacteria are divided between the α- and β-Proteobacteria.

Representative genera: *Rhodospirillum, Rhodopseudomonas*

Nitrifying Proteobacteria

This group comprises aerobic Gram-negative chemolithoautotrophs that derive their energy from the oxidation of inorganic nitrogen compounds (either ammonia or nitrite), and their carbon from CO_2. The nitrifying bacteria are represented in both the α- and β-Proteobacteria.

The oxidation of ammonia through to nitrate is a two-stage process, with specific bacteria carrying out each stage (ammonia to nitrite and nitrite to nitrate). This is reflected in the generic names of the bacteria, bearing the prefix *Nitroso-* or *Nitro-* according to whether they carry out the first or second reaction. Nitrifying bacteria play an essential role in the cycling of nitrogen in terrestrial, marine and freshwater habitats. Nitrite, which is toxic to many forms of life, rarely accumulates in the environment, due to the activity of the Nitrobacteria. As was the case with the purple photosynthetic bacteria, several cell forms are represented among the nitrifiers.

Representative genera: *Nitrosomonas* ($NH_4^+ \longrightarrow NO_2^-$)
Nitrobacter ($NO_2^- \longrightarrow NO_3^-$)

Iron- and sulphur-oxidising Proteobacteria

Two further groups of environmentally significant chemolithoautotrophs derive their energy through the oxidation of reduced iron and sulphur respectively.

Among the sulphur oxidisers, perhaps the best studied are members of the genus *Acidithiobacillus* *, which includes extreme acidophiles such as *A. thiooxidans* that are capable of growth at a pH as low as 1.0! These may utilise sulphur in its elemental form, as H_2S, metal sulphides, or other forms of reduced sulphur such as thiosulphate:

$$2S^0 + 3O_2 + 2H_2O \longrightarrow 2H_2SO_4$$
$$S^{2-} + 4O_2 \longrightarrow 2SO_4^{2-}$$
$$S_2O_3^{2-} + 2O_2 + H_2O \longrightarrow 2SO_4^{2-} + 2H^+$$

The result of all these reactions is the production of sulphuric acid and a lowering of the environmental pH. Bacteria such as these are responsible for the phenomenon of *acid mine drainage*. Their environmental impact is discussed in Chapter 16, whilst Chapter 17 describes their use in the extraction of valuable metals from intractable mineral ores. A particularly valuable organism in the latter context is *A. ferrooxidans*, due to its ability to use not only reduced sulphur compounds as energy sources, but also reduced iron (see below).

A second group of sulphur oxidisers are bacteria that exist not as single cells, but join to form *filaments*, the best known of which is *Beggiotoa*. These are typically found in sulphur springs, marine sediments and hydrothermal vents at the bottom of the sea.

Acidithiobacillus ferrooxidans is also an example of an *iron oxidiser*. At normal physiological pH values and in the presence of oxygen, reduced iron (iron II, Fe^{2+}) is spontaneously oxidised to the oxidised form (iron III, Fe^{3+}). Under very acidic conditions, the iron remains in its reduced form, unless acted on by certain bacteria. *A. ferrooxidans*; is an obligate aerobe able to use iron II as an energy source, converting it to iron III at an optimum pH range of around 2:

$$2Fe^{2+} + {}^1/_2O_2 + 2H^+ \longrightarrow Fe^{3+} + H_2O$$

Gallionella ferruginea, on the other hand, grows around neutrality in oxygen-poor environments such as bogs and iron springs. Ferric hydroxide is excreted from the cell and deposited on a stalk-like structure projecting from, and much bigger than, the cell itself. This gives the macroscopic impression of a mass of red/brown twisted filaments.

Representative genera: *Acidithiobacillus*, *Beggiotoa* (sulphur oxidisers)
Leptospirillum, *Gallionella* (iron oxidisers)

* Note: *A. thiooxidans* and *A. ferrooxidans* formerly belonged to the genus *Thiobacillus*. In 2000, several species were assigned to new genera, however you may still find them referred to by their old names.

Hydrogen-oxidising Proteobacteria

This diverse group of bacteria are united by their ability to derive energy by using hydrogen gas as a donor of electrons, and oxygen as an acceptor:

$$2H_2 + O_2 \longrightarrow 2H_2O$$

Nearly all the members of this group are facultative chemolithotrophs, i.e. they can also grow as heterotrophs, utilising organic compounds instead of CO_2 as their carbon source, and indeed most grow more efficiently in this way.

Representative genera: *Alcaligenes*, *Ralstonia*

Nitrogen-fixing Proteobacteria

The α-Proteobacteria includes certain genera of **nitrogen-fixing bacteria.** These are able to fix (reduce) atmospheric N_2 as NH_4^+ for subsequent incorporation into cellular materials, a process that requires a considerable input of energy in the form of ATP:

> Nitrogen fixation is limited to a few species of bacteria and cyanobacteria. No eucaryotes are known to have this property.

$$N_2 + 8e^- + 8H^+ + 16ATP \longrightarrow, 2NH_4^+ + 16ADP + 16Pi$$

Nitrogen-fixing bacteria may be free-living in the soil (e.g. *Azotobacter*), or form a symbiotic relationship with cells on the root hairs of leguminous plants such as peas, beans and clover (e.g. *Rhizobium*). The nitrogenase responsible for the reaction (actually a complex of two enzymes) is highly sensitive to oxygen; many nitrogen fixers are anaerobes, while others have devised ways of keeping the cell interior oxygen-free. Nitrogen fixation is discussed further in Chapters 15 and 16.

Closely related to *Rhizobium*, but unable to fix nitrogen, are members of the genus *Agrobacterium*. Like *Rhizobium*, these enter the tissues of plants, but instead of forming a mutually beneficial association, cause cell proliferation and tumour formation. *A. tumefaciens* has proved to be a valuable tool in the genetic engineering of plants, and is discussed further in Chapter 12.

Representative genera: *Rhizobium, Azotobacter*

Methanotrophic Proteobacteria

In discussing the Archaea earlier in this chapter, we encountered species capable of the production of methane, a gas found widely in such diverse locations as marshes, sewage sludge and animal intestines. Certain proteobacteria are able to utilise this methane as a carbon and energy source and are known as *methanotrophs*.

> Methane is one of the so-called greenhouse gases, responsible for the phenomenon of global warming. Its effects would be much more pronounced if it were not for the activity of methanotrophic bacteria.

Methanotrophs are strict aerobes, requiring oxygen for the oxidation of methane. The methane-generating bacteria, however, as we've seen are anaerobes; methanotrophs are consequently to be found at aerobic/anaerobic interfaces such as topsoil, where they can find both the oxygen and the methane they require. The methane is firstly oxidised to methanol, then to formaldehyde, by means of separate enzyme systems. Some of the carbon in formaldehyde is assimilated into organic cellular material, while some is further oxidised to carbon dioxide.

Bacteria able to utilise other single-carbon compounds such as methanol (CH_3OH) or methylamine (CH_3NH_2) are termed *methylotrophs*. Depending on whether they possess the enzyme methane monooxygenase (MMO), they may also be methanotrophs.

Representative genera: *Methylomonas, Methylococcus*

> The methylotroph *Methylophilus methylotrophus* was once produced in huge quantities as a source of 'single cell protein' for use as animal feed, until the low price of alternatives such as soya and fish meal made it commercially unviable.

Sulphate- and sulphur-reducing Proteobacteria

Some 20 genera of anaerobic δ-Proteobacteria reduce either elemental sulphur or oxidised forms of sulphur such as sulphate to hydrogen sulphide. Organic compounds such as pyruvate, lactate or certain fatty acids act as electron donors:

$$\underset{\text{Lactate}}{2CH_3CHOHCOO^-} + SO_4^{2-} \xrightarrow{\textit{Desulfovibrio}} \underset{\text{Acetate}}{2CH_3COO^-} + H_2S + 2HCO_3^-$$

Sulphate- and sulphur-reducers are found in anaerobic muds and play an important role in the global sulphur cycle.

Representative genera: *Desulfovibrio* (sulphate), *Desulfuromonas* (sulphur)

Enteric Proteobacteria

This is a large group of rod-shaped bacteria, mostly motile by means of peritrichous flagella, which all belong to the γ-Proteobacteria. They are facultative aerobes, characterised by their ability in anaerobic conditions to carry out fermentation of glucose and other sugars to give a variety of products. The nature of these products allows division into two principal groups, the mixed acid fermenters and the butanediol fermenters (Figure 6.23). All the enteric bacteria test negative for cytochrome *c* oxidase (see *Vibrio* and related genera below). In view of their similar appearance, members of the group are distinguished from one another largely by means of their biochemical characteristics. An unknown isolate is subjected to a series of tests including its ability to utilise substrates such as lactose and citrate, convert tryptophan to indole, and hydrolyse urea. On the basis of its response to each test, a characteristic profile can be built up for the isolate, and matched against those of known species (see Table 7.3).

The most thoroughly studied of all bacteria, *Escherichia coli* (*E. coli*) is a member of this group, as are a number of important pathogens of humans such as *Salmonella*, *Shigella* and *Yersinia* (the causative agent of plague).

Representative genera: *Escherichia, Enterobacter*

Table 7.3 Identification of enteric bacteria on the basis of their biochemical and other properties

Some of the tests used to identify isolates of enteric bacteria are listed below. The table on the next page indicates typical results obtained for common genera; note, however, that for many cases, the result of a test may vary for different species within a genus. The symbols + and – indicate that most or all species within a genus give a positive or negative result, whilst +/– denotes that results are more variable within a genus.

Test	Description
Indole	Tests for ability to produce indole from the amino acid tryptophan.
Methyl Red	Acid production causes methyl red indicator to turn red.
Voges-Proskauer	Tests for ability to ferment glucose to acetoin.
Citrate utilisation	Demonstrates ability to utilise citrate as sole carbon source.
Urease	Demonstrates presence of the enzyme urease by detecting rise in pH due to urea being converted to ammonia and CO_2.
Gas from sugars	Production of gas from sugars such as glucose is demonstrated by collection in a Durham tube (a small inverted tube placed in a liquid medium).
H_2S production	Production of H_2S from sulphate reduction or from sulphur-containing amino acids is demonstrated by the formation of black iron sulphide in an iron-rich medium.
Ornithine decarboxylase	Growth on medium enriched in ornithine leads to pH change when enzyme is present.
Motility	Diffusion through soft agar demonstrates cellular movement.
Gelatin liquefaction	Demonstrates presence of proteolytic enzymes capable of liquefying a medium containing gelatin.
% age GC	Nucleotide composition determined by melting point measurements.

(*Continued*)

Table 7.3 Identification of enteric bacteria on the basis of their biochemical properties (*Continued*)

	Escherichia	*Salmonella*	*Shigella*	*Citrobacter*	*Proteus*	*Serratia*	*Klebsiella*	*Enterobacter*	*Erwinia*
Indole	+	−	+/−	+/−	+/−	−	+/−	−	−
Methyl Red	+	+	+	+	+	+/−	+	+/−	+
Voges-Proskauer	−	−	−	−	+/−	+	+	+	+
Citrate utilisation	−	+/−	−	+	+/−	+	+	+	−
Urease	−	−	−	+	+	−	+/−	−	−
Gas from glucose	+	+	−	+	+	+/−	+	+	−
H_2S production	−	−	−	+/−	+	−	−	−	+
Ornithine decarboxylase	+	+	+/−	+	+/−	−	+/−	+	−
Motility	+	+	−	+	+	+	−	+	+
Gelatin liquefaction	−	−	−	−	+	+	−	+/−	+/−
% GC	48–52	50–53	49–53	50–52	38–41	53–59	53–58	52–60	50–58

Table 7.4 Differentiation between enteric bacteria, vibrios and pseudomonads

	Enteric bacteria	Vibrios	Pseudomonads
Oxidase test	−ve	+ve	+ve
Glucose fermentation	+ve	+ve	−ve
Flagella	Peritrichous	Polar*	Polar

*When grown on solid media, some *Vibrio* species also develop lateral flagella, a unique arrangement termed mixed flagellation.

Vibrio and related genera

A few other genera, including *Vibrio* and *Aeromonas,* are also facultative anaerobes able to carry out the fermentative reactions described above, but are differentiated from the enteric bacteria by being oxidase-positive (Table 7.4). *Vibrio* and *Photobacterium* both include examples of marine *bioluminescent* species; these are widely found both in seawater and associated with fish and other marine life. The luminescence, which requires the presence of oxygen, is due to an oxidation reaction carried out by the enzyme *luciferase*.

Bioluminescence is the production of light by living systems

Vibrio cholerae is the causative agent of cholera, a debilitating and often fatal form of acute diarrhoea transmitted in faecally contaminated water. It remains a major killer in many third world countries. Several species of *Vibrio*, including *V. cholerae*, have been shown to possess two circular chromosomes instead of the usual one.

Representative genera: *Vibrio, Aeromonas*

The Pseudomonads

Members of this group of proteobacteria, the most important genus of which is *Pseudomonas*, are straight or curved rods with polar flagella. They are chemoheterotrophs that generally utilise the Entner–Doudoroff pathway (see Chapter 6) rather than glycolysis for the oxidation of hexoses. They are differentiated from the enteric bacteria (Table 7.4) by being oxidase-positive and incapable of fermentation. A characteristic of many pseudomonads is the ability to utilise an extremely wide range of organic compounds (maybe over 100!) for carbon and energy, something that makes them very important in the recycling of carbon in the environment. Several species are significant pathogens of animals and plants; *Pseudomonas aeruginosa* is an effective coloniser of wounds and burns in humans, while *P. syringae* causes chlorosis (yellowing of leaves) in a range of plants. Because of their ability to grow at low temperatures, a number of pseudomonads are important in the spoilage of food.

Burkholderia cepacia can utilise an exceptionally wide range of organic carbon sources, including sugars, carboxylic acids, alcohols, amino acids, aromatic compounds and amines, to name but a few!

Although most species carry out aerobic respiration with oxygen as the terminal electron acceptor, a few are capable of substituting nitrate (anaerobic respiration, see Chapter 6).

Representative genera: *Pseudomonas, Burkholderia*

Acetic acid bacteria

Acetobacter and *Gluconobacter* are two genera of the α-Proteobacteria that convert ethanol into acetic acid, a highly significant reaction in the food and drink industries (see Chapter 17). Both genera are strict aerobes, but unlike *Acetobacter*, which can oxidise the acetic acid right through to carbon dioxide and water, *Gluconobacter* lacks all the enzymes of the TCA cycle, and cannot oxidise it further.

Acetobacter species also have the ability, rare in bacteria, to synthesise cellulose; the cells become surrounded by a mass of extracellular fibrils, forming a pellicle at the surface of an unshaken liquid culture.

Representative genera: *Acetobacter, Gluconobacter*

Stalked and budding Proteobacteria

The members of this group of aquatic Proteobacteria differ noticeably in their appearance from typical bacteria by their possession of extracellular extensions known as *prosthecae*; these take a variety of forms but are always narrower than the cell itself. They are true extensions of the cell, containing cytoplasm, rather than completely extracellular appendages.

In the stalked bacteria such as *Caulobacter* (Figure 7.5), the prostheca serves both as a means of attaching the cell to its substratum, and to enhance nutrient absorption by

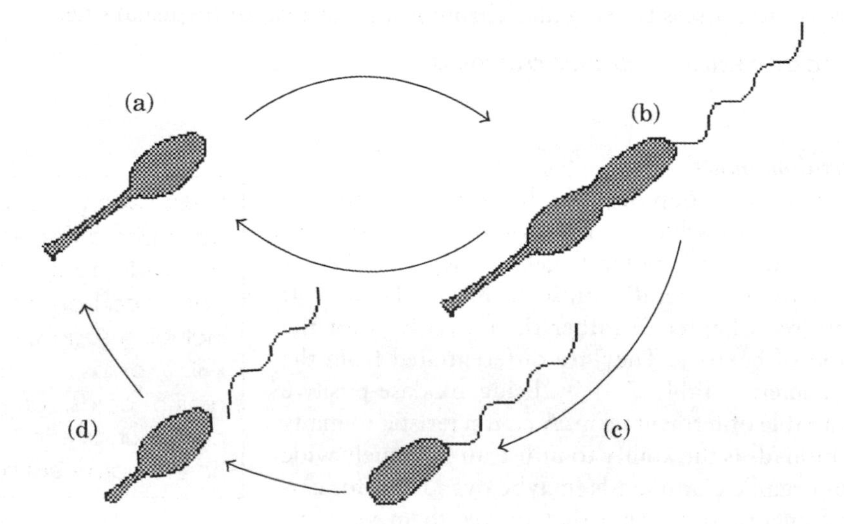

Figure 7.5 The life cycle of *Caulobacter*, a stalked bacterium. The stalked 'mother' cell attaches to a surface by means of a holdfast (a). It grows in length and develops a flagellum (b), before undergoing binary fission. The flagellated swarmer cell swims away (c), and on reaching a suitable substratum, loses its flagellum and develops a stalk or prostheca (d). Reproduced by permission of Dr James Brown, North Carolina State University

increasing the surface area-to-volume ratio of the cell. The latter enables such bacteria to live in waters containing extremely low levels of nutrient. *Caulobacter* lives part of its life cycle as a free-swimming swarmer cell with no prostheca but instead a flagellum for mobility.

The iron oxidiser *Gallionella* (see Nitrifying Proteobacteria above) may be regarded as a stalked bacterium, however it is not truly prosthecate, as its stalk does not contain cytoplasm.

In the budding bacteria, the prostheca is involved in a distinctive form of reproduction, in which two cells of unequal size are produced (c.f. typical binary fission, which results in two identical daughter cells). The daughter cell buds off from the mother cell, either directly, or as *Hyphomicrobium* spp. at the end of a hypha (stalk) (Figure 7.6). Once detached, the daughter cell grows to full size and eventually produces its own buds. *Hyphomicrobium* is a methanotroph and a methylotroph, so it also belongs to the methanotrophs described earlier.

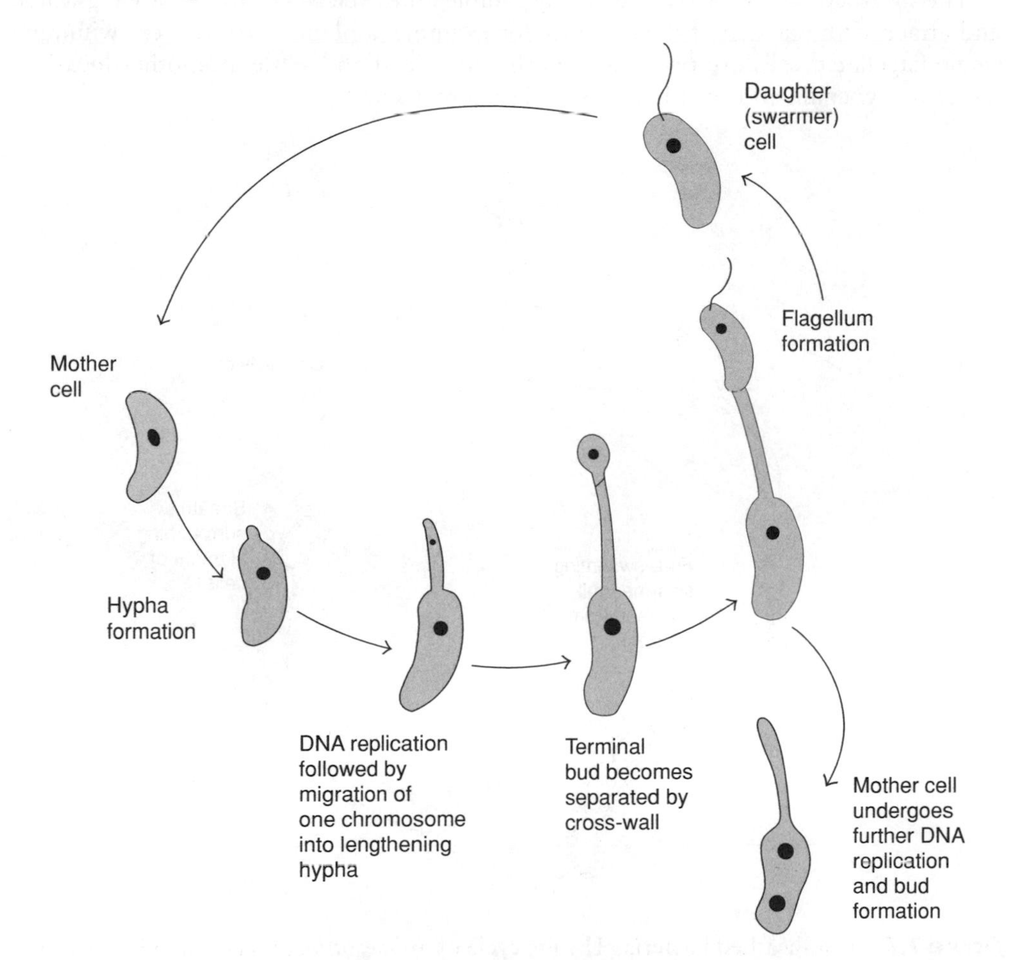

Figure 7.6 The budding bacteria: reproduction in *Hyphomicrobium*. Before reproduction takes place, the vegetative cell develops a stalk or hypha, at the end of which a bud develops. This produces a flagellum, and separates to form a motile swarmer cell

In some bacteria, more than one prostheca is found per cell; these *polyprosthecate* forms include the genus *Stella*, whose name ('a star') derives from its six symmetrically arranged buds.

Representative genera: *Caulobacter, Hyphomicrobium*

Sheathed Proteobacteria

Some genera of β Proteobacteria exist as chains of cells surrounded by a tube-like *sheath*, made up of a carbohydrate/protein/lipid complex. In some cases, the sheath contains deposits of manganese oxide or ferric hydroxide, which may be the product of chemical or biological oxidation. Empty sheaths encrusted with oxides may remain long after the bacterial cells have died off or been released. As with the stalked bacteria (see above) the sheath helps in the absorption of nutrients, and may also offer protection against predators.

The sheathed bacteria have a relatively complex life cycle. They live in flowing water, and attach with one end of the chain to, for example, a plant or rock. Free-swimming single flagellated cells are released from the distal end and settle at another location, where a new chain and sheath are formed (Figure 7.7).

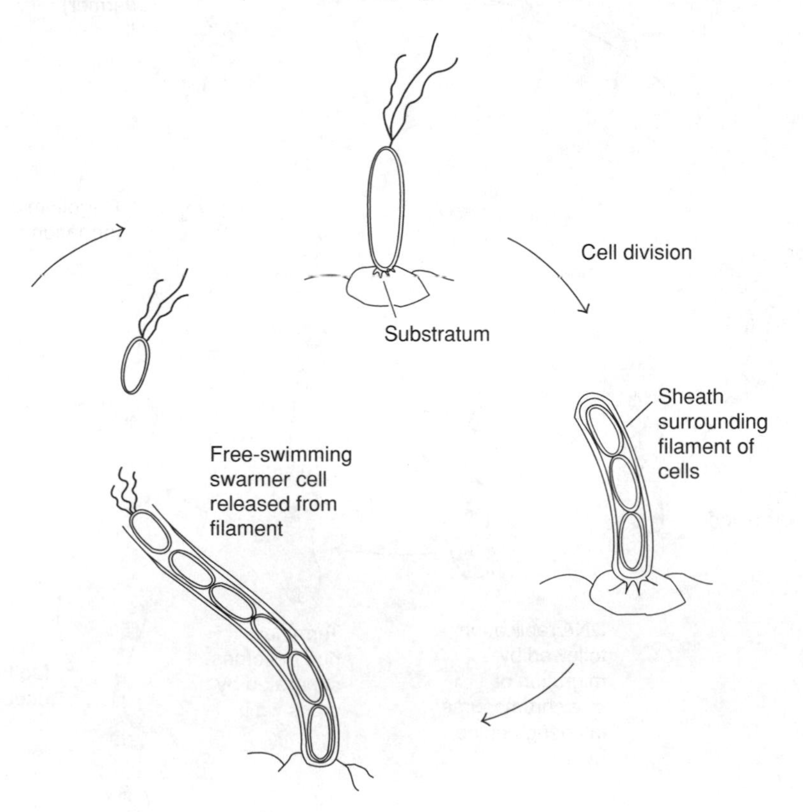

Figure 7.7 The sheathed bacteria. The life cycle of *Sphaerotilus*. Free-swimming swarmer cells settle on an appropriate substratum and give rise to long filaments contained within a sheath. New locations become colonised when flagellated cells are released into the water to complete the cycle

Sphaerotilus forms thick 'streamers' in polluted water, and is a familiar sight around sewage outlets.

Representative genera: *Sphaerotilus, Leptothrix*

Predatory Proteobacteria

Bdellovibrio is a unique genus belonging to the δ-Proteobacteria. It is a very small comma-shaped bacterium, which actually attacks and lives inside other Gram-negative bacteria (Figure 7.8). Powered by its flagellum, it collides with its prey at high speed and penetrates even thick cell walls by a combination of enzyme secretion and mechanical boring. It takes up residence in the periplasmic space, between the plasma membrane and cell wall. The host's nucleic acid and protein synthesis cease, and its macromolecules are degraded, providing nutrients for the invader, which grows into a long

> The recently-sequenced genome of *Bdellovibrio bacterivorans* has been shown to encode a huge number of lytic enzymes. Its ability to metabolise amino acids, however, is limited, necessitating its unusual mode of existence.

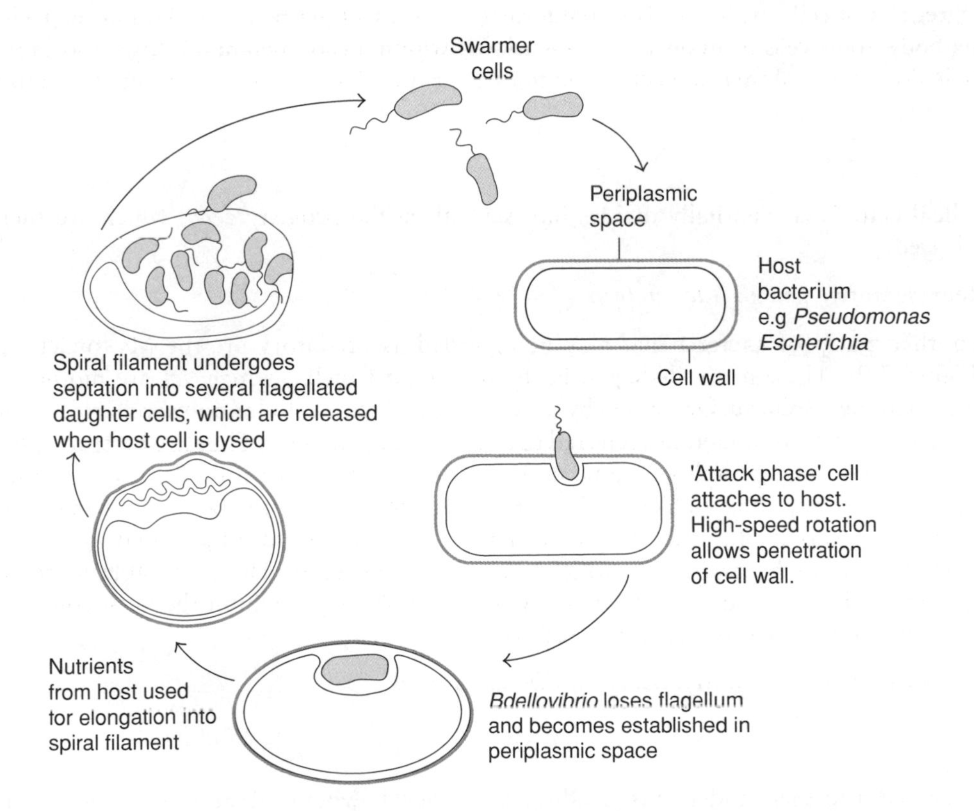

Figure 7.8 The life cycle of *Bdellovibrio*, a bacterial predator. Once *Bdellovibrio* has taken up residence in the periplasmic space of its host, it loses its flagellum and becomes non-motile. In nutrient-rich environments, *Bdellovibrio* is also capable of independent growth

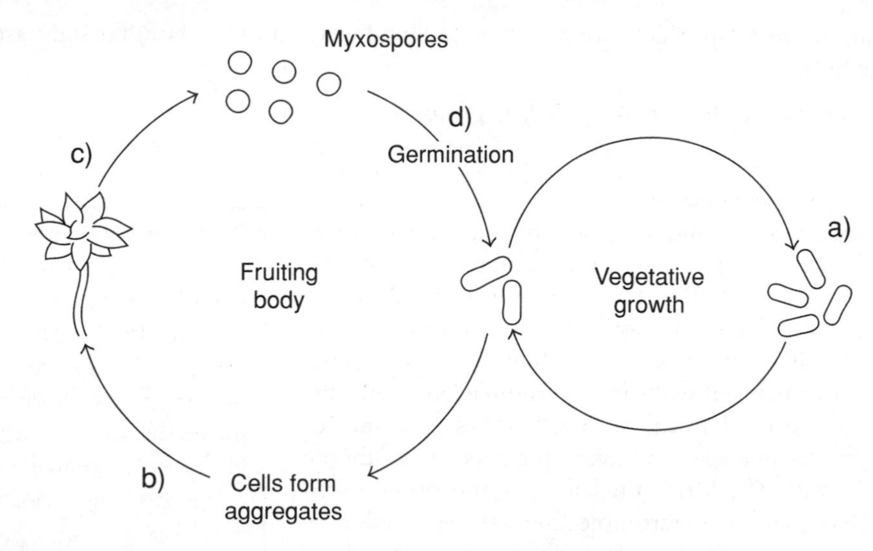

Figure 7.9 The Myxobacteria: a complex bacterial life cycle. When nutrients are in plentiful supply, myxobacteria divide by binary fission (a). On depletion of nutrients, they form aggregates of cells, which leads to the formation of a fruiting body (b). Within the fruiting body, some cells form myxospores, enclosed within a sporangium (c). Myxospores remain dormant until environmental conditions are favourable, then germinate into vegetative cells (d)

helical cell. This eventually divides into several motile progeny cells, which are then released.

Representative genus: *Bdellovibrio*

Another group of bacteria that may be regarded as predatory are the Myxobacteria (Figure 7.9). These are rod-shaped bacteria lacking flagella, which yet are motile by gliding along a solid surface, aided by the excretion of extracellular polysaccharides. For this reason they are sometimes referred to as the *gliding bacteria*. They are heterotrophs, typically requiring complex organic nutrients, which they obtain by the lysis of other types of bacteria. Thus, unlike *Bdellovibrio*, they digest their prey before they ingest it. When a rich supply of nutrients is not available, many thousands of cells may aggregate to form *fruiting bodies*, inside which *myxospores* develop. These are able to resist drought and lack of nutrients for many years. Myxobacteria exhibit the most complex life cycles of any procaryote so far studied.

Representative genera: *Myxococcus, Chondromyces*

Spirilla

Collected together under this heading are several genera of aerobic (mostly microaerophilic) spiral-shaped bacteria with polar flagella. These include free-living, symbiotic and parasitic types.

Spirilla such as *Aquaspirillum* and *Magnetospirillum* contain *magnetosomes*, intracellular particles of iron oxide (magnetite, Fe_3O_4). Such magnetotactic bacteria have the remarkable ability to orientate themselves with respect to the earth's magnetic field (*magnetotaxis*).

Two important pathogens of humans are included in the spirilla; *Campylobacter jejuni* is responsible for foodborne gastroenteritis, while *Helicobacter pylori* has in recent times been identified as the cause of many cases of peptic ulcers.

Representative genera: *Magnetospirillum, Campylobacter*

Rickettsia

This group comprises arthropod-borne intracellular parasites of vertebrates, and includes the causative agents of human diseases such as *typhus* and *Rocky Mountain spotted fever*. The bacteria are taken up by host phagocytic cells, where they multiply and eventually cause lysis.

The Rickettsia are aerobic organotrophs, but some possess an unusual mode of energy metabolism, only being able to oxidise intermediate metabolites such as glutamate and succinate, which they obtain from their host. *Rickettsia* and *Coxiella*, the two main genera, are not closely related phylogenetically and are placed in the α- and γ-Proteobacteria, respectively.

Representative genera: *Rickettsia, Coxiella*

Neisseria and related Proteobacteria

All members of this loose collection of bacteria are aerobic non-motile cocci, typically seen as pairs, with flattened sides where they join. Some however only assume this morphology during stationary growth phase. Many are found in warm-blooded animals, and some species are pathogenic. The genus *Neisseria* includes species responsible for gonorrhoea and meningitis in humans.

Other Gram-negative phyla

The following section considers those Gram-negative bacteria not included in the Proteobacteria. These phyla are not closely related in the phylogenetic sense, either to each other or to the Proteobacteria.

Phylum Cyanobacteria: the blue-green bacteria

The Cyanobacteria are placed in volume 1 of the second edition of *Bergey*, along with the Archaea (see above), the deeply branching bacteria, the 'Deinococcus–Thermus' group, and the green sulphur and green non-sulphur bacteria (see below). See also Box 7.2.

Members of the Cyanobacteria were once known as blue-green algae because they carry out the same kind of *oxygenic photosynthesis* as algae and green plants

Box 7.2 Prochlorophyta – a missing link?

It has long been thought that the chloroplasts of eucaryotic cells arose as a result of incorporating unicellular cyanobacteria into their cytoplasm. The photosynthetic pigments of plants and green algae, however, are not the same as those of the blue greens; both contain chlorophyll *a*, but while the former also have chlorophyll *b*, the blue greens have a unique group of pigments called phycobilins. In the mid-1970s a group of bacteria was discovered which seemed to offer an explanation for this conundrum. Whilst definitely procaryotic, the Prochlorophyta possess chlorophylls *a* and *b* and lack phycobilins, making them a more likely candidate for the origins of eucaryotic chloroplasts. Initially, these marine bacteria were placed by taxonomists in a phylum of their own, but in the second edition of *Bergey*, they are included among the Cyanobacteria.

(Chapter 6). They are the only group of procaryotes capable of carrying out this form of photosynthesis; all the other groups of photosynthetic bacteria to be discussed in this chapter carry out an anoxygenic form. When it became possible to examine cell structure in more detail with the electron microscope, it became clear that the cyanobacteria were in fact procaryotic, and hence quite distinct from the true algae. Old habits die hard, however, and the term 'blue-green algae' is still encountered, particularly in the popular press. Being procaryotic, cyanobacteria do not possess chloroplasts; however they contain lamellar membranes called *thylakoids*, which serve as the site of photosynthetic pigments and as the location for both light-gathering and electron transfer processes.

Early members of the Cyanobacteria evolved when the oxygen content of the earth's atmosphere was much lower than it is now, and these organisms are thought to have been responsible for its gradual increase, since photosynthetic eucaryotes did not arise until many millions of years later.

Cyanobacteria are Gram-negative bacteria which may be unicellular or filamentous; in spite of the name by which they were formerly known, they may also appear variously as red, black or purple, according to the pigments they possess. A characteristic of many cyanobacteria is the ability to fix atmospheric nitrogen, that is, to reduce it to ammonium ions (NH_4^+) for incorporation into cellular constituents (see above). In filamentous forms, this activity is associated with specialised, enlarged cells called *heterocysts* (Figure 7.10).

The tiny unicellular cyanobacterium *Prochlorococcus* is found in oceans throughout the tropical and temperate regions and is thought to be the most abundant photosynthetic organism on our planet. It has several strains adapted to different light conditions. Some cyanobacteria are responsible for the production of unsightly (and smelly!) 'algal' blooms in waters rich in nutrients such as phosphate. When they die, their decomposition by other bacteria leads to oxygen depletion and the death of other aquatic life forms. Bloom-forming species contain gas vacuoles to aid their buoyancy.

Representative genera: *Oscillatoria, Anabaena*

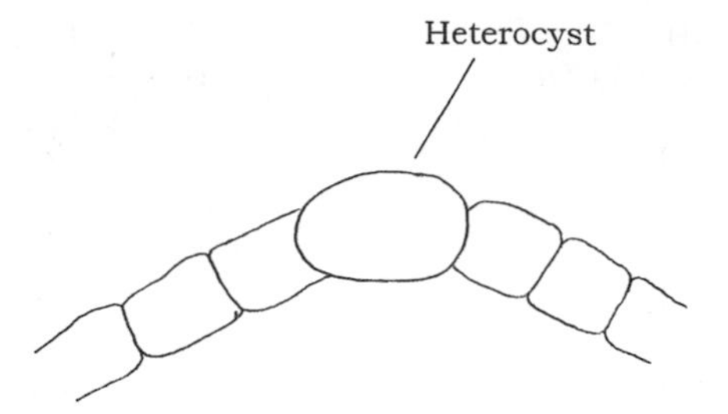

Figure 7.10 Cyanobacteria. Nitrogen fixation takes place in specialised cells called heterocysts, which develop from ordinary cells when supplies of available nitrogen (e.g. ammonia) are depleted. The heterocyst loses its ability to photosynthesise and, therefore, to produce oxygen. This is essential because oxygen is highly inhibitory to the nitrogenase enzyme complex

Phylum Chlorobi (green sulphur bacteria) and phylum Chloroflexi (green non-sulphur bacteria)

We have already come across three distinct groups of photosynthetic bacteria in this chapter: the purple sulphur and purple nonsulphur bacteria and the Cyanobacteria: here we consider the remaining two groups, the green sulphur and green non sulphur bacteria.

The green sulphur bacteria (phylum Chlorobi), like their purple counterparts (see above), are anaerobic photolithotrophs that utilise reduced sulphur compounds instead of water as an electron donor, and generate elemental sulphur. They differ, however, in a number of respects. The sulphur is deposited *outside* the cell, and CO_2 is assimilated not by the Calvin cycle, but by a reversal of the steps of the TCA cycle (see Chapter 6). The photosynthetic pigments in the green sulphur bacteria are contained in sac-like structures called *chlorosomes* that are associated with the inside of the plasma membrane.

Most members of the green non-sulphur bacteria (phylum Chloroflexi) are filamentous thermophiles, living in non-acid hot springs, where they form thick bacterial mats. Like the purple non-sulphur bacteria, they are photoheterotrophs, but can also grow in the dark as chemoheterotrophs.

Representative genera: *Chlorobium* (green sulphur), *Chloroflexus* (green non-sulphur).

Phylum Aquificae and phylum Thermotogae: the deeply branching bacteria

These two phyla are regarded as the two deepest (= oldest) branches in the evolution of the Bacteria and both comprise highly thermophilic Gram-negative rods. They are the only members of the Bacteria that can compare with the Archaea in their ability to live at high temperatures (optimal growth $>80\,^{\circ}C$). The two phyla differ in their mode of nutrition: the Aquificae are autotrophs capable of oxidising hydrogen or sulphur, while the Thermotogae are anaerobic heterotrophs, fermenting carbohydrates.

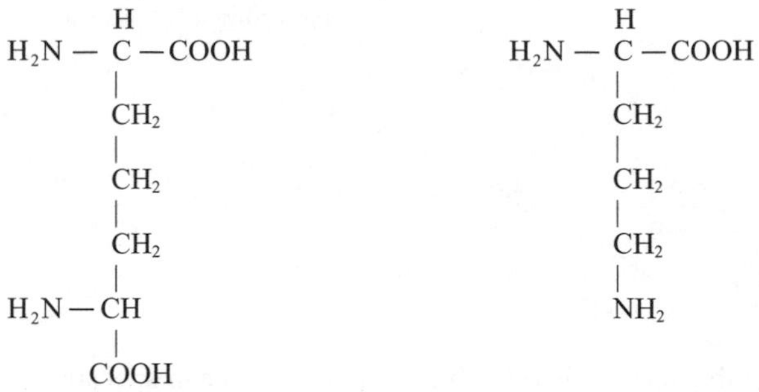

```
        H                          H
H2N — C — COOH           H2N — C — COOH
        |                          |
       CH2                        CH2
        |                          |
       CH2                        CH2
        |                          |
       CH2                        CH2
        |                          |
H2N — CH                          NH2
        |
      COOH
```

(a) Diaminopimelic acid (b) Ornithine

Figure 7.11 Members of the Deinococcus–Thermus group have an unusual form of peptidoglycan. The diaminopimelic acid at position three on the amino acid chain attached to *N*-acetylmuramic acid (see Figure 3.6) is replaced by ornithine

Members of the Thermotogae are characteristically surrounded, sometimes in a chain, by a proteinaceous sheath (or 'toga').

Representative genera: *Aquifex* (phylum Aquificae), *Thermotoga* (phylum Thermotogae)

Phylum 'Deinococcus–Thermus'*

Only three genera are included in this phylum, but two of these are of particular interest because of remarkable physiological properties. *Thermus* species are thermophiles whose best known member is *T. aquaticus*; this is the source of the thermostable enzyme *Taq* polymerase, used in the polymerase chain reaction (Chapter 12). *Deinococcus* species show an extraordinary degree of resistance to radiation, due, it seems, to an unusually powerful DNA repair system. This also enables *Deinococcus* to resist chemical mutagens (see Chapter 12). Both *Deinococcus* and *Thermus* species have an unusual refinement to the structure of their peptidoglycan, with ornithine replacing diaminopimelic acid (Figure 7.11).

Representative genera: *Deinococcus*, *Thermus*

The remaining phyla of Gram-negative bacteria are grouped together in Volume 5 of the second edition of *Bergey*.

Phylum Planctomycetes

This very ancient group of bacteria has a number of unusual properties, including cell division by budding, the lack of any peptidoglycan in their cell walls and the presence

* This phylum has not yet been assigned a formal name.

of a degree of internal compartmentalisation. (Recall from Chapter 3 that membrane-bound compartments are regarded as a quintessentially *eucaryotic* feature.)

The recently discovered reaction known as *anammox* (anaerobic ammonium oxidation), whereby ammonium and nitrite are converted to nitrogen gas:

$$NH_4^+ + NO_2^- = N_2 + 2H_2O$$

has been attributed to certain members of the Planctomycetes. It is thought that this reaction may be responsible for much of the nitrogen cycling in the world's oceans. The bacteria concerned have only been identified by means of their rRNA gene sequences, and have as yet been assigned only provisional generic and specific names. They are anaerobic chemolithoautotrophs, however this is not typical of the Planctomycetes, most of which are aerobic chemoorganoheterotrophs.

Recent studies propose that the Planctomycetes should be placed much closer to the root of any proposed phylogenetic tree than had previously been proposed.

Representative genera: *Planctomyces, Pirellula*

Phylum Chlamydiae

Formerly grouped with the Rickettsia (see above), these non-motile obligate parasites of birds and mammals are now assigned a separate phylum comprising only five genera, of which *Chlamydia* is the most important. Like the Rickettsia, members of the Chlamydiae have extremely small cells, and very limited metabolic capacities, and depend on the host cell for energy generation. Unlike that group, however, they are not dependent on an arthropod vector for transmission from host to host.

Chlamydia trachomatis is the causative agent of trachoma, a major cause of blindness in humans. Different strains of this same species are responsible for one of the most common forms of sexually transmitted disease. *C. psittaci* causes the avian disease psittacosis, and *C. pneumoniae* causes chlamydial pneumonia in humans as well as being linked to some cases of coronary artery disease.

Representative genus: *Chlamydia*

Phylum Spirochaetes

The Spirochaetes are distinguished from all other bacteria by their slender helical morphology and corkscrew-like movement. This is made possible by *endoflagella* (axial filaments), so-called because they are enclosed in the space between the cell and a flexible sheath that surrounds it.

Spirochaetes comprise both aerobic and anaerobic bacteria that inhabit a wide range of habitats, including water and soil as well as the gut and oral cavities of both vertebrate and invertebrate animals. Some species are important pathogens of humans, including *Treponema pallidum* (syphilis) and *Leptospira interrogans* (leptospirosis).

Representative genera: *Treponema, Leptospira*

Phylum Bacteroidetes

No unifying phenotypic feature characterises this diverse group, but their phylogenetic closeness causes them to be placed together. In light of this, we can only consider examples, without claiming them to be in any way representative.

The genus *Flavobacterium* takes its name from the yellow carotenoid pigments secreted by its members. These are aerobic, free-living, aquatic forms, although they are also associated with food spoilage.

In contrast, *Bacteroides* species are obligate anaerobes found in the human gut, where they ferment undigested food to acetate or lactate. Here they outnumber all other microbial forms, and are responsible for a significant percentage of the weight of human faeces. Some species can also be pathogenic, and may cause peritonitis in cases where the large intestine or appendix has become perforated.

Representative genera: *Bacteroides, Flavobacterium*

Phylum: Verrucomicrobia

Members of the Verrucomicrobia form several prosthecae per cell; these are similar to those described for certain Proteobacteria (see above). Although widespread in terrestrial, freshwater and marine environments, only a handful of representatives have been isolated in pure culture.

Representative genera: *Verrucomicrobium, Prosthecobacter*

The Gram-positive bacteria: phylum Firmicutes and phylum Actinobacteria

In the second edition of *Bergey*, the Gram-positive bacteria are divided into two large phyla, the Firmicutes and the Actinobacteria. Some 2500 species are known, but a substantial proportion of these belong to just a handful of genera. Gram-positive bacteria mostly have a chemoheterotrophic mode of nutrition and include among their number several important human pathogens, as well as industrially significant forms.

The base composition of an organism's DNA can be expressed as the percentage of cytosine and guanine residues (per cent GC content); the technique is used widely in microbial taxonomy, and the Gram-positive bacteria are divided into those whose GC content is significantly over or under 50 per cent. It is convenient to consider groupings within the high GC and low GC forms as follows:

Phylum Firmicutes (low GC):	spore-forming
	non-spore forming
	mycoplasma
Phylum Actinobacteria (high GC):	actinomycetes
	coryneform bacteria

Phylum Firmicutes: The low GC Gram-positive bacteria

The low GC Gram-positive bacteria form volume 3 of the second edition of *Bergey*.

The *spore-forming* Gram-positive bacteria include two large genera, *Clostridium* and *Bacillus*. Although not particularly close in phylogenetic terms, they are both capable of propagation by *endospores*.

Clostridium species are obligate anaerobes, and common inhabitants of soil. Sugars are fermented to various end-products such as butyric acid, acetone or butanol. Lacking an electron transport system, they obtain all their ATP from substrate-level phosphorylation.

Several species of *Clostridium* are serious human pathogens including *C. botulinum* (botulism) and *C. tetani* (tetanus). *C. perfringens* causes gas gangrene, and if ingested, can also result in gastroenteritis. All these conditions are due to the production of bacterial exotoxins. The resistance of spores to heating is thus highly relevant both in medicine and in the food industry. Related to *Clostridium* are the heliobacteria, two genera of anaerobic photoheterotrophic rods, some of which produce endospores. They are the only known photosynthetic Gram-positive bacteria.

So-called 'Botox' injections, much in vogue in certain circles as a cosmetic treatment, involve low doses of *C. botulinum* exotoxin. By acting as a muscle relaxant, they are intended to reduce the facial wrinkles that develop with the passing of time! The toxin is also used to treat medical conditions in which abnormal muscle contractions make it impossible for patients to open their eyes properly.

Bacillus species are aerobes or facultative anaerobes. They are chemoheterotrophs and usually motile by means of peritrichous flagella. Only a few species of *Bacillus* are pathogenic in humans, notably *B. anthracis*, the causative agent of anthrax. This is seen by many as a potential agent of bioterrorism, and here again the relative indestructibility of its spores is a crucial factor. Other species, conversely, are positively beneficial to humans; antibiotics such as bacitracin and polymixin are produced by *Bacillus* species, whilst the toxin from *B. thuringiensis* has been used as a natural insecticide (see Chapter 12).

Representative genera: *Bacillus*, *Clostridium*

The non-spore-forming low GC Gram-positive bacteria include a number of medically and industrially significant genera, a few of which are discussed below.

The lactic acid bacteria are a taxonomically diverse group containing both rods (*Lactobacillus*) and cocci (*Streptococcus*, *Lactococcus*), all characterised by their fermentative metabolism with lactic acid as end-product. Although they are able to tolerate oxygen, these bacteria do not use it in respiration. They are said to be *aerotolerant*. Like the clostridia, they lack cytochromes, and are therefore unable to carry out electron transport phosphorylation. The lactic acid bacteria have limited synthetic capabilities, so they are dependent on a supply of nutrients such as amino acids, purines/pyrimidines and vitamins. There has been growing interest in recent years in the use of certain lactic acid bacteria as *probiotics*.

Probiotics are living organisms that are deliberately ingested by humans with the aim of promoting health.

The genus *Streptococcus* remains a large one, although some members have been assigned to new genera in recent years, e.g. *Enterococcus, Lactococcus*. Streptococci are classified in a number of ways on the basis of phenotypic characteristics, but these do not correspond to phylogenetic relationships. Many species produce *haemolysis* when grown on blood agar, due to the production of toxins called *haemolysins*. In α-haemolysis, haemoglobin is reduced to methaemoglobin, resulting in a partial clearance of the medium and a characteristic green colour. β-Haemolysis causes a complete lysis of the red blood cells, leaving an area of clearing in the agar. A few species are non-haemolytic. Streptococci are also classified on the basis of carbohydrate antigens found in the cell wall; this system, which assigns each organism to a lettered group, is named after its devisor, Rebecca Lancefield.

Haemolysis is the lysis (bursting) of red blood cells. It may be brought about by bacterial toxins called haemolysins.

Pathogenic species of *Streptococcus* include *S. pyogenes* ('strep' sore throat, as well as the more serious rheumatic fever), *S. pneumoniae* (pneumococcal pneumonia) and *S. mutans* (tooth decay). Cells of *Streptococcus* exist mostly in chains, but in *S. pneumoniae* they are characteristically paired.

Lactobacillus is used very widely in the food and drink industry in the production of such diverse foodstuffs as yoghurt, cheeses, pickled foods (e.g. sauerkraut) and certain beers. This is discussed further in Chapter 17.

The cells of staphylococci occur in irregular bunches rather than ordered chains. They also produce lactic acid but can additionally carry out aerobic respiration involving cytochromes, and lack the complex nutritional requirements of the lactic acid bacteria. They are resistant to drying and able to tolerate relatively high concentrations of salt. These properties allow *Staphylococcus aureus* to be a normal inhabitant of the human skin, where it can sometimes give rise to dermatological conditions such as acne, boils and impetigo. It is also found in the respiratory tract of many healthy individuals, to whom it poses no threat, but in people whose immune system has been in some way compromised, it can cause serious respiratory infections. *S. aureus* can also cause a type of food poisoning and is the causative agent of toxic shock syndrome. Widespread antibiotic use has been largely responsible for the development of resistant forms of *S. aureus*, which have become ubiquitous inhabitants of hospitals (methicillin-resistant *Staphylococcus aureus*: MRSA). The problem of antibiotic resistance is discussed at greater length in Chapter 14.

Representative genera: *Streptococcus, Staphylococcus*

The Mycoplasma (Class Mollicutes) lack a cell wall and hence have a fluid shape (*pleomorphic*). Since the Gram test is based on the peptidoglycan content of a cell wall, why are these organisms grouped with the Gram-positive bacteria? The answer is that although they do not give a positive Gram test, they are clearly related at the genetic level to other members of the low GC Gram-positive group. The membranes of mycoplasma contain sterols; these help in resisting osmotic lysis, and are often essential as a growth requirement. Saprophytic, commensal and parasitic forms are known, and some species are associated with respiratory diseases in animals. Mycoplasma frequently occur as contaminants in the culture of animal cells, because their small size allows them to pass through filters, and they are resistant to antibiotics directed at cell wall synthesis.

Members of the Mycoplasma are among the smallest of all known cells and have some of the smallest genomes (just over half a million base pairs).

Representative genera: *Mycoplasma, Ureoplasma*

Phylum Actinobacteria: The high GC Gram-positive bacteria

The high GC gram-positive bacteria make up volume 4 of the second edition of *Bergey.*

The actinomycetes are aerobic, filamentous bacteria that form branching *mycelia* superficially similar to those of the Fungi (Chapter 8). Remember, however, that the actinomycetes are procaryotes and the fungi are eucaryotes, so the mycelia formed by the former are appreciably smaller. In some cases, the mycelium extends clear of the substratum and bears asexual *conidiospores* at the hyphal tips. These are produced by the formation of cross-walls and pinching off of spores, which are often coloured. The best known actinomycete genus is *Streptomyces*, which contains some 500 species, all with a characteristically high GC content (69–73 per cent). *Streptomyces* are very prevalent in soil, where they saprobically degrade a wide range of complex organic substrates by means of extracellular enzymes. Indeed, the characteristic musty smell of many soils is due to the production of a volatile organic compound called *geosmin*. A high proportion of therapeutically useful antibiotics derive from *Streptomyces* species, including well-known examples such as streptomycin, erythromycin and tetracycline (see Chapter 14).

> It has been suggested that camels are able to locate water sources in the desert by detecting the smell of geosmins given off by *Streptomyces* spp. in the damp earth!

Most actinomycetes, including *Streptomyces*, are aerobic; however, members of the genus *Actinomyces* are facultative anaerobes.

Representative genus: *Streptomyces*

The coryneform bacteria are morphologically half way between single celled bacilli and the branching filamentous actinomycetes. They are rods that show rudimentary branching, giving rise to characteristic 'V' and 'Y' shapes. Among the genera in this group are *Corynebacterium*, *Mycobacterium, Propionibacterium* and *Nocardia.*

Corynebacterium species are common in soil, and are also found in the mouths of a variety of animals. *C. diphtheriae* is the causative agent of diphtheria; it only becomes pathogenic when it has been infected by a bacteriophage that carries the gene for the diphtheria exotoxin.

Members of the genus *Mycobacterium* are characterised by their unusual cell wall structure; they include unusual complex lipids called mycolic acids. This causes the cells to be positive for the *acid-fast* staining technique, a useful way of identifying the presence of these bacteria. Mycobacteria are rod shaped, sometimes becoming filamentous; when filaments are formed, propagation is by means of fragmentation. *M. leprae* and *M. tuberculosis* cause, respectively, leprosy and tuberculosis in humans.

> The acid-fast test assesses the ability of an organism to retain hot carbol fuchsin stain when rinsed with acidic alcohol.

Propionibacterium species ferment lactic acid to propionic acid. Some species are important in the production of Swiss cheeses, whilst *P. acnes* is the main cause of acne in humans.

Representative genera: *Corynebacterium, Propionibacterium*

Bacteria and human disease

Although a detailed discussion of bacterial diseases falls outside the remit of this introductory text, their effect on the human race is too huge to fail to mention them completely. Some important examples have been mentioned briefly in the preceding text and Table 7.5 summarises the principal bacterial diseases of humans. This chapter about bacteria concludes with a brief discussion of four bacterial diseases, each providing an example of a different mode of transmission.

Table 7.5 Some bacterial diseases of humans

	Genus	Disease
Gram-positive	*Staphylococcus*	Impetigo, food poisoning, endocarditis, bronchitis, toxic shock syndrome
	Streptococcus	Pneumonia, pharyngitis, meningitis, scarlet fever, dental caries
	Enterococcus	Enteritis
	Listeria	Listeriosis
	Bacillus	Anthrax
	Clostridium	Tetanus, botulism, gangrene
	Corynebacterium	Diphtheria
	Mycobacterium	Leprosy, tuberculosis
	Propionibacterium	Acne
	Mycoplasma	Pneumonia, vaginosis
Gram-negative	*Salmonella*	Salmonellosis
	Escherichia	Gastroenteritis
	Shigella	Dysentery
	Neisseria	Gonorrhoea, meningitis
	Bordetella	Whooping cough
	Legionella	Legionnaires' disease
	Pseudomonas	Infections of burns
	Vibrio	Cholera
	Campylobacter	Gastroenteritis
	Helicobacter	Peptic ulcers
	Haemophilus	Bronchitis, pneumonia
	Treponema	Syphilis
	Chlamydia	Pneumonia, urethritis, trachoma

The table shows the genera responsible for some important bacterial diseases of humans. Note that many of the diseases listed may be caused only by a particular species within the genus.

Waterborne transmission: cholera

The causative agent *Vibrio cholerae* is ingested in faecally contaminated water or food. The bacteria attach by means of adhesins to the intestinal mucosa, where, without actually penetrating the cells, they release the cholera exotoxin. This comprises an 'A' and several 'B' subunits; the former is the active ingredient, while the latter attach to epithelial cells by binding to a specific glycolipid in the membrane. This allows the passage of the 'A' subunit into the cell, where it causes the activation of an enzyme called adenylate cyclase (Figure 7.12). This results in uncontrolled production of cyclic AMP, causing active secretion of chloride and water into the intestinal lumen. The outcome of this is huge fluid loss (10 l or more per day) through profuse and debilitating diarrhoea. In the young, old and sick, death through dehydration and salt depletion can follow within a very short time. If proper liquid and electrolyte replacement therapy is available, recovery rates can be very high. Although it is now very rare in the developed world (in 2002, only two cases were reported in the USA), cholera is a major killer in the third world. It is easily preventable by means of clean water supplies and improved

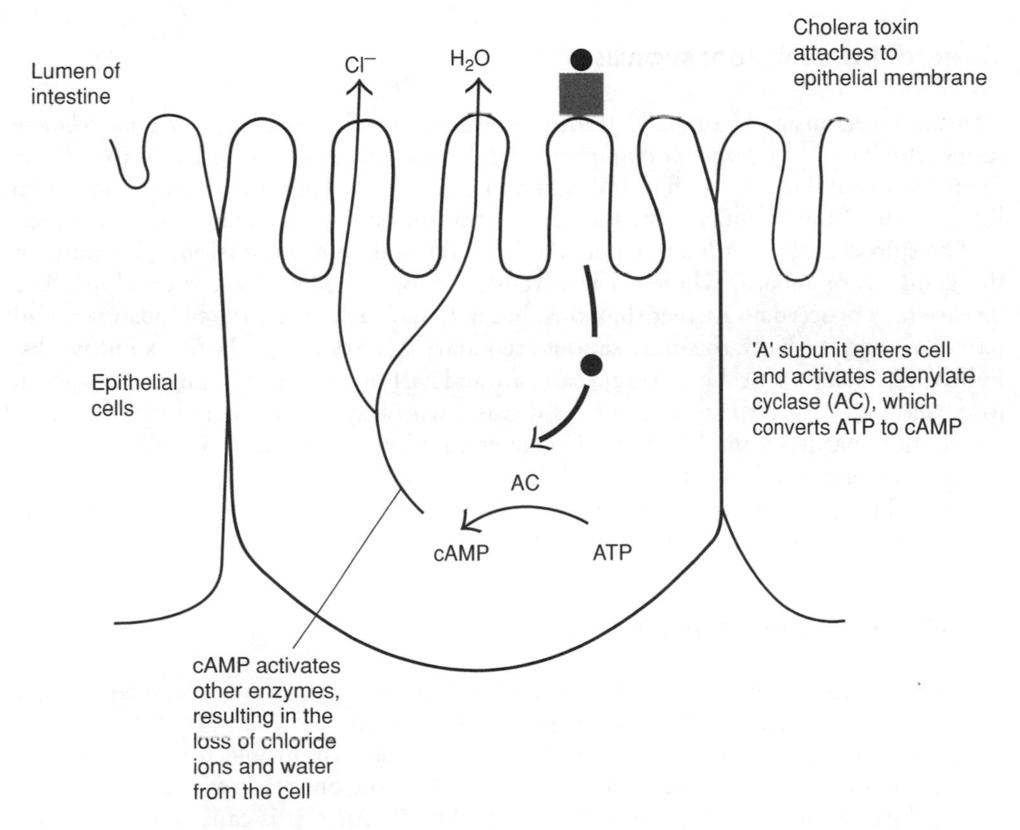

Figure 7.12 The action of the cholera exotoxin. Activation of the enzyme adenylate cyclase results in elevated levels of cyclic AMP and the secretion of electrolytes out of the epithelial cells lining the intestine and into the lumen. This is followed by water loss, resulting in debilitating dehydration

sanitation, however when these services break down, for example during war or after an earthquake, cholera outbreaks quickly follow.

Airborne transmission: 'strep' throat

Streptococcal pharyngitis, commonly known as strep throat, is one of the commonest bacterial diseases of humans, being particularly common in children of school age. The primary means of transmission is by the inhalation from coughs and sneezes of respiratory droplets containing *Streptococcus pyogenes* (β-haemolytic type A streptococci), although other routes (kissing, infected handkerchiefs) are possible. The primary symptoms are a red and raw throat (and/or tonsils), accompanied by headaches and fever. *S. pyogenes* attaches to the throat mucosa, stimulating an inflammatory response and secreting virulence factors that destroy host blood cells. Although self-limiting within a week or so, strep throat should be treated with penicillin or erythromycin as more serious streptococcal diseases such as scarlet fever and rheumatic fever may follow if it is left untreated.

Contact transmission: syphilis

Causative organisms of sexually transmitted diseases such as *Neisseria gonorrhoeae* (gonorrhoea) and *Treponema pallidum* (syphilis) are extremely sensitive to the effects of environmental factors such as UV light and desiccation. They are therefore unable to live outside of their human host, and rely for transmission on intimate human contact.

The spirochaete *T. pallidum* enters the body through minor abrasions, generally on the genitalia or mouth, where a characteristic lesion called a *chancre* develops. The disease may proceed no further than this, but if *T. pallidum* enters the bloodstream and passes around the body, the more serious secondary stage develops, lasting some weeks. Following a latent period of several years, around half of secondary syphilis cases go on to develop into the tertiary stage of the disease, whose symptoms may include mental retardation, paralysis and blindness. Congenital syphilis is caused by *T. pallidum* being passed from a mother to her unborn child.

The primary and secondary stages of syphilis are readily treated by penicillin; however, the tertiary stage is much less responsive to such therapy.

Vector-borne transmission: plague

A limited number of bacterial diseases reach their human hosts via an insect intermediary from their main host, usually another species of mammal.

Plague (bubonic plague, the Black Death) has been responsible for the deaths of untold millions of people in terrible epidemics such as the ones that wiped out as much as one-third of the population of Europe in the Middle Ages. It is caused by the Gram-negative bacterium *Yersinia pestis*, whose normal host is a rat, but can be spread to humans by fleas. The bacteria pass to the lymph nodes, where they multiply, causing the swellings known as bubos. *Y. pestis* produces an exotoxin, which prevents it from being destroyed by the host's macrophages; instead, it is able to multiply inside them.

From the lymph nodes, the bacteria spread via the bloodstream to other tissues such as the liver and lungs.

Once established in the lungs (pneumonic plague), plague can spread from human to human by airborne transmission in respiratory droplets. Untreated, plague has a high rate of fatality, particularly for the pneumonic form of the disease. Early treatment with streptomycin or tetracycline, however, is largely successful. Improved public health measures and the awareness of the dangers of rats and other rodents have meant that confirmed cases of plague are now relatively few.

Test yourself

1 The second edition of *Bergey's Manual of Systematic Bacteriology* is based on the ________ relationships between microorganisms, as determined by a comparison of their ________ sequences.

2 The three domains into which all living things are divided are the ________, the ________ and the ________ .

3 Archaean membrane lipids contain branched ________ instead of fatty acids.

4 Certain members of the Euryarchaeota have the unique ability to produce methane. Such organisms are termed ________ .

5 Members of the Crenarchaeota are mostly extreme ________ .

6 The biggest single phylum of bacteria is the ________ , which includes many of the best-known Gram-negative species.

7 The form of photosynthesis carried out by bacteria such as the purple sulphur bacteria is fundamentally different to that carried out by plants and algae because it is ________ .

8 The genus *Nitrosomonas* carries out the first step of the nitrification process, converting ________ to ________ .

9 Acid mine drainage is largely due to the activities of bacteria that are able to oxidise ________ compounds.

10 Bacteria such as *Rhizobium* are able to ________ atmospheric ________ nitrogen in association with the roots of ________ plants.

11 The genus *Desulfovibrio* is found in ________ conditions, where it ________ sulphur compounds to ________ ________ .

12 Two main forms of fermentation are found in the enteric bacteria: ________ ________ fermentation and ________ fermentation.

13 The pseudomonads can be distinguished from the enteric bacteria because they are ________ -positive and incapable of carrying out ________ .

14 The extracellular extension possessed by members of the stalked bacteria is called a ________ .

15 *Coxiella* is a member of the ________ , arthropod-borne intracellular parasites of vertebrates.

16 The Cyanobacteria are the only procaryotes capable of carrying out ________ photosynthesis.

17 Some members of the Planctomycetes can carry out ________ reactions, in which ammonia and nitrite are converted to nitrogen gas.

18 *Treponema pallidum*, the causative agent of syphilis, belongs to the ________ , a group characterized by their ________ shape and ________ movement.

19 ________ is an obligate anaerobe, and the most numerous microorganism in the human gut.

20 *Bacillus* and *Clostridium* are Gram-positive genera characterised by their ability to form ________ .

21 The Lancefield system is used to classify the ________ on the basis of cell surface antigens.

22 The ________ are exceptionally small bacteria, which lack a cell wall.

23 The ________ are branching, filamentous bacteria. One genus, ________ , is the source of many useful antibiotics.

24 Cell walls of the genus *Mycobacterium* contain ________ acids.

25 The diphtheria exotoxin is encoded by a ________ of *Corynebacterium diphtheriae*.

8
The Fungi

As we saw in the introduction to this section on microbial diversity, fungi were for many years classified along with bacteria, algae and the slime moulds (Chapter 9) as members of the Kingdom Plantae. As recently as the 1960s it was possible to find fungi being discussed under this heading, but in more recent times there has been universal agreement that they should be assigned their own kingdom. This is because fungi differ from plants in two quite fundamental respects:

- plants obtain energy from the sun, fungi do not
- plants utilise CO_2 as a carbon source, fungi do not.

To use the terminology we introduced in Chapter 4, plants are photoautotrophs, whereas fungi are chemoheterotrophs. In fact it now seems on the basis of molecular evidence that fungi are more closely related to animals than they are to plants!

We may define true fungi as primarily terrestrial, spore-bearing organisms, lacking chlorophyll and having a heterotrophic, absorptive mode of nutrition. Some 80 000 species are known and it is thought possible that at least a million more remain to be described! True fungi are a monophyletic group; that is, they are all thought to descend from a common ancestor, some 550 million years ago.

Fungi are of great importance economically and socially, and may have beneficial or detrimental effects. Many fungi, particularly yeasts, are involved in industrial fermentation processes (Chapter 17). These include, for example, the production of bread and alcohol, while other fungi are essential to the cheese-making process. Many antibiotics, including penicillin, derive from fungi, as does the immunosuppressive drug cyclosporin. Along with bacteria, fungi are responsible for the decomposition and reprocessing of vast amounts of complex organic matter; some of this is recycled to the atmosphere as CO_2, while much is rendered into a form that can be utilised by other organisms (Chapter 16). The other side of this coin is seen in the activity of fungi that degrade and destroy materials of economic importance such as wood, paper and leather, employing essentially the same biochemical processes. Additionally, some fungi may cause disease; huge damage is caused to crops and other commercially valuable plants, while a number of human diseases, particularly of the skin and scalp, are also caused by fungi.

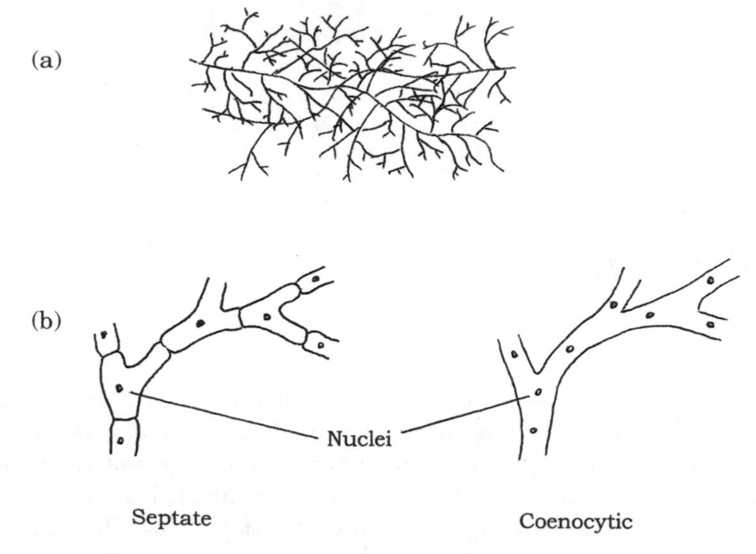

Figure 8.1 Hyphae and mycelia. (a) Individual hyphae branch and aggregate to form a mycelium. (b) Hyphae may or may not contain cross-walls (septa)

General biology of the Fungi

Morphology

Fungi range in form and size from unicellular yeasts to large mushrooms and puffballs. *Yeasts* are unicellular, do not have flagella and reproduce asexually by budding or transverse fission, or sexually by spore formation. Multicellular forms such as moulds have long, branched, threadlike filaments called *hyphae*, which aggregate together to form a tangled *mycelium* (Figure 8.1a). In some fungi the hyphae have crosswalls or septa (sing: septum) separating cells, which may nevertheless be joined by one or more pores, which permit cytoplasmic streaming, a form of internal transport. Such hyphae are said to be *septate*; others have no crosswalls and are therefore *coenocytic* (i.e. many nuclei within a single plasma membrane, Figure 8.1b).

Many fungi are *dimorphic*, that is, they exist in two distinct forms. Some fungi that cause human infections can change from the yeast form in the human to a mycelial form in the environment in response to changes in nutrients, and environmental factors such as CO_2 concentration and temperature. This change in body form is known as the *YM shift*; in fungi associated with plants, the shift often occurs the other way round, i.e. the mycelial form exists in the plant and the yeast form in the environment.

One of the features that caused taxonomists finally to remove fungi from the plant kingdom was the distinctive chemical nature of the fungal cell wall. Whereas plant and algal cells have walls composed of cellulose, the cell wall of fungi is made up principally of chitin (Figure 3.17b), a strong but flexible polysaccharide that is also found in the exoskeleton of insects. It is a polymer whose repeating subunit is *N*-acetylglucosamine, a compound we encountered when discussing peptidoglycan structure (Chapter 3).

Nutrition

Most fungi are *saprobic* (although some have other modes of nutrition), that is, they obtain their nutrients from decaying matter, which they grow over and through, frequently secreting enzymes extracellularly to break down complex molecules to simpler forms that can then be absorbed by the hyphae. Most fungi are able to synthesise their own amino acids and proteins from carbohydrates and simple nitrogenous compounds. Although fungi are unable to move, they can swiftly colonise new territory as a result of the rapid rate at which their hyphae grow. All energy is concentrated on adding length rather than thickness; this growth pattern leads to an increase in surface area and is an adaptation to an absorptive way of life. Carbohydrates are stored mainly in the form of glycogen (c.f. starch in higher plants, green algae). Metabolism is generally aerobic, but some yeasts can function as facultative anaerobes.

> The term saprobe describes an organism that feeds on dead and decaying organic materials. The older term *saprophyte* is no longer used, since the name perpetuates the idea of fungi being plants (*phyton* = a plant). Saprobes contribute greatly to the recycling of carbon and other elements.

Reproduction

Although, as we will see, there is a good deal of variety among the patterns of reproduction among the fungi, all share in common the feature of reproducing by *spores*; these are non-motile reproductive cells that rely on being carried by animals or the wind for their dispersal. The hyphae that bear the spores usually project up into the air, aiding their dispersal. One of the main reasons that we have to practise aseptic techniques in the laboratory is that fungal spores are pretty well ubiquitous, and will germinate and grow if they find a suitable growth medium. Spores of the common black bread mould, *Rhizopus*, (see below) have been found in the air over the North Pole, and hundreds of miles out to sea. In some fungi the aerial spore-bearing hyphae are developed into large complex structures called fruiting bodies. The most familiar example of a fruiting body is the mushroom. Many people think that the mushroom itself is the whole fungus but it only represents a part of it; most is buried away out of sight below the surface of the soil or rotting material, a network of nearly invisible hyphae.

Classification of the Fungi

The Fungi are arranged into four major phyla on the basis of differences in their sexual reproduction. These are:

Zygomycota } 'Lower Fungi'
Chytridiomycota }

Ascomycota } 'Higher Fungi'
Basidiomycota

The designation 'higher' and 'lower' Fungi, is an unofficial distinction based on the fact that the latter two (much larger) groups possess septate hyphae. In some books, you may come across references to a group called the Deuteromycota or Fungi Imperfecti. This is not a taxonomic grouping, but a 'holding area' for species in which only an asexual stage has been recognised, and which cannot therefore be assigned to any of the above groups. It is now possible, however, by means of DNA analysis, to place such species with their nearest relatives, mostly in the Ascomycota or Basidiomycota.

> Members of the different fungal phyla are given the suffix -*cetes*, e.g. ascomycetes.

In the following pages, we shall look at each of the phyla in turn, concentrating particularly on their life cycles.

Zygomycota

The Zygomycota is a relatively small phylum, comprising less than a thousand species. Its members are typically found in soil, or on decaying organic matter, including animal droppings. Some members of the group are of great importance in the formation of a mutualistic association with plant roots known as a *mycorrhiza*. This will be discussed in more detail in Chapter 15.

Members of the Zygomycota are characterised by the formation of a dormant form, the *zygospore*, which is resistant to unfavourable environmental conditions. Hyphae are coenocytic, with numerous haploid nuclei, but few dividing walls or septa.

Familiar examples of this group are *Mucor* and the black bread mould *Rhizopus*. The life cycle of *Rhizopus* is shown in Figure 8.2. Hyphae spread rapidly over the surface of the substrate (bread, fruit etc.) and penetrate it, absorbing soluble nutrients such as sugars. Upright hyphae develop, carrying at their tip *sporangia*, full of black haploid spores. The spores give the characteristic colour to the mould; they are the asexual reproductive structures, and are released when the thin wall of the sporangium ruptures. When conditions are favourable, *Rhizopus* reproduces in this way; each spore, upon finding a suitable substratum for growth, is capable of germinating and initiating a new mycelium.

> A sporangium is a structure inside which spores develop. It is held aloft on an aerial hypha called a sporangophore.

Sexual reproduction occurs when environmental conditions are unfavourable. Most species of *Rhizopus* are *heterothallic*; that is, there exist two distinct mating strains known as + and −. Sexual reproduction is only possible between a member of the + -strain and a member of the − -strain. Although reproductively distinct,

> A dicaryon is a structure formed by two cells whose contents, but not nuclei, have fused.

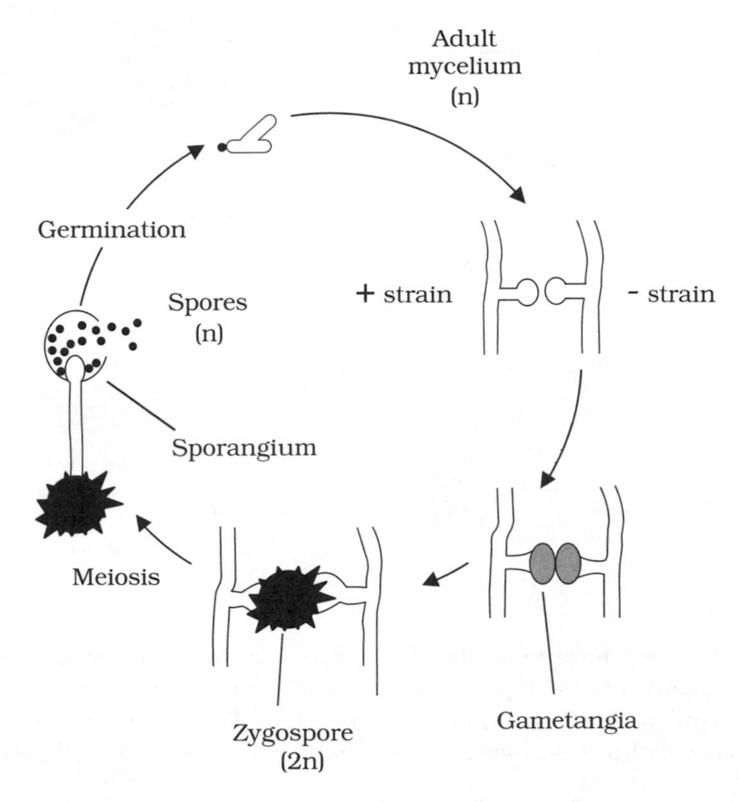

Figure 8.2 Zygomycota: the life cycle of *Rhizopus*. Both sexual and asexual cycles involve the production of sporangiospores. In sexual reproduction, hyphae from different mating strains fuse to form a diploid zygospore, via a short-lived dicaryotic intermediate. Germination of the zygospore gives rise to an aerial sporangium; this contains many haploid sporangiospores, which give rise to another vegetative mycelium

these two types are morphologically identical; because of this it is not appropriate to refer to them as 'male' and 'female'. When hyphae of opposite mating types come into contact, a cross-wall develops a short distance behind each tip, and the regions thus isolated swell to produce *gametangia* (Figure 8.2). These fuse to form a single large multinucleate cell. Note that at this stage, the nuclei from each parent have paired up *but not fused*, forming a *dicaryon*. Dicaryon formation is found in all fungal phyla apart from the chytrids, and may be regarded as an intermediate stage between the haploid and diploid conditions (Figure 8.3). The proportion of the life cycle it occupies varies considerably. A thick protective covering develops around the dicaryon in *Rhizopus*, forming the zygospore, which can survive extremes of draught and temperature and may remain dormant for months. When conditions are favourable again, the nuclei from each strain fuse in pairs, to give a fully diploid zygote. Just before germination, meiosis occurs, then an aerial sporangiophore emerges, terminating in a sporangium. Production and dispersal of haploid spores then occur as in the asexual life cycle and a new mycelium forms when a spore germinates.

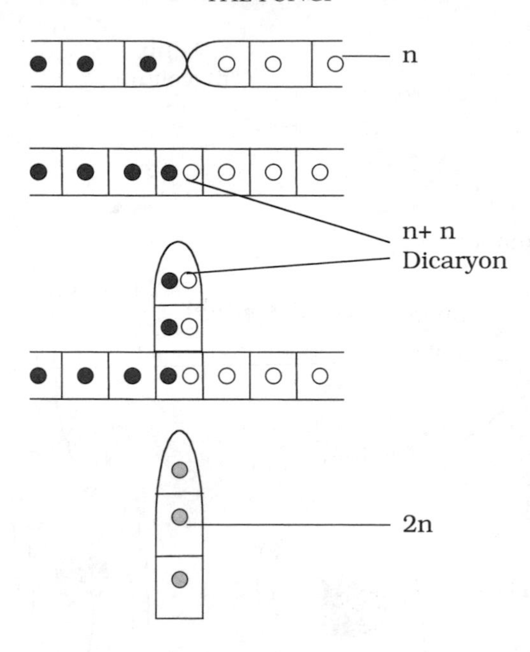

Figure 8.3 Dicaryon formation. The dicaryon ($n + n$) represents an intermediate stage between the haploid (n) and diploid ($2n$) states. The participating cells have undergone plasmogamy (fusion of cytoplasm) but not caryogamy (fusion of nuclei). Each dicaryotic cell contains one nucleus from each parent cell. Fusion of nuclei leads to the formation of a true diploid

In *Rhizopus*, therefore, we have a life cycle in which the haploid form predominates (the zygospore is the only diploid stage), and we have sexual reproduction without the involvement of motile gametes.

Chytridiomycota

The chytrids are believed to have been the first of the fungal groups to diverge from a common ancestor many millions of years ago. They differ from all other fungal groups by possessing flagellated zoospores. At one time, the Fungi were defined by their lack of flagella, so the chytrids were assigned to the Protista (see Chapter 9). However, molecular evidence, including the possession of a chitinous cell wall, suggests that it would be more appropriate to place them among the Fungi.

Some members of the chytrids may live saprobically on decaying plant and animal matter, while others are parasites of plants and algae. Another group live anaerobically in the rumen of animals such as sheep and cattle. In recent years there has been evidence that a parasitic species of chytrid is at least partially responsible for the dramatic decline in frog populations in certain parts of the world. Some chytrids are unicellular, while others form mycelia of coenocytic hyphae. Reproduction may be asexual by means of motile zoospores or sexual. The latter may involve fusion of gametes to produce a diploid zygote, but there is no dicaryotic stage in the life cycle.

Ascomycota

A lichen is formed by the symbiotic association of a fungus (usually an ascomycete) and an alga or cyanophyte.

The Ascomycota are characterised by the production of haploid *ascospores* through the meiosis of a diploid nucleus in a small sac called an *ascus*. For this reason they are sometimes called the sac fungi or cap fungi. Many of the fungi that cause serious plant diseases such as Dutch elm disease and powdery mildew belong to this group. They include some 30 000 species, among them yeasts, food spoilage moulds, brown fruit rotting fungi and truffles. Note that the latter, often regarded as the most prized type of mushrooms by gourmets, are assigned to a completely different group to the true mushrooms, which belong to the Basidiomycota. Around half of ascomycote species exist in associations with algae to form *lichens*; these will be discussed more fully in Chapter 15. Most ascomycetes produce mycelia that superficially resemble those of zygomycetes, but differ in that they have distinct, albeit perforated cross walls (septa) separating each cell.

Asexual reproduction in most ascomycetes involves the production of airborne spores called *conidia*. These are carried on the ends of specialised hyphae called *conidiophores*, where they may be pinched off as chains or clusters (Figure 8.4). Note that the conidia are not contained within sporangia; they may be naked or protected by a flask-like structure called the pycnidium. Asexual reproduction by conidia formation is a means of rapid propagation for the fungus in favourable conditions. The characteristic green,

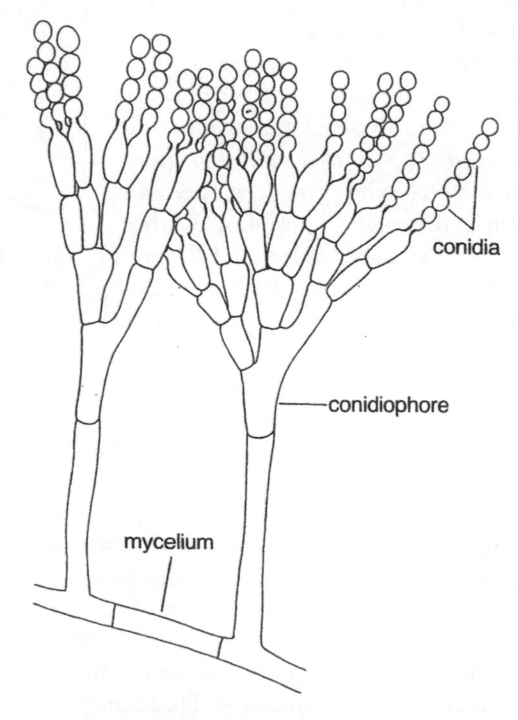

Figure 8.4 Asexual reproduction in the Ascomycota. Chains of conidia develop at the end of specialised hyphae called conidiophores

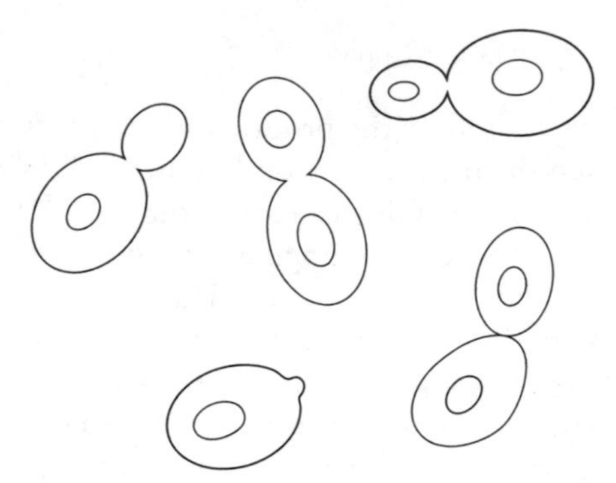

Figure 8.5 Yeast cells in various stages of budding. A protuberance or bud develops on the parent yeast; the nucleus undergoes division and one copy passes into the bud. Eventually the bud is walled off and separated to form a new cell

pink or brown colour of many moulds is due to the pigmentation of the conidia, which are produced in huge numbers and dispersed by air currents. The conidia germinate to form another mycelium (haploid).

In the case of the unicellular yeasts, asexual reproduction occurs as the result of *budding*, a pinching off of a protuberance from the cell, which eventually grows to full size (Figure 8.5).

Plasmogamy is the fusion of the cytoplasmic content of two cells. Caryogamy is the fusion of nuclei from two different cells.

Sexual reproduction in some ascomycetes involves separate + and − mating strains similar to those seen in zygomycetes, whilst in other cases an individual will be self-fertile, and thus able to mate with itself. When there are separate strains, the hyphae involved in reproduction are termed the *ascogonium* (− -strain) and *antheridium* (+ -strain). In either case, two hyphae grow together and there is a fusion of their cytoplasm (Figure 8.6). Within this fused structure, rather like the zygospore of *Rhizopus*, nuclei pair, but do not fuse; thus the resulting structure is a dicaryon. Following cytoplasmic fusion (*plasmogamy*), branching hyphae develop. These hyphae are septate, i.e. partitioned off into separate cells, but each cell is dicaryotic, having a nucleus from each parental type.

As we have seen, a key characteristic of ascomycetes is the production of sexual spores in sac-like structures called asci. These develop in distinct macroscopic fruiting bodies called *ascocarps*, which arise from the aggregation of dicaryotic hyphae (Figure 8.6). At the tip of each dicaryotic hypha, pairs of nuclei fuse to give a diploid zygote; this is followed by one meiotic and one mitotic division, giving rise to eight haploid ascospores. An ascocarp may contain thousands of asci, each with eight ascospores. *Tetrad analysis* of ascospores has proved a valuable technique in genetic mapping.

When the ascus is mature, it splits open at its tip and the ascospores are released. They are dispersed, often over long distances, by air currents. If a mature ascocarp is disturbed, it may release smoke-like puffs containing thousands of ascospores. The germinating ascospore forms a new mycelium. Frequently there are many rounds of asexual reproduction between successive rounds of sexual reproduction by ascospore production.

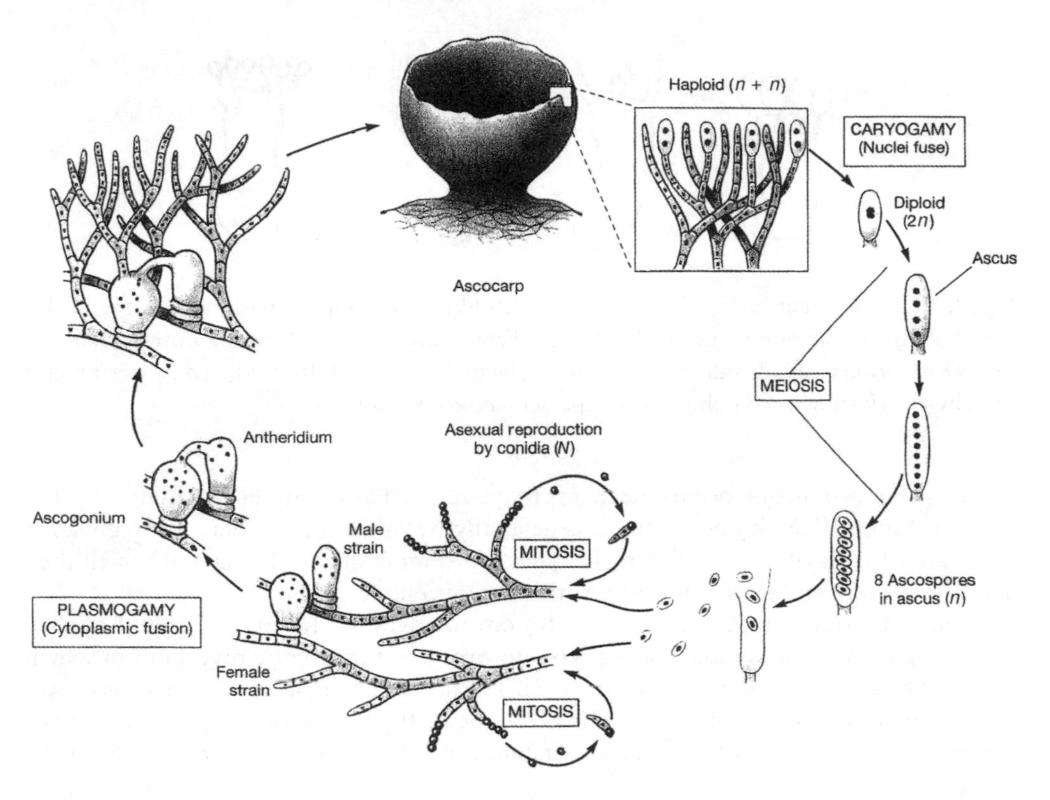

Figure 8.6 Sexual reproduction in the Ascomycota. Nuclei in the cells at the tips of the ascogenous hyphae fuse to give a diploid zygote. Meiotic and mitotic divisions result in the formation of eight haploid ascospores inside a tubular ascus. On germination, each ascospore is capable of giving rise to a new mycelium. From Black, JG: Microbiology: Principles and Explorations, 4th edn, John Wiley & Sons Inc., 1999. Reproduced by permission of the publishers

As with the zygomycetes, the diploid stage in ascomycetes is very brief. The life cycles of the two groups differ, however, in the greater role played by the dicaryotic form in ascomycetes.

Basidiomycota

This large group of some 25 000 species contains the true mushrooms and toadstools as well as other familiar fungi such as puffballs and bracket fungi. In fact the great majority of the fungi that we see in fields and woodlands belong to the Basidiomycota. They are of great economic importance in the breakdown of wood and other plant material (Chapter 16). The group derives its common name of the club fungi from the way that the spore-bearing hyphae involved in reproduction are swollen at the tips, resembling clubs (the *basidia*: Figure 8.7).

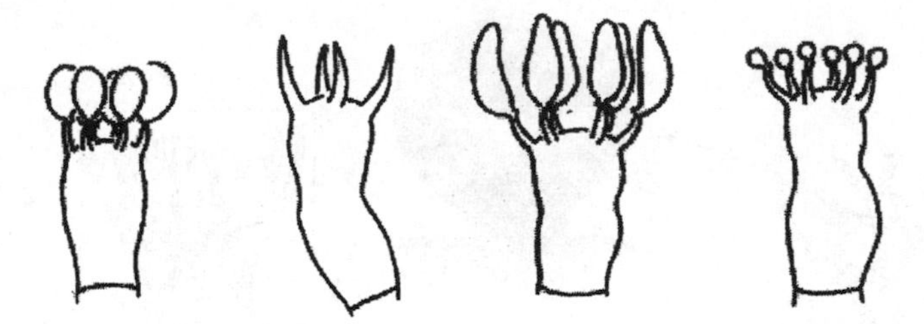

Figure 8.7 Different forms of basidia. The club-like appearance of basidia gives the Basidiomycota their common name of 'club Fungi'. From Langer, E: Die Gattung Botryobasidium DONK (Corticiaceae, Basidiomycetes). Bibl Mycol 158, 1–459. Reproduced by permission of Schweizerbart Science Publishers (http://schweizerbart.de)

Asexual reproduction occurs much less frequently in basidiomycetes than in the other types of fungi. When it does occur, it is generally by means of conidia, although some types are capable of fragmenting their hyphae into individual cells, each of which then acts like a spore and germinates to form a new mycelium.

Sexual reproduction in a typical mushroom involves the fusion of haploid hyphae belonging to two compatible mating types to produce a dicaryotic mycelium in which each cell has two haploid nuclei (Figure 8.8). The most striking feature of this secondary mycelium is the *clamp connection*; this is unique to the Basidiomycota and is a device for ensuring that as growth continues, each new cell has one nucleus from each of the

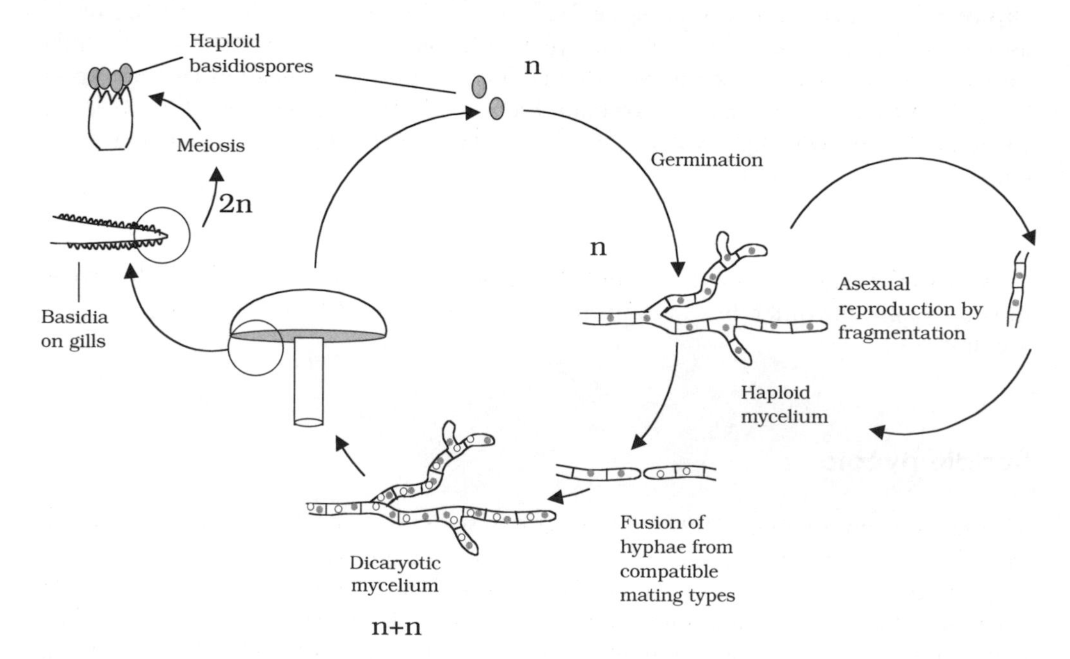

Figure 8.8 Life cycle of the Basidiomycota. Most of the fungus exists as a mycelial mass underground. The mushroom is an aerial fruiting body that facilitates the dispersal of spores

Box 8.1 Clamp connections

Imagine the situation in a dicaryotic cell when mitosis took place. The most likely outcome would be that as the two nuclei divided, both of one type would end up in one daughter cell and both of the other type in the other. As growth and cell division proceeds in the secondary mycelium, it is therefore necessary to ensure that the dicaryotic state is maintained, i.e. that each new cell inherits one nucleus of each type from its parent. This is achieved by means of *clamp connections*.

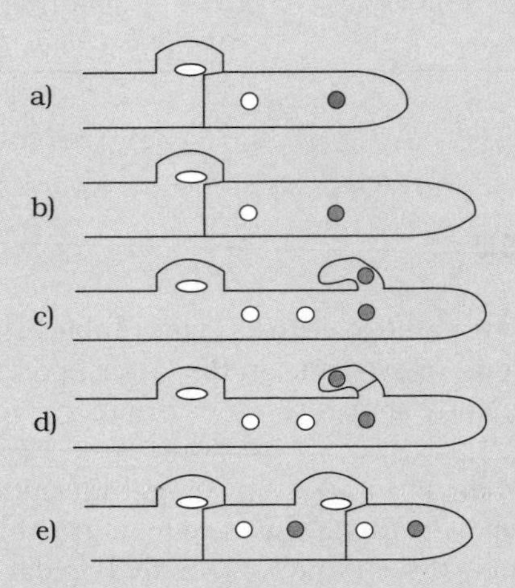

As the terminal hyphal cell elongates (b), a tube-like clamp connection grows out backwards (c). As this happens, one nucleus enters the clamp and mitosis occurs simultaneously in both parental nuclei. A septum is formed to separate the first pair of nuclei (d), then, as the loop of the clamp connection is completed, a second septum separates the second pair (e). The result is two new daughter cells, each with one copy of each nuclear type.

parent mating strains (see Box 8.1). This dicaryotic secondary mycelium continues to grow, overwhelming any remaining haploid hyphae from the parent fungi.

When the secondary mycelium has been developing for some time, it forms a dense compact ball or button, which pushes up just above the surface and expands into a *basidiocarp*; this is the mushroom itself. Stalk formation and upward growth is extremely rapid; a stalk or stipe of 10 cm can be formed in only about 6–9 hours. The growth is initially towards light (positive phototropism) and then upward (negative geotropism). As the cap expands, fleshy flaps radiating from the centre of its underside open up. These are the *gills*, made up of compacted hyphae with numerous basidia arranged at right angles. As each basidium matures, its two nuclei finally fuse, and then undergo meiosis to produce four haploid basidiospores. A single large mushroom can produce millions of basidiospores in the space of a few days. They are discharged from the end of the basidia and then fall by gravity from the gills. Air currents then carry them away for dispersal. Upon finding a suitable substratum, the spores germinate into a haploid mycelium just below the surface of the soil, thus completing the life cycle.

Table 8.1 Some fungal diseases of humans

Disease	Fungus
Histoplasmosis	*Histoplasma capsulatum*
Blastoplasmosis	*Blastomyces dermatitidis*
Cryptococcosis	*Cryptococcus neoformans*
Cutaneous mycoses	*Trichophyton* spp.
Pneumocystis pneumonia	*Pneumocystis carinii*
Candidiasis ('thrush')	*Candida albicans*
Aspergillosis	*Aspergillus fumigatus*

Fungi and disease

A limited number of fungi are pathogenic to humans (Table 8.1). *Mycoses* (sing: mycosis) in humans may be cutaneous, or systemic; in the latter, spores generally enter the body by inhalation, but subsequently spread to other organ systems via the blood, causing serious, even fatal disease.

Cutaneous mycoses are the most common fungal infections found in humans, and are caused by fungi known as *dermatophytes*, which are able to utilise the keratin of skin, hair or nails by secreting the enzyme *keratinase*. Popular names for such infections include ringworm and athletes' foot. They are highly contagious, but not usually serious conditions.

Systemic mycoses can be much more serious, and include conditions such as histoplasmosis and blastomycosis. The former is caused by *Histoplasma capsulatum*, and is associated with areas where there is contamination by bat or bird excrement. It is thought that the number of people displaying clinical symptoms of histoplasmosis represents only a small proportion of the total number infected. If confined to the lungs, the condition is generally self-limiting, but if disseminated to other parts of the body such as the heart or central nervous system, it can be fatal. The causative agents of both diseases exhibit dimorphism; they exist in the environment as mycelia but convert to yeast at the higher temperature of their human host.

Aspergillus fumigatus is an example of an *opportunistic pathogen*, that is, an organism which, although usually harmless, can act as a pathogen in individuals whose resistance to infection has been lowered. Other opportunistic mycoses include candidiasis ('thrush') and *Pneumocystis* pneumonia. The latter is found in a high percentage of acquired immune deficiency syndrome (AIDS) patients, whose immune defences have been compromised. The causative organism, *Pneumocystis carinii*, was previously considered to be a protozoan, and has only been classed as a fungus in the last decade, as a result of DNA/RNA sequence evidence. It lives as a commensal in a variety of mammals, and is probably transmitted to humans through contact with dogs.

The incidence of opportunistic mycoses has increased greatly since the introduction of antibiotics, immunosuppressants and cytotoxic drugs. Each of these either suppresses the individual's natural defences, or eliminates harmless microbial competitors, allowing the fungal species to flourish.

Box 8.2 Ergot

Members of the genus *Claviceps* may infect a variety of grains, particularly rye, when they come into flower, giving rise to the condition called *ergot*. No great damage is caused to the crop, but as the fungus develops in the maturing grain, powerful hallucinatory compounds are produced, which cause ergotism in those who consume bread made from the affected grain. This was relatively common in the Middle Ages, when it was known as St Anthony's Fire. The hallucinatory effects of ergotism have been put forward by some as an explanation for outbreaks of mass hysteria such as witch hunts and also for the cause of the abandonment of the ship, the *Mary Celeste*. The effects can go beyond the psychological causing convulsions and even death. In small controlled amounts, the drugs derived from ergot can be medically useful in certain situations such as the induction of childbirth and the relief of migraine headaches.

Many fungi produce natural *mycotoxins*; these are *secondary metabolites*, which, if consumed by humans, can cause food poisoning that can sometimes be fatal. Certain species of mushrooms ('toadstools') including the genus *Amanita* contain substances that are highly poisonous to humans. Other examples of mycotoxin illnesses include ergotism (see Box 8.2) and aflatoxin poisoning. Aflatoxins are *carcinogenic* toxins produced by *Aspergillus flavus* that grows on stored peanuts. In the early 1960s, the turkey industry in the UK was almost crippled by 'Turkey X disease', caused by the consumption of feed contaminated by *A. flavus*.

Secondary metabolites are produced by a microorganism after the phase of active growth has ceased. Such substances are usually not required for essential metabolic or cell maintenance purposes. Examples include toxins pigments and most antibiotics. Carcinogenic means cancer-causing.

It is thought likely that all animals are parasitised by one fungus or another. Extraordinary though it may seem, there are even fungi that act as predators on small soil animals such as nematode worms, producing constrictive hyphal loops that tighten, immobilising the prey.

Fungi also cause disease in plants, and can have a devastating effect on crops of economic importance, either on the living plant or in storage subsequent to harvesting. Rusts, smuts and mildews are all examples of common plant diseases caused by fungi. The effects of fungi on materials such as wood and textiles will be considered in Chapter 16.

Test yourself

1 Categorised according to their carbon and energy sources, all fungi are ________.

2 Fungi whose cells are separated by cross-walls are described as ________.

3 Fungi whose cells lack cross-walls are described as ________.

4 The YM shift describes an alternation between ________ and ________.

5 The principal component of cell walls is ________.

6 The ________ ________ (or Deuteromycota) is a term used to describe those species in which no sexual reproductive stage has been observed.

7 The Chytridiomycota and the Zygomycota are sometimes termed the ________ fungi.

8 The Ascomycota and the Basidiomycota are sometimes termed the ________ fungi.

9 A resistant form, the ________, enables members of the Zygomycota to overcome adverse environmental conditions.

10 An aerial hypha that gives rise to spore formation is termed a ________.

11 The chytrids are thought to be the most primitive fungal group. Unlike other fungi, they possess ________.

12 Members of the Ascomycota reproduce asexually by means of spores called ________.

13 Unicellular yeasts reproduce asexually by the process of ________.

14 In sexual reproduction in the Ascomycota, haploid ascospores develop in fruiting bodies called ________.

15 ________ cells contain two nuclei, one inherited from each parent.

16 The appearance of spore-bearing hyphae called basidia gives the Basidiomycota their common name: the ________ ________.

17 The ________ connection is a morphological feature unique to basidiomycetes.

18 A mushroom is a fruiting body of a member of the Basidiomycota; its scientific name is a ________.

19 A ________ is a fungal disease of humans. It may be ________ or ________.

20 *Aspergillus flavus* produces carcinogenic toxins called ________.

9

The Protista

Although not such an all-embracing a term as originally envisaged by Haeckel, the Protista represents a very diverse group of organisms, united by their possession of eucaryotic characteristics, and failure to fit satisfactorily into the animal, plant or fungal kingdoms. Some scientists limit use of the name to unicellular organisms, while others also include organisms such as the macroscopic algae, which are not accommodated conveniently elsewhere.

It has become clear from molecular studies that some members of the Protista bear only a very distant relationship to each other, making it an unsatisfactory grouping in many respects. In an introductory text such as this, however, it is convenient and, it is hoped, less confusing, to retain the name Protista as a chapter heading, as long as the student understands that it does not represent a coherent taxonomic grouping of phylogenetically related organisms. At the end of the chapter, we shall look at how members of the Protista are placed in modern, phylogenetic taxonomic schemes.

The Protista is a grouping of convenience, containing organisms not easily accommodated elsewhere. It includes all unicellular and colonial eucaryotic organisms, but is often expanded to include multicellular algae.

It has been found helpful in the past to think of Protists as being divided into those with characteristics that are plant–like (the Algae), animal-like (the Protozoa) and fungus-like (the water moulds and slime moulds). We shall discuss each of these groups in turn in the following pages. It should be borne in mind, however, that molecular evidence suggests such a division to be artificial; on the basis of molecular and cytological comparison, the 'animal-like' protozoan *Trypanosoma*, for example, is closely related to the photosynthetic (and therefore 'plant-like') *Euglena*.

'The Algae'

The Algae is a collective name traditionally given to several phyla of primitive, and mostly aquatic plants, making up a highly diverse group of over 30 000 species. They display a wide variety of structure, habitat and life-cycle, ranging from single-celled forms to massive seaweeds tens of metres in length. Most algae share a number of common features which caused them to be grouped together. Among these are:

- possession of the pigment chlorophyll
- deriving energy from the sun by means of oxygenic photosynthesis
- fixing carbon from CO_2 or dissolved bicarbonate (see Chapter 6).

Modern taxonomy attempts to reflect more accurately the relationship between organisms with an assumed common ancestor. Thus, in the following pages, the unicellular 'algae' are discussed in relation to other unicellular eucaryotes. Multicellular forms, including the Phaeophyta (brown algae) and Rhodophyta (red algae), are not discussed at great length and are included for the sake of completeness.

Structural characteristics of algal protists

All algal types are eucaryotic, and therefore contain the internal organelles we encountered in Chapter 3, that is, nuclei, mitochondria, endoplasmic reticulum, ribosomes, Golgi body, and in most instances, chloroplasts. With the exception of one group (the Euglenophyta), all have a cellulose cell wall, which is frequently modified with other polysaccharides, including pectin and alginic acids. In some cases, the cell wall may be fortified with deposits of calcium carbonate or silica. This is permeable to small molecules and ions, but impermeable to macromolecules. To the exterior of the cell may be one or two flagella, with the typical eucaryotic 9 + 2 microstructure (see Figure 3.18), which may allow unicellular types to move through the water; cilia are not found in any algae.

The characteristics used to place algal protists into different taxa include the type of chlorophyll present, the form in which carbohydrate is stored, and the structure of the cell wall (Table 9.1). A group not considered here are the cyanophytes, previously known as the blue-green algae; although they carry out oxygenic photosynthesis, they are procaryotes, and as such are more closely related to certain bacteria. They are therefore discussed in Chapter 7.

Euglenophyta

> A pellicle is a semi-rigid structure composed of protein strips found surrounding the cell of many unicellular protozoans and algae

This is a group of unicellular flagellated organisms, which probably represent the most ancient group of algal protists. Individuals range in size from 10–500 μm. Euglenophytes are commonly found in fresh water, particularly that with a high organic content, and to a lesser extent, in soil, brackish water and salt water. Members of this group have a well-defined nucleus, and chloroplasts containing chlorophylls *a* and *b* (Figure 9.1). The storage product of photosynthesis is a β-1,3-linked glucan called *paramylon*, found almost exclusively in this group. Euglenophytes lack a cellulose cell wall but have instead, situated within the plasma membrane, a flexible *pellicle* made up of interlocking protein strips, a characteristic which links them to certain protozoan species. A further similarity is the way in which locomotion is achieved by the undulation of a terminal flagellum. Movement towards a light source is facilitated in many euglenids by two structures

Table 9.1 Characteristics of major algal groups

	Common name	Morphology	Pigments	Storage compound	Cell wall
Euglenophyta	Euglenids	Unicellular	Chlorophyll *a* & *b*	Paramylon	None
Pyrrophyta	Dinoflagellates	Unicellular	Chlorophyll *a* & *c*, xanthophylls	Starch	Cellulose/ none
Chrysophyta	Golden-brown algae, diatoms	Unicellular	Chlorophyll *a* & *c*	Lipids	Cellulose, silica, CaCO3, etc.
Chlorophyta	Green algae	Unicellular to multicellular	Chlorophyll *a* & *b*	Starch	Cellulose
Phaeophyta	Brown algae	Multicellular	Chlorophyll *a* & *c*, xanthophylls	Laminarin	Cellulose
Rhodophyta	Red algae	Mostly multicellular	Chlorophyll *a* & *d*, phycocyanin, phycoery-thrin	Starch	Cellulose

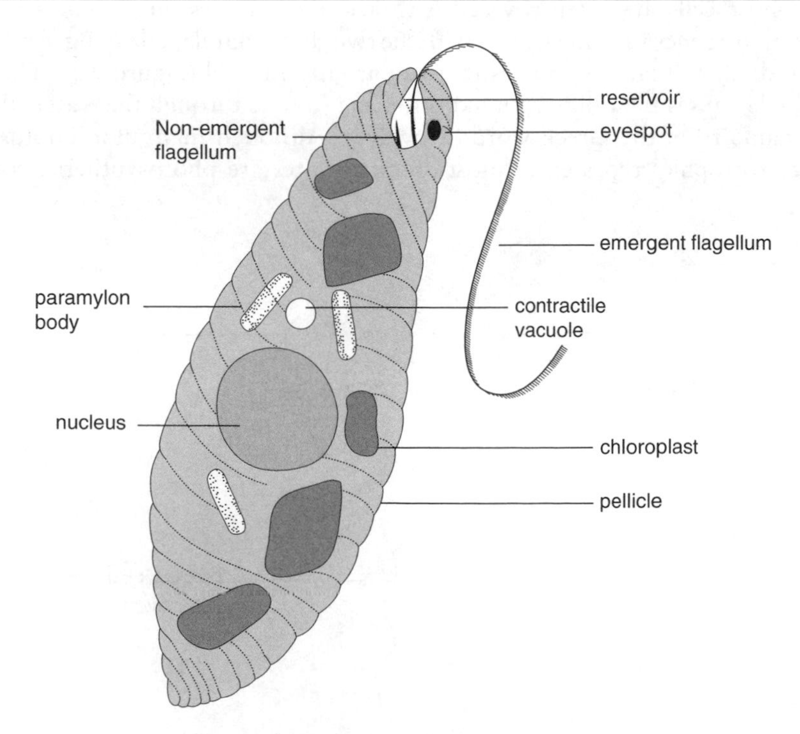

Figure 9.1 *Euglena. Euglena* has a number of features in common with the zooflagellates (see Figure 9.11), but its possession of chloroplasts has meant that it has traditionally been classified among the algae

situated near the base of the flagellum; these are the *paraflagellar body* and the *stigma* or *eyespot*. The latter is particularly conspicuous, as it is typically an orange-red colour, and relatively large.

Reproduction is by binary fission (i.e. by asexual means only). Division starts at the anterior end, and proceeds longitudinally down the length of the cell, giving the cell a characteristic 'two-headed' appearance. During mitosis, the chromosomes within the nucleus replicate, forming pairs that split longitudinally. Since the euglenophyte is usually haploid, it thus becomes diploid for a short period. As fission proceeds, one daughter cell retains the old flagellum, while the other one generates a new one later. As in the binary fission of bacteria, the progeny are genetically identical, i.e. clones. When conditions are unfavourable for survival due to failing nutrient supplies, the cells round up to form cysts surrounded by a gelatinous covering; these have an increased complement of paramylon granules, but no flagella. An important respect in which euglenids may be at variance with the notion of 'plant-like protists' is their ability to exist as heterotrophs under certain conditions. When this happens, they lose their photosynthetic pigments and feed saprobically on dead organic material in the water.

Dinoflagellata

The dinoflagellates (also known variously as Pyrrophyta, or 'fire algae') are chiefly marine planktonic types, comprising some 2000 species. This is another unicellular group, but one whose cells are often covered with armoured plates known as *thecae* (sing: theca). They are generally biflagellate, with the two dissimilar flagella lying in part within the longitudinal and lateral grooves that run around the cell (Figure 9.2). The beating of the flagella causes the cell to spin like a top as it moves through the water (the group takes its name from the Greek word 'to whirl'). Although many non-photosynthetic (chemoheterotrophic) types exist, most dinoflagellates are photosynthetic, containing

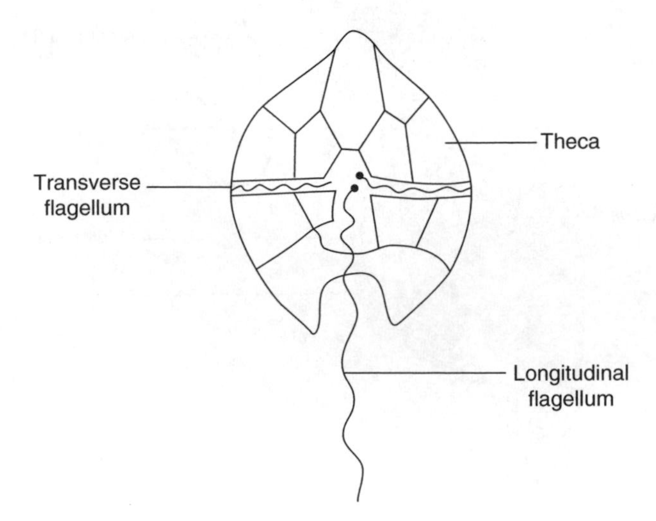

Figure 9.2 A dinoflagellate. Note the two flagella in perpendicular grooves. Each plays its part in the organism's locomotion

Box 9.1 Why would an alga want to glow in the dark?

The production of bioluminescence by several dinoflagellate species is thought to have a protective function. Such algae are the natural prey of copepods, tiny crustaceans found in astronomical numbers as part of the zooplankton. The bioluminescence could have an effect directly, by acting as a warning signal to the copepods, or indirectly, by making those crustaceans which had consumed glowing algae much more conspicuous to their own predators.

chlorophyll *a* and *c* plus certain carotenoids and xanthophylls, which give them a red/golden appearance. As a group, they are second only to the diatoms (see below) as the primary photosynthetic producers in the marine environment. Some dinoflagellates form endosymbiotic relationships with marine animals such as corals and sea anemones; these are termed *zooxanthellae*. An unusual feature of dinoflagellate ultrastructure is that the chromosomes contain little, if any, histone protein, and exist almost permanently in the condensed form.

Some tropical species of dinoflagellate emit light, the only algae to do so (see Box 9.1). Due to an enzyme–substrate (luciferin–luciferase) interaction, this can cause a spectacular glow in the water at night, especially when the water is disturbed, for example by a ship. *Bioluminescence* of this kind has proved to be a useful 'tagging' system for cells in biological research. Other marine dinoflagellates can produce metabolites that act as nerve toxins to higher animals. Shellfish such as mussels and oysters can concentrate these with no harm to themselves, but they can be fatal to humans who consume them. Sometimes, when conditions are highly favourable, an explosion of growth results in the development of huge 'red tides' of dinoflagellates in coastal waters. This produces a build-up of toxins, and may lead to the death of massive numbers of fish and other marine life. The greatly increased incidence of these blooms in recent decades is probably due to pollution by fertilisers containing nitrates and phosphates.

Reproduction by asexual means involves binary fission. In armoured forms, the theca may be shed before cell division, or split along suture lines; in either case, daughter cells must regenerate the missing sections. Sexual reproduction is known to occur in some dinoflagellates, and is probably more widespread. Gametes produced by mitosis fuse to produce a diploid zygote; this undergoes meiosis to reinstate the haploid condition in the offspring. In some species we see *isogamy*, the fusion of identical, motile gametes, while in others, *anisogamy* occurs, in which gametes of dissimilar size fuse. Fusion may occur between genetically identical gametes, or only when the gametes come from genetically distinct populations.

Diatoms

The diatoms, which belong to the division Chrysophyta (the golden-brown algae), make up the majority of phytoplankton in marine food chains, and as such are the most important group of algal protists in terms of photosynthetic production. Over 10 000 species of diatom are recognised, but some experts feel that the real number is many times greater than this.

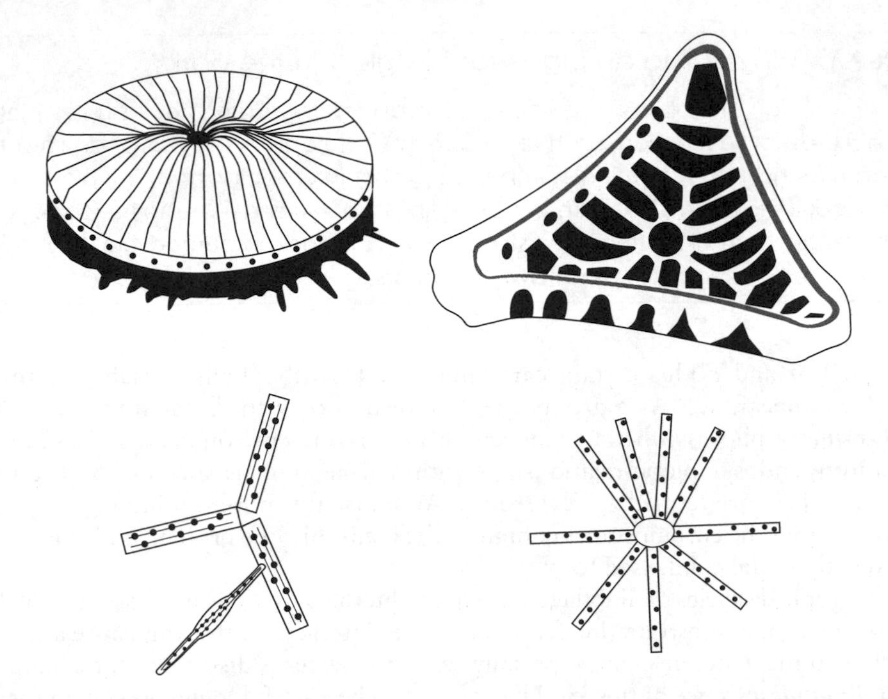

Figure 9.3 Diatoms are covered by an intricate two-part silicaceous shell called a frustule, whose often highly striking appearance make them among the most beautiful of microorganisms

As with the dinoflagellates, chlorophylls *a* and *c* are present, but not chlorophyll *b*. Their colour is due to carotenoids and xanthophylls (particularly fucoxanthin) masking the chlorophyll.

Diatoms have their cells surrounded by a silica-based shell known as a *frustule*, composed of two overlapping halves (the epitheca and the hypotheca). Microbiologists are rarely able to resist the temptation to liken this structure to that of a petri dish, and with good reason. With the electron microscope it can be seen that the frustule is perforated with numerous tiny pores that connect the protoplast of the cell with the outside environment. Diatom classification is based almost entirely on the shape and pattern of these shells, which are uniform for a particular species, and often have a very striking appearance (Figure 9.3). When diatoms die, their shells fall to the bottom of the sea, and can accumulate in thick layers where they represent a valuable mineral resource. This fine, light material (diatomaceous earth) has a number of applications, for example in filtration systems, and also as a light abrasive in products such as silver polish or toothpaste.

Reproduction is usually asexual by binary fission, but a sexual phase with the production of haploid gametes can occur. Chrysophytes are unusual among the three primitive groups of algae in that they are diploid. In diatoms, asexual reproduction involves mitotic cell division, with each daughter cell receiving one half of the parental frustule, and synthesizing a new one to complement it. The newly formed half, however, always acts as the hypotheca (lower half) of the new cell; consequently, one in two daughter cells will be slightly smaller than the parent, an effect which is heightened over a number

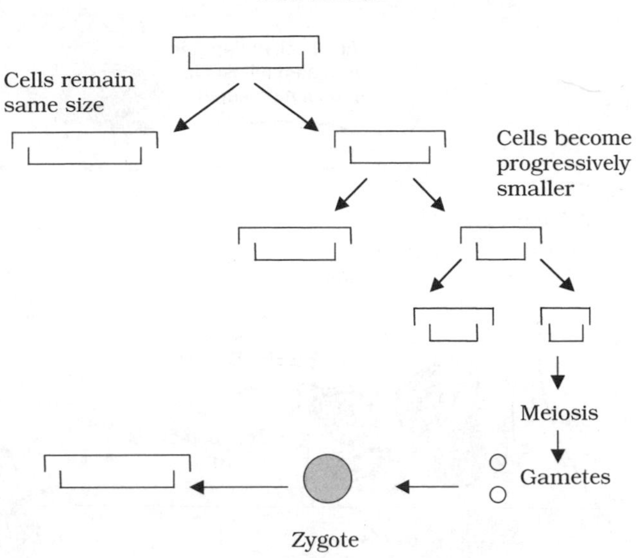

Figure 9.4 Asexual reproduction in diatoms. Two halves of the parent cell separate following mitosis, and a new half develops to fit inside it. Thus, the parental shell always forms the upper (larger half) of the new individual. For half of the population, this means a gradual diminution of size over a number of generations. Eventually, a minimum size is reached, and a sexual cycle is entered. Gametes are produced and the resulting zygote develops into a full-size cell

of generations (Figure 9.4). This process continues until a critical size is reached, and the diatoms undergo a phase of sexual reproduction, which re-establishes the normal frustule size. In species whose frustules have a degree of elasticity, the daughter cells are able to expand, and the problem of cell diminution does not arise. In bilaterally symmetrical (long, thin) diatoms, meiosis in parental cells produces identical, non-motile gametes, which fuse to form a zygote. The radially symmetrical (round) forms provide an example of the third pattern of gamete fusion found in the algae: *oogamy*. Here, there is a clear distinction between the small, motile sperm cell and the larger, immobile egg cell. Both are produced by meiosis in the parental cell, followed, in the case of the male, by several rounds of mitosis, to give a large number of sperm cells.

Chlorophyta

The green algae have always attracted a lot of interest because, as a group, they share a good deal in common with the higher plants in terms of ultrastructure, metabolism and photosynthetic pigments, pointing to the likelihood of a common ancestor. They possess both chlorophyll *a* and *b* and certain carotenoids, store carbohydrate in the form of starch, and generally have a rigid cell wall containing cellulose. The starch is stored in structures called *pyrenoids*, which are found within chloroplasts. There are two phylogenetically distinct lines of green algae, the Charophyta and the Chlorophyta; the latter are much the bigger group, but the charophytes seem to be more closely related to green plants (see Figure 9.18).

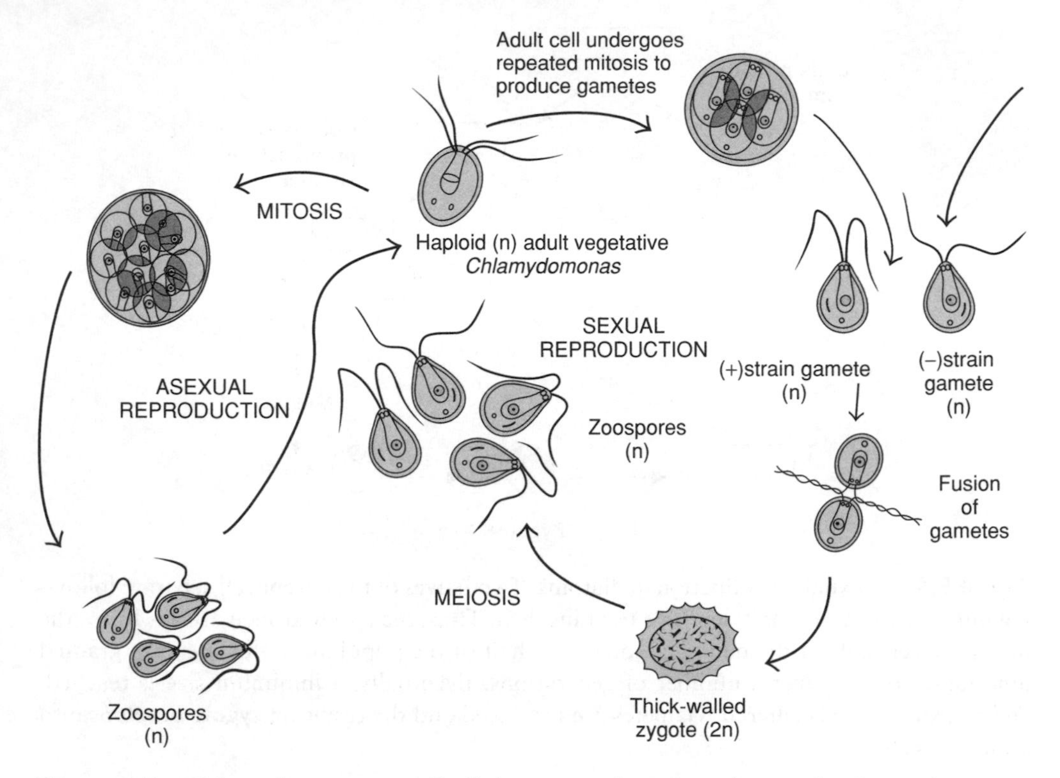

Figure 9.5 *Chlamydomonas*, a unicellular green alga. Sexual reproduction only occurs under adverse conditions, resulting in the production of a resistant zygote. When conditions are favourable, asexual reproduction by means of zoospores predominates

Chlorophytes demonstrate a wide variety of body forms, ranging from unicellular types to colonial, filamentous, membranous and tubular forms. The vast majority of species are freshwater aquatic, but a few marine and pseudoterrestrial representatives exist.

The genus usually chosen to illustrate the unicellular condition in chlorophytes is *Chlamydomonas* (Figure 9.5). This has a single chloroplast, similar in structure and shape to that of a higher plant, and containing a pyrenoid. Situated together at the anterior end is a pair of smooth or whiplash flagella, whose regular, ordered contractions propel it through the water. A further structural feature found in *Chlamydomonas* and other motile forms of green algae is the stigma or eye-spot; this is made up of granules of a carotenoid pigment and is at least partially responsible for orienting the cell with respect to light.

Reproduction in *Chlamydomonas* and other unicellular types under favourable conditions of light, temperature and nutrients, occurs asexually by the production of *zoospores*. A single haploid adult loses its flagella and undergoes mitosis to produce several daughter cells, which then secrete cell walls and flagella and take up an independent existence of their own. This can result in a tremendous increase in numbers; a single cell can divide as many as eight times in one day. Sexual reproduction in *Chlamydomonas*, which occurs when conditions are less favourable, differs

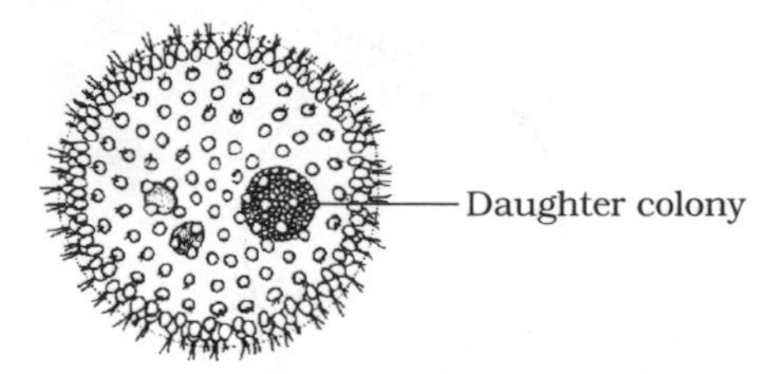

Figure 9.6 *Volvox*, a colonial green alga. The colony comprises thousands of biflagellated cells embedded in mucilage. Note the presence of daughter colonies, produced by asexual reproduction; these are eventually liberated and assume an independent existence

in detail according to the species (Figure 9.5). Any one of the three variants of gamete production seen in the algae may be seen: isogamy, anisogamy and oogamy. In all cases, two haploid gametes undergo a fusion of both cytoplasm and nuclei to give a diploid zygote. The gametes may simply be unmodified haploid adult cells, or they may arise through mitotic cleavage of the adult, depending on the species. The process of isogamy, where the two gametes are morphologically alike and cannot be differentiated visually, only occurs in relatively lowly organisms such as *Chlamydomonas*. In some species we see the beginnings of sexual differentiation – there are two mating strains, designated + and −, and fusion will only take place between individuals of opposite strains. The diploid zygote, once formed, often develops into a tough-walled protective spore called a *zygospore*, which tides the organism over conditions of cold or drought. At an appropriate time the zygospore is stimulated to recommence the life-cycle, and meiosis occurs, to produce haploid cells, which then mature into adult individuals.

In *C. braunii*, sexual reproduction is anisogamous; a + strain produces eight microgametes and a − strain produces four macrogametes. In *C. coccifera*, simple oogamy occurs, in which a vegetative cell loses its flagella, rounds off and enlarges; this acts as the female gamete or ovum, and is fertilised by male gametes formed by other cells.

The next level of organization in the green algae is seen in the *colonial* types, typified by *Volvox*. These, like the unicellular types, are motile by means of flagella, and exist as a number of cells embedded in a jelly-like matrix (Figure 9.6). Both the number of cells and the way they are arranged is fixed and characteristic of a particular species. During growth, the number of cells does not increase. In simpler types, all cells seem to be identical but in more complex forms there are distinct anterior and posterior ends, with the stigma more prominent at the anterior, and the posterior cells becoming larger. Reproduction can occur asexually or sexually.

The diversity of body forms in multicellular chlorophytes referred to earlier is matched by that of their life cycles. Two examples are described here.

Oedogonium is a filamentous type. When young it attaches to the substratum by a basal holdfast, but unless it lives in flowing water the adult form is free-floating. Asexual reproduction occurs by means of motile zoospores, which swim free for around an hour before becoming fixed to a substratum and developing into a new filament. In sexual reproduction, the process of sexual differentiation is carried a step further than we have seen so far, with two separate filaments producing gametes from specialised cells called *gametangia* (Figure 9.7). These are morphologically distinct, with the male being termed an *antheridium* and the female an *oogonium*. Gametes (morphologically

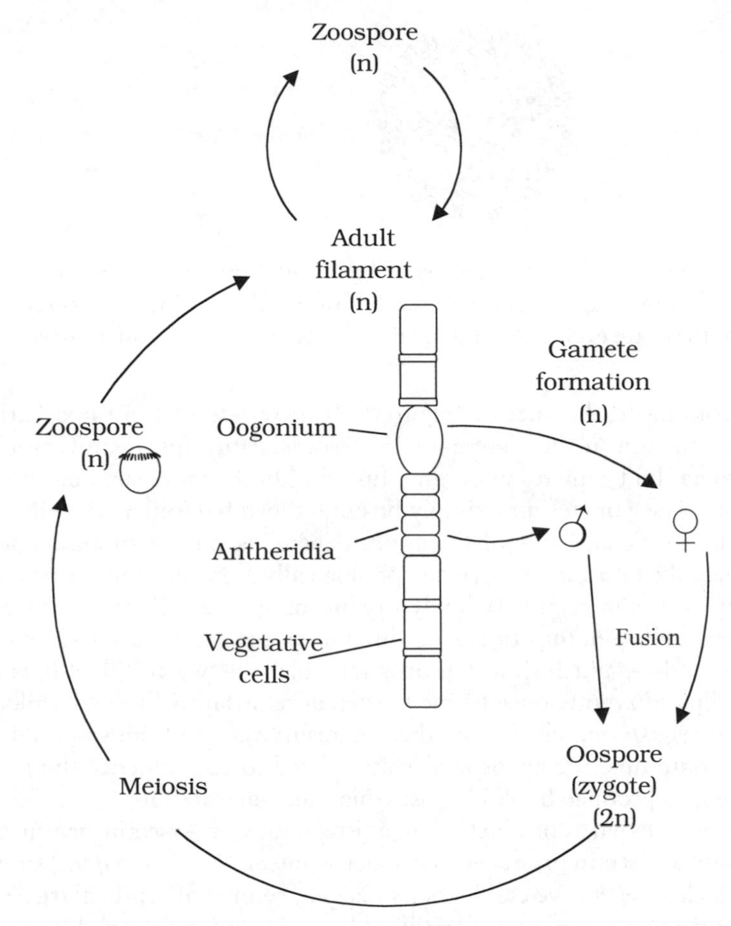

Figure 9.7 The life cycle of *Oedogonium*, a filamentous green alga. In the example shown, gametangia of both sexes are present on the same filament. The oospore (zygote) that results from the fusion of gametes is the only diploid part of the life cycle. Some species of *Oedogonium* have male and female gametangia on separate filaments

distinct: anisogamy) fuse to form a resistant zygote or *oospore*, which, when conditions are favourable, undergoes meiosis to produce four haploid zoospores, each of which can germinate into a young haploid filament. The oospore is thus the only diploid phase in the life cycle. In *Oedogonium* there are species with separate male and female filaments (*dioecious*) as well as ones with both sexes on the same filament (*monoecious*).

A second main form of multicellularity in green algae is the *parenchymatous* state, by which we mean that the cells divide in more than one plane, giving the plant thickness as well as length and width. An example of this is *Ulva*, the sea lettuce, a familiar sight at the seaside in shallow water, attached to rocks or other objects. *Ulva* has a flat, membranous structure, comprising two layers of cells. Reproductively it is of interest because it features alternation of generations, a feature of all the higher green plants. This means that both haploid and diploid mature individuals exist in the life cycle. Gametes are released from one haploid adult and fuse with gametes similarly released from another to form a zygote (Figure 9.8). In most species of *Ulva*, the male and

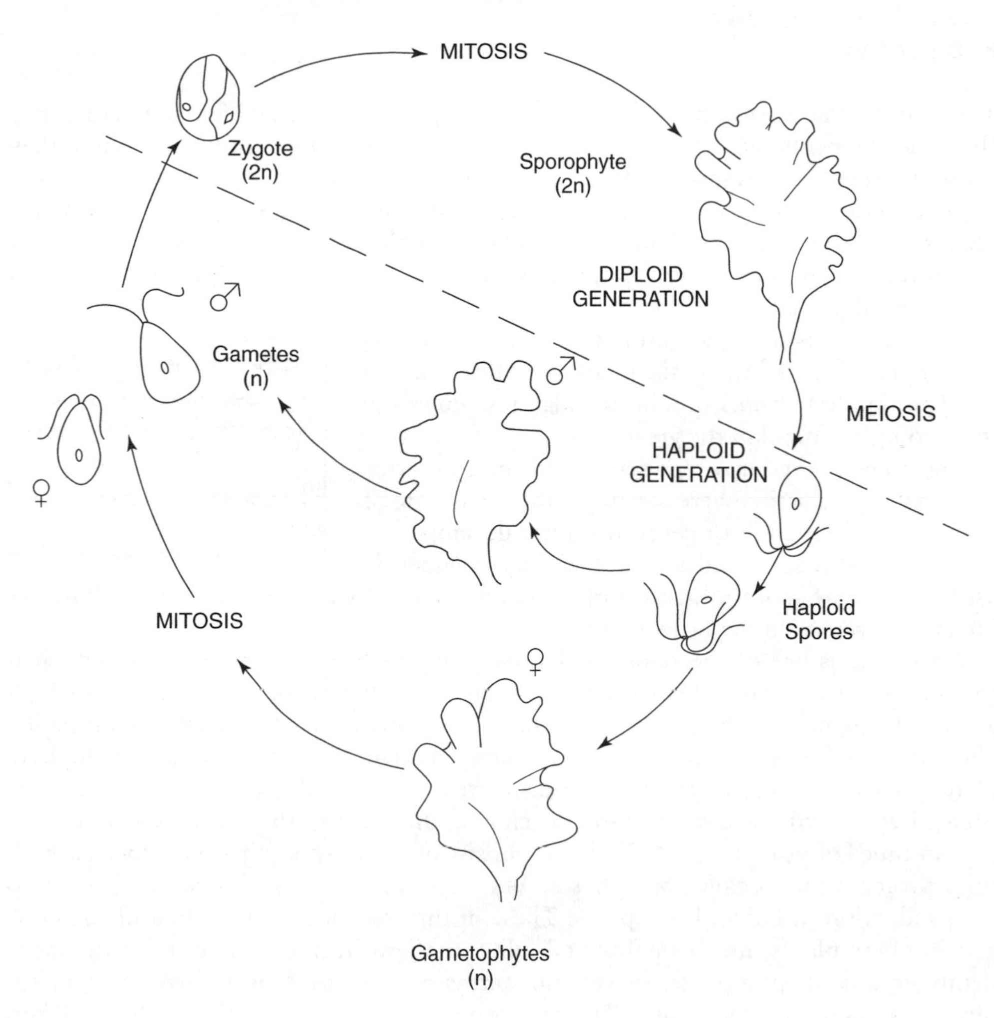

Figure 9.8 Isomorphic alternation of generations: *Ulva*. The life cycle of *Ulva* involves morphologically identical haploid and diploid plants. Fusion of gametes forms a zygote, which grows into the mature diploid plant. Meiosis produces haploid zoospores, which give rise to separate male and female haploid plants

female gametes are morphologically identical (isogamy). The zygote germinates to form a diploid plant, indistinguishable from the plant that produced the gametes, except for its complement of chromosomes. When the diploid plant is mature, it produces haploid zoospores by meiosis, which settle on an appropriate substratum and develop into haploid *Ulva* plants. This form of alternation of generations is called *isomorphic*, because both haploid and diploid forms look alike and each assumes an equal dominance in the life cycle. It is, however, more usual for alternation of generations to be *heteromorphic*, with the *sporophyte* and the *gametophyte* being physically dissimilar, and with one form or other dominating.

> The sporophyte is the diploid, spore-forming stage in a life cycle with alternation of generations.
>
> The haploid, gamete-forming stage is called the gametophyte

Phaeophyta

The brown algae are multicellular, large and complex seaweeds, which dominate rocky shores in temperate and polar regions. Apart from one or two freshwater types, they are all marine. The presence of fucoxanthin masks the presence of chlorophylls *a* and *c*. (In this context it must be stated here that not all 'brown' seaweeds look brown, nor indeed do all the 'red' ones look red). Unlike the higher plants and green algae, which use starch as a food reserve, the phaeophytes use an unusual polysaccharide called laminarin (β-1,3-glucan).

The level of tissue organisation in the brown algae is greatly in advance of any of the types we've discussed so far. The simplest *thallus* of a brown alga resembles the most complex found in the greens.

Thallus is the word used to describe a simple vegetative plant body showing no differentiation into root, stem and leaf.

The phaeophytes also represent an advance in terms of sexual reproduction; here oogamy is the usual state of affairs and alternation of generations has developed to such an extent that diploid and haploid stages frequently assume separate morphological forms. Again, we shall use two examples to illustrate life cycle diversity in the brown algae.

Laminaria is one of the kelps, the largest group of brown algae. It grows attached to underwater rocks or other objects by means of holdfasts, root-like structures which anchor the plant. The thallus is further subdivided into a stalk-like stipe and a broader, blade-like lamina. Reproduction in *Laminaria* involves sporophyte and gametophyte plants that are morphologically quite distinct; (heteromorphic alternation of generations). Reproductive areas called *sori* develop on the blade of the diploid sporophyte at certain times of year (Figure 9.9). These consist of many sporangia, interspersed with thick protective hairs called paraphyses. As the sori develop, meiosis occurs, leading to the production of haploid zoospores. These in turn develop into haploid filamentous gametophyte plants, much smaller and quite different in morphology from the more highly organised sporophyte. Indeed, in contrast to the large sporophyte the gametophyte is a microscopic structure. The gametophytes are dioecious, that is the male and female reproductive structures are borne on separate individuals. The female plant bears a number of *oogonia*, each of which produces a single egg, which escapes through a pore at the apex of the oogonium, but remains attached in a sort of cup, formed by the surrounds of the pore. In similar fashion the male plant bears several antheridia, each liberating a single antherozoid; this however is motile by means of flagella and fertilises the egg. The diploid zygote so produced grows immediately into a new sporophyte plant.

In our second example of a phaeophyte life cycle, there is no alternation of generations at all, the gametophyte generation having been completely lost. The wracks are familiar seaweeds found in the intertidal zone, and *Fucus vesiculosus*, known commonly as the bladder wrack, is one of the best known (Figure 9.10). It gets its name from the air bladders distributed on its surface, which assist buoyancy.

The adult has reproductive structures called receptacles, slight swellings situated at the tip of the thallus; within these are flask-like invaginations called *conceptacles* which contain the male or female gametangia, again interspersed with sterile paraphyses. *F. vesiculosus* is monoecious but some other *Fucus* species are dioecious. Each antheridium

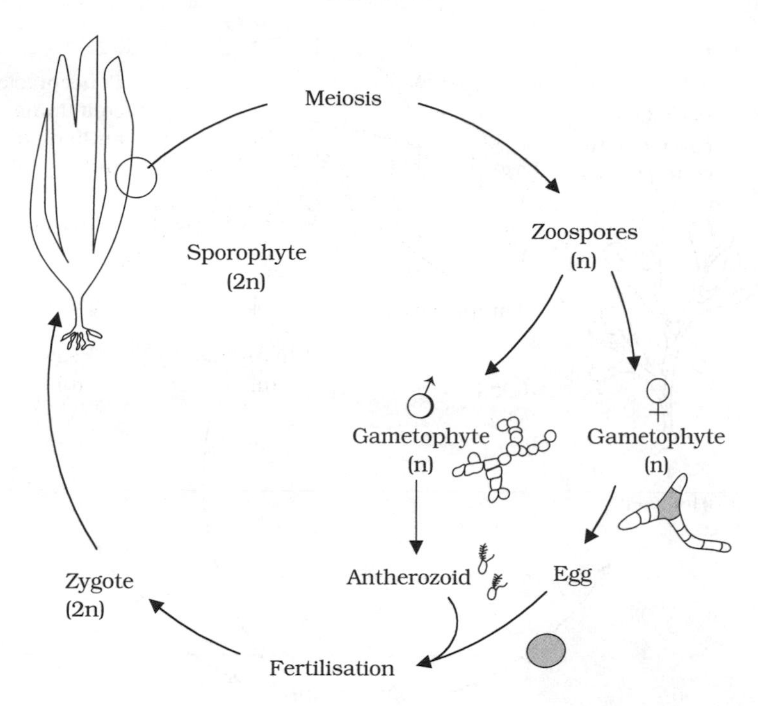

Figure 9.9 Heteromorphic alternation of generations: *Laminaria*. The haploid and diploid generations are morphologically quite distinct: in contrast to the large robust sporophyte, the gametophyte is a tiny filamentous structure

undergoes meiosis, followed by mitosis to produce 64 antherozoids or sperm, while by the same means eight eggs are produced in the oogonium. At high tide these gametes are released into the open water. Fertilisation results in a diploid zygote, which continues to drift quite free, while secreting a mucilaginous covering. It eventually settles, becoming anchored by the mucilage, and germinates into an adult individual. Here then, we have a life cycle in which there is no gametophyte generation, and no specialised asexual reproduction (although in certain conditions fragments may regenerate to form adults).

Rhodophyta

The red coloration of the rhodophytes is due to the pigments phycoerythrin and phycocyanin, which mask the chlorophylls present, in this case *a* and *d*. The biggest single difference between the red algae and the other groups we have looked at so far is that they lack flagella at any stage of their life cycle. Thus they are completely lacking in any motile forms, even in the reproductive stages; the gametes rely on being passively dispersed. Almost all the red algae are multicellular marine species, inhabiting habitats ranging from shallow rock pools to the ocean's deeps.

Life cycles vary considerably, and may be quite complex, with variations on the alternation of generations theme. Several species of the more primitive red algae reproduce

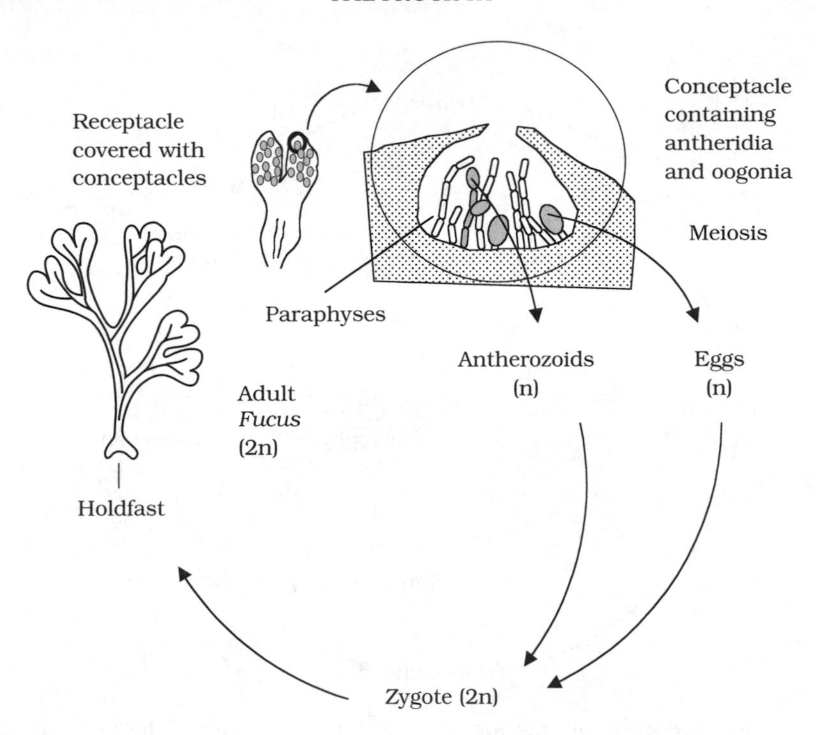

Figure 9.10 *Fucus*: a life cycle with no haploid plant. The only representatives of the haploid state in the life cycle of *Fucus* are the gametes produced by meiosis within the male and female gametangia

asexually by releasing spores into the water. These attach to an appropriate substrate and mature into an adult.

Red algae are the source of several complex polysaccharides of commercial value. Agar and agarose are used in the laboratory in microbial growth media and electrophoresis gels respectively, whilst carrageenan is an important thickening agent in the food industry. In addition, *Porphyra* species are cultivated in Japan for use in sushi dishes.

'The Protozoa'

The name Protozoa comes from the Greek, meaning 'first animal', and was originally applied to single-celled organisms regarded as having animal-like characteristics (multicellular animals were termed Metazoa). Protozoans as a group have evolved an amazing range of variations on the single-celled form, particularly with respect to the different means of achieving movement. They are a morphologically diverse group of well over 50 000 species; although the majority are free-living, the group also includes *commensal* forms and some extremely important parasites of animals and humans.

A commensal lives in or on another organism, deriving some benefit from the association but not harming the other party.

Most protozoans are found in freshwater or marine habitats, where they form a significant component of *plankton*, and represent an important link in the food chain. Although water is essential for the survival of protozoans, many are terrestrial, living saprobically in moist soil.

Plankton are the floating microscopic organisms of aquatic systems.

Remember that a protozoan needs to pack all the functions of an entire eucaryotic organism into a single cell; consequently a protozoan cell may be much more complex than a single animal cell, which is dedicated to a single function. Thus, protozoans display most of the typical features of a eucaryotic cell discussed in Chapter 3, but they may also have evolved certain specialised features. The single cell is bounded by the typical bilayer membrane discussed earlier, but depending on the type in question, this may in turn be covered by a variety of organic or inorganic substances to form an envelope or shell.

One of the most characteristic structural features of protozoans is the *contractile vacuole*, whose role is to pump out excess amounts of water that enter the cell by osmosis. The activity of the contractile vacuole is directly related to the osmotic potential differential between the cell and its surroundings. This is vitally important for freshwater protozoans, since the hypotonic nature of their environment means that water is continually entering the cell. The contractile vacuole often has a star-shaped appearance, the radiating arms being canals that drain water from the cytoplasm into the vacuole.

The contractile vacuole is a fluid-filled vacuole involved in the osmoregulation of certain protists.

Most protozoans have a heterotrophic mode of nutrition, typically ingesting particulate food such as bacteria, and digesting them in phagocytic vacuoles. Since they actively 'hunt' their food rather than simply absorbing it across the cell surface, it is not surprising that the majority of protozoans are capable of movement. The structural features used to achieve locomotion (e.g. cilia, flagella) are among the characteristics used to classify the protozoans.

We shall now examine the characteristics of the principal groupings into which the protozoans have traditionally been divided. It should be repeated, however, that the Protozoa do not represent a coherent taxonomic grouping with a common ancestor, but rather a phylogenetically diverse collection of species with certain features in common. Indeed, each of the four groups is now regarded as having a closer evolutionary relationship with certain 'algal' groups than with each other. See Figure 9.18 for a modern view of how the various taxonomic groupings of protozoans are related.

The zooflagellates (Mastigophora)

Members of this, the biggest and most primitive group of protozoans, are characterised by the long flagellum (*mastigos* = 'a whip'), by which they propel themselves around. Although typical zooflagellates have a single flagellum, some types possess several. The prefix 'zoo-' distinguishes them from plant-like flagellates such as *Euglena*, but as we have already mentioned, such a distinction is not necessarily warranted on molecular and structural grounds (see the end of this chapter).

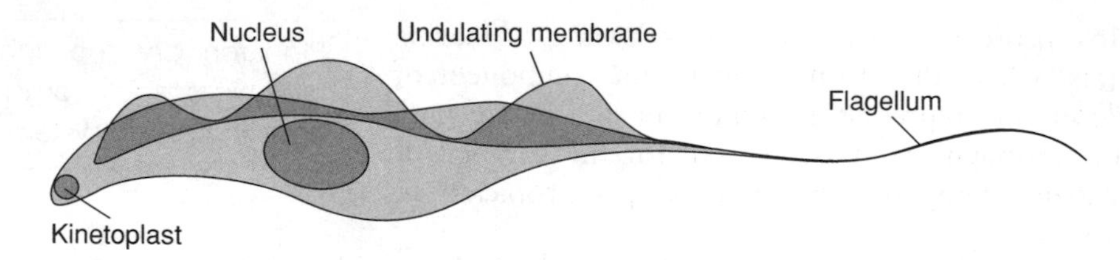

Figure 9.11 *Trypanosoma*, a zooflagellate. The kinetoplast is contained within a single large mitochondrion that runs almost the entire length of the cell (not shown). The flagellum is continuous with the undulating membrane. From Baron, EJ, Chang, RS, Howard, DH, Miller, JN & Turner, JA: Medical Microbiology: A Short Course, John Wiley & Sons Inc., 1994. Reproduced by permission of the publishers

> A *kinetoplast* is a specialised structure within the mitochondria of certain flagellated protozoans, and is the site of their mitochondrial DNA.

Zooflagellates may be free-living, symbiotic or parasitic. An example of the latter is the causative agent of African sleeping sickness in humans, *Trypanosoma brucei* (Figure 9.11). This belongs to the kinetoplastids, a group characterised by the possession of a unique organelle called the *kinetoplast*, found within the cell's single, large, tubular mitochondrion, and containing its own DNA. The flagellum extends back to form the edge of a long, *undulating membrane* that gives *Trypanosoma* its characteristic locomotion.

The infectious form of *T. brucei* develops in the salivary glands of the intermediate host, the tsetse fly, and is passed to the human host when a bite punctures the skin. Here, it eventually reaches the central nervous system by way of the blood or lymphatic systems. Inflammation of the brain and spinal cord results in the characteristic lethargy, coma and eventual death of the patient.

Reproduction in the zooflagellates is generally by binary fission. In the case of *T. brucei*, this occurs both in the human host and in the gut of the tsetse fly, from which it migrates to the salivary glands to complete its life-cycle.

The choanoflagellates (Figure 9.12) are a group of zooflagellates of particular interest, as it is thought that they represent the closest single-celled relatives of animals. They possess a 'collar' of microvilli that surrounds the base of a single flagellum, an arrangement that is also seen in the simplest multicellular animals, the sponges. This connection is made even more apparent in colonial forms of choanoflagellates, and is supported by molecular evidence. Also, both choanoflagellates and animals share the flat, lamellar type of cristae in their mitochondria.

A third group of zooflagellates worthy of note are the diplomonads. These have two nuclei per cell and multiple flagella, but their most remarkable feature is that they do not apparently possess any mitochondria (but see Chapter 3). *Giardia lamblia*, the causative agent of the intestinal disease giardiasis, is a member of this group. Unlike *Trypanosoma*, *Giardia* does not have a secondary host, but survives outside the body as a resistant cyst, before it is taken up again in infected water. The diplomonads occupy a very distant branch of suggested phylogenetic trees from the kinetoplastids and choanoflagellates (see Figure 9.18). The parabasalians are another group of amitochondriate flagellates, whose best-known member is *Trichomonas vaginalis*, a cause of infections of the female urogenital tract.

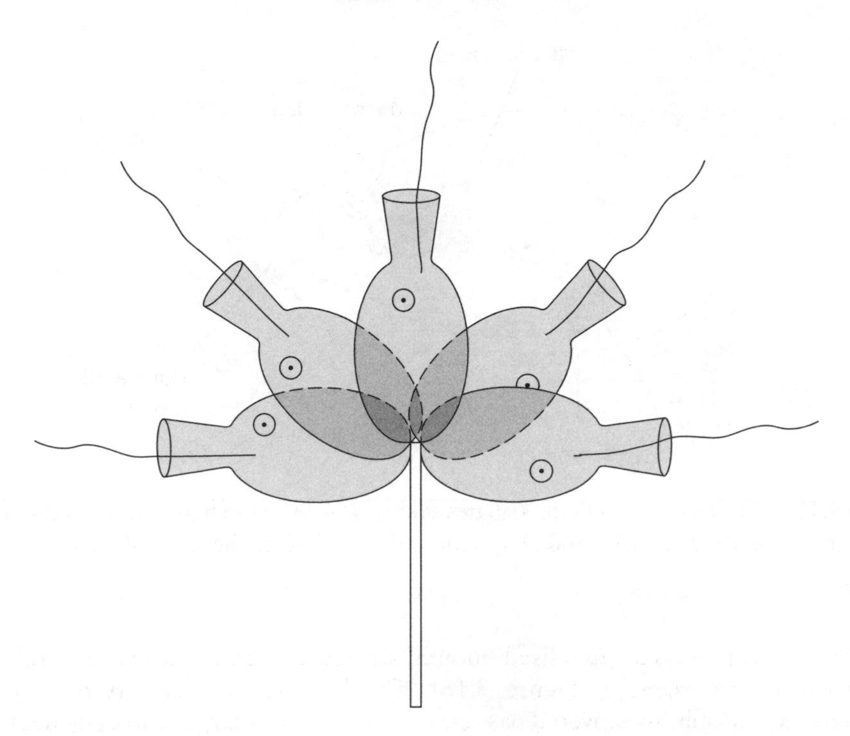

Figure 9.12 Choanoflagellates are free-living zooflagellates characterised by a collar of tentacles or microvilli that surround the single flagellum. Choanoflagellates are often colonial, as in the example shown

The ciliates (Ciliophora)

The largest group of protozoans, the ciliates, are also the most complex, showing the highest level of internal organisation in any single-celled organism. Most are free-living types such as *Paramecium* (Figure 9.13), and as the name suggests, they are characterised by the possession of cilia, which may be present all over the cell surface or arranged in rows or bands. They beat in a co-ordinated fashion to propel the organism, or assist in the ingestion of food particles.

A unique feature of the ciliates is that they possess two distinct types of nuclei (Box 9.3):

- *macronuclei* are concerned with encoding the enzymes and other proteins required for the cell's essential metabolic processes. They are polyploid, containing many copies of the genome.

- *micronuclei*, of which there may be as many as 80 per cell, are involved solely in sexual reproduction by conjugation.

As might be expected, removal of the macronucleus leads quickly to the death of the cell; however, cells lacking micronuclei can continue to live, and reproduce asexually by binary fission.

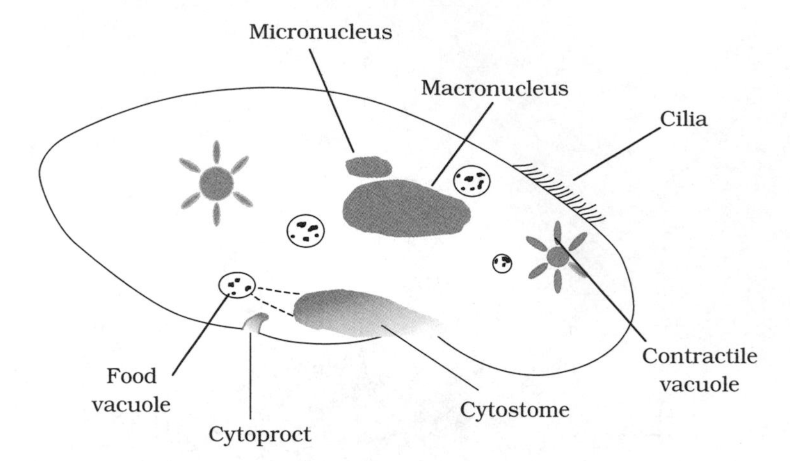

Figure 9.13 *Paramecium*, a ciliate. Ciliates such as *Paramecium* have specialised structures for the ingestion food particles and elimination of waste. Note the rows of cilia covering the surface

Most ciliates possess a specialised 'mouth' structure, the *cytostome*, through which food particles are ingested (Figure 9.13). The beating of cilia directs them to a cytopharynx, a membrane-covered passage or tube, which enlarges and detaches to form a food vacuole. Fusion with lysosomes and digestion by enzymes occurs as described earlier. Undigested particles are ejected from a region on the surface (the anal pore or *cytoproct*).

As well as cilia, some members of the group have *trichocysts* projecting from the cell surface, harpoon-like structures that can be used for attachment or defence.

Some ciliated protozoans are anaerobic, such as those found in the rumen of cattle. The only ciliate known to cause disease in humans is *Balantidium coli*, which causes a form of dysentery.

The amoebas (Sarcodina)

The amoebas are characterised by the possession of *pseudopodia* (='false feet'), temporary projections from the cell into which cytoplasm flows until the organism has moved forward (Figure 9.14). This means that amoebas are continually changing their body shape and the position of their internal organelles. Pseudopodia are also used to capture and engulf food, forming a vacuole around it. Once again, digestive enzymes are released from lysosomes and the food particle dissolved. Once absorption of soluble nutrients has taken place, undigested waste is ejected by the vacuole moving back to the cell surface.

Reproduction in the amoebas is by simple binary fission. Most amoebas are free-living, in aquatic environments; their mode of movement and feeding makes them well adapted to life on the bottom of ponds and lakes, where there is a good supply of prey organisms and suspended organic matter. Also included in the group, however, are some important parasites, including *Entamoeba histolytica*, which causes amoebic dysentery

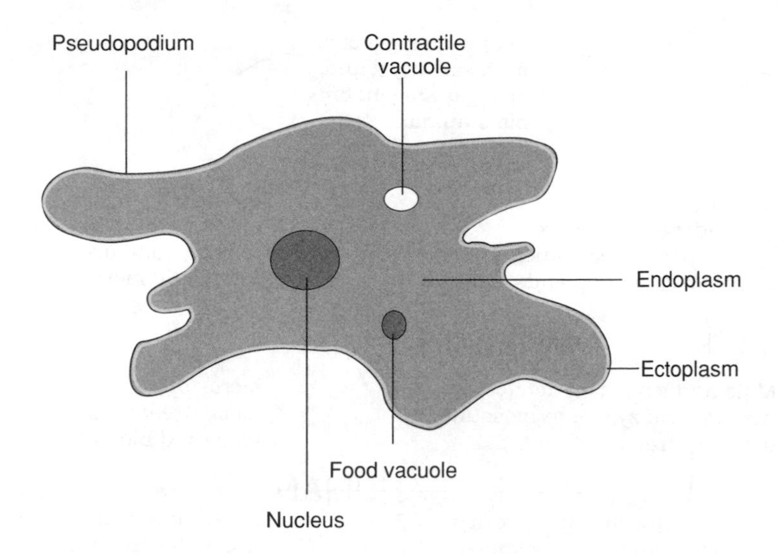

Figure 9.14 Amoeboid structure. The internal features of an amoeba change position as the cell changes its shape by cytoplasmic streaming

in humans. Ingested in faecally contaminated water, it is responsible for some 50 000 –100 000 deaths world-wide every year. Unlike its free-living relatives, *Entamoeba* is unable to reproduce outside of its host.

Amoebas with external shells: Foraminifera and Radiolaria

Some types of amoeba have an external shell covering the cell. The Foraminifera secrete a shell of protein coated with calcium carbonate; their shells are covered with pores through which their long, filamentous pseudopodia project. Some foraminiferans are zooplankton, microscopic organisms living at the surface of the sea, while others are bottom-dwellers. It is the discarded shells of countless long-dead creatures such as these that make up deposits of limestone many hundreds of metres in depth. Many thousands of shells are needed to form just 1 g of such a deposit! The White Cliffs of Dover are an example of a limestone deposit made up largely of foraminiferan shells.

The outer surface of radiolarians is composed of silica, which again is perforated to allow the passage of many very fine pseudopodia.

The sporozoans (Apicomplexa)

Members of this group are all parasitic, infecting a range of vertebrates and invertebrates. They have complex life cycles involving both haploid and diploid phases and infecting more than one host. Probably the best known is *Plasmodium*, the causative agent of malaria, which spends part of its life in a species of mosquito (Figure 9.15). Sporozoans are characterised by a spore-like stage called a *sporozoite*, which is involved

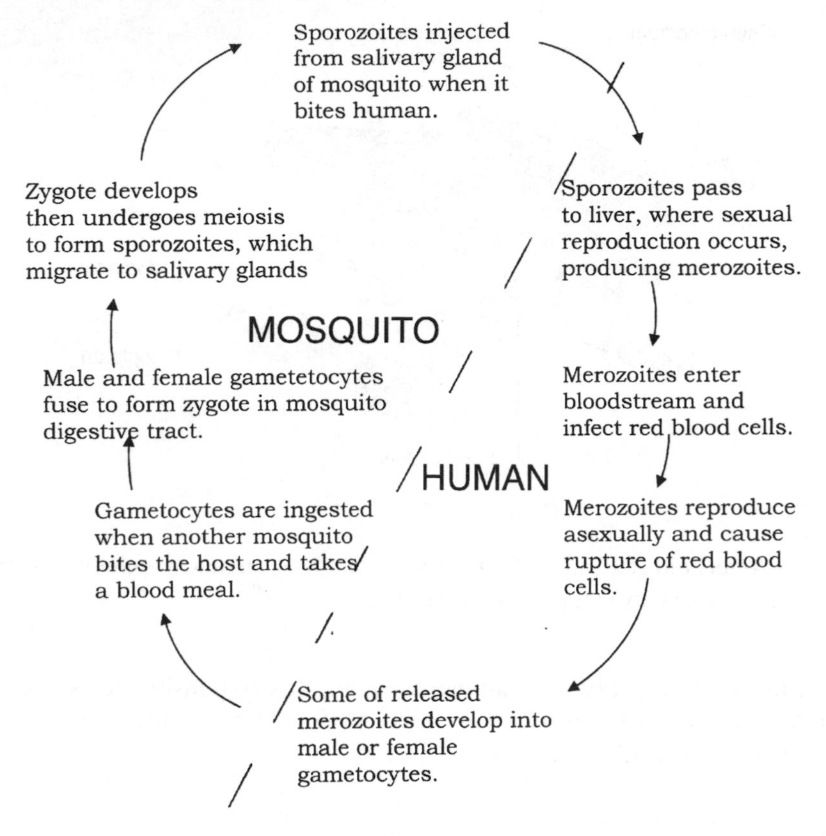

Figure 9.15 *Plasmodium*, the causative agent of malaria, has two hosts. Four different species of *Plasmodium* cause malaria, the most widespread human infectious disease. Each is transmitted by the *Anopheles* mosquito. Asexual reproduction (schizogeny) takes place in the human host; the sexual cycle occurs in the mosquito, following ingestion of blood containing gametocytes

in the transmission of the parasite to a new host. The tip of the sporozoite contains a complex of structures that assist in the penetration of the host's tissues. Unlike the protozoans discussed above, sporozoans are generally non-motile, and absorb soluble nutrients across the cell surface rather than ingesting particulate matter.

A sporozoite is a motile infective stage of members of the Sporozoa that gives rise to an asexual stage within the new host.

The slime moulds and water moulds (the fungus-like protists)

The final group to consider in this chapter are the so-called 'fungus-like' protists. Its members are phylogenetically diverse, and as we'll see towards the end of this chapter, its two principal groupings, the slime moulds and the water moulds, are placed far apart from each other in modern classification systems.

Oomycota (water moulds)

Water moulds resemble true fungi in their gross structure, comprising a mass of branched hyphae. At the cellular and molecular level however, they bear very little resemblance, and are not at all closely related.

The Oomycota derive their name from the single large egg cell that is fertilised to produce a diploid zygote as part of the sexual reproduction cycle.

Many water moulds play an important role in the decomposition of dead plants and animals in freshwater ecosystems, while others are parasitic on the gills of fish. Terrestrial members of the Oomycota include a number of important plant pathogens, such as rusts and mildews, which can have a devastating effect on crops such as tobacco and potatoes.

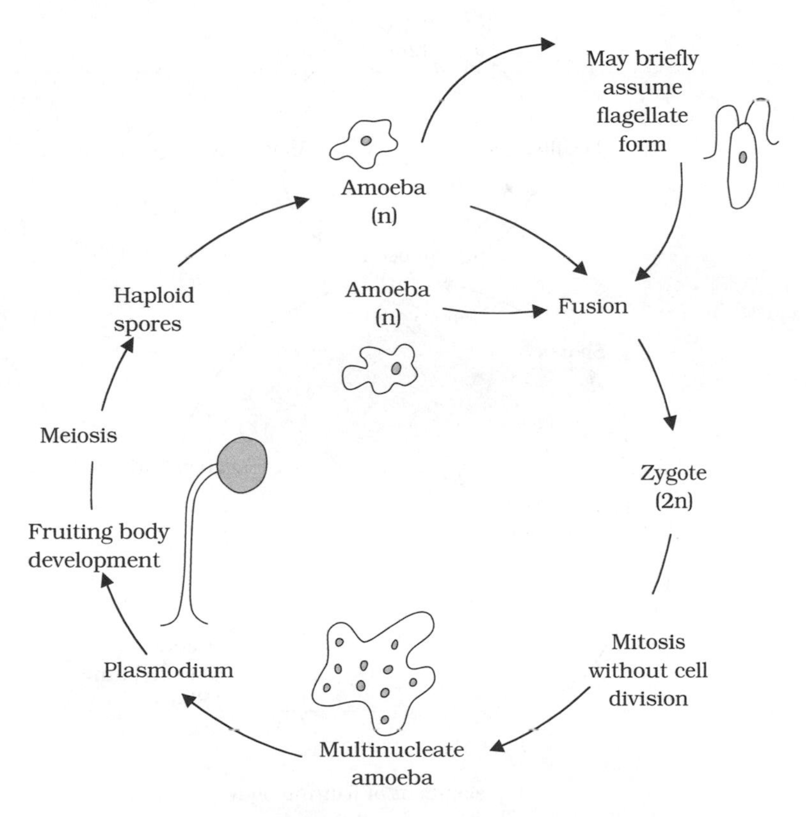

Figure 9.16 The plasmodial slime moulds. Acellular slime moulds such as *Physarum* produce amorphous coenocytic plasmodia which move by amoeboid movement and phagocytically engulf particles of food. Fruiting bodies bearing sporangia release haploid spores, which germinate to form new amoebas

Myxomycota (plasmodial slime moulds)

At one stage in their life cycle, the plasmodial or acellular slime moulds exist as a single-celled amoeboid form. Two of these haploid amoebas fuse to give a diploid cell, which then undergoes repeated divisions of the nucleus, without any accompanying cell division; the result is a *plasmodium*, a mass of cytoplasm that contains numerous nuclei surrounded by a single membrane (Figure 9.16).This retains the amoeboid property of cytoplasmic streaming, so the whole multinucleate structure is able to move in a creeping fashion. This 'feeding plasmodium', which may be several centimetres in length and often brightly coloured, feeds phagocytically on rotting vegetation. Fruiting bodies develop from the plasmodium when it is mature or when conditions are unfavourable, and a cycle of sexual reproduction is entered. When favourable conditions return, meiosis gives rise to haploid spores, which germinate to produce the amoeboid form once more.

> A plasmodium is a mass of protoplasm containing several nuclei and bounded by a cytoplasmic membrane.

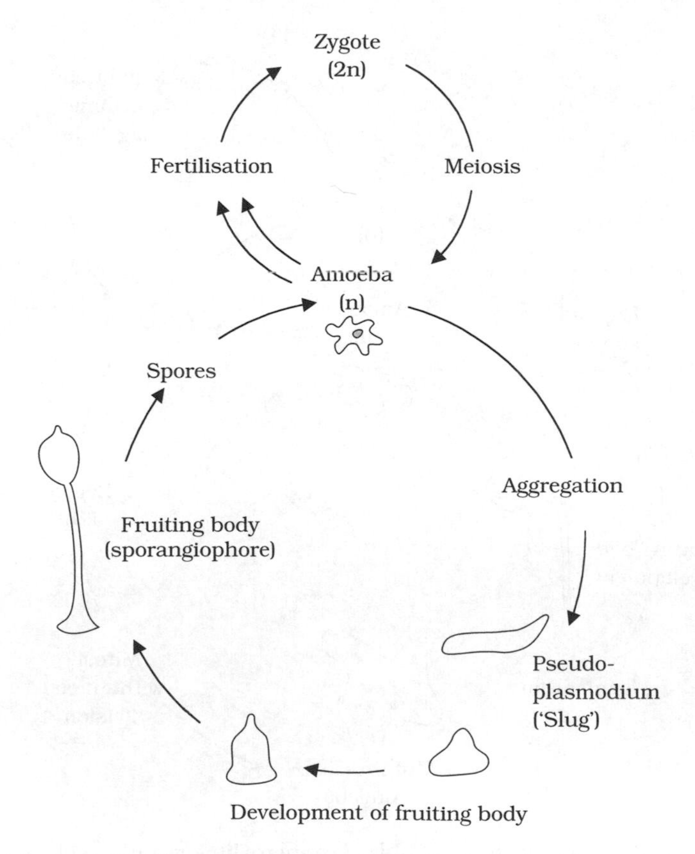

Figure 9.17 The cellular slime moulds. Fruiting bodies develop from the pseudoplasmodium or 'slug' and release haploid spores that develop into individual amoebas. Only haploid forms participate in this cycle, which is therefore asexual. Sexual reproduction can also occur, involving the production of dormant diploid spores called macrocysts. Note that the pseudoplasmodium of cellular slime moulds is entirely cellular

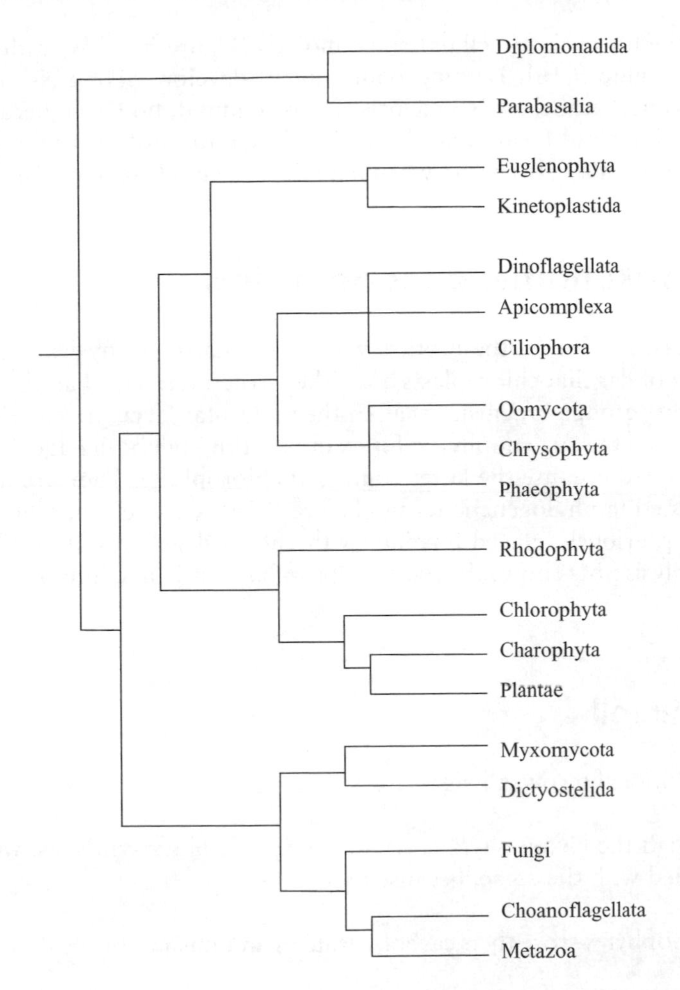

Figure 9.18 A modern view of eucaryotic taxonomy. A possible scheme for the relationship between protistan groups based on 18S RNA data. The positions of the fungi, plants and multicellular animals (Metazoa) are also shown. Note that some protistan groups placed together in traditional schemes (e.g. kinetoplastids and choanoflagellates) are very distant in phylogenetic terms. The diplomonads and parabasilians are shown as diverging from the main stock before the acquisition of mitochondria through endosymbiosis with bacteria. This hypothesis may need to be revised in light of recent evidence that these organisms did once possess mitochondria but have since lost them

Dictyostelida (cellular slime moulds)

A unicellular amoeboid form also figures in the life cycle of the other group of slime moulds, the Dictyostelida (Figure 9.17). This haploid amoeba is the main vegetative form, but when food supplies are scarce, large numbers aggregate to form a slug-like blob, superficially not unlike the plasmodium described above. Unlike the plasmodium, however, this aggregate is fully cellular, so each component cell retains its plasma membrane.

Compare the life cycle of cellular slime moulds (Figure 9.17) with that of the plasmodial kind (Figure 9.16). Fruiting bodies again develop, giving rise to spores that germinate into new amoebas. No meiosis step is required, however, because the whole cycle comprises haploid forms, and this is therefore a form of asexual reproduction. A simple sexual cycle may also occur, when haploid amoebas fuse to give a diploid zygote.

Protistan taxonomy: a modern view

Traditional classification of the protists was on the basis of physical features such as the possession of flagella, chloroplasts and other structures. This had the effect of placing into separate groups organisms that, at the molecular level, are closely related. The kinetoplastids and the euglenophytes, for example, although both flagellate forms, were distantly separated because the latter contained chloroplasts. They are now known to be closely related in phylogenetic terms (Figure 9.18). On the other hand, the kinetoplastids were previously placed together with choanoflagellates as zooflagellates, but molecular analysis now shows the two groups to have little in common in evolutionary terms.

Test yourself

1 In nutritional terms, all algae are ________.

2 Although the Cyanophyta carry out oxygenic photosynthesis, they are not classified with the algae, because they are ——.

3 Euglenophytes store their carbohydrate in an unusual form called ________.

4 The ________ ________ is a reddish structure situated near the flagellum, and is responsible for reacting to light stimuli.

5 Members of the Pyrrophyta are more commonly known as the ________.

6 Diatoms are surrounded by a ________ made of silica, which is perforated by numerous ________.

7 During asexual reproduction in the diatoms, one daughter cell at each division gets progressively smaller. This is stopped by entering a phase of ________ ________.

8 The group of algae thought to be phylogenetically the closest to the green plants is the ________.

9 Multicellular algae are not differentiated into root, stem and leaf; their vegetative body form is called a ________.

10 Sexual reproduction involving clearly differentiated sperm and egg cells is termed ________.

11 Species which have reproductive structures of the two sexes on separate individuals are said to be ________.

12 A ________ alternation of generations involves morphologically distinct sporophyte and gametophyte adult phases.

13 The coloration of the red algae is due to ________ and ________, which mask the presence of ________.

14 Protozoans are a ________ diverse group, which do not share a ________ ________.

15 Osmoregulation in protozoans is carried out by the ________ ________.

16 The kinetoplastids include ________, the causative agent of African sleeping sickness. They show little evolutionary relationship to the other group of flagellated protozoans, the ________.

17 ________ such as *Giardia* lack conventional mitochondria.

18 Ciliates possess two kinds of ________.

19 The ________ are amoebas that have a calcium-based shell.

20 The acellular slime moulds are characterised by the formation of a ________, a mass of cytoplasm containing numerous nuclei.

10
Viruses

In Chapter 1, we saw how, in the late 19th century, one disease after another, in plants as well as in animals, was shown to have a bacterial cause. In 1892, however, the Russian Dimitri Iwanowsky made a surprising discovery concerning the condition known as tobacco mosaic disease. He showed that an extract from an infected leaf retained the ability to transmit the disease to another plant even after being passed through a porcelain filter. This recently developed device was believed to remove even the smallest bacteria, and it was therefore proposed that perhaps the cause of the disease was not an organism, but a filterable toxin. The work of the Dutch botanist Martinus Beijerinck and others around the turn of the century, however, (see Table 10.1) led to the idea of viruses, filterable entities much smaller than bacteria, that were responsible for a wide range of diseases in plants, animals and members of the microbial world.

What are viruses?

All viruses are obligate intracellular parasites; they inhabit a no-man's-land between the living and the non-living worlds, and possess characteristics of both. They are now known to differ radically from the simplest true organisms, bacteria, in a number of respects:

- they cannot be observed using a light microscope
- they have no internal cellular structure
- they contain either DNA or RNA, but not both*
- they are incapable of replication unless occupying an appropriate living host cell
- they are incapable of metabolism
- individuals show no increase in size.

When inside a host cell, viruses show some of the features of a living organism, such as the ability to replicate themselves, but outside the cell they are just inert chemical

* Some viruses have DNA and RNA at different phases of their growth cycle. See p. 253

Table 10.1 Some milestones in the history of virology

1892	Tobacco Mosaic Disease (TMD) shown to be caused by a filterable agent.	Iwanowsky
1898	Proposal that TMD is due to a novel type of infectious agent.	Beijerinck
	Demonstration of first viral disease in animals (foot and mouth).	Loeffler & Frosch
1901	Demonstration of first human viral disease (yellow fever).	Reed
1915/1917	Discovery of bacterial viruses (bacteriophages).	Twort, d'Herelle
1918	Spanish influenza pandemic	
1935	TMV is first virus to be crystallised.	Stanley
1937	Separation of TMV into protein and nucleic acid fractions.	Bawden & Pirie
1939	Viruses visible under electron microscope	Kausche, Pfankuch & Ruska
1955	Spontaneous reassembly of TMV from protein and RNA components.	Fraenkel-Conrat & Williams
1971	Discovery of viroids.	Diener
1980	Sequencing of first complete viral genome (CaMV)	Frank
1982	Sequencing of first RNA genome (TMV)	
	Recombinant Hepatitis B vaccine	
	Discovery of prions	Prusiner
1983	Discovery of HIV, thought to be causative agent of AIDS	Montaigner and Gallo
1990	Retrovirus used as vector in first human gene therapy trial.	Anderson
2001	BSE outbreak in UK	
2003	Outbreak of new human viral disease (SARS) in SE Asia	

TMV, Tobacco mosaic virus; CaMV, Cauliflower mosaic caulimovirus.

structures, thus fuelling the debate as to whether they can be considered to be life forms. A particular virus has a limited host range, that is, it is only able to infect certain cell types. Nobody is sure how viruses evolved; Box 10.1 describes some current ideas.

Viral structure

The demonstration by Wendel Stanley in 1935 that a preparation of tobacco mosaic virus could be crystallised was an indication of the relative chemical homogeneity of viruses, and meant that they could not be thought of in the same terms as other living things. Compared to even the most primitive cellular organism, viruses have a very simple structure (Figure 10.1). An intact viral particle, or *virion*, has in essence just two components: a core of nucleic acid, surrounded and protected by a protein coat or

Box 10.1 Where do viruses come from?

Three major mechanisms have been proposed for the evolution of viruses:

- *'Escaped gene' theory*: Viruses derive from normal cellular nucleic acids and 'gain independence' from the cell. DNA viruses could come from plasmids or transposable elements (see Chapter 12), while RNA viruses could derive from mRNA.
- *Regressive theory*: Gradual degeneration of procaryotes living parasitically in eucaryotic cells. Enveloped forms such as poxviruses are most likely to have been formed in this way.
- *Coevolution theory*: Independent evolution alongside cellular forms from primordial soup.

Some scientists consider it unlikely that the same mechanism could account for the diversity of viruses we see today, and therefore propose that viruses must have evolved many times over. A study published in 2004 conversely proposes that all viruses share a common ancestor and may even have developed before cellular life forms.

capsid, the combination of the two being known as the *nucleocapsid*. In certain virus types, the nucleocapsid is further surrounded by a membranous *envelope*, partly derived from host cell material. Most viruses are smaller than even the smallest bacterial cells; Figure 10.2 shows the size of some viruses compared to that of typical bacterial and eucaryotic cells.

The viral genome

The genetic material of a virus may be either RNA or DNA, and either of these may be single-stranded or double-stranded (Figure 10.3). As shown in Figure 10.4, the genome may furthermore be circular or linear. An additional variation in the viral genome is

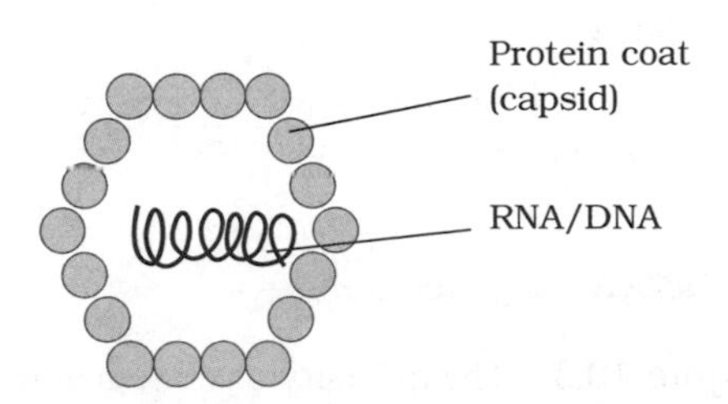

Figure 10.1 Viral structure. Viruses comprise a nucleic acid genome surrounded by a protein coat (capsid). Both naked and enveloped forms are shown

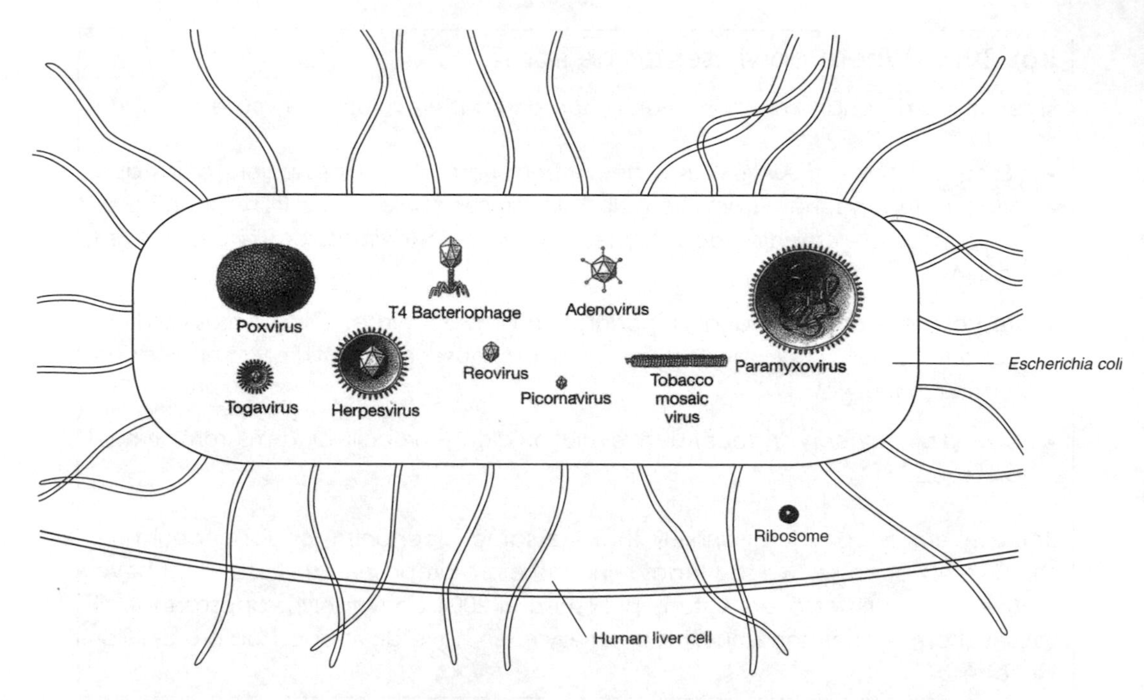

Figure 10.2 Viruses are much smaller than cells. The viruses shown are drawn to scale and compared to an *E. coli* cell and a human liver cell. As a guide, *E. coli* cells are around 2 μm in length. From Black, JG: Microbiology: Principles and Explorations, 4th edn, John Wiley & Sons Inc., 1999. Reproduced by permission of the publishers

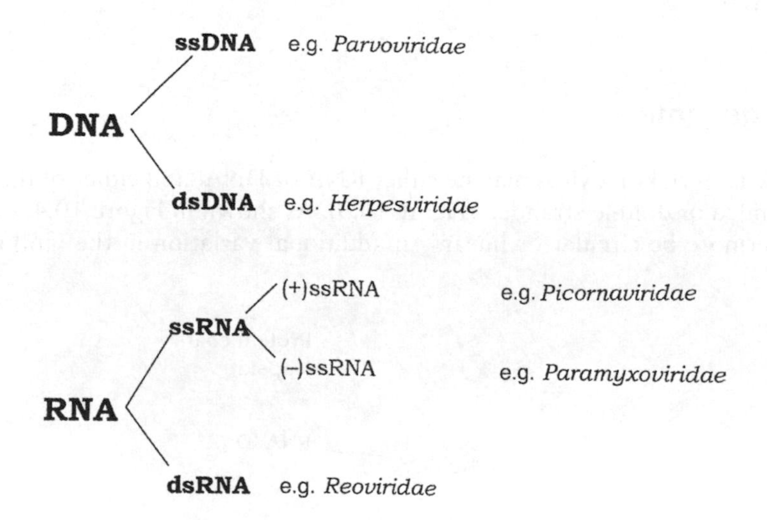

Figure 10.3 The diversity of viral genomes

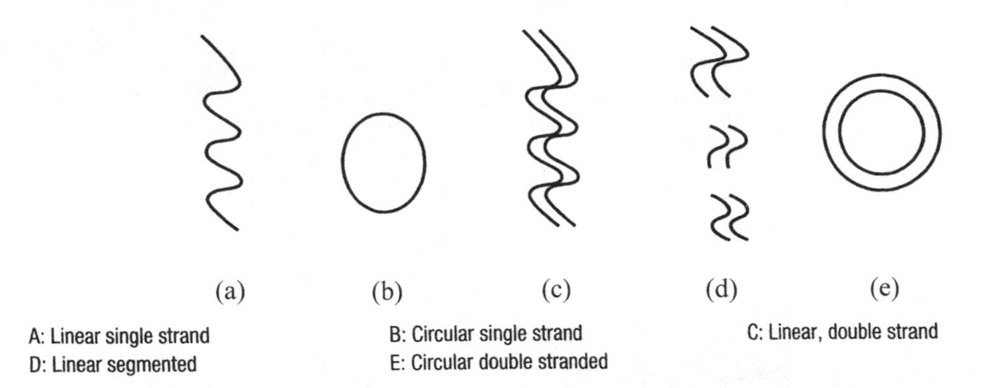

A: Linear single strand
B: Circular single strand
C: Linear, double strand
D: Linear segmented
E: Circular double stranded

Figure 10.4 Viral genomes. Viral genomes may be circular or linear. Some RNA viruses have their genome broken up into segments, each encoding a separate protein (a) Linear single strand; (b) Circular single strand; (c) Linear double strand; (d) Linear segmented; (e) Circular double stranded. From Hardy, SP: Human Microbiology, Taylor and Francis, 2002. Reproduced by permission of Thomson Publishing Services

seen in certain RNA viruses, such as the influenza virus; here, instead of existing as a single molecule, it is segmented, existing as several pieces, each of which may encode a separate protein. In some plant viruses, the segments may be present in separate particles, so in order for replication to occur, a number of virions need to co-infect a cell, thereby complementing each other (multipartite genomes)! Double-stranded RNA is always present in the segmented form.

The size of the genome varies greatly; it may contain as few as four genes or as many as over 200 (see Box 10.2). These genes may code for both structural and non-structural proteins; the latter include enzymes such as RNA/DNA polymerases required for viral replication.

Single-stranded RNA viral genomes can be divided into two types, known as *(+) sense* and *(−) sense RNA*. The former is able to act as mRNA, attach to ribosomes and become translated into the relevant proteins within the host cell. As such, it is infectious in its own right. Minus (−) sense RNA, on the other hand, is only infectious in the presence of a capsid protein possessing RNA polymerase activity. This is needed to convert the (−) RNA into its complementary (+) strand, which then acts as a template for protein production, as described above.

Box 10.2 The mother of invention

A gene in most organisms comprises a discrete linear sequence of DNA with a distinct starting point, which codes for a specific protein product. Some viruses however use the same stretch of DNA for more than one gene. By beginning at different points and using different reading frames, the same code can have a different meaning! These overlapping genes, which are also found in some bacteria, provide an ingenious solution to the problem of having such a small genome size.

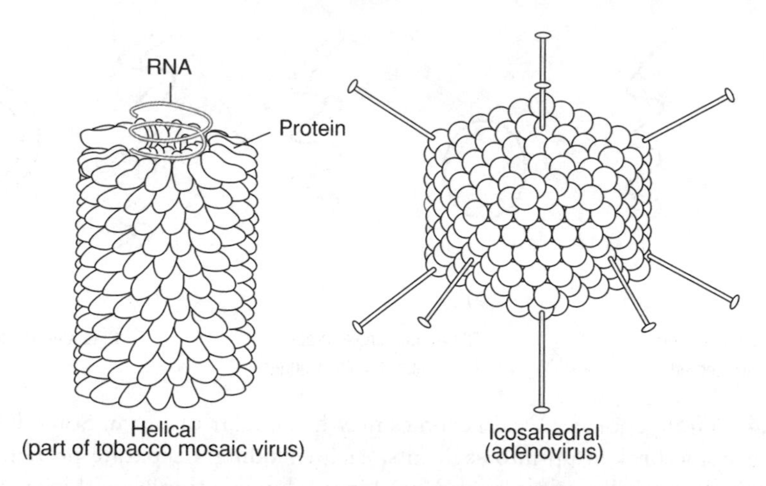

Figure 10.5 Viral capsids have two basic forms, helical and icosahedral. Complex viruses represent a fusion of both forms. From Harper, D: Molecular Virology, 2nd edn, Bios Scientific Publishers, 1998. Reproduced by permission of Thomson Publishing Services

When DNA forms the genome of viruses, it is usually double-stranded (dsDNA), although some of the smaller ones such as the parvoviruses have ssDNA (Figure 10.3).

Capsid structure

The characteristic shape of a virus particle is determined by its protein coat or capsid. In the non-enveloped viruses, the capsid represents the outermost layer, and plays a role in attaching the virus to the surface of a host cell. It also acts to protect the nucleic acid against harmful environmental factors such as UV light and desiccation, as well as the acid and degradative enzymes encountered in the gastrointestinal tract.

The capsid is made up of a number of subunits called *capsomers* (Figure 10.5), and may comprise a few different protein types or just one. The number of capsomers is constant for a particular viral type. This repetitive subunit construction is necessitated by the small amount of protein-encoding RNA/DNA in the viral genome. The capsomers have the ability to interact with each other spontaneously to form the completed capsid by a process of self-assembly. This would be less easily achieved if there were large numbers of different protein types. Capsomers are arranged symmetrically, giving rise to two principal capsid shapes, *icosahedral* and *helical* (Figure 10.5). Both shapes can be found in either enveloped or non-enveloped viruses. Complex viruses, such as certain bacteriophages, contain elements of both helical and icosahedral symmetry.

Helical capsids

A number of plant viruses, including the well-studied tobacco mosaic virus, have a rod-like structure when viewed under the electron microscope (Figure 10.5a). This is caused by a helical arrangement of capsomers, resulting in a tube or cylinder, with room in the centre for the nucleic acid element, which fits into a groove on the inside. The diameter of the helix is determined by the nature of the protein(s) making up the capsomers; its length depends on the size of the nucleic acid core.

Icosahedral capsids
An icosahedron is a regular three-dimensional shape with 20 triangular faces, and 12 points or corners (Figure 10.5). The overall effect is of a roughly spherical structure.

The icosahedron has a low surface-area to volume ratio, allowing for the maximum amount of nucleic acid to be packaged.

The viral envelope

Envelopes are much more common in animal viruses than in those of plants. The lipid bilayer covering an enveloped virus is derived from the nuclear or cytoplasmic membrane of a previous host. Embedded in this, however, are proteins (usually glycoproteins) encoded by the virus's own genome. These may project from the surface of the virion as spikes, which may be instrumental in allowing the virus to bind to or penetrate its host cell (Figure 10.6). The envelope is more susceptible than the capsid to environmental pressures, and the virus needs to remain moist in order to survive. Consequently, such viruses are transmitted by means of body fluids such as blood (e.g. hepatitis B virus) or respiratory secretions (e.g. influenza virus).

Classification of viruses

As we saw at the beginning of this chapter, viruses are not considered to be strictly living, and their classification is a complex issue. As with true organisms we have species, genera, families and orders of viruses, but none of the higher groupings (class, phylum, kingdom). Latin binomials (e.g. *Homo sapiens, Escherichia coli*), familiar

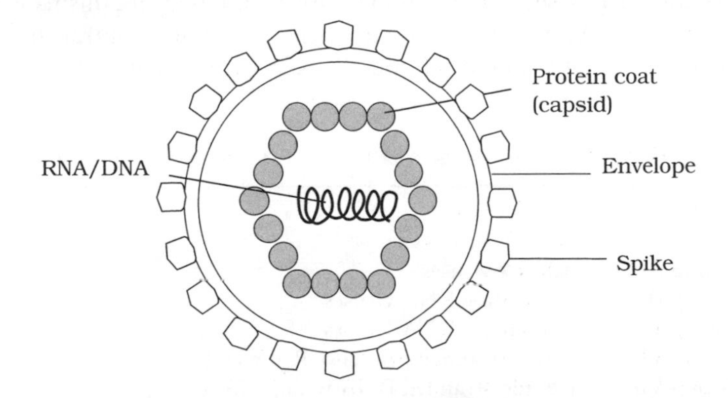

Figure 10.6 An enveloped virus. The envelope derives from the host cell's membrane, and includes virus-encoded proteins. Some viral envelopes contain projecting spikes, which may assist in attachment to the host

from conventional biological taxonomy, are not used for viruses; however, a proposal for non-Latinised viral binomials has been proposed. Originally, no attempt was made to draw up any sort of phylogenetic relationship between the viruses, but more recent developments in sequencing of viral genomes has meant that insights are being gained in this area.

Factors taken into account in the classification of viruses include:

- host range (vertebrate/invertebrate, plant, algae/fungi, bacteria)
- morphology (capsid symmetry, enveloped/non-enveloped, capsomer number)
- genome type/mode of replication (see Figure 10.3).

In 1971, David Baltimore proposed a scheme that orders the viruses with respect to the strategies used for mRNA production. This results in seven major groupings (Table 10.2). The most recent meeting (2005) of the International Commission on Taxonomy of Viruses (ICTV, established in 1973) produced a report which recognises three orders, 73 families, 287 genera and more than 1900 species of virus. Countless others, undiscovered or insufficiently characterised, also exist.

Virus families always end in '*-viridae*', subfamilies in '*-virinae*' and genera in '*-virus*'. Such names are italicised and capitalised, whereas this is not done for species. e.g. Order: *Mononegavirales*, Family: *Paramyxoviridae*, Subfamily: *Paramyxovirinae*, Genus: *Morbillivirus*, Species: measles virus. For informal usage, we would talk about, for example, 'the picornovirus family', or the 'enterovirus genus'.

An indication of just how complex the taxonomy of viruses can be is given by the fact that in 1999, a paper was published in a leading virology journal, entitled: '*How to write the name of virus species*'!

Viral replication cycles

One characteristic viruses share in common with true living organisms is the need to reproduce themselves*. As we have seen, all viruses are obligate intracellular parasites, and so in order to replicate, a host cell must be successfully entered. It is the host cell

Table 10.2 Major groupings of viruses based on the Baltimore system

Group I	dsDNA viruses
Group II	ssDNA viruses
Group III	dsRNA viruses
Group IV	(+) sense ssRNA viruses
Group V	(−) sense ssRNA viruses
Group VI	Single-stranded (+) sense RNA with DNA intermediate
Group VII	Double-stranded DNA with RNA intermediate

* Since the processes involved proceed at the molecular rather than the organismal level, it is more appropriate to speak of viral *replication* than of reproduction.

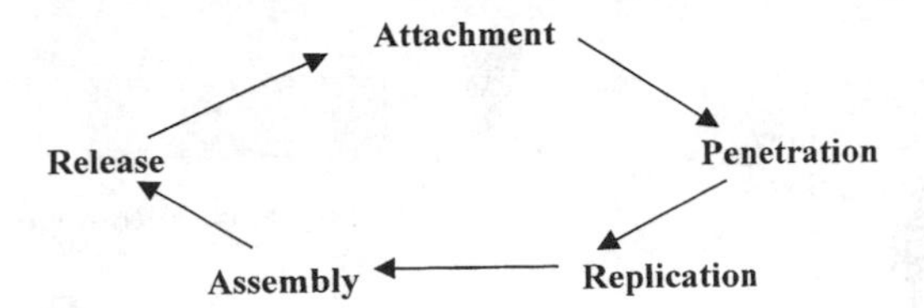

Figure 10.7 Main stages in a viral replication cycle. The replication cycle of all viruses is based on this generalised pattern

that provides much of the 'machinery' necessary for viral replication. All viral growth cycles follow the same general sequence of events (Figure 10.7), with some differences from one type to another, determined by viral structure and the nature of the host cell.

Replication cycles in bacteriophages

Viruses that infect bacterial cells are called *bacteriophages* (phages for short), which means, literally, 'bacteria eaters'. Perhaps the best understood of all viral replication cycles are those of a class of bacteriophages which infect *E. coli*, known as the *T-even phages*. These are large, complex viruses, with a characteristic head and tail structure (Figure 10.8). The double-stranded, linear DNA genome contains over 100 genes, and is contained within the icosahedral head. The growth cycle is said to be *lytic*, because it culminates in the lysis (=bursting) of the host cell. Figure 10.9 shows the lytic cycle of phage T4, and the main stages are described below.

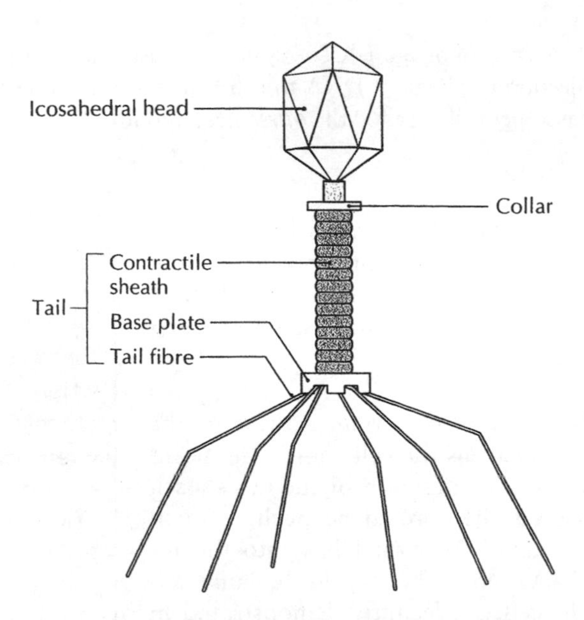

Figure 10.8 A T-even bacteriophage. Note the characteristic 'head plus tail' structure. The tail fibres and base plate are involved in the attachment of the phage to its host cell's surface. Reproduced by permission of Professor Michael J Pelczar, University of Maryland

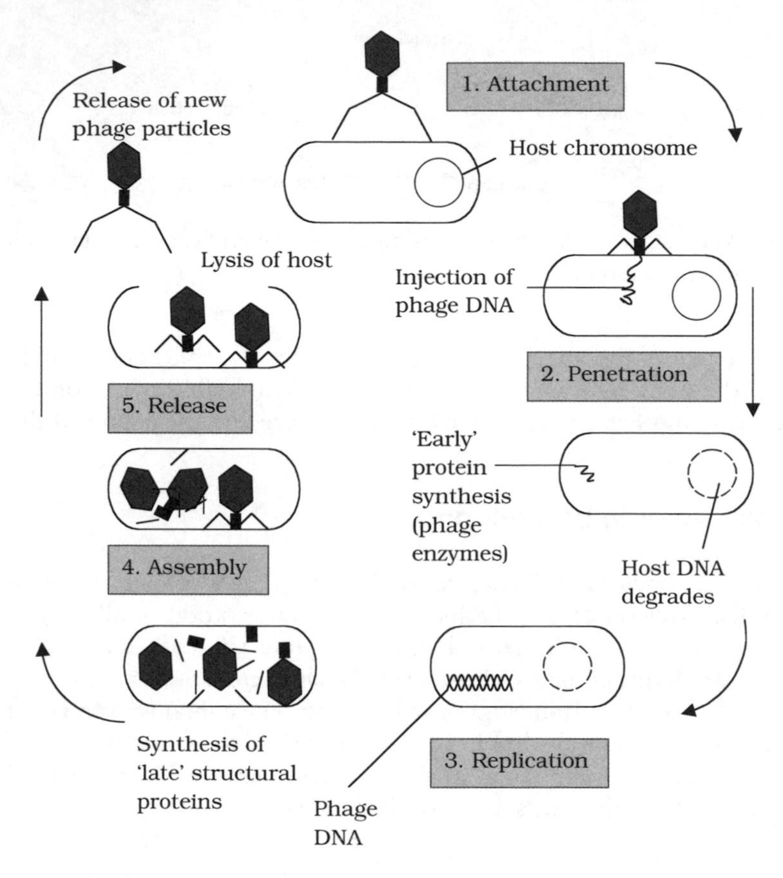

Figure 10.9 The lytic cycle of phage T4. The cycle comprises the five main stages described in the text; from injection of phage of DNA to cell lysis takes 22 minutes. The number of phage particles released per cell is called the *burst size*, and for T4 it ranges from 50 to 200

1 *Adsorption* (*attachment*): T4 attaches by means of specific tail fibre proteins to complementary receptors on the host cell's surface. The nature of these receptors is one of the main factors in determining a virus's host specificity.

2 *Penetration*: The enzyme *lysozyme*, present in the tail of the phage, weakens the cell wall at the point of attachment, and a contraction of the tail sheath of the phage causes the core to be pushed down into the cell, releasing the viral DNA into the interior of the bacterium. The capsid remains entirely outside the cell, as elegantly demonstrated in the famous experiment by Hershey and Chase (see Chapter 11).

You might reasonably ask yourself why cells would evolve receptor molecules for viruses, when the outcome is clearly not in their interests. The answer is, of course, that they haven't; the receptors have other biological properties, but the viruses have 'taken them over'.

3 *Replication*: Phage genes cause host protein and nucleic acid synthesis to be switched off, so that all of the host's metabolic machinery becomes dedicated to the synthesis of phage DNA and proteins. Host nucleic acids are degraded by phage-encoded enzymes, thereby providing a supply of nucleotide building blocks. Host enzymes are employed to replicate phage DNA, which is then transcribed into mRNA and translated into protein.

'Early' proteins are viral enzymes including DNA polymerase, used to synthesise more phage DNA, and others that disrupt normal host processes. Later, production switches to 'late', structural proteins, required for the construction of capsids; these are produced in much greater quantities.

4 *Assembly*: Once synthesised in sufficient quantities, capsid and DNA components assemble spontaneously into viral particles. The head and tail regions are synthesised separately, then the head is filled with the DNA genome, and joined onto the tail.

5 *Release*: Phage-encoded lysozyme weakens the cell wall, and leads to lysis of the cell and release of viral particles;these are able to infect new host cells, and in so doing recommence the cycle. During the early phase of infection, the host cell contains components of phage, but no complete particles. This period is known as the *eclipse period*. The time which elapses between the attachment of a phage particle to the cell surface and the release of newly-synthesised phages is the *latent period* (sometimes known as the *burst time*); for T4 under optimal conditions, this is around 22 minutes. This can be seen in a one-step growth curve, as shown in Figure 10.10.

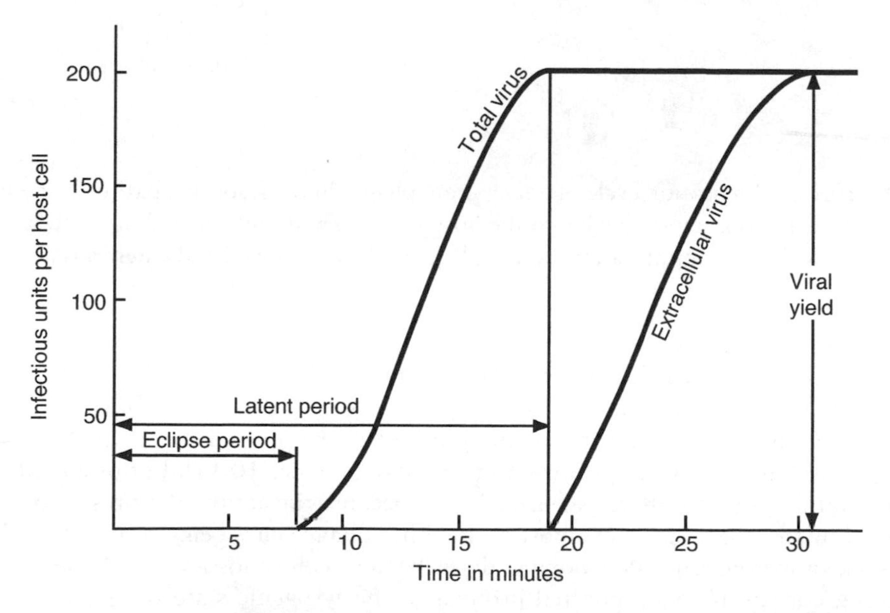

Figure 10.10 The one-step growth curve. During the eclipse period the host cell does not contain complete phage particles. Following synthesis of new particles, they are released, signalling the end of the latent period. The left-hand curve represents the number of phage particles, while the number of free (extracellular) particles is shown by the right-hand curve

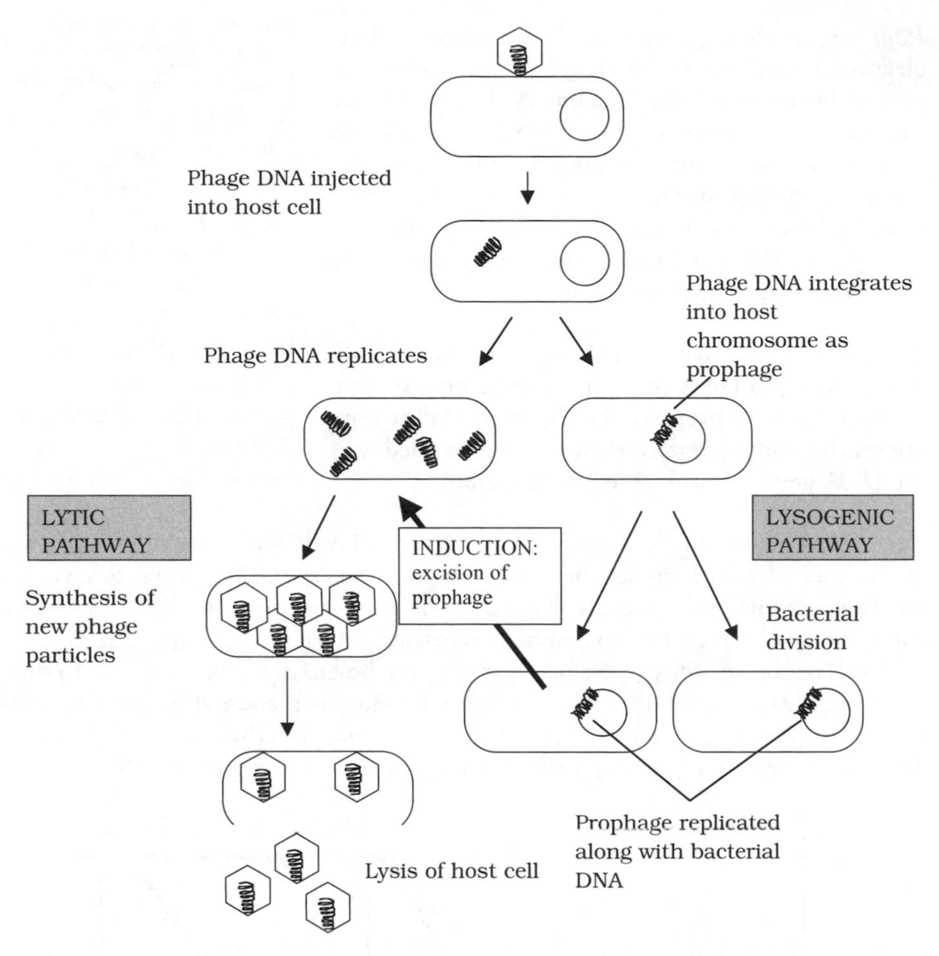

Figure 10.11 Replication cycle of a temperate phage. In the lysogenic pathway, the phage DNA is integrated as a prophage into the host genome, and replicated along with it. Upon induction by an appropriate stimulus, the phage DNA is removed and enters a lytic cycle

Lysogenic replication cycle

Phages such as T4, which cause the lysis of their cells, are termed *virulent* phages. *Temperate* phages, in addition to following a lytic cycle as outlined above, are able to undergo an alternative form of growth cycle. Here, the phage DNA actually becomes incorporated into the host's genome as a *prophage* (Figure 10.11). In this condition of *lysogeny*, the host cell suffers no harm. This is because the action of repressor proteins, encoded by the phage, prevents most of the other phage genes being transcribed. These genes are, however, replicated along with the bacterial chromosome, so all the bacterial offspring contain the incorporated prophage. The lysogenic state is ended when the survival of the host cell is threatened, usually by an environmental factor such as UV light or a chemical mutagen. Inactivation of the repressor protein allows the phage DNA to be excised, and adopt a circular form in the cytoplasm. In this form, it initiates a lytic cycle, resulting in destruction of the host cell. An example of a temperate phage is bacteriophage λ (Lambda), which infects certain strains of *E. coli*. Bacterial strains that can incorporate phage DNA in this way are termed *lysogens*.

Replication cycles in animal viruses

Viruses that infect multicellular organisms such as animals may be specific not only to a particular organism, but also to a particular cell or tissue type. This is known as the *tissue tropism* of the virus, and is due to the fact that attachment occurs via specific receptors on the host cell surface.

The growth cycles of animal viruses have the same main stages as described for bacteriophages (Figure 10.7), but may differ a good deal in some of the details. Most of these variations are a reflection of differences in structure between bacterial and animal host cells.

Adsorption and penetration

Animal viruses do not have the head and tail structure of phages, so it follows that their method of attachment is different. The specific interaction with a host receptor is made via some component of the capsid, or, in the case of enveloped viruses, by special structures such as spikes (*peplomers*). Viral attachment sites can frequently be blocked by host antibody molecules; however some viruses (e.g. the rhinoviruses) have overcome this by having their sites situated in deep depressions, inaccessible to the antibodies.

Whereas bacteriophages inject their nucleic acid component from the outside, the process in animal viruses is more complex, a fact reflected in the time taken for completion of the process. Animal viruses do not have to cope with a thick cell wall, and in many such cases the entire virion is internalised. This necessitates the extra step of uncoating, a process carried out by host enzymes. Many animal viruses possess an envelope; such viruses are taken into the cell either by fusion with the cell membrane, or by endocytosis (Figure 10.12). While some non-enveloped types release only their nucleic acid component into the cytoplasm, others require additionally that virus-encoded enzymes be introduced to ensure successful replication.

Replication (DNA viruses)

The DNA of animal cells, unlike that of bacteria, is compartmentalised within a nucleus, and it is here that replication and transcription of viral DNA generally occur*. Messenger RNA then passes to ribosomes in the cytoplasm for translation (Figure 10.13). In the case of viruses with a ssDNA genome, a double-stranded intermediate is formed, which serves as a template for mRNA synthesis.

Assembly

Translation products are finally returned to the nucleus for assembly into new virus particles.

* Poxviruses are an exception. Both replication and assembly occur in the cytoplasm.

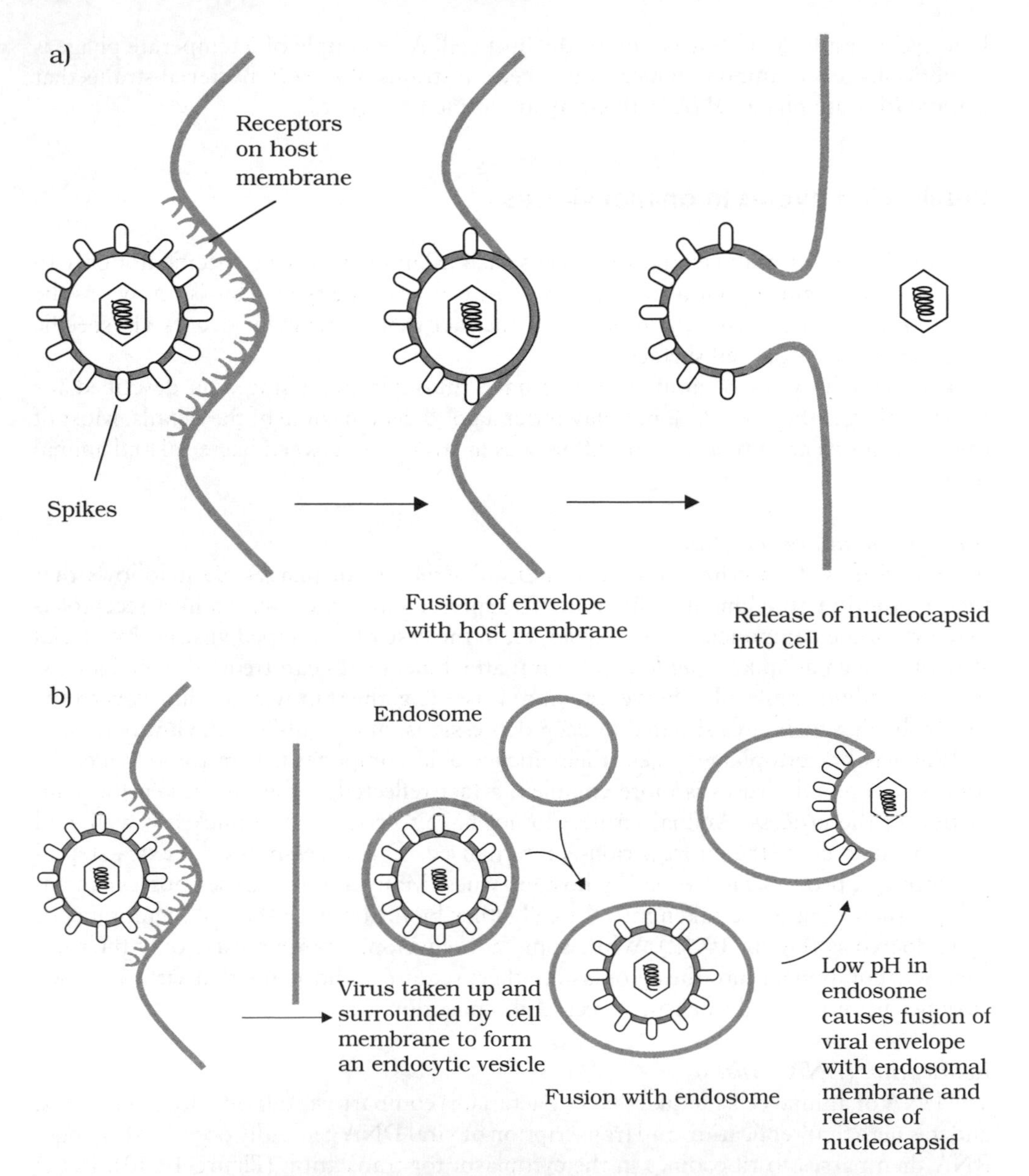

Figure 10.12 Enveloped viruses enter the host cell by fusion or endocytosis. (a) Fusion between viral envelope and host membrane results in release of nucleocapsid into the cell. Fusion depends on the interaction between spikes in the envelope and specific surface receptors. (b) Viral particles bound to the plasma membrane are internalised by endocytosis. Acidification within the endosome allows the release of the nucleocapsid into the cell

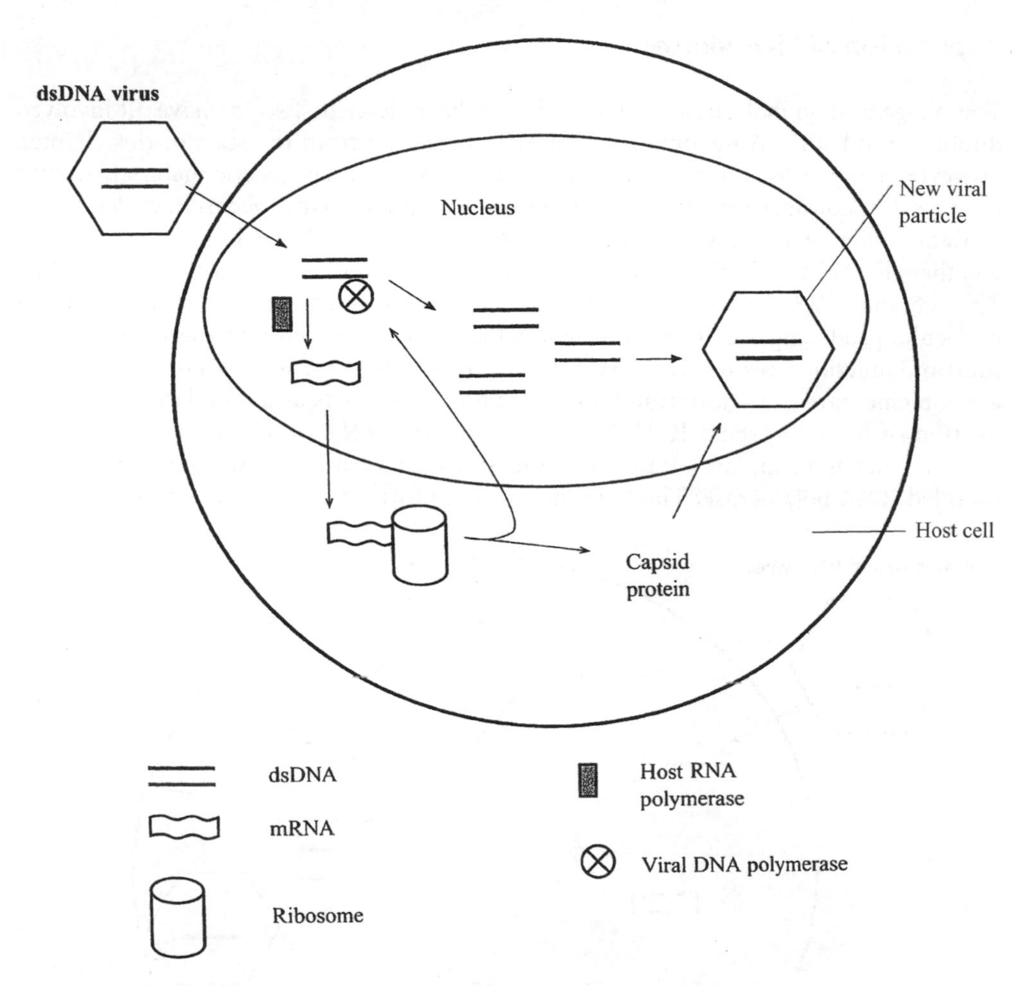

Figure 10.13 Replication in dsDNA viruses. Replication of viral DNA and transcription to mRNA take place in the nucleus of the host cell. The mRNA then passes out into the cytoplasm, where protein synthesis occurs on ribosomes. The capsid protein so produced returns to the nucleus for assembly into new viral particles. Newly synthesised DNA polymerase also returns to the nucleus, for further DNA replication. From Hardy, SP: Human Microbiology, Taylor and Francis, 2002. Reproduced by permission of Thomson Publishing Services

Release

Naked (non-enveloped) viruses are generally released by lysis of the host cell. In the case of enveloped forms, release is more gradual. The host's plasma membrane is modified by the insertion of virus-encoded proteins, before engulfing the virus particle and releasing it by a process of *budding*. This can be seen as essentially the reverse of the process of internalisation by fusion (Figure 10.12a).

> Herpesviruses are unusual in deriving their envelope from the nuclear, rather than cytoplasmic membrane.

Replication of RNA viruses

The phage and animal virus growth cycles we have described so far have all involved double-stranded DNA genomes. As you will remember from the start of this chapter, however, many viruses contain RNA instead of DNA as their genetic material, and we now need to consider briefly how these viruses complete their replication cycles.

Replication of RNA viruses occurs in the cytoplasm of the host; depending on whether the RNA is single- or double-stranded, and (+) or (−) sense, the details differ. The genome of a (+) *sense single stranded RNA virus* functions directly as an mRNA molecule, producing a giant polyprotein, which is then cleaved into the various structural and functional proteins of the virus. In order for the (+) sense RNA to be replicated, a complementary (−) sense strand must be made, which acts as a template for the production of more (+) sense RNA (Figure 10.14). The RNA of a (−) *sense RNA virus* must first act as a template for the formation of its complementary sequence by a virally encoded RNA polymerase. The (+) sense RNA so formed has two functions: (i) to act

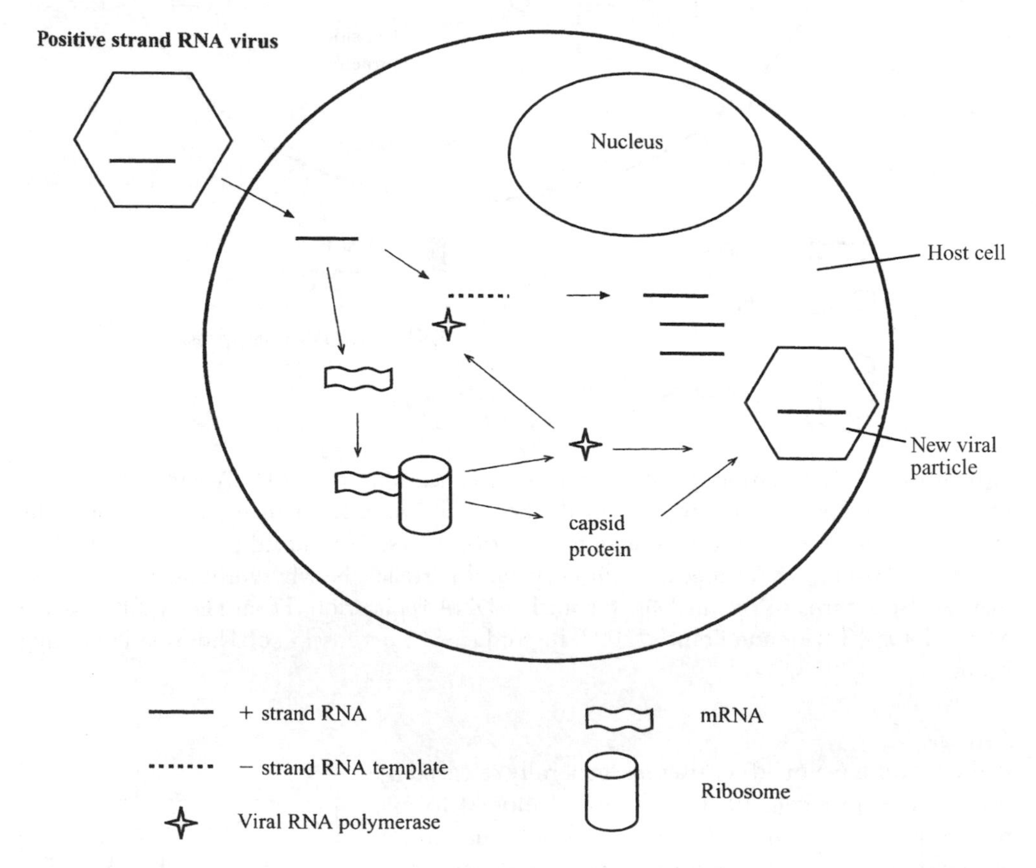

Figure 10.14 Replication in (+) sense single-stranded RNA viruses. On entering the cell, the (+) sense ssRNA genome is able to act directly as mRNA, directing the synthesis of capsid protein and RNA polymerase. In addition, it replicates itself, being converted firstly into a (−) sense ssRNA intermediate. All steps take place outside of the nucleus. From Hardy, SP: Human Microbiology, Taylor and Francis, 2002. Reproduced by permission of Thomson Publishing Services

as mRNA and undergo translation into the virus's various proteins, and (ii) to act as template for the production of more genomic (−) sense RNA (Figure 10.15).

Double-stranded RNA viruses are all segmented. They form separate mRNAs for each of their proteins by transcription of the (−) strand of their genome. These are each translated, and later form an aggregate (*subviral particle*) with specific proteins, where they act as templates for the synthesis of a double-stranded RNA genome, ready for incorporation into a new viral particle.

Two final, rather complicated, variations on the viral replication cycles involve the enzyme *reverse transcriptase*, first discovered in 1970 (see Box 10.3)

Retroviruses

These viruses, which include some important human pathogens, have a genome that exists as RNA and DNA at different part of their replication cycle. Retroviruses have a (+) sense ss-RNA genome which is unique among viruses in being diploid. The two copies of the genome serve as templates for the enzyme reverse transcriptase to produce

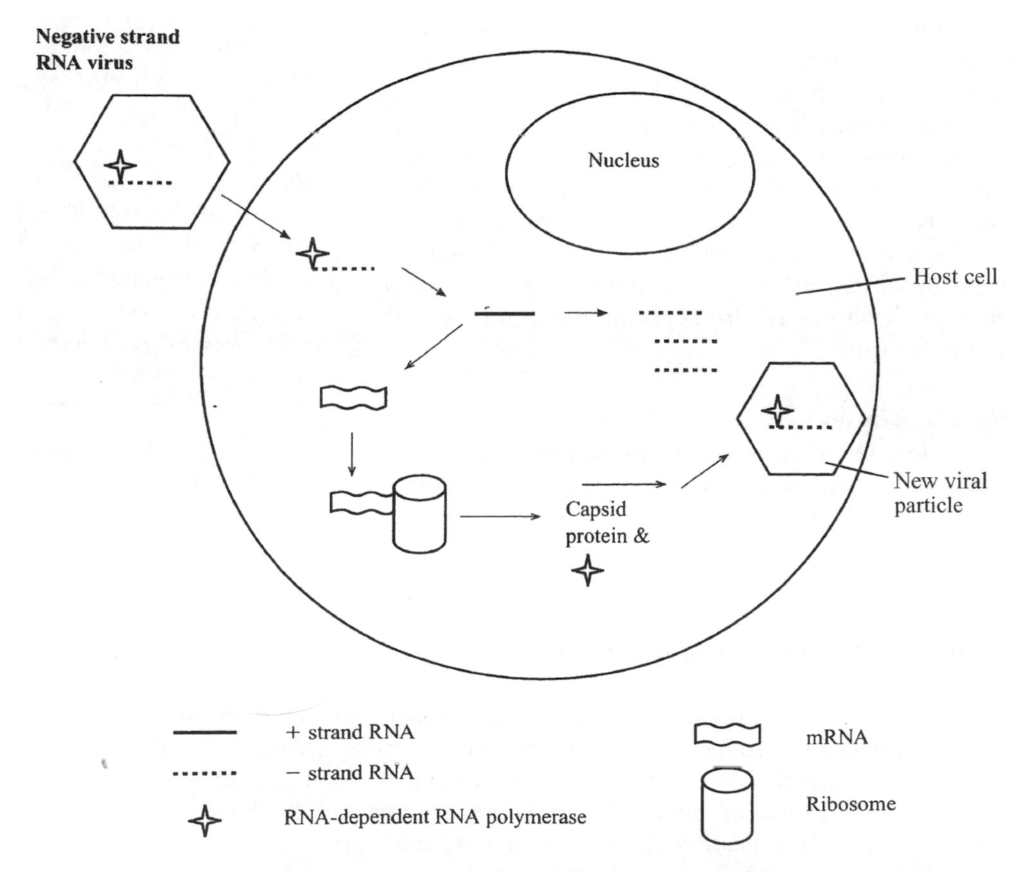

Figure 10.15 Replication in (−) sense single-stranded RNA viruses. Before it can function as mRNA, the (−) sense ssRNA must be converted to its complementary (+) sense sequence. This serves both as mRNA and as template for the production of new (−) sense ssRNA genomes. From Hardy, SP: Human Microbiology, Taylor and Francis, 2002. Reproduced by permission of Thomson Publishing Services

Box 10.3 The enzyme that breaks the rules

The discovery in 1970 of an enzyme that could convert an RNA template into DNA caused great surprise in the scientific world. The action of this reverse transcriptase or RNA-dependent DNA polymerase is a startling exception to the 'central dogma' of molecular biology, that the flow of genetic information is unidirectional, from DNA to RNA to protein (see Chapter 11). This in its revised form can be represented:

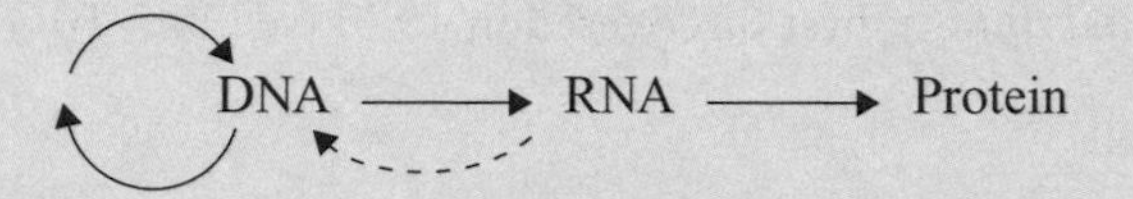

(The circular arrow to the left denotes DNA's ability to be replicated; the dotted arrow represents the action of reverse transcriptase.)

a complementary strand of DNA. The RNA component of this hybrid is then degraded, allowing the synthesis of a second strand of DNA. This *proviral DNA* passes to the nucleus, where it is incorporated into the host's genome (Figure 10.16). Transcription by means of a host RNA polymerase results in mRNA, which is translated into viral proteins and also serves as genomic material for the new retrovirus particles. The human immunodeficiency virus (HIV), the causative agent of acquired immune deficiency syndrome, is an important example of a retrovirus.

Incorporation of retroviral DNA into the host genome parallels the integration seen in lysogenic phage growth cycles. Unlike the prophage, however, the provirus is not capable of a separate existence away from the host chromosome.

Hepadnaviruses

In two families of viruses (hepadnaviruses and caulimoviruses), DNA and RNA phases again alternate, but their order of appearance is reversed, so that a dsDNA genome is produced. This is made possible by reverse transcriptase occurring at a later stage, during the maturation of the virus particle.

Replication cycles in plant viruses

Viral infections of plants can be spread by one of two principal pathways. *Horizontal transmission* is the introduction of a virus from the outside, and typically involves insect vectors, which use their mouthparts to penetrate the cell wall and introduce the virus. This form of transmission can also occur by means of inanimate objects such as garden tools. In *vertical transmission*, the virus is passed from a plant to its offspring, either by asexual propagation or through infected seeds.

The majority of plant viruses discovered so far have an RNA genome, although DNA forms such as the caulimoviruses (see above) are also known. Replication is similar to that of animal viruses, depending on the nature of the viral genome. An infection only becomes significant if it spreads throughout the plant (a *systemic* infection). Viral

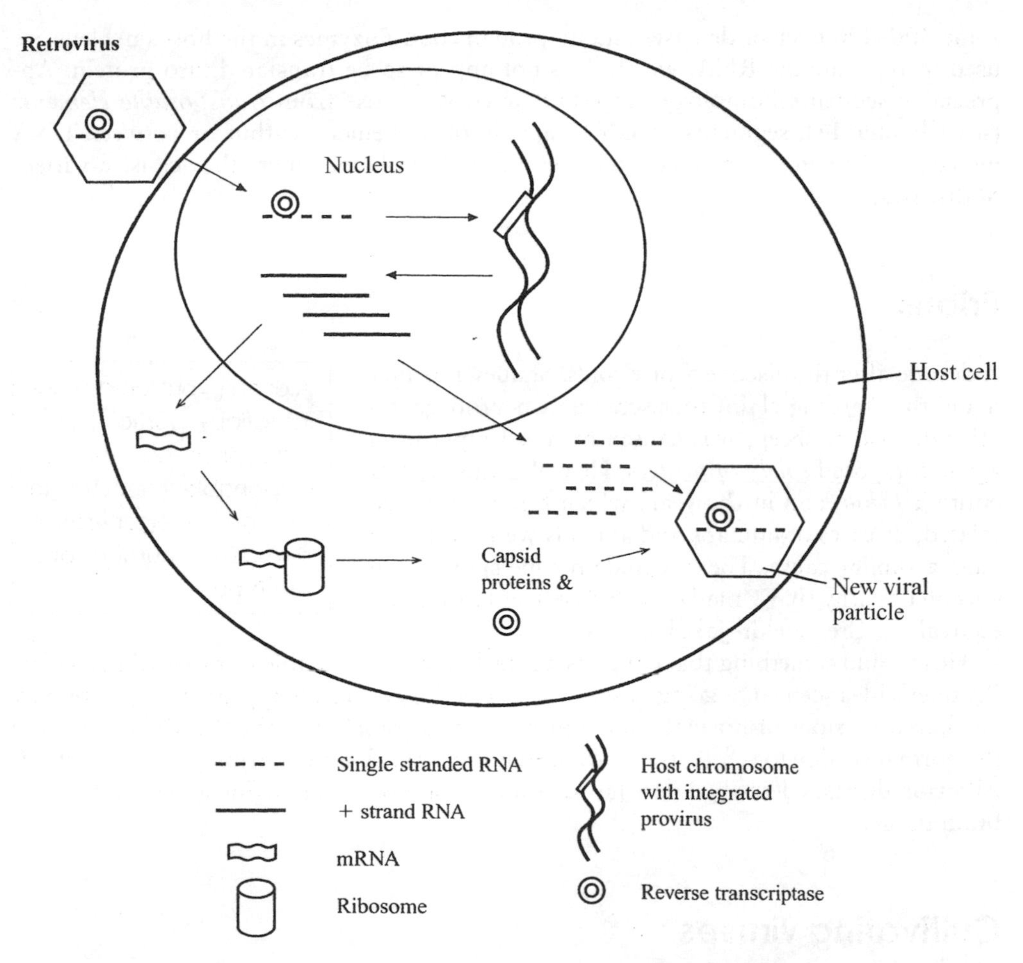

Figure 10.16 Replication in retroviruses. Reverse transcriptase makes a DNA copy of the single-stranded RNA retroviral genome. This is integrated into the host genome and is transcribed by the cellular machinery. Messenger RNA passes out to the ribosomes, where translation into viral coat proteins and more reverse transcriptase occurs. Retrovirus packaging takes place outside the nucleus. From Hardy, SP: Human Microbiology, Taylor and Francis, 2002. Reproduced by permission of Thomson Publishing Services

particles do this by moving through the plasmodesmata, naturally occurring cytoplasmic strands linking adjacent plant cells.

Viroids

A viroid is a plant pathogen that comprises only ssRNA and does not code for a protein product.

In 1971, Theodor Diener proposed the name *viroid* to describe a newly discovered pathogen of potatoes. Viroids are many times smaller than the smallest virus, and consist solely of a small circle of ssRNA containing some 300–400 nucleotide bases and no protein coat. Enzymes in the host's nucleus are used to replicate the RNA, which does not appear to be translated into protein. Appreciable sequence homology suggests that viroids arose from *transposable elements* (see Chapter 11), segments of DNA capable of movement within or between DNA molecules. To date, viroids have only been found in plants, where they cause a variety of diseases.

Prions

A prion (=proteinaceous infectious particle) is a self-replicating protein responsible for a range of neurodegenerative disorders in humans and mammals.

A decade after the discovery of viroids, Stanley Prusiner made the startling claim that scrapie, a neurodegenerative disease of sheep, was caused by a self-replicating agent composed *solely of protein*. He called this type of entity a *prion*, and in the years which followed, other, related, diseases of humans and animals were shown to have a similar cause. These include bovine spongiform encephalopathy (BSE, 'mad cow disease') and its human equivalent, Creutzfeldt–Jakob disease.

How could something that contains no nucleic acid be capable of replicating itself? – Prusiner's idea seemed to go against the basic rules of biology. It appears that prions may be altered versions of normal animal proteins, and somehow have the ability to cause the normal version to refold itself into the mutant form. Thus the prion propagates itself. All prion diseases described thus far are similar conditions, involving a degeneration of brain tissue.

Cultivating viruses

Whilst the growth of bacteria in the laboratory generally demands only a supply of the relevant nutrients and appropriate environmental conditions, maintaining viruses presents special challenges. Think back to the start of this chapter, and you will realise why this is so; all viruses are obligate intracellular parasites, and therefore need an appropriate host cell if they are to replicate.

Bacteriophages, for example, are grown in culture with their bacterial hosts. Stock cultures of phages are prepared by allowing them to infect a broth culture of the appropriate bacterium. Successful propagation of phages results in a clearing of the culture's turbidity; centrifugation removes any remaining bacteria, leaving the phage particles in the supernatant. A quantitative measure of phages, known as the *titre*, can be obtained by mixing them with a much greater number of bacteria and immobilising them in agar. Due to their numbers, the bacteria grow as a confluent *lawn*. Some become infected by phage, and when new viral particles are released following lysis of their host, they infect more host cells. Because they are immobilised in agar, the phages are only able to infect cells in the immediate vicinity. As more and more cells in the same area are lysed, an area of clearing called a *plaque* appears in the lawn of bacteria (Figure 10.17). Quantification is based on the assumption that each visible plaque arises from

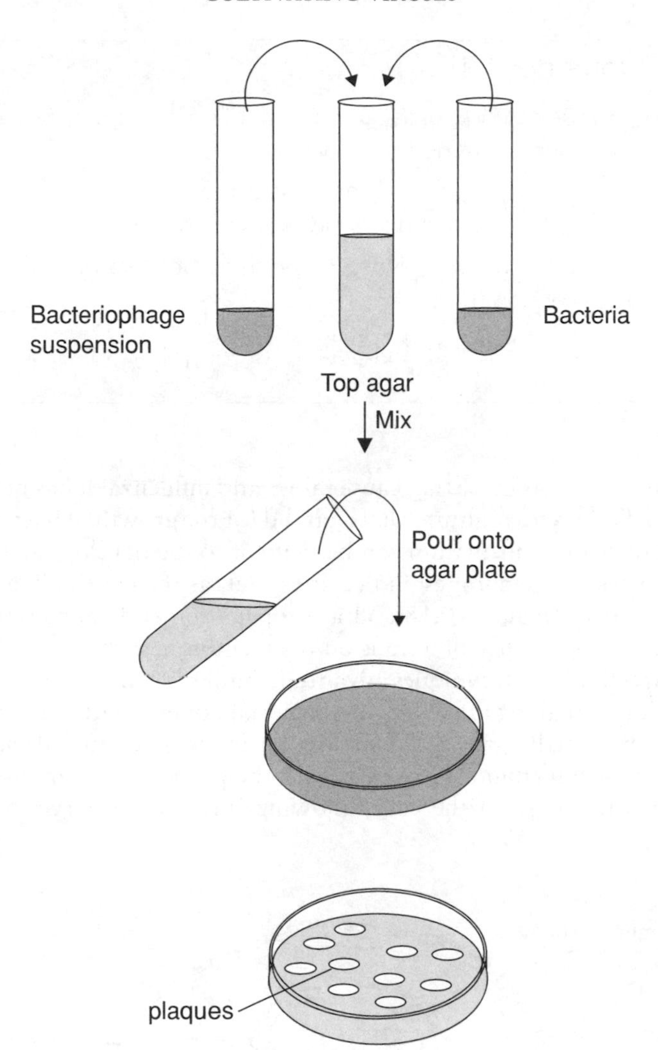

Figure 10.17 The plaque assay for bacteriophages. The number of particles in a phage preparation can be estimated by means of a plaque assay. Phages and bacteria are mixed in a soft agar then poured onto the surface of an agar plate. Bacteria grow to develop a confluent lawn, and the presence of phage is indicated by areas of clearing (plaques) where bacteria have been lysed. From Reece, RJ: Analysis of Genes and Genomes, John Wiley & Sons, 2003. Reproduced by permission of the publishers

infection by a single phage particle. Thus, we speak of *plaque-forming units* (pfu: see Box 10.4).

Animal viruses used to be propagated in the host animal; clearly there are limitations to this, not least when the host is human! One of the major breakthroughs in the field of virus cultivation was made in 1931 when it was shown by Alice Woodruff and Ernest Goodpasture that fertilised chicken's eggs could serve as a host for a number of animal

> **Box 10.4** Plaque counts
>
> The number of plaque forming units (pfu) per millilitre of suspension can be calculated by a plaque count by using the equation:
>
> $$\text{pfu/ml} = \frac{\text{no. of plaques}}{\text{dilution factor} \times \text{volume (ml)}}$$
>
> e.g. 100μl of a 1 in 10,000 dilution gives 53 colonies when plated out.
> Thus the original suspension had:
>
> $$\frac{53}{10^{-4} \times 0.1} = 5.3 \times 10^{6} \text{ pfu/ml}$$

and human viruses, such as those that cause rabies and influenza. It has been said that the chicken embryo did for virus culture what agar did for the growth of bacteria. Depending on the virus in question, inoculation can be made into the developing embryo itself or into one of the various membranes and cavities such as the chorioallantoic membrane or the allantoic cavity (Figure 10.18). Viral propagation is demonstrated by death of the embryo, or the appearance of lesions on the membranes.

In the 1950s, cell culture techniques advanced, thanks in part to the widespread availability of antibiotics, making the control of bacterial contamination much more readily achieved. Cells are usually grown as *monolayers* in tissue culture flasks containing a suitable liquid growth medium. Treatment with the protease trypsin dissolves the connective tissue matrix between the cells, allowing them to be harvested, and used to

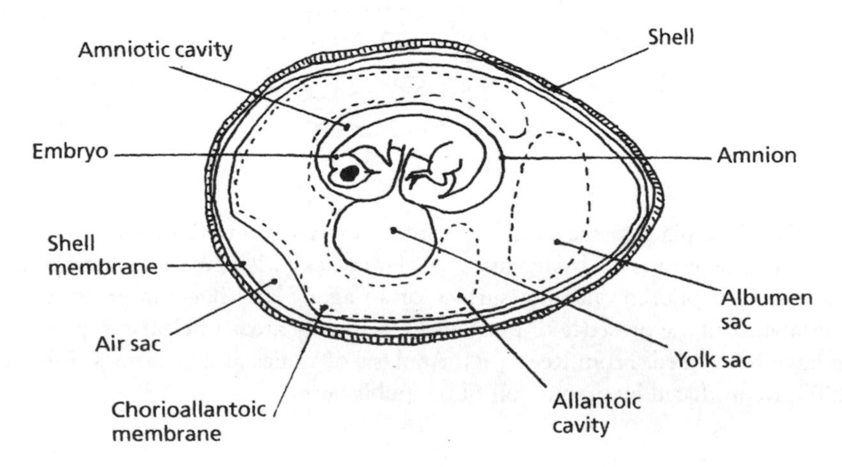

Figure 10.18 The culture of animal viruses in embryonated eggs. Viruses such as the influenza virus are cultured more effectively in eggs than in cell culture. The chorioallantoic membrane provides epithelial cells that act as host to the virus. From Heritage, J, Evans, EGV & Killington, RA: Introductory Microbiology, Cambridge University Press, 1996. Reproduced by permission of the publishers

seed new cultures. Changes in cell morphology, known generically as *cytopathic effects*, are indicators of viral infection, and may be used diagnostically in the identification of specific viral types.

Plant viruses need to overcome the barrier presented by the cellulose cell wall of the plant; in nature this is often achieved by the piercing mouthparts of an insect vector or by entering areas of damaged tissue. Experimentally, viruses can be introduced into an appropriate host by rubbing the surface of a leaf with the virus together with a mild abrasive to create a minor wound.

Viral diseases in humans

Viruses are responsible for causing some of the most serious infectious diseases to affect humans. Some important examples are listed in Table 10.3, and some are discussed in a little more detail below.

Airborne transmission: influenza

Influenza is a disease of the respiratory tract caused by members of the *Orthomyxoviridae*. Transmission occurs as a result of inhaling airborne respiratory droplets from an infected individual. Infection by the influenza virus results in the destruction of epithelial cells of the respiratory tract, leaving the host open to secondary infections from bacteria such as *Haemophilus influenzae* and *Staphylococcus aureus*. It is these secondary infections that are responsible for the great majority of fatalities caused by influenza. Generally, sufferers from influenza recover completely within 10–14 days, but some people, notably the elderly and those with chronic health problems, may develop complications such as pneumonia.

The influenza virus has an envelope, and a segmented (−) sense ssRNA genome (Figure 10.19). The envelope contains two types of protein spike, each of which plays a crucial role in the virus's infectivity:

- Neuraminidase is an enzyme which hydrolyses sialic acid, thereby assisting in the release of viral particles.
- Haemagglutinin enables the virus to attach to host cells by binding to epithelial sialic acid residues. It also helps in the fusion of the viral envelope with the cell membrane.

Both types of spike act as antigens, proteins that stimulate the production of antibodies in a host. One of the reasons that influenza is such a successful virus is that the 'N' and 'H' antigens are prone to undergoing changes (*antigenic shift*) so that the antigenic 'signature' of the virus becomes altered, and host immunity is evaded. Different strains of the influenza virus are given a code denoting which variants of the antigens they carry; the strain that caused the 1918 pandemic, for example, was N_1H_1, while the one responsible for the outbreak of 'bird flu' in SE Asia in 2003/4 was H_5N_1.

Table 10.3 Some important viruses of humans

Virus	Family	Disease	Genome type
Adenovirus	*Adenoviridae*	Respiratory infections	dsDNA
Ebola virus	*Filoviridae*	Haemorrhagic fever	(−)ssRNA
Epstein–Barr virus	*Herpesviridae*	Infectious mononucleosis	dsDNA
Hepatovirus A	*Picornaviridae*	Hepatitis A	(+)ssRNA
Herpes simplex Type I	*Herpesviridae*	Cold sores	dsDNA
Herpes simplex Type II	*Herpesviridae*	Genital warts	dsDNA
Human immunodeficiency virus (HIV)	*Retroviridae*	Acquired immune deficiency syndrome (AIDS)	(+)ssRNA*
Human papillomavirus	*Papovaviridae*	Warts	dsDNA
Influenza virus	*Orthomyxoviridae*	Influenza	(−)ssRNA
Lassa virus	*Arenaviridae*	Lassa fever	(−)ssRNA
Morbillivirus	*Paramyxoviridae*	Measles	(−)ssRNA
Norwalk virus	*Calciviridae*	Enteritis	(+)ssRNA
Paramyxovirus	*Paramyxoviridae*	Mumps	(−)ssRNA
Polio virus	*Picornaviridae*	Poliomyelitis	(+)ssRNA
Rabies virus	*Rhabdoviridae*	Rabies	(−)ssRNA
Rhinovirus	*Picornaviridae*	Common cold	(+)ssRNA
Rotavirus	*Reoviridae*	Enteritis	dsRNA
Rubella virus	*Togaviridae*	German measles	(+)ssRNA
Smallpox virus	*Poxviridae*	Smallpox	dsDNA
Varicella-Zoster	*Herpesviridae*	Chicken pox, shingles	dsDNA
Yellow fever virus	*Flaviviridae*	Yellow fever	(+)ssRNA

*The genome of HIV, like that of other retroviruses also has a DNA phase. See the text.

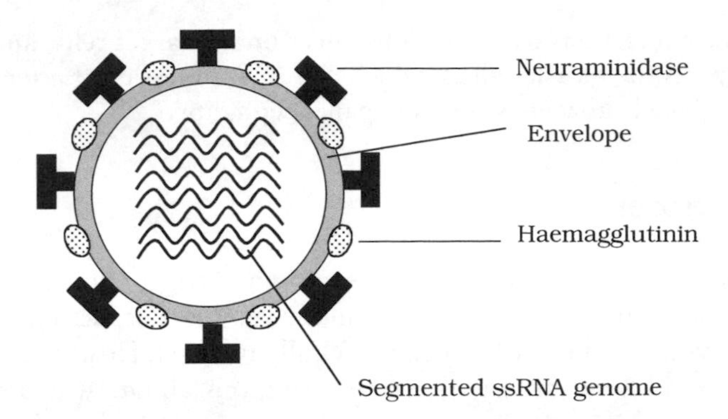

Figure 10.19 The influenza virus. The RNA segments are bound to protein, forming a nucleocapsid, and are surrounded by further protein. The two types of spike assist in the attachment and penetration of the virus into its host

Transmission by water or food: viral gastroenteritis

rota = 'a wheel', describing the distinctive appearance of this type of virus.

Everybody must surely be familiar with the symptoms of gastroenteritis – sickness, diarrhoea, headaches and fever. The cause of this gastroenteritis may be bacterial (e.g. *Salmonella*) or viral. The major cause of the viral form is the human *rotavirus* , which, together with the Norwalk virus, is responsible for the majority of reported cases. The rotavirus has a segmented, dsRNA genome, and is a non-enveloped virus.

The virus damages the villi in the upper part of the intestinal tract, affecting normal ion transport, and resulting in the characteristic water loss. Transmission of gastroenteritis is via the faecal–oral route, that is, by the ingestion of faecally contaminated food or water. Poor hygiene practice or contaminated water supplies are usually to blame for the perpetuation of the cycle. Normally, the condition is self-limiting, lasting only a couple of days; the normal treatment is fluid replacement therapy. In areas where clean water supplies are not available, however, the outcome can be much more serious. In the Third World, the condition is a major killer; it is the principal cause of infant mortality, and the cause of some five to ten million deaths per year.

Latent and slow (persistent) viral infections

After an infection has passed, a virus may sometimes remain in the body for long periods, causing no harm. It may be reactivated, however, by stress or some change in the individual's health, and initiate a disease state. Well known examples of *latent* viral infections are cold sores and shingles, both caused by members of the herpesvirus family. A virus of this sort will remain with an individual throughout their lifetime.

Whereas latent virus infections are characterised by a sudden increase in virus production, in *persistent* (slow) infections the increase is more gradual, building up over several years. Such infections have a serious effect on the target cells, and are generally fatal. An example is the measles virus, which can re-manifest itself after many years in a rare condition called subacute sclerosing panencephalitis.

Viruses and cancer

A number of chemical and physical agents are known to trigger the uncontrolled proliferation of cells that characterise cancers, but in the last two decades it has become clear that at least six types of human cancer can be virally induced. How do cells lose control of their division, and how are viruses able to bring this about? It is now known that cells contain genes called *protooncogenes*, involved in normal cell replication. They are normally under the control of other, *tumour-suppressor genes*, but these can be blocked by proteins encoded by certain DNA viruses. When this happens, the protooncogene functions as an *oncogene*, and cell division is allowed to proceed uncontrolled. Retroviruses have a different mechanism; they carry their own, altered, version of the cellular oncogene, which becomes integrated into the host's genome and leads to uncontrolled cell growth. Retrovirus oncogenes are thought to have been acquired originally from human (or animal) genomes, with the RNA transcript becoming incorporated into the retrovirus particle.

An oncogene is a gene associated with the conversion of a cell to a cancerous form.

Emerging and re-emerging viral diseases

As a result of changes in the pathogen or in the host population, completely new infectious diseases may arise, or we may experience the reappearance of diseases previously considered to be under control. These are known as, respectively, *emerging* and *re-emerging* infections. Changed patterns of human population movement are often responsible for the development of such infections, with the spread of smallpox to the New World by European colonisers being a famous example. Frequently emerging virus infections are *zoonotic* in origin, that is, they are transferred to humans from animal reservoirs. HIV, for example, is thought to have developed from a similar virus found in monkeys.

A zoonosis is a disease normally found in animals, but transmissible to humans under certain circumstances.

While this book was in preparation, there was a sharp reminder of the ever-present threat of emerging viral diseases, in the form of a new viral disease called severe acute respiratory syndrome (SARS). The outbreak of this disease began in Guangdong province in southern China in November 2002. The Chinese authorities were heavily criticised for not reporting the extent of the outbreak until some 3 months later, by which time cases were appearing in many parts of the world, illustrating the role of increased intercontinental travel in the spread of such a disease. At its peak in April 2003, over 1000 new cases of SARS were being reported per week. The cause of SARS was quickly identified as a member of the *Coronaviridae* (single-stranded RNA viruses). Transmitted by droplets from coughs and sneezes, it produces flu-like symptoms, but has a mortality rate of around 4 percent. Strict public health measures were brought into force, including restrictions in flights to and from affected areas, and the number of reported cases began to subside. In July 2003, the World Health Organisation announced that the final country, Taiwan, had been removed from its list of SARS-infected countries. By this time, SARS had claimed over 800 lives, mostly in China and Hong Kong, but with a number of deaths occurring as far afield as Canada and South Africa. Although apparently under control, isolated cases of SARS infections emerged in late 2003, and early 2004, most of which could be linked to laboratory workers. In January 2004 the Chinese authorities announced that a SARS vaccine was about to enter human clinical trials.

Virus vaccines

Smallpox, once the scourge of millions, was in 1979 the first infectious disease to be declared successfully eradicated. This followed a worldwide campaign of vaccination by the World Health Organisation over the previous decade, and was made feasible by the fact that humans are the only reservoir for the virus. *Vaccination* is a preventative strategy that aims to stimulate the host immune system, by exposing it to the infectious agent in question in an inactivated or incomplete form.

There are four main classes of virus vaccines:

- *Attenuated* (='weakened') vaccines contain 'live' viruses, but ones whose pathogenicity has been greatly reduced. The aim is to mimic an infection in order to stimulate an immune response, but without bringing about the disease itself. A famous example of this type of vaccine is the polio vaccine developed by Albert Sabin in the 1960s. The cowpox virus used by Edward Jenner in his pioneering vaccination work in the late 18th century was a naturally occurring attenuated version of the smallpox virus.

- *Inactivated* vaccines contain viruses which have been exposed to a denaturing agent such as formalin. This has the effect of rendering them non-infectious, while at the same time retaining their ability to stimulate an immune response. Vaccines directed against influenza are of this type.

- *Subunit* vaccines depend on the stimulation of an immune response by just a part of the virus. Since the complete virus is not introduced, there is no chance of infection, so vaccines of this type have the attraction of being very safe. Subunit vaccines are often made using recombinant DNA technology (Chapter 12); the first example to be approved for human use was the hepatitis B vaccine, which consists of part of the protein coat of the virus produced in specially engineered yeast cells.

- *DNA* vaccines are also the product of modern molecular biology techniques. DNA coding for virus antigens is directly injected into the host, where it is expressed and triggers a response by the immune system. Vaccines of this type have not so far been approved for use in humans.

Test yourself

1 Unlike bacteria, viruses are able to pass through a ________.

2 Viruses are incapable of ________ except when occupying a host cell.

3 A ________ is another name for an intact virus particle.

4 The nucleic acid of a virus is surrounded by a protein coat or ________. In some forms, this may in turn be surrounded by an ________.

5 In some types of ________ virus, the genome is segmented.

6 Parvoviruses are unusual among DNA viruses in having a ________ ________ genome.

7 Capsids may be either ________ or ________ in shape.

8 The ________ scheme of classification orders viruses in terms of their method of mRNA production.

9 The replication cycle of the T-even phages is described as ________ because it results in the ________ of the host cell.

10 The time between infection of a host cell and the release of newly-synthesised viral particles is called the ________ ________.

11 The DNA of ________ phages becomes integrated into the host genome, where it is known as a ________.

12 Enveloped viruses are taken into the host cell either by ________ with the cell membrane or by a process of ________.

13 The viral envelope is formed by the viral particle being engulfed in the host ________ ________, into which proteins encoded by the ________ are inserted.

14 In order for the genome of ss-RNA virus to be replicated, a ________ strand must be formed to act as a template.

15 Retroviruses contain the enzyme reverse ________, which can synthesise ________ from an ________ template.

16 The genome of most plant viruses is made of ________.

17 Subviral particles found in plants and comprising RNA but no ________ are known as ________.

18 When bacteriophages infect bacterial cells, their presence can be detected as ________ on an agar plate.

19 The envelope of the influenza virus has two types of protein spike made of ________ and ________ that play a part in the infectious process.

20 The surface antigens of viruses often undergo a gradual modification known as ________ ________, which helps to protect them against the host immune system.

21 Most cases of viral gastroenteritis in humans are caused by the ________.

22 Shingles and cold sores, caused by herpesviruses, are examples of ________ viral infections.

23 ________ infections of humans are transmitted from an animal source.

24 ________ vaccines contain 'live' viruses whose ________ has been reduced.

25 The hepatitis B vaccine is an example of a ________ vaccine, which depends on stimulating the immune response without introducing the whole virus particle.

Part IV

Microbial Genetics

11

Microbial Genetics

When any living organism reproduces, it passes on genetic information to its offspring. This information takes the form of *genes*, linear sequences of DNA that can be thought of as the basic units of heredity. The total complement of an organism's genetic material is called its *genome*.

A gene is a sequence of DNA that usually encodes a polypeptide.

How do we know genes are made of DNA?

The concept of the gene as an inherited physical entity determining some aspect of an organism's phenotype dates back to the earliest days of genetics. The question of what genes are actually made of was a major concern of molecular biologists (not that they would have described themselves as such!) in the first half of the 20th century. Since it was recognised by this time that genes must be located on chromosomes, and that chromosomes (in eucaryotes) comprised largely protein and DNA, the reasonable assumption was made that genes must be made up of one of these substances. In the early years, protein was regarded as the more likely candidate, since, from what was known of molecular structure at the time, it offered far more scope for the variation which would be essential to account for the thousands of genes that any organism must possess. The road to proving that DNA is in fact the 'stuff of life' was a long and hard one, which can be read about elsewhere; we shall mention below just some of the key experiments which provided crucial evidence.

In 1928 the Englishman Fred Griffith carried out a seminal series of experiments which not only demonstrated for the first time the phenomenon of genetic transfer in bacteria (a subject we shall consider in more detail later in this chapter), but also acted as the first step towards proving that DNA was the genetic material. As we shall see, Griffith showed that it was possible for heritable characteristics to be transferred from one type of bacterium to another, but the cellular component responsible for this phenomenon was not known at this time.

Attempts were made throughout the 1930s to isolate and identify the *transforming principle*, as it became known, and in 1944 Avery, MacLeod and McCarty published a paper, which for the first time, proposed DNA as the genetic material. Avery and his colleagues demonstrated that when DNA was rendered inactive by enzymatic treatment, transforming ability was lost from a cell extract, but if proteins,

Box 11.1 Hershey and Chase: the Waring blender experiment

In 1952 Alfred Hershey and Martha Chase provided further experimental evidence that DNA was the genetic material. In their experiment, the bacteriophage T2 was grown with *E. coli* cells in a medium with radiolabelled ingredients, so that their proteins contained ^{35}S, and their DNA ^{32}P. The phages were harvested, and mixed with a fresh culture of *E. coli*. They were left long enough for the phage particles to infect the bacteria, (but not long enough to produce new phage particles and lyse the cells). The culture was then subjected to mechanical agitation in a Waring blender, which, it was hoped, would remove the 'shell' of the phages from the outside of the bacteria, but leave the injected genetic material inside.

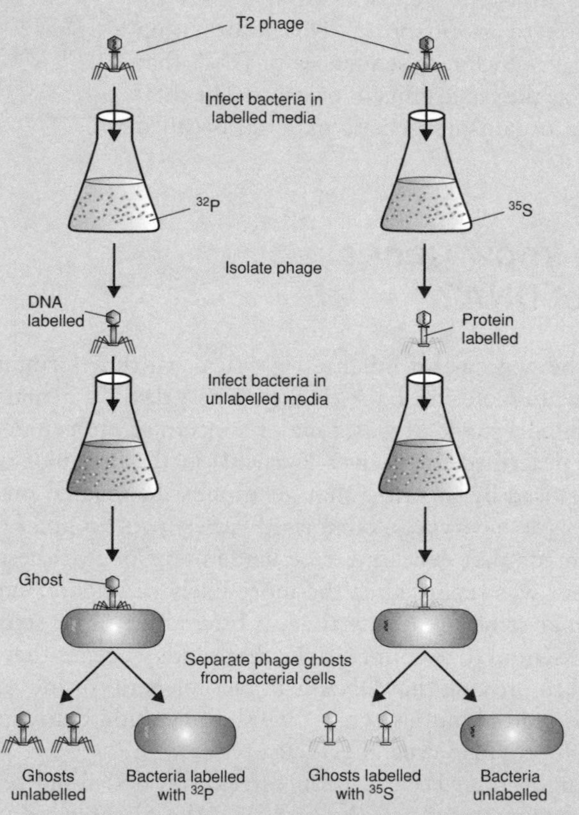

From Reece, RJ: Analysis of Genes and Genomes, John Wiley & Sons, 2003. Reproduced by permission of the publishers

The bacteria were sedimented by centrifugation, leaving the much lighter phage 'shells' in the supernatant. When solid and liquid phases were analysed for ^{32}P and ^{35}S, it was found that nearly all the ^{32}P was associated with the bacterial cells, while the great majority of the ^{35}S remained in the supernatant. The conclusion drawn from these results is that it was the ^{32}P-labelled DNA which had been injected into the bacteria, and which was therefore the genetic material.

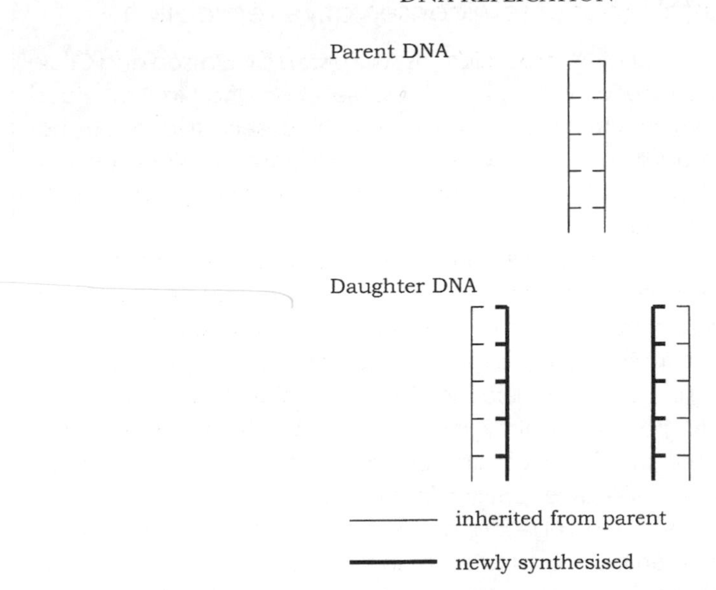

Figure 11.1 DNA replication is semiconservative. Following replication, each new DNA molecule comprises one strand from the parent DNA and one newly synthesised strand

carbohydrates or any other cellular component was similarly inactivated, the ability was retained. In spite of this apparently convincing proof, the pro-protein lobby was not easily persuaded. It was to be several more years before the experimental results of Alfred Hershey and Martha Chase (Box 11.1) coupled with Watson and Crick's model for DNA structure (Figure 2.23) finally cemented the universal acceptance of DNA's central role in genetics.

DNA replication

The elucidation of the structure of DNA by Watson and Crick in 1953 stands as one of the great scientific breakthroughs of the 20th century. An important implication of their model lay in the complementary nature of the two strands, as they commented themselves in their famous paper in *Nature*:

It has not escaped our notice that the specific pairing we have postulated immediately suggests a possible copying mechanism for the genetic material.

The mechanism they proposed became known as the *semiconservative* replication of DNA, so called because each daughter molecule comprises one parental strand and one newly synthesised strand (Figure 11.1). The parental double helix of DNA unwinds and each strand acts as a template for the production of a new complementary strand, with new nucleotides being added according to the rules of base-pairing. In 1957, 4 years after the publication of Watson and Crick's model, Matthew Meselson and Franklin Stahl provided experimental proof of the semiconservative nature of DNA replication (Box 11.2). The exact way in which replication takes place in procaryotic and eucaryotic cells differs in some respects, but we shall take as our model replication in *E. coli*, since it has been studied so extensively.

Box 11.2 Experimental proof of semiconservative replication

The semiconservative model of DNA replication, as suggested by Watson and Crick, was not the only one in circulation during the 1950s. Meselson and Stahl provided experimental evidence which not only supported the semiconservative model, but also showed the other models to be unworkable. Their elegantly designed experiment used newly developed techniques to differentiate between parental DNA and newly synthesised material.

First, they grew *E. coli* in a medium with ammonium salts containing the heavy isotope ^{15}N as the only source of nitrogen. This was done for several generations of growth, so that all the cells contained nitrogen exclusively in the heavy form.

The bacteria were then transferred to a medium containing nitrogen in the normal, ^{14}N form. After a single round of replication, DNA was isolated from the culture and subjected to *density gradient centrifugation*. This is able to differentiate between DNA containing the two forms of nitrogen, as ^{15}N has a greater buoyant density and therefore settles at a lower position in the tube.

Meselson and Stahl found just a single band of DNA after centrifugation, with a density intermediate between that of ^{14}N and ^{15}N, indicating a hybrid molecule, as predicted by the semiconservative model. After a second round of *E. coli* replication in a medium containing ^{14}N, two bands of DNA were produced, one hybrid and one containing exclusively ^{14}N, exactly as predicted by the semiconservative model, but inconsistent with other hypotheses.

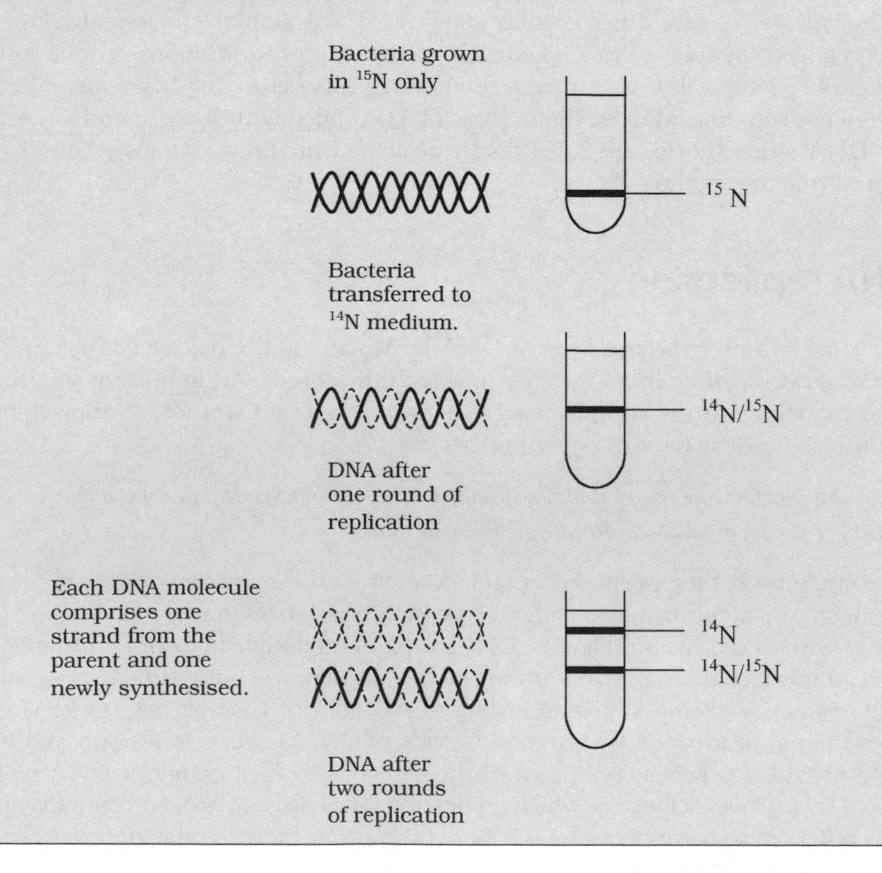

Bacteria grown in 15N only
15 N
Bacteria transferred to 14N medium.
14N/15N
DNA after one round of replication
Each DNA molecule comprises one strand from the parent and one newly synthesised.
14N
14N/15N
DNA after two rounds of replication

DNA replication in procaryotes

You may recall from Chapter 4 that bacteria multiply by a process of binary fission; before this occurs, each cell must duplicate its genetic information so that each daughter cell has a copy.

DNA replication involves the action of a number of specialised enzymes:

- helicases
- DNA topoisomerases
- DNA polymerase I
- DNA polymerase III
- DNA primase
- DNA ligase.

DNA replication takes place at a *replication fork*, a Y-shaped structure formed by the separating strands. The fork moves along the DNA as replication proceeds.

A primer is a short sequence of single-stranded DNA or RNA required by DNA polymerase as a starting point for chain extension.

Replication begins at a specific sequence called an *origin of replication*. The two strands of DNA are caused to separate by *helicases* (Figure 11.2), while *single-stranded DNA binding proteins* (SSB) prevents them rejoining. Opening out part of the double helix causes increased tension (supercoiling) elsewhere in the molecule, which is relieved by the enzyme *DNA topoiosomerase* (sometimes known as *DNA gyrase*). As the 'zipper' moves along, and more single stranded DNA is exposed, *DNA polymerase III* adds new nucleotides to form a complementary second strand, according to the rules of base-pairing. DNA polymerases are not capable of initiating the synthesis of an entirely new strand, but can only extend an existing one. This is because they require a free 3′-OH group onto which to attach new nucleotides. Thus, DNA polymerase III *can only work in the 5′–3′ direction*. A form of RNA polymerase called *DNA primase* synthesises a short single strand of RNA, which can be used as a *primer* by the DNA polymerase III. (Figure 11.2).

When replication occurs, complementary nucleotides are added to one of the strands (the *leading* strand) in a continuous fashion (Figure 11.2). The other strand (the *lagging* strand), however, runs in the opposite polarity; so how is a complementary sequence synthesised here? The answer is that DNA polymerase III allows a little unwinding to take place and then, starting at the fork, works back over from a new primer, in the 5′–3′ direction. Thus, the second strand is synthesised discontinuously, in short bursts, about 1000–2000 nucleotides at a time. These short stretches of DNA are called *Okazaki fragments*, after their discoverers.

On the lagging strand, a new RNA primer is needed at the start of every Okazaki fragment. These short sequences of RNA are later removed by *DNA polymerase I*, which

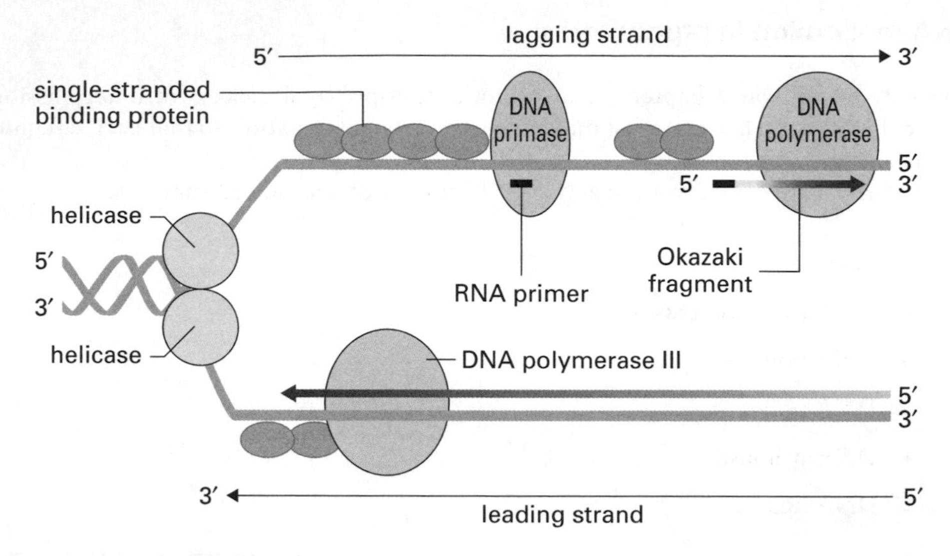

Figure 11.2 DNA replication takes place at a replication fork. Strands of DNA are separated and unwound by helicase and DNA gyrase, and prevented from rejoining by the attachment of single-strand binding proteins. Starting from an RNA primer, DNA polymerase III adds the complementary nucleotides to form a second strand. On the leading strand, a single primer is set down (not shown), and replication proceeds uninterrupted in the same direction as the fork. Because DNA polymerase III only works in the $5' \rightarrow 3'$ direction, on the lagging strand a new primer must be added periodically as the strands open up. Replication here is thus discontinuous, as a series of Okazaki fragments (see the text). From Bolsover, SR, Hyams, JS, Jones, S, Shepherd, EA & White, HA: From Genes to Cells, John Wiley & Sons, 1997. Reproduced by permission of the publishers

then replaces them with DNA nucleotides. Finally, the fragments are joined together by the action of *DNA ligase*. Replication is bidirectional (Figure 11.3), with two forks moving in opposite directions; when they meet, the whole chromosome is copied and replication is complete*.

DNA ligase repairs breaks in DNA by re-establishing phosphodiester bonds in the sugar-phosphate backbone.

What happens when replication goes wrong?

It is clearly important that synthesis of a complementary second strand of DNA should occur with complete accuracy, but occasionally, a non-complementary nucleotide is inserted; this may happen as frequently as once in every 10 000 nucleotides. Cells, however, are able to have a second attempt to incorporate the correct base because of the proofreading activity of the enzymes DNA polymerase I and III. These are able to

* This rather long-winded description hardly does justice to a process that incorporates new nucleotides at a rate of around 1000 per second!

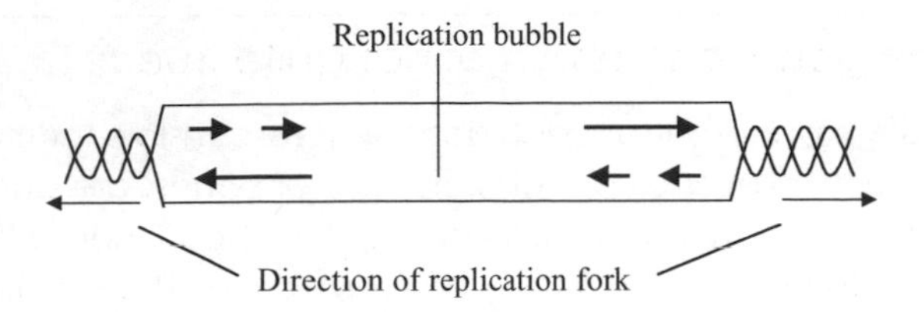

Figure 11.3 DNA replication is bidirectional. Two replication forks form simultaneously, moving away from each other and developing a replication bubble

cut out the 'wrong' nucleotide and replace it with the correct one. As a result of this monitoring, mistakes are very rare; they are thought to occur at a frequency of around one in every billion (10^9) nucleotides copied. Mistakes that do slip through the net result in *mutations*, which are discussed later in this chapter.

DNA replication in eucaryotes

In eucaryotic organisms, the DNA is linear, not circular, so the process of replication differs in some respects. Genome sizes are generally much bigger in eucaryotes, and replication much slower, so numerous replication forks are active simultaneously on each chromosome. Replication takes place bidirectionally, creating numerous replication bubbles (Figure 11.4) which merge with one another until the whole chromosome has been covered.

What exactly do genes do?

At the start of the 20th century Archibald Garrod had proposed that inherited disorders such as alkaptonuria may be due to a defect in certain key metabolic enzymes, thus offering for the first time an explanation of how genetic information is expressed. His ideas were not really developed however, until the work of George Beadle and Edward Tatum in the 1940s, whose experiments with the bread mould *Neurospora* led to the formulation of the *one gene, one enzyme* hypothesis. Although now acknowledged to be

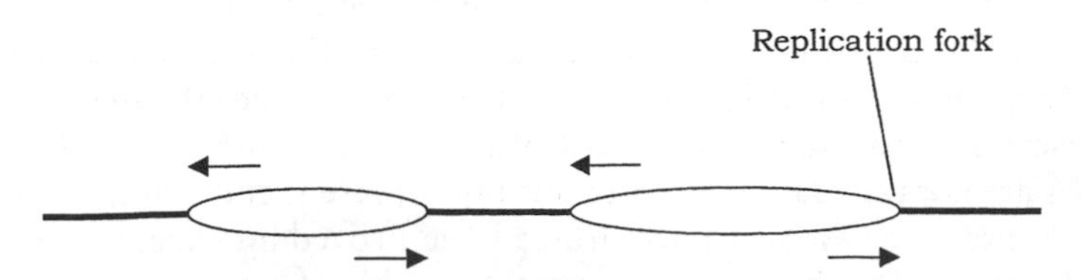

Figure 11.4 DNA replication in eucaryotes. Many replication bubbles develop simultaneously; they extend towards each other and eventually merge. The arrows denote the direction of the replication forks

Box 11.3 One gene, one enzyme: not quite true

Beadle and Tatum proposed that each gene was responsible for the production of a specific enzyme. However all proteins, not just enzymes, are encoded by DNA, and furthermore, some have a quaternary structure (see Chapter 2) with different polypeptide subunits being encoded by different genes. The hypothesis was therefore modified to *one gene, one polypeptide*. Later still, it emerged that even this is not always the case, as some genes do not encode proteins at all, but forms of RNA.

somewhat over-generalised (see Box 11.3), this model proved useful in the years when the molecular basis of gene action was being elucidated.

Having established that genes are made of DNA, and having a model for the structure of DNA that explained how it was able to copy itself, the way was open in the 1950s for scientists to work out the mechanism by which the information encoded in a DNA sequence was converted into a specific protein.

How does a gene direct the synthesis of a protein?

You may recall from Chapter 2 that both DNA and proteins are polymers whose 'building blocks' (nucleotides and amino acids respectively) can be put together in an almost infinite number of sequences. The sequence of amino acids making up the primary structure of a protein is determined by the sequence of nucleotides in the particular gene responsible for its production. It does this not directly, but through an intermediary molecule, now known to be a form of RNA called *messenger RNA* (mRNA). It is this intermediary that carries out the crucial task of passing the information encoded in the DNA sequence to the site of protein synthesis. This unidirectional flow of information can be summarised:

DNA ⟶ mRNA ⟶ Protein

DNA replication

and is often referred to as the Central Dogma of biology, because of its applicability to all forms of life. Proposed by Crick in the late 1950s, this is still accepted as being a true model of the basic events in protein synthesis. Sometimes the message encoded in DNA is transcribed into either *ribosomal RNA* (rRNA) or *transfer RNA* (tRNA); these types of RNA are not translated into proteins but represent end-products in themselves. (In the last chapter, we saw that retroviruses have proved an exception to one part of the Central Dogma, since they possess enzymes capable of forming DNA from an RNA template.)

The conversion of information encoded as DNA into the synthesis of a polypeptide chain occurs in two distinct phases (Figure 11.5); first the 'message' encoded in the

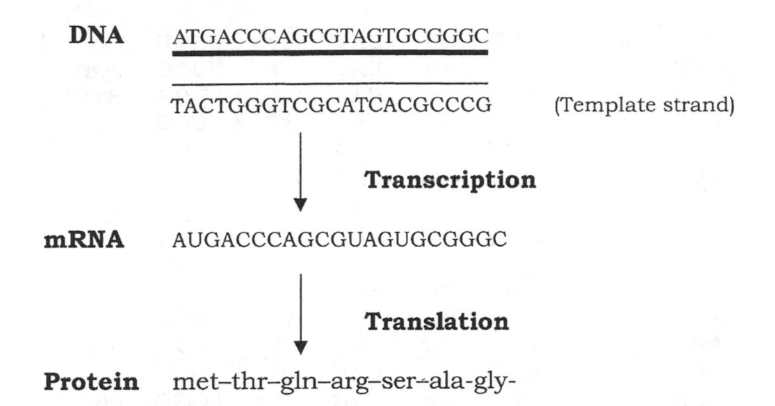

Figure 11.5 Transcription and translation. The sequence on one strand of DNA is transcribed as a molecule of mRNA (with uracil replacing thymine). The triplet code on the mRNA is then translated at ribosomes as a series of amino acids

DNA sequence of a gene is converted to mRNA by *transcription*, then this directs the assembly of a specific sequence of amino acids during *translation*. We shall discuss how this happens shortly, but first we need to consider the question: how does the sequence of nucleotides in a gene serve as an instruction in the synthesis of proteins?

> Messenger RNA (mRNA) is formed from a DNA template (transcription) and carries its encoded message to the ribosome, where it directs synthesis of a polypeptide (translation).

The genetic code

Messenger RNA carries information (copied from a DNA template) in the form of a *genetic code* that directs the synthesis of a particular protein. The nature of this code was worked out in the early 1960s by Marshall Nirenberg, Har Gobind Khorana and others (Box 11.4). The message encoded in a gene takes the form of a series of triplets (*codons*). Of the 64 possible three-letter combinations of A, C, G and U, 61 correspond to specific amino acids, while the remaining three act as 'stop' messages, indicating that reading of the message should cease at that point (Figure 11.6). It is essential that the reading of the message starts at the correct place, otherwise the *reading frame* (groups of

Box 11.4 The genetic code is (almost) universal

The genetic code was first worked out using *E. coli*, but soon found to apply to other organisms too. It seemed reasonable to assume that the code was universal, i.e. applicable to all life forms. It has been shown, however, that certain genes such as those found in the mitochondria of some eucaryotes employ a slight variation of the code. Mitochondria have their own transcription enzymes, ribosomes and tRNAs and so are able to use a modified system.

UUU	phe	**UCU**	ser	**UAU**	tyr	**UGU**	cys
UUC	phe	**UCC**	ser	**UAC**	tyr	**UGC**	cys
UUA	leu	**UCA**	ser	**UAA**	**STOP**	**UGA**	**STOP**
UUG	leu	**UCG**	ser	**UAG**	**STOP**	**UGG**	trp
CUU	leu	**CCU**	pro	**CAU**	his	**CGU**	arg
CUC	leu	**CCC**	pro	**CAC**	his	**CGC**	arg
CUA	leu	**CCA**	pro	**CAA**	gln	**CGA**	arg
CUG	leu	**CCG**	pro	**CAG**	gln	**CGG**	arg
AUU	ile	**ACU**	thr	**AAU**	asn	**AGU**	ser
AUC	ile	**ACC**	thr	**AAC**	asn	**AGC**	ser
AUA	ile	**ACA**	thr	**AAA**	lys	**AGA**	arg
AUG	**met**	**ACG**	thr	**AAG**	lys	**AGG**	arg
GUU	val	**GCU**	ala	**GAU**	asp	**GGU**	gly
GUC	val	**GCC**	ala	**GAC**	asp	**GGC**	gly
GUA	val	**GCA**	ala	**GAA**	glu	**GGA**	gly
GUG	val	**GCG**	ala	**GAG**	glu	**GGG**	gly

Figure 11.6 The genetic code. Apart from methionine and tryptophan, all amino acids can be coded for by more than one triplet codon (for some, e.g. leucine, there may be as many as six). The code is thus said to be degenerate. Three of the triplet sequences are stop codons, and represent the point at which translation of the mRNA message must end. Translation always begins at an AUG codon, meaning that newly-synthesised proteins always begin with a methionine residue. See Box 2.4 for full names of amino acids

three nucleotides) may become disrupted. This would lead to a completely inappropriate sequence of amino acids being produced. *Frameshift mutations* (see below) have this effect. Since there are only 20 amino acids to account for, it follows that the genetic code is degenerate, that is, a particular amino acid may be coded for by more than one triplet. Amino acids such as serine and leucine are encoded by as many as six alternatives each, whilst tryptophan and methionine are the only amino acids to have just a single codon (Figure 11.6).

Transcription in procaryotes

In the first phase of gene expression, one strand of DNA acts as a template for the production of a complementary strand of RNA. In the outline that follows, we shall describe how mRNA is synthesised, but remember that sometimes the product of transcription is rRNA or tRNA. An important point to note is that the coding strand is not the same for all genes; some are encoded on one strand, some on the other. Whereas in DNA replication the whole molecule is copied, an RNA transcript is made only of specific sections of DNA, typically single genes. The enzyme *RNA polymerase*, unlike DNA polymerase, is able to use completely single-stranded material, that is, no primer is required. It is able to synthesise an mRNA chain from scratch, according to the coded sequence on the template, and working in the 5′ to 3′ direction (Figure 11.7). In order to do this the RNA polymerase needs instructions for when to start and finish. First, it recognises a short sequence of DNA called a *promoter*, which occurs upstream of a gene. A protein cofactor called *sigma* (σ) assists in attachment to this, and is released

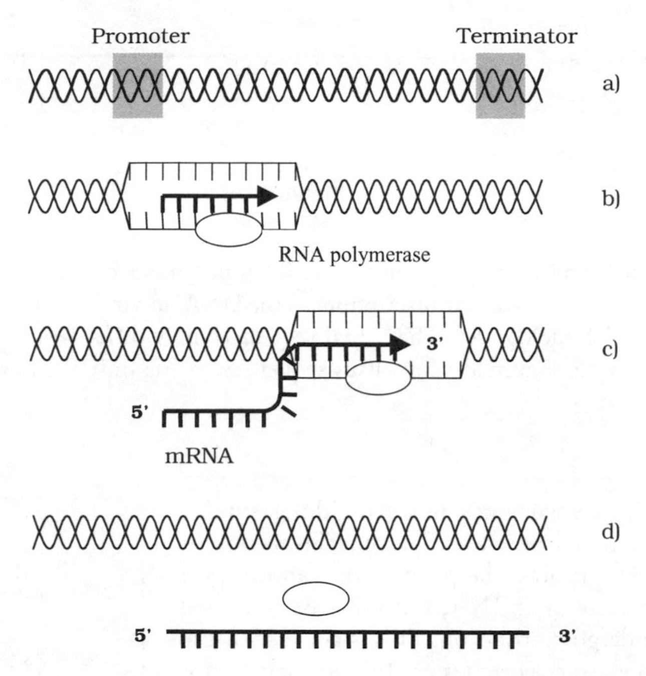

Figure 11.7 Transcription in procaryotes. (a) The gene to be transcribed is flanked by a promoter and a terminator sequence. (b) Following localised unwinding of the double-stranded DNA, one strand acts as a template for RNA polymerase to make a complementary copy of mRNA. (c) The mRNA is extended; only the most recently copied part remains associated with the DNA. (d) After reaching the termination sequence, both RNA polymerase and newly synthesised mRNA detach from the DNA, which reverts to its fully double-stranded state

again shortly after transcription commences. The promoter tells the RNA polymerase where transcription should start, and also on which strand. The efficiency with which a promoter binds the RNA polymerase determines how frequently a particular gene will be transcribed. The promoter comprises two parts, one 10 bases upstream (known as the *Pribnow box*), and the other 35 bases upstream (Figure 11.8). RNA polymerase binds to the promoter, and the double helix of the DNA is caused to unwind a little at a time, exposing the coding sequence on one strand. Ribonucleotides are added one by one to form a growing RNA chain, according to the sequence on the template; this occurs at a rate of some 30–50 nucleotides per second. Remember from Chapter 2 that RNA has uracil rather than thymine, so that a 'U' is incorporated into the mRNA whenever an 'A' appears on the template. Transcription stops when a *terminator* sequence is recognised by the RNA polymerase; both the enzyme and the newly synthesised mRNA are released. Unlike the promoter sequence, the terminator is not transcribed. Some termination sequences are dependent on the presence of a protein called the rho factor (ρ). Groups of bacterial proteins having related functions may have their genes grouped together. Only the last one has a termination sequence, so a single, contiguous mRNA is produced, encoding several proteins (*polycistronic* mRNA).

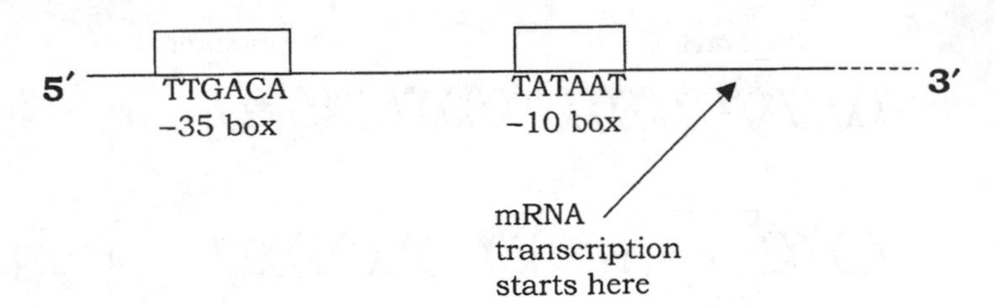

Figure 11.8 The promoter sequence in *E. coli*. RNA polymerase attaches at a point 35 nucleotides upstream of the start of transcription; as the DNA unwinds, it binds to the Pribnow box, situated at −10 nucleotides. The actual sequences may differ from gene to gene; the sequences shown in the figure are consensus sequences. Note: only the non-template DNA strand is shown

Eucaryotic genes generally contain non-coding sequences (introns) in between the coding sequences (exons).

Transcription in eucaryotes proceeds along similar lines, but with certain differences. The most important of these is that in eucaryotes, the product of transcription does not act directly as mRNA, but must be modified before it can undergo translation. This is because of the presence within eucaryotic genes of DNA sequences not involved in coding for amino acids. These are called *introns* (c.f. coding sequences = *exons*), and are removed to give the final mRNA by a process of *RNA splicing* (Figure 11.9).

Translation

The message encoded in mRNA is translated into a sequence of amino acids at the ribosome. The ribosomes are not protein-specific; they can translate any mRNA to synthesise its protein. Amino acids are brought to the ribosome by a transfer RNA (tRNA) molecule. Each tRNA acts as an adaptor, bearing at one end the complementary sequence for a particular triplet codon, and at the other the corresponding amino acid (Figure 11.10). It recognises a specific codon and binds to it by complementary base pairing, thus ensuring that the appropriate amino acid is added to the growing peptide chain at that point. Enzymes called *aminoacyl-tRNA synthetases* ensure that each tRNA is coupled with the correct amino acid in an ATP-dependent process.

There is at least one type of tRNA for each amino acid, each with a three base *anticodon*, enabling it to bind to the complementary triplet sequence on the mRNA. However, there is not a different anticodon for each of the 61 possible codons, in fact there are less than 40. To explain this, the *wobble* hypothesis proposed that certain non-standard pairings are allowed between the third nucleotide of the codon and the first of the anticodon (Table 11.1). This means that a single anticodon may pair with more than one codon (Figure 11.11).

Translation starts when the small ribosomal subunit binds to a specific sequence on the mRNA upstream of where translation is to begin. This is the *ribosome binding site*; in procaryotes this sequence is AGGAGG (the *Shine–Dalgarno* sequence).

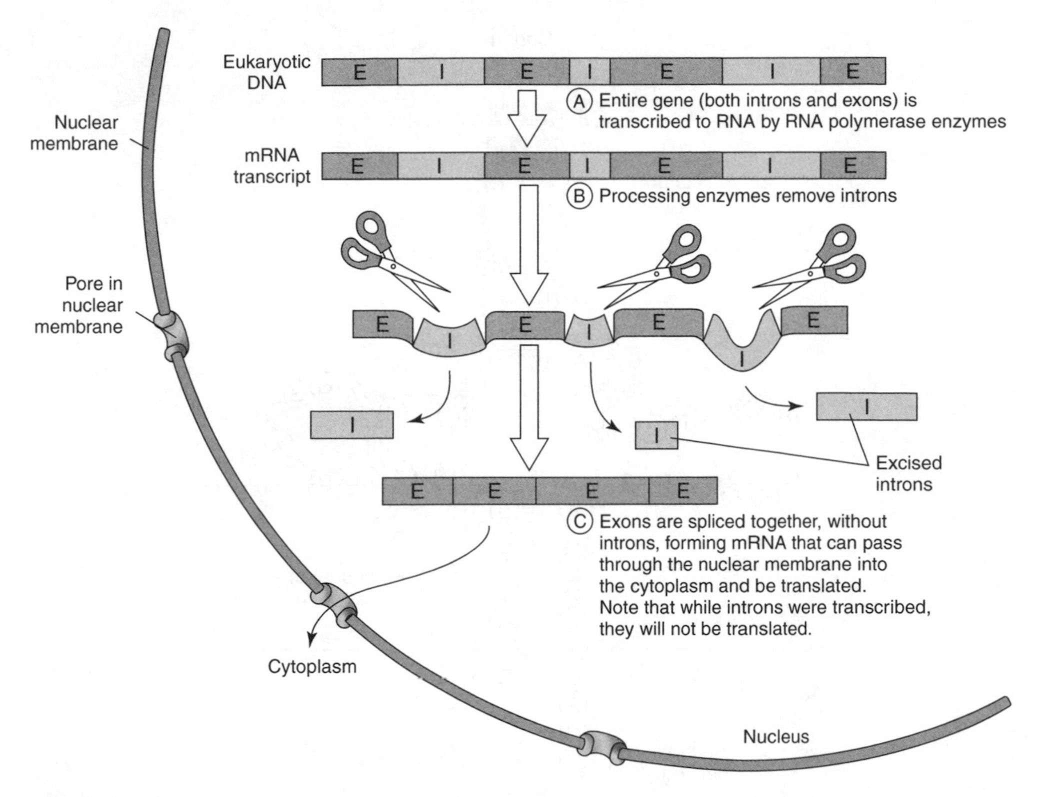

Figure 11.9 Eucaryotic genes contain non-coding sequences. The product of transcription (primary RNA transcript) cannot act as mRNA because it contains sequences that do not code for the final polypeptide. These introns must be removed and the remaining coding sequences (exons) spliced together to give the mature mRNA. From Black, JG: Microbiology: Principles and Explorations, 4th edn, John Wiley & Sons Inc., 1999. Reproduced by permission of the publishers

This sets the ribosome in the correct reading frame to read the message encoded on the mRNA. A tRNA carrying a formylmethionine then binds to the AUG start codon on the mRNA. The large ribosomal subunit joins, and the initiation complex is complete (Figure 11.12). Proteins called *initiation factors* help to assemble the initiation complex, with energy provided by GTP.

The initial amino acid in the chain is always a methionine, corresponding to the AUG start codon. In procaryotes there is a modified form called formylmethionine (fMet); a special tRNA carries it to the initiation site.

The positioning of the large subunit means that the initiation codon (AUG) fits into the P-site, and the next triplet on the mRNA is aligned with the *A-site*. *Elongation* of the peptide chain (Figure 11.13) starts when a second tRNA carrying an amino acid is added at the A-site. *Peptidyl transferase* activity breaks the link between the first amino acid and its tRNA, and forms a peptide bond with the second amino acid. The catalytic action is due partly to the ribozyme activity of the large subunit rRNA. The ribosome moves along

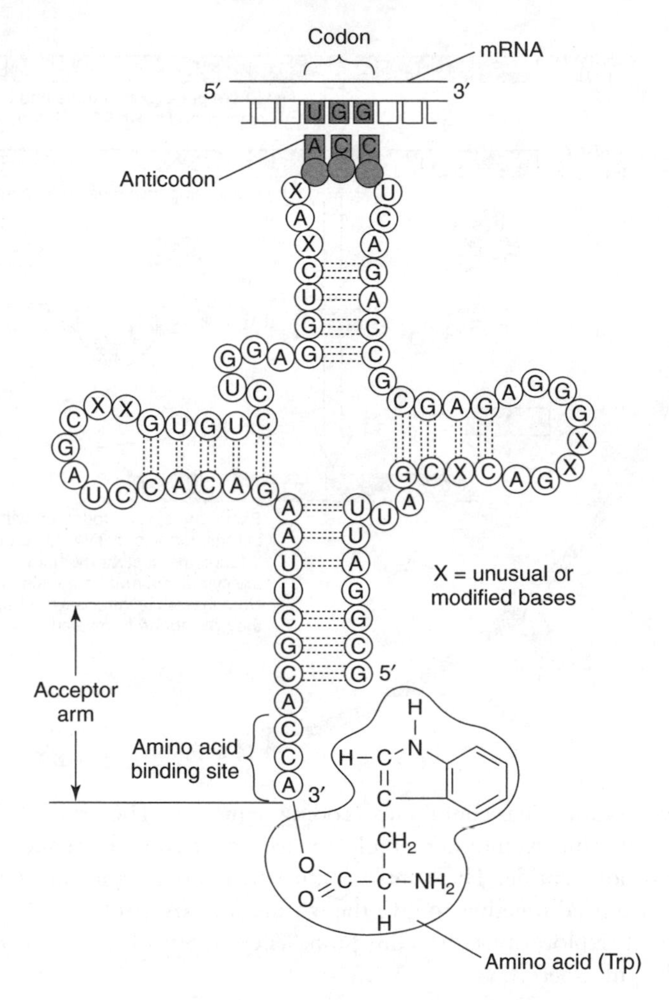

Figure 11.10 Transfer RNA. The single-stranded tRNA adopts its characteristic cloverleaf appearance due to partial base pairing between complementary sequences. The molecule contains some modified nucleotides such as inosine and methylguanosine. Transfer RNA acts as an adaptor between the triplet codon on the mRNA and the corresponding amino acid (tryptophan in the example shown). It base-pairs with the mRNA via a complementary anticodon, thus bringing the amino acid into position for incorporation into the growing peptide chain. From Black, JG: Microbiology: Principles and Explorations, 4th edn, John Wiley & Sons Inc., 1999. Reproduced by permission of the publishers

Table 11.1 The Wobble Hypothesis: Permitted Pairings and Mispairings

1st Anticodon Base	3rd Codon Base
C	G
A	U
U	A or G
G	U or C
I (inosine)	U, C or A

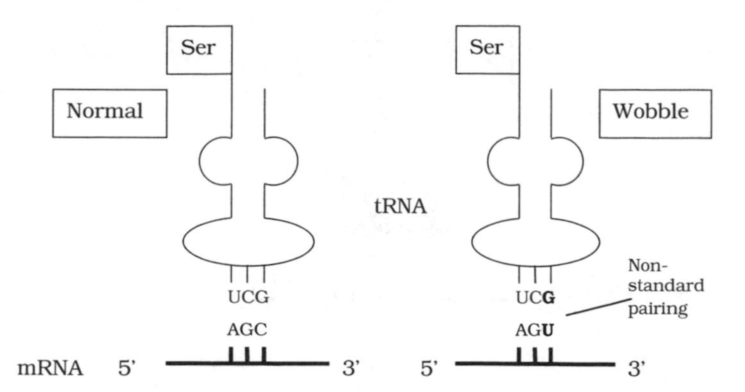

Figure 11.11 The wobble hypothesis. The codons AGC and AGU each encode the amino acid serine. Both can be 'read' by the same tRNA anticodon due to the non-standard 'wobble' pairing allowed at the codon's final base.

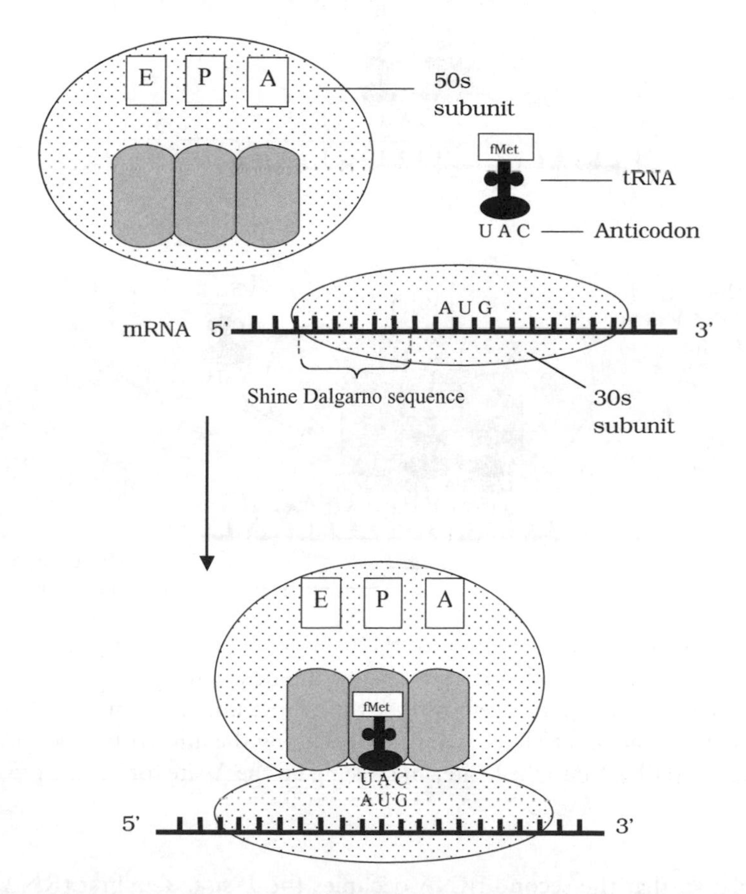

Figure 11.12 Translation: formation of the initiation complex. The small (30S) ribosomal subunit binds to the Shine–Dalgarno sequence, and a formylmethionine initiator tRNA attaches to the AUG codon just downstream. The large ribosomal subunit then attaches so that the fMet-tRNA enters the P-site

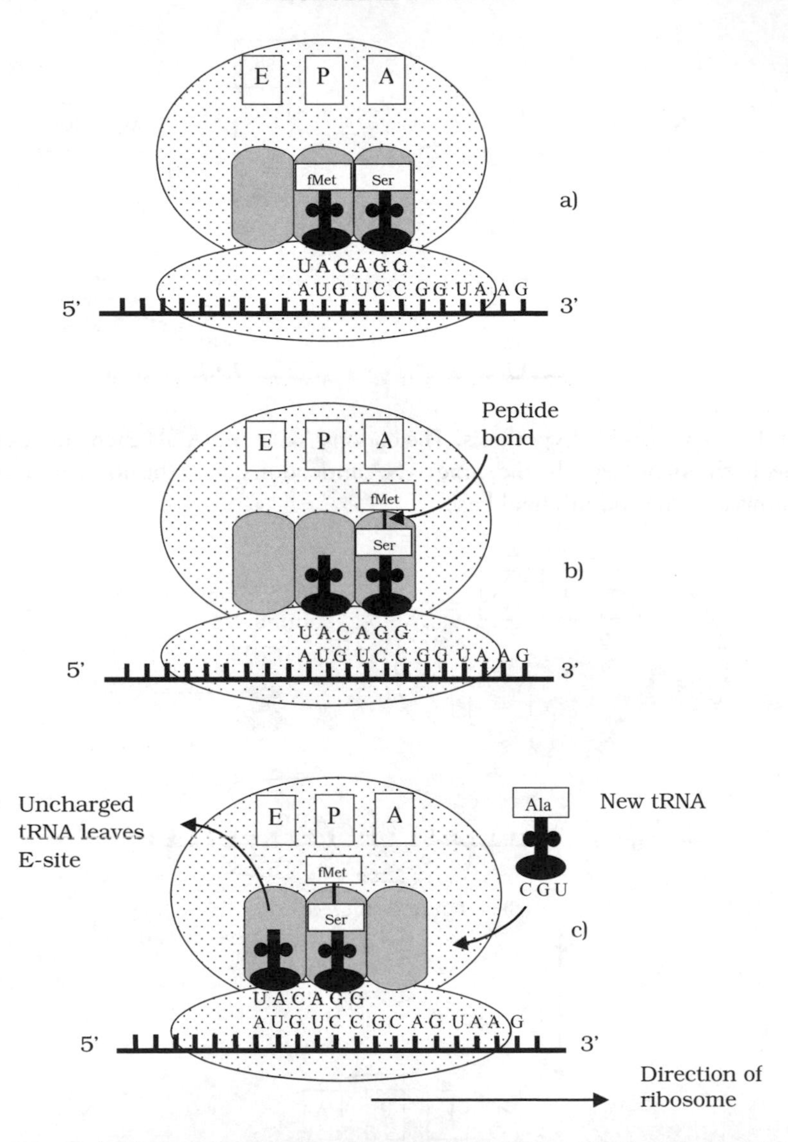

Figure 11.13 Translation: elongation of the peptide chain. (a) Transfer RNA enters the A-site, carrying the amino acid corresponding to the next triplet on the mRNA (serine in diagram). (b) Peptidyl transferase removes the formylmethionine from its tRNA and joins it to the second amino acid. (c) The ribosome moves along the mRNA by one triplet, pushing the first uncharged tRNA into the E-site, and freeing up the A-site for the next tRNA to enter

by one triplet so that the second tRNA occupies the *P-site*. The first tRNA is released from its amino acid, and passes to the *E-site* before being released from the ribosome. A third aminoacyl tRNA moves into the A-site, corresponding to the next codon on the mRNA. Elongation continues in this way until a stop codon is encountered (UAG, UAA, UGA). Release factors cleave the polypeptide chain from the final tRNA and the ribosome dissociates into its subunits.

Regulation of gene expression

The proteins synthesised by microorganisms may broadly be divided into structural proteins required for the fabric of the cell, and enzymes, used to maintain the essential metabolic processes of the cell. It would be terribly wasteful (and possibly harmful) to produce all these proteins incessantly, regardless of whether or not the cell actually required them, so microorganisms, in common with other living things, have mechanisms of *gene regulation*, whereby genes can be switched on and off according to the cell's requirements.

Induction of gene expression

The synthesis of many enzymes required for the catabolism (breakdown) of a substrate is regulated by enzyme induction. If the substrate molecule is not present in the environment, there is little point in synthesising the enzyme needed to break it down. In terms of conservation of cellular energy, it makes more sense for such enzymes to be produced only when they are needed, that is, when the appropriate substrate molecule is present. The substrate itself therefore acts as the *inducer* of the enzyme's synthesis. An example of this that has been studied in great depth concerns the enzyme β-galactosidase, used by *E. coli* to convert the disaccharide lactose into its constituent sugars:

> Enzymes whose production can be switched on and off are called *inducible enzymes*. This distinguishes them from *constitutive enzymes*, which are always produced regardless of prevailing conditions.

$$\text{Lactose} \xrightarrow[\beta\text{-galactosidase}]{} \text{Glucose} + \text{Galactose}$$

β-galactosidase has been intensively studied as part of the *lac operon* of *E. coli*. This is made up of three structural genes designated *Z*, *Y* and *A*, which are clustered together and share a common promoter and terminator (Figure 11.14). The genes code for, respectively, β-galactosidase, a permease and a transacetylase. The permease is necessary for the transport of lactose into the cell, while the role of the transacetylase is not entirely clear, although it is essential for the metabolism of lactose. Grouping the three genes together in this way ensures an 'all-or-nothing' expression of the three proteins. Transcription of these structural genes into their respective mRNAs is initiated by the enzyme RNA polymerase binding to the promoter sequence. However, this is only possible in the presence of lactose; in its absence, a *repressor protein* binds to an operator site, adjacent to the promoter, preventing RNA polymerase binding to the promoter, and therefore preventing mRNA

> A group of functionally related genes involved in the regulation of enzyme synthesis and positioned together at the same locus is called an operon. It contains both structural and regulatory genes.

> A repressor protein regulates the transcription of a gene by binding to its operator sequence. It is encoded by a regulatory gene.

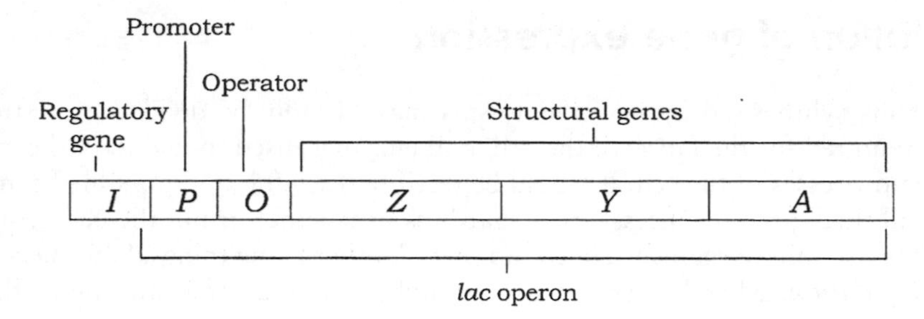

Figure 11.14 The *lac* operon comprises three structural genes under the control of a single promoter and operator sequence. The role of the regulatory gene *I* is described in Figure 11.15

production. Production of the repressor protein is encoded by a *regulator gene* (*I*), situated slightly upstream from the operon (Figure 11.15a).

How then, does the presence of lactose overcome this regulatory mechanism? Allolactose, an isomer of lactose and an intermediate in its breakdown, attaches to a site on the *lac* repressor, thereby reducing the latter's affinity for the operator, and neutralising its blocking effect (Figure 11.15b). The structural genes are then transcribed into mRNA, which is subsequently translated into the three proteins described above, and the lactose is broken down. In the absence of lactose, there are only trace amounts of β-galactosidase present in an *E. coli* cell; this increases some 1000-fold in its presence.

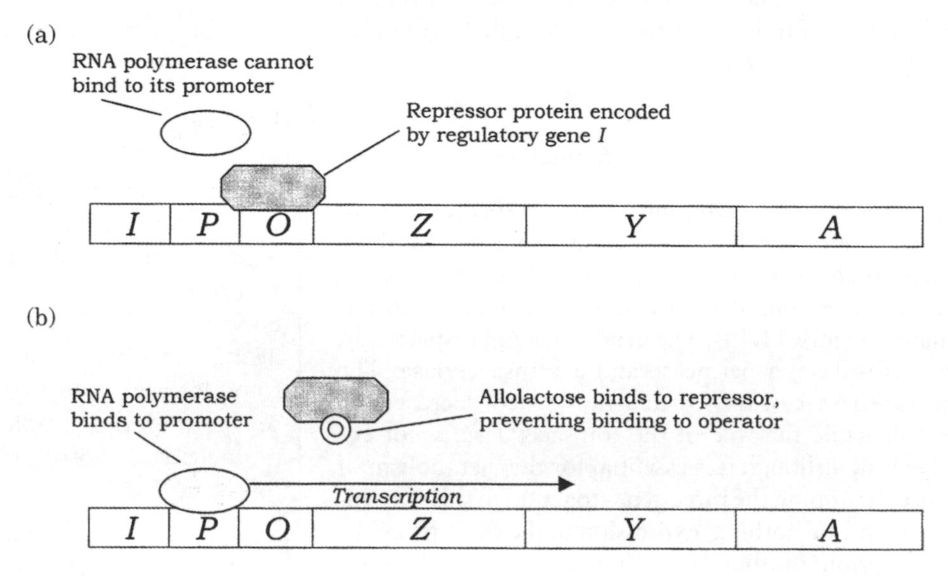

Figure 11.15 The *lac* operon is inducible. (a) In the absence of the substrate lactose, the *lac* operon is 'switched off', due to a repressor protein encoded by the regulatory gene *I*. The repressor binds to the operator site, preventing the binding of RNA polymerase to the promoter and therefore blocking transcription. (b) Allolactose acts as an inducer by binding the repressor protein and preventing it from blocking the promoter site. Transcription of the three structural genes is able to proceed unhindered

When all the lactose has been consumed, the repressor protein is free to block the operator gene once more, and the needless synthesis of further β-galactosidase ceases.

The *lac* operon can also be induced by isopropyl β-thiogalactoside (IPTG); *E. coli* is not able to break this down, so the genes remain permanently switched on. IPTG is utilised as an inducer in cloning systems involving the expression of the *lacZ* gene on pUC plasmids (Chapter 12).

The *lac* operon is subject to control by positive as well as negative regulator proteins. Transcription of the operon only occurs if another regulatory protein called *catabolite activator protein* (CAP) is bound to the promoter sequence (see Box 11.5). This is dependent on a relatively high concentration of the nucleotide cAMP which only occurs when glucose is scarce. The activation of the *lac* operon thus occurs only if lactose is present and glucose is (almost) absent.

Repression of gene expression

The induction of gene expression, such as we have just described for the *lac* operon, generally relates to catabolic (breakdown) reactions. Anabolic (synthetic) reactions, such as those leading to the production of specific amino acids, by contrast, are often controlled by the *repression* of key genes.

Enzyme repression mechanisms operate along similar lines to induction mechanisms, but the determining factor here is not the substrate of the enzymes in question (lactose in our example), but the end-product of their action. The *trp* operon contains a cluster of genes encoding five enzymes involved in the synthesis of the amino acid tryptophan. (Figure 11.16) In the presence of tryptophan, the cell has no need to synthesise its

Box 11.5 A choice of substrates

Glucose is central to the reactions of glycolysis (Chapter 6), and is utilised by *E. coli* with high efficiency, because the enzymes involved are permanently switched on or *constitutive*. The β-galactosidase required for lactose breakdown, however, must be induced. What happens then, when *E. coli* is presented with a mixture of both glucose and lactose? It would be more efficient to metabolise the glucose, with the ready-to-use enzymes, but from what you have learnt elsewhere in this section (see Figure 11.15b), the presence of lactose would induce formation of β-galactosidase and subsequent lactose breakdown, a less energy-efficient way of going about things. In fact, *E. coli* has a way of making sure that while the readily utilised glucose is present, it takes precedence. It does this by repressing the formation of β-galactosidase, a phenomenon known as *catabolite repression*. Thus, the presence of a 'preferred' nutrient prevents the synthesis of enzymes needed to metabolise a less favoured one.

This is because glucose inhibits the formation of cAMP, which is required for the binding of the CAP to its site on the *lac* promoter. When glucose levels drop, more cAMP forms and causes CAP to bind to the CAP binding site. Thus, after a delay, the enzymes needed for lactose catabolism are synthesised, and the lactose is utilised, leading to a diauxic growth curve (see Chapter 5).

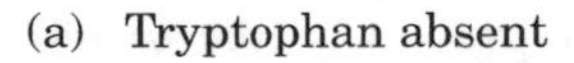

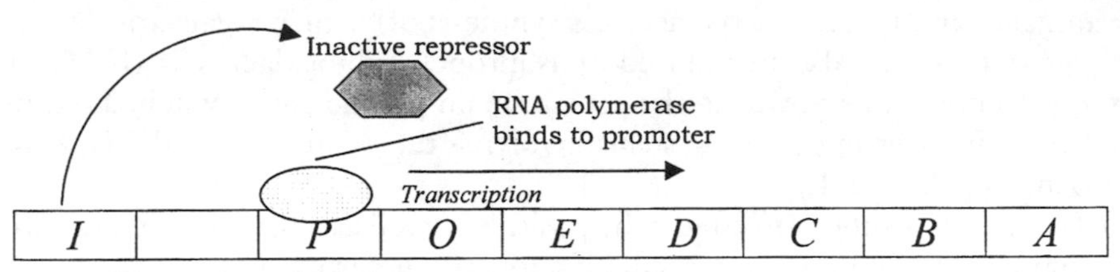

All five enzymes for tryptophan synthesis produced

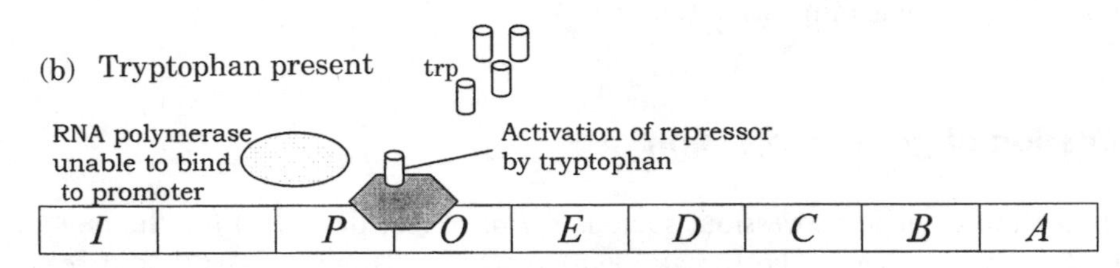

No enzymes produced for tryptophan synthesis

Figure 11.16 The *trp* operon. Five structural genes *A-E* encode enzymes necessary for the synthesis of tryptophan. (a) In the absence of tryptophan, transcription of the operon proceeds unhindered. Although a repressor protein is produced, it is inactive and unable to bind to the operator sequence. (b) Tryptophan activates the repressor by binding to it. This prevents RNA polymerase from binding to the promoter, and transcription is blocked

own, so the operon is switched off. This is achieved by tryptophan binding to and activating a repressor protein, which in turn binds to the operator of the *trp* operon and prevents transcription of the synthetic enzymes. The tryptophan here is said to act as a *corepressor*. As tryptophan is used up and its level in the cell falls, the repressor reverts to its inactive form, allowing transcription of the tryptophan-synthesising enzymes to go ahead unhindered.

Global gene regulation

The systems of gene regulation discussed above apply to individual operons; sometimes, however, a change in environmental conditions necessitates the regulation of many genes at once. These global regulatory systems respond to stimuli such as oxygen depletion and temperature change, and utilise a number of different mechanisms.

The molecular basis of mutations

Any alteration made to the DNA sequence of an organism is called a *mutation*. This may or may not have an effect on the phenotype (physically manifested properties) of the organism. It may for example enable bacteria to grow without the need for a particular growth supplement, or confer resistance to an antibiotic. Just how this happened was for many years a source of debate. Since the mutant forms only became apparent after the change in conditions (e.g. withdrawal of a nutrient, addition of antibiotic), have some of the bacteria been *induced* to adapt to the new conditions, or are mutant forms arising all the time at a very low frequency, and merely *selected* by the environmental change? In 1943, Salvador Luria and Max Delbrück devised the *fluctuation test* to settle the matter (see Box 11.6). Results of the fluctuation test together with other evidence led to the understanding that mutations occur spontaneously in nature at a very low frequency. As we shall see later on in this section, however, they can also be induced by a variety of chemical and physical agents. Any change to the DNA sequence is heritable, thus mutations represent a major source of evolutionary variation. Bacteria make marvellous tools for the study of mutations because of their huge numbers and very short generation times.

Since the DNA sequence of a gene represents highly ordered coded information, most mutations have a neutral or detrimental effect on the organism's phenotype, but occasionally a mutation occurs which confers an advantage to an organism, making it better able to survive and reproduce in a particular environment. Mutants that are favoured in this way may eventually become the dominant type in a population, and, by steps like this, evolution gradually takes place.

Mutations occur spontaneously in any part of an organism's genome. Spontaneous mutations causing an inactivation of gene function occur in bacteria at the rate of about one in a million for a given gene at each round of cell division. Most genes within a given organism show similar rates of mutation, relative to their gene size; clearly a larger 'target' will be 'hit' more often than a small one.

How do mutations occur?

Figure 11.17(a) reminds us how the code in DNA is transcribed into messenger RNA and then translated into a sequence of amino acids. Each time the DNA undergoes replication, this same sequence will be passed on, coding for the same sequence of amino acids. Occasionally, mistakes occur during replication. Cells have repair mechanisms to minimise these errors, but what happens if a mistake still slips through? In Figure 11.17(b), we can see the effect of one nucleotide being inserted into the strand instead of another. When the next round of replication occurs, the modified DNA will act as a template for a newly synthesised strand, which at this position will be made complementary to the new, 'wrong' base, instead of the original one, and thus the mistake will be perpetuated.

This is an example of the simplest type of mutation, a *point* mutation, where one nucleotide has been substituted by another. The example shown is a *missense* mutation, which has resulted in the affected triplet coding for a different amino acid; this may or may not have an effect on the phenotype of the organism. RNA polymerase, which transcribes the DNA sequence into mRNA, is unable to tell that an error has occurred, and faithfully

> A missense mutation alters the sense of the message encoded in the DNA, and results in an incorrect amino acid being produced at the point where it occurs.

Box 11.6 Settling an argument: the fluctuation test

Luria and Delbrück designed the fluctuation test to show whether resistance in *E. coli* to the bacteriophage T1 was *induced* or occurred *spontaneously*.

Suppose we divide a broth culture of bacteria into two, then inoculate half into a flask of fresh broth, and divide the other half between a large number of smaller cultures in tubes

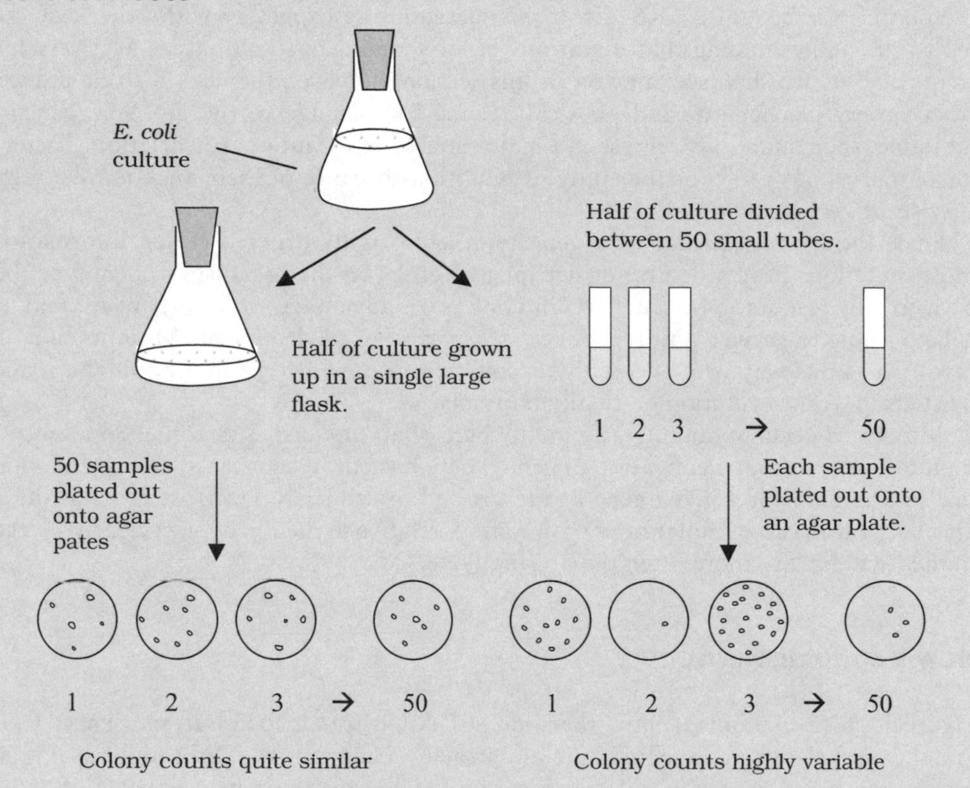

After allowing the bacteria to grow, samples are taken from tubes and flask, and spread onto a selective agar medium (one covered with T1 phage). All the plates deriving from the single bulk culture have roughly the same number of resistant colonies, but the plates resulting from the smaller, tube cultures show very variable colony counts. Although the *average* number of colonies across the 50 tubes is similar to that obtained from the bulk culture, the *individual* plate counts varies greatly, from none on several plates to over a hundred on another.

If the phage-resistance was *induced* by the presence of the phage, we would expect all plates in the experiment to give rise to approximately the same number of resistant colonies, as all cultures experienced the same exposure. If however, the resistant forms were spontaneously arising all the time at a low level in the population, the numbers of resistant colonies would be dependent on when, if at all, a mutation had taken place in a particular tube culture. A mutant arising early in the incubation period would give rise to more resistant offspring and therefore more colonies than one that arose later.

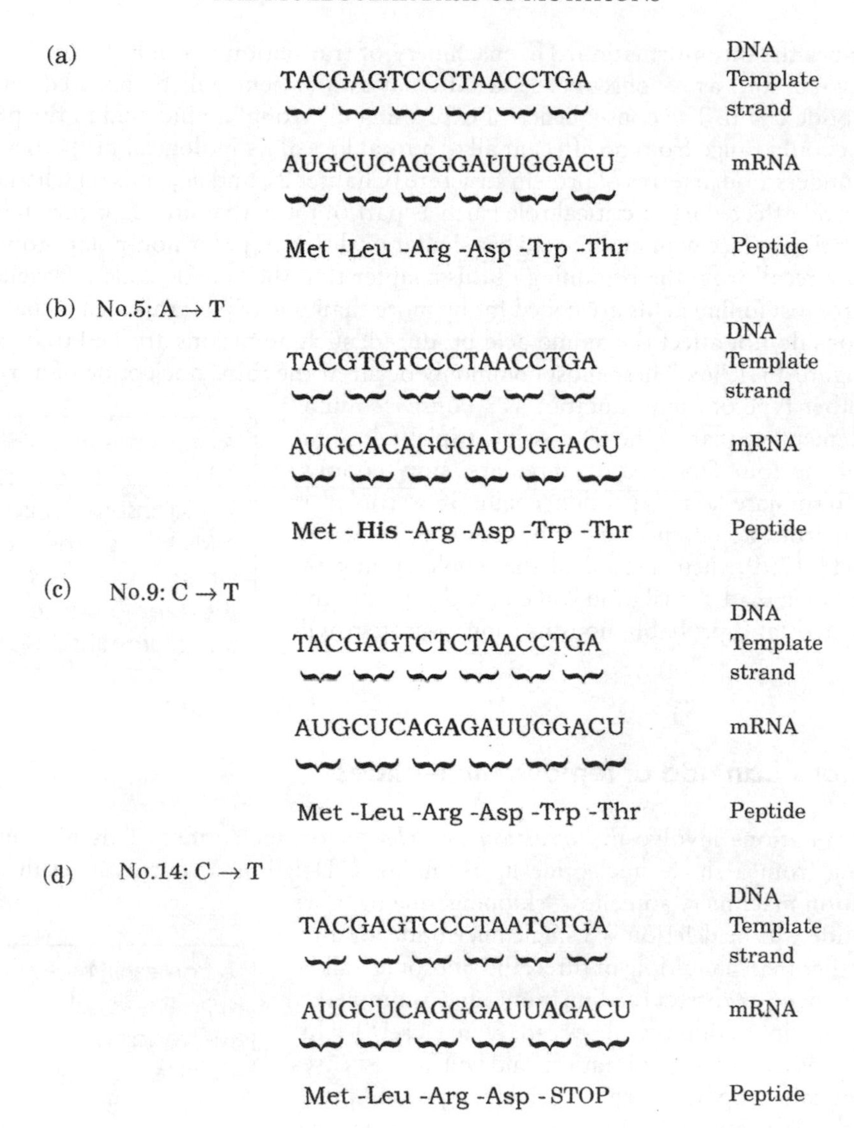

Figure 11.17 Mutations can alter the sense of the DNA message. (a) A short sequence of DNA is transcribed into mRNA and then transcribed into the corresponding amino acids. (b) A single base change from T to A results in a missense mutation, as a histidine is substituted for a leucine. (c) A silent mutation has altered the DNA (and therefore mRNA) sequence, but has not changed its sense. Both AGG and AGA are mRNA triplets that code for arginine. (d) A nonsense mutation has replaced a tryptophan codon with a STOP codon, hence bringing the peptide chain to a premature end

transcribes the misinformation. The machinery of translation is similarly 'unaware' of the mistake, and as a consequence, a different amino acid will be inserted into the polypeptide chain. The consequence of expressing a 'wrong' amino acid in the protein product could range from no effect at all to a total loss of its biological properties. This can be understood in terms of protein structure (Chapter 2), and depends on whether the amino acid affected has a critical role (such as part of the active site of an enzyme), and whether the replacement amino acid has similar or different polar/non-polar properties. You may recall from the beginning of this chapter that the genetic code is degenerate, and that most amino acids are coded for by more than one triplet; this means that some mutations do not affect the amino acid produced; such mutations are said to be *silent*, as in Figure 11.17(c). These most commonly occur at the third nucleotide of a triplet.

Another type of point mutation is a *nonsense* mutation. Remember that of the 64 possible triplet permutations of the four DNA bases, three are 'stop' codons, which terminate a polypeptide chain. If a triplet is changed from a coding to a 'stop' codon as shown in Figure 11.17(d), then instead of the whole coding sequence being read, translation will end at this point, and a truncated (and probably non-functional) protein will result.

> A nonsense mutation results in a 'stop' codon being inserted into the mRNA at the point where it occurs, and the premature termination of translation.

Mutations can add or remove nucleotides

Other mutations involve the *insertion* or *deletion* of nucleotides. This may involve anything from a single nucleotide up to millions. Deletions occur as a result of the replication machinery somehow 'skipping' one or more nucleotides. If the deletion is a single nucleotide, or anything other than a multiple of three, the ribosome will be thrown out of its correct reading frame, and a completely new set of triplet codons will be read (Figure 11.18). This is known as a *frameshift* mutation, and will in most cases result in catastrophic changes to the final protein product. If the deletion is a multiple of three nucleotides, the reading frame will be preserved and the effect on the protein less drastic.

> A frameshift mutation results in a change to the reading frame, and an altered sequence of amino acids results downstream of the point where it occurs.

Mutations can be reversed

Just as it is possible for a mutation to occur spontaneously, so it is possible for the nucleotide change causing it to be spontaneously reversed – in other words, a mutant can mutate back to being a *wildtype*. This is known for obvious reasons as a *reverse* or back mutation. When this happens, the original genotype and phenotype are restored. Whereas a forward mutation results from any change that inactivates a gene, a back mutation is more specific; it must restore function to a protein damaged by a specific

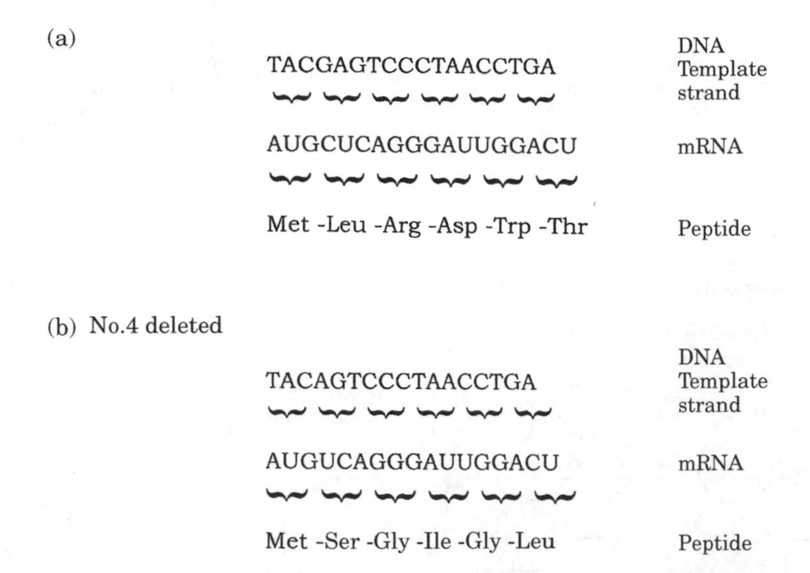

Figure 11.18 Frameshift mutations. (a) A short sequence of DNA is transcribed into mRNA and then transcribed into the corresponding amino acids. (b) The fourth nucleotide in the sequence is deleted, upsetting the groups of triplets or reading frame. This alters the sense of the remainder of the message, leading to a completely different sequence of amino acids

mutation. Not surprisingly, given this specificity, the rate of back mutations is much less frequent.

It is possible for the wildtype phenotype to be restored, not through a reversal of the original base change, but due to a second mutation at a different location. The effect of this second mutation is to suppress the effects of the first one. These are called *suppressor* or second site mutations. They are double mutants that produce a *pseudowildtype*; the phenotype appears to be wildtype, but the genotype differs.

Mutations have a variety of mechanisms

Why are mistakes made every so often during DNA replication? One source of erroneous base incorporation is a phenomenon called *tautomerism*. Some of the nucleotide bases occur in rare alternative forms, which have different base-pairing properties. For example, as you will recall from Chapter 2, cytosine normally has an amino group that provides a hydrogen atom for bonding with the complementary keto group of guanine. Occasionally, however (once in about every 10^4–10^5 molecules), the cytosine may undergo a rearrangement called a tautomeric shift, which results in the amino group changing to an imino group (=NH), and this now behaves in pairing terms as if it were thymine, and therefore pairs with adenine (Figure 11.19a). Similarly, thymine might undergo a tautomeric shift, changing its usual keto form (C=O) to the rare enol form (COH). This then takes on the pairing properties of cytosine and pairs with guanine (Figure 11.19b). The result of such mispairing is that at subsequent rounds of replication,

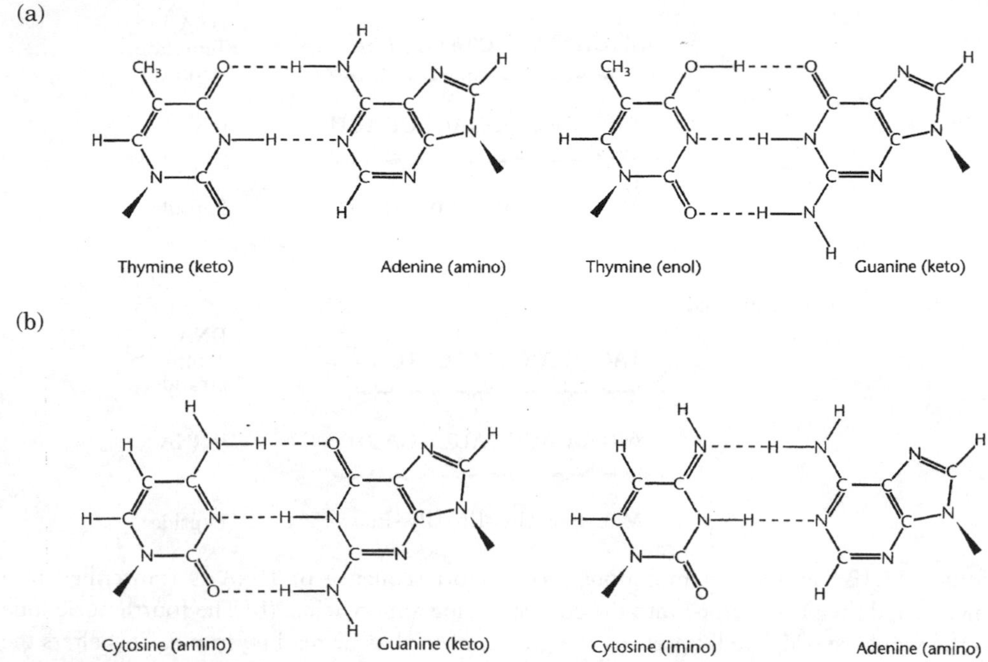

(a)

Thymine (keto) Adenine (amino) Thymine (enol) Guanine (keto)

(b)

Cytosine (amino) Guanine (keto) Cytosine (imino) Adenine (amino)

Figure 11.19 Rare forms of nucleotides may cause mispairing. If DNA replication takes place whilst a base is in its rare alternative form, an incorrect base will be incorporated into the newly synthesised second strand. (a) In its normal keto form, thymine pairs with adenine in keeping with the rules of base-pairing. In its rare enol form, however, it preferentially pairs with guanine. (b) Cytosine can likewise assume the rare imino form and mispair with adenine

one half of the DNA molecules will contain a wrong base pair at that point. The fact that spontaneous mutations occur much less frequently than the rate just quoted is due to the DNA repair mechanisms discussed earlier in this chapter.

Mutations also occur in viruses

As we saw in Chapter 10, the genome of viruses can be composed of either DNA or RNA, and these are subject to mutation just like cellular genomes. The rate of mutation in RNA viruses is much higher than in those containing DNA, about a thousand times higher, in fact. This is significant, because by frequently changing in this way, pathogenic viruses are able to stay one step ahead of host defence systems.

Mutagenic agents increase the rate of mutations

In the early part of the 20th century, the existence of mutations was appreciated, but attempts to gain a fuller understanding of them were hampered by the fact that they occurred so rarely. In the 1920s, however, it was shown that the mutation rate in both barley plants and the fruit fly *Drosophila* was greatly increased as a result of exposure to X-rays. In the following decades, a number of chemical and physical agents were shown also to cause mutations. *Mutagens* such as these raise the general level of mutations in a population, rather than the incidence of mutation in at particular location in the genome. They are extremely useful tools for the microbial geneticist, but need to be treated with great care because they are also mutagenic (and in most cases carcinogenic) towards humans.

Chemical or physical agents capable of inducing mutations are termed mutagens.

Chemical mutagens can be divided into five classes:

Base analogues

A base analogue is able to 'mimic' one of the four normal DNA bases by having a chemical structure sufficiently similar for it to be incorporated into a DNA molecule instead of that base during replication. One such base is 5-bromouracil (5-BU); in its usual keto form, it acts like thymine (which it closely resembles – see Figure 11.20), and therefore pairs with adenine, and is not mutagenic. However 5-BU is capable of tautomerising to the enol form, which, remember, pairs with G, not A. Whereas for thymine the enol form is a rarity, for 5-BU it is more common, hence mispairing with guanine occurs more frequently. So, we get the same outcome as with spontaneous mutation – the TA has been replaced by a CG – but its frequency is much increased. Several other base analogues, such as 2-aminopurine, which mimics adenine, have similar effects to 5-BU. Mutations brought about by base analogues all result in a purine being replaced by another purine, or a pyrimidine being replaced by another pyrimidine; this type of mutation is called a *transition*.

Alkylating agents

As the name suggests, this group of mutagens act by adding an alkyl group (e.g. methyl, ethyl) at various positions on DNA bases. Ethylethanesulphonate (EES) and ethylmethanesulphonate (EMS) are alkylating agents, used as laboratory mutagens. They act by alkylating guanine or thymine at the oxygen atom involved in hydrogen bonding. This leads to an impairment of the normal base-pairing properties, and causes guanine for example to mispair with thymine.

Substitutions brought about by alkylating agents can be either transitions or *transversions*.

Deaminating agents

Nitrous acid is a potent mutagen, which alters the base-pairing affinities of cytosine and adenine by replacing an amino group with an oxygen atom. Transitions occur in both directions, AT $\rightarrow$ GC and GC $\rightarrow$ AT. Note that although guanine is deaminated, its base pairing properties are not affected.

Intercalating agents

These mutagens exert their effect by inserting themselves between adjacent nucleotides in a single strand of DNA. They distort the strand at this site and cause either the addition or, less commonly, deletion of a nucleotide. This leads to frameshift mutations, and often

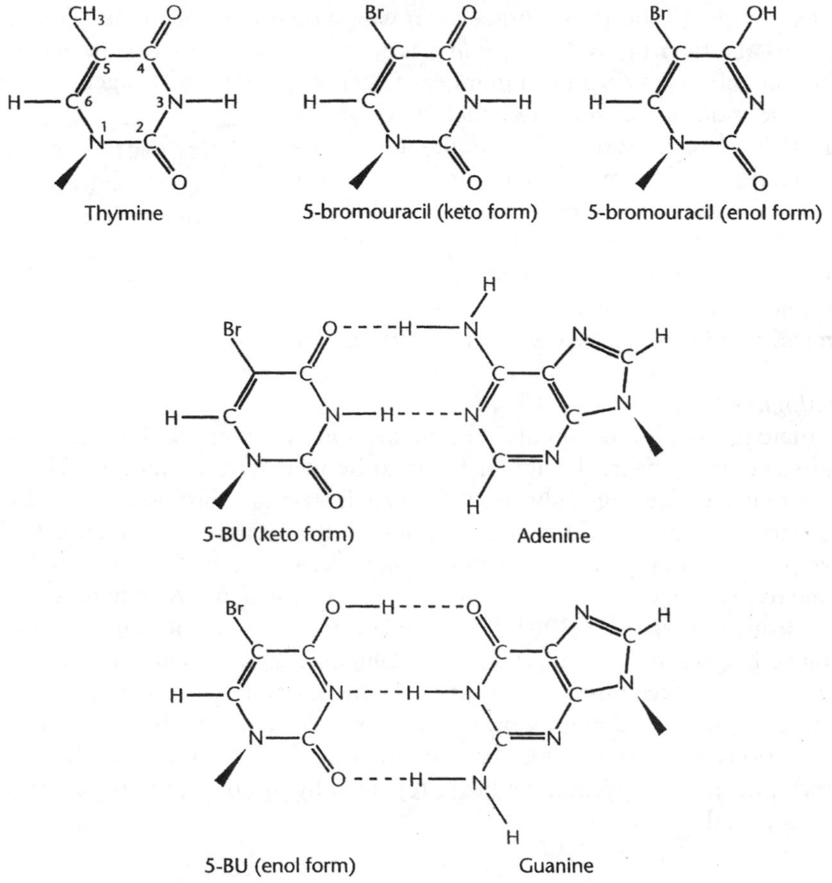

Thymine 5-bromouracil (keto form) 5-bromouracil (enol form)

5-BU (keto form) Adenine

5-BU (enol form) Guanine

Figure 11.20 Base analogues mimic DNA base structure. In its keto form, the base analogue 5-bromouracil (5-BU) forms base pairs with adenine, but in the enol form it mispairs with guanine. 5-BU differs from thymine only in the substitution of a bromine atom in place of a methyl group. This makes the occurrence of a tautomeric shift more likely

during replication the bound mutagen interferes with new strand synthesis, leading to a gross deletion of nucleotides. Ethidium bromide, commonly used for the visualisation of small amounts of DNA in the molecular biology laboratory, is an intercalating agent; it has a planar structure with roughly the same dimensions as a purine–pyrimidine pairing.

Hydroxylating agents
Hydroxylamine has a specific mutagenic effect, hydroxylating the amino group of cytosine to cause the transition of GC → AT.

Base analogues and certain intercalating agents can only exert their mutagenic effects if they are incorporated into DNA while it is replicating. Others, which depend on altering base pairing by modifying the structure of DNA bases, are effective on both replicating and non-replicating DNA.

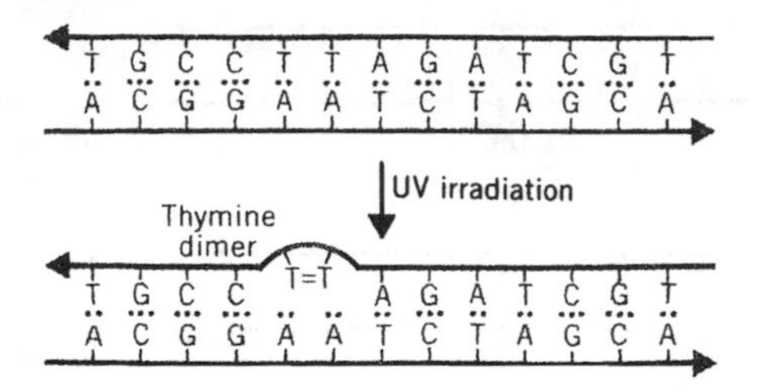

Figure 11.21 Thymine dimer formation. Absorption of UV light results in the formation of dimers between adjacent thymine residues in the same strand. This can lead to errors being introduced into the DNA sequence. From Gardner, EJ, Simmons MJ and Snustad, DP: Principles of Genetics, 8th edn, John Wiley & Sons Inc., 1991. Reproduced by permission of the publishers.

Physical mutagens

As mentioned at the start of this section, the first mutagenic agent to be demonstrated were X-rays; along with ultraviolet light, they are the most commonly used physical mutagen. X-rays, like other forms of ionising radiation, cause the formation of highly reactive free radicals. These can bring about changes in base structure as well as gross chromosomal alterations. DNA strands can be broken and reannealed incorrectly, to produce errors in the sequence.

In recent years, we have all become much more aware of the possible perils of excessive exposure to sunlight. Ultraviolet light (UV) from the sun damages DNA in skin cells, which can lead to them becoming cancerous. The specific action is on adjacent pyrimidine bases (usually thymines) on the same strand, which are cross-linked to form a dimer (Figure 11.21). This results in a distortion of the helix and interferes with replication. UV light is most effective at wavelengths around 260 nm, as this is the wavelength most strongly absorbed by the DNA bases.

DNA damage can be repaired

All organisms have developed the means to repair damage to their DNA, in addition to the proofreading mechanism described earlier

The most common way of dealing with mutations is by means of a method called *excision repair*, in which enzymes recognise and cut out the altered region of DNA and then fill in the missing bases, using the other strand as a template (Figure 11.22). *Mismatch repair* is used to repair the incorporation of a base that has escaped the proofreading system. In a situation such as this, it is not immediately obvious which is the correct base and which is the mistake, so how does the cell know which strand to replace? In *E. coli*, the old strand is distinguished from the new by the fact that some of its bases are methylated. This only occurs some time after replication, so newly synthesised strands will not have the methyl groups and can thus be recognised.

Less frequently, the alteration in DNA structure is simply reversed by *direct repair* mechanisms, the best known of which is *photoreactivation*. This involves an enzyme

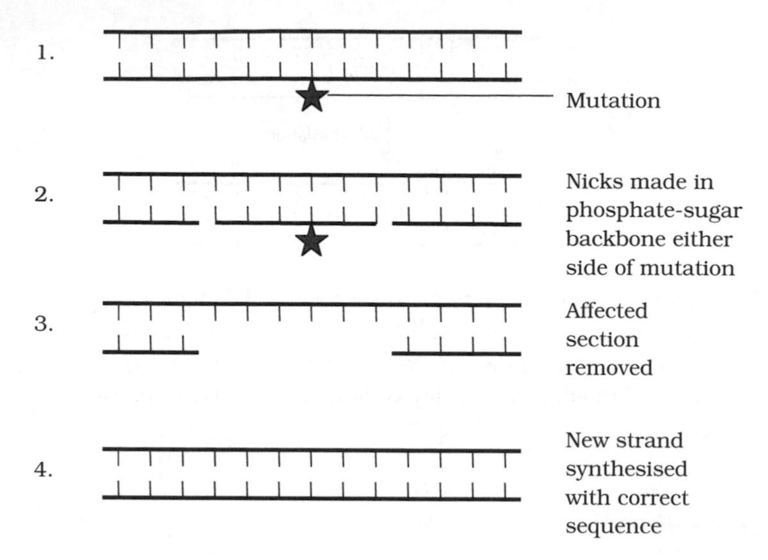

Figure 11.22 Excision repair. A section of the affected strand either side of the mutation is removed and replaced by the action of DNA polymerase and DNA ligase

called *DNA photolyase*, which breaks the bonds formed between adjacent thymine bases by UV light (see above) before replication takes place. Unusually, the photolyase is dependant on visible light (>300 nm) for activation. Another form of direct repair, involving the enzyme methylguanine transferase allows the reversal of the effects of alkylating agents.

Loss of repair mechanisms can allow mutations to become established which would normally be corrected. A harmful strain of *E. coli* which emerged in the 1980s was shown to have developed its pathogenicity due to a deficiency in its repair enzymes.

Carcinogenicity testing: the Ames test

The great majority of *carcinogenic* substances, that is, substances that cause cancer in humans and animals, are also *mutagenic* in bacteria. This fact has been used to develop an initial screening procedure for carcinogens; instead of the expensive and time-consuming process of exposing laboratory animals (not to mention the moral issues involved), a substance can be tested on bacteria to see if it induces mutations.

The *Ames Test* assesses the ability of a substance to cause reverse mutations in auxotrophic strains of *Salmonella* that have lost the ability to synthesise the amino acid histidine (*his*$^-$). Rates of back mutation (assessed by the ability to grow in a histidine-free medium) are compared in the presence and absence of the test substance (Figure 11.23). A reversion to *his*$^+$ at a rate higher than that of the control indicates a mutagen. Many substances are procarcinogens, only becoming mutagenic/carcinogenic after metabolic conversion by mammals; in order to test for these, an extract of rat liver is added to the experimental system as a source of the necessary enzymes.

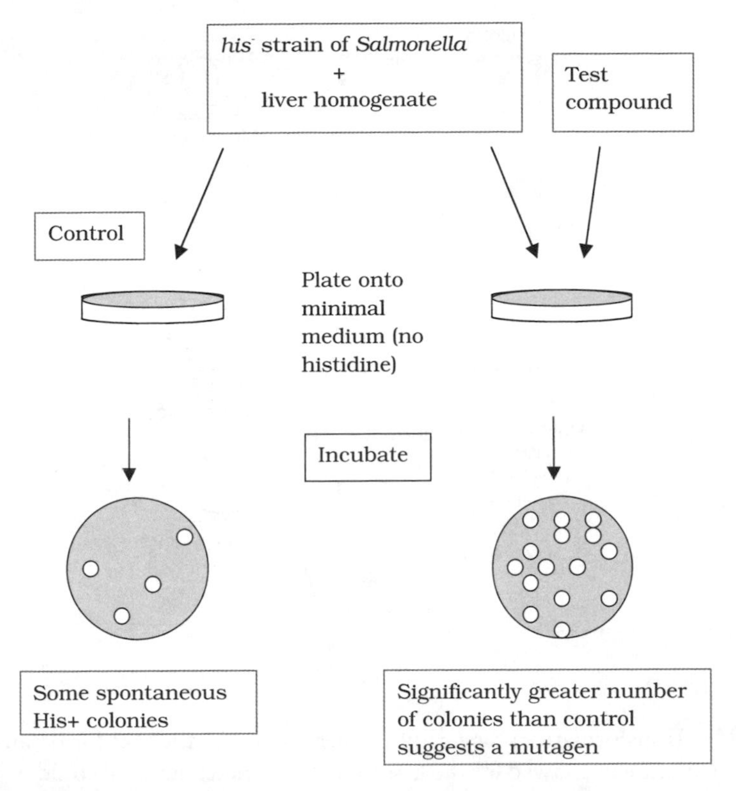

Figure 11.23 The Ames test. The test assesses the ability of a mutagen to bring about reverse mutations in a strain of *S. typhimurium* auxotrophic for histidine. Since it is possible that a small number of revertants may occur due to spontaneous mutation, results are compared with a control plate with no added mutagen

Genetic transfer in microorganisms

The various mechanisms of mutation described above result in an alteration in the genetic make-up of an organism. This can also occur by *recombination*, in which genetic material from two cells combines to produce a variant different to either parent cell.

In bacteria, this involves the transfer of DNA from one cell (*donor*) to another (*recipient*). Because transfer occurs between cells of the same generation (unlike the genetic variation brought about by sexual reproduction in eucaryotes, where it is passed to the next generation), it is sometimes referred to as *horizontal transfer*. There are three ways in which gene transfer can occur in bacteria, which we shall explore in the following sections.

Transformation

Transformation is the simplest of these, and also the first to have been described. We have already referred to the classic experiment of Fred Griffith in 1928, the first demonstration

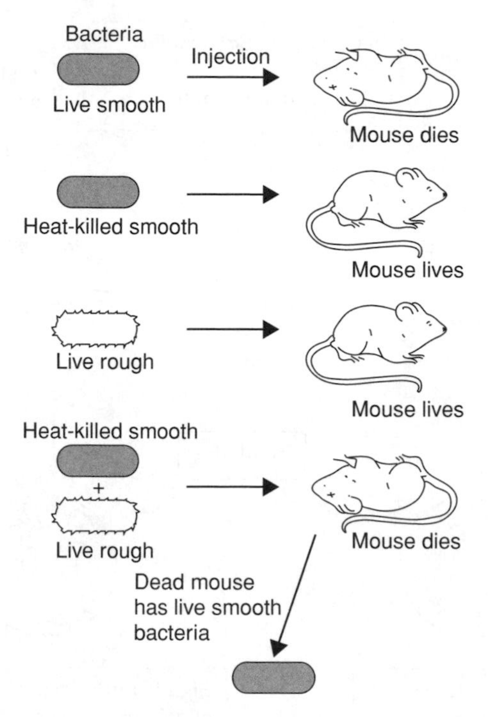

Figure 11.24 Transformation: the Griffith experiment. See the text for details. The result of the fourth experiment showed for the first time that genetic material could be passed from one bacterium to another. From Reece, RJ: Analysis of Genes and Genomes, John Wiley & Sons, 2003. Reproduced by permission of the publishers.

that genetic transfer can occur in bacteria. Griffith had previously demonstrated the existence of two strains of the bacterium *Streptococcus pneumoniae*, which is one of the causative agents of pneumonia in humans, and is also extremely virulent in mice. The *S (smooth)-form* produced a polysaccharide capsule, whilst the *R (rough)-form* did not. These formed recognisably different colonies when grown on a solid medium, but more importantly, differed in their ability to bring about disease in experimental animals. The R-form, lacking the protective capsule, was easily destroyed by the defence system of the host. Griffith observed the effects of injecting mice with bacterial cells of both forms; these are outlined below and in Figure 11.24. The results of experiments 1–3 were predictable, those of experiment 4 startlingly unexpected:

1 Mice injected with live cells of the R-form were unaffected.

2 Mice injected with live cells of the S-form died, and large numbers of S-form bacteria were recovered from their blood.

3 Mice injected with cells of the S-form that had been killed by heating at 60 °C were unharmed and no bacteria were recovered from their blood.

4 Mice injected with a mixture of living R-form and heat-killed S-form cells died, and living S-form bacteria were isolated from their blood.

The S-form bacteria recovered from the mice in the crucial fourth experiment possessed a polysaccharide capsule like other S-forms, and, critically, *were able to pass on this characteristic to subsequent generations*. This finding went against the prevailing view that bacteria simply underwent binary fission, a completely asexual process involving no genetic transfer. Griffith deduced that some as yet unknown substance had passed from the heat-killed S-form cells to some of the living R-forms and conferred on them the ability to make capsules (see Box 11.7). Not long afterwards it was shown that this process of *transformation* could happen in the test tube, without the involvement of a host animal, and, as we have seen, it was eventually shown that Griffith's 'transforming principle' was DNA.

How does transformation occur?

The uptake of foreign DNA from the environment is known to occur naturally in a number of bacterial types, both Gram-positive and Gram-negative, by taking up fragments of *naked* DNA released from dead cells in the vicinity. Being linear and very fragile, the DNA is easily broken into fragments, each carrying on average around 10 genes. Transformation will only happen at a specific stage in the bacterial life cycle, when cells are in a physiological state known as *competence*. This occurs at different times in different bacteria, but is commonly during late log phase. One of the reasons why only a low percentage of recipient cells become transformed is that only a small proportion of them are at any one time in a state of competence. The expression of proteins essential to the transformation process is dependant on the secretion of a *competence factor*.

The exact mechanism of transformation varies somewhat according to species; the process for *Bacillus subtilis* is shown in Figure 11.25. Mere uptake of exogenous DNA is not enough to cause transformation; it must also be integrated into the host genome, displacing a single strand, which is subsequently degraded. Upon DNA replication and cell division, one daughter cell will inherit the parent genotype, and the other

Box 11.7 Transformation is not due to reverse mutation

At the time of Griffith's experiment, it was already known that the wildtype S-form could mutate to the R-form and vice versa. It might therefore be argued that the results of his fourth experiment could easily be explained away in this way. Griffith's experiment was sufficiently well designed to refute this argument, however, because the bacteria he used were of two different serotypes, meaning that they produced different *types* of capsule (types II and III). Even if the IIR-form cells had mutated back to the wildtype (IIS), they could not have produced the type III capsule Griffith observed in the bacteria he recovered from the mice. The only explanation is that the ability to make type III capsules had been acquired from the heat-killed type III cells.

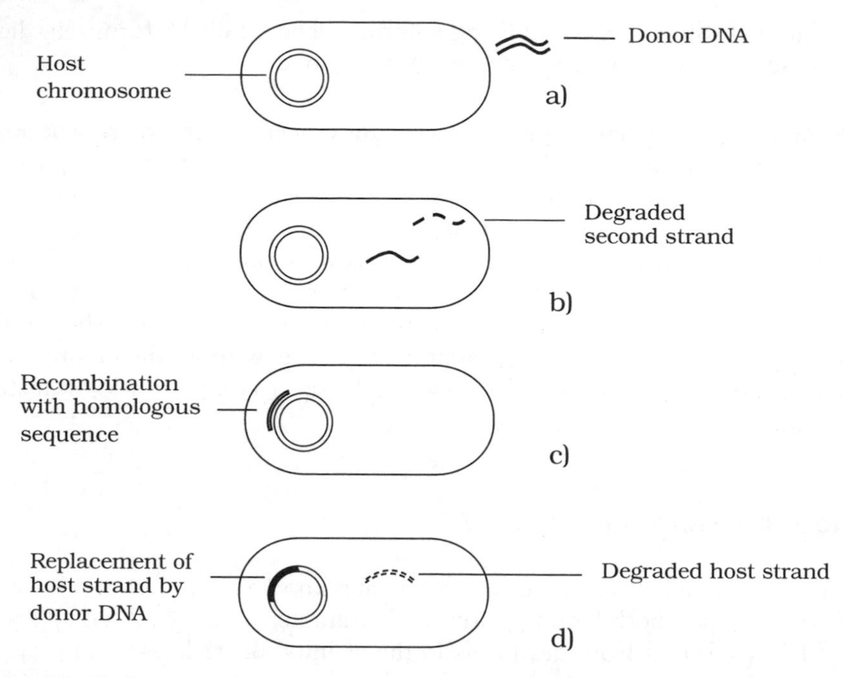

Figure 11.25 Transformation in *Bacillus subtilis*. (a) A fragment of donor DNA become bound to the recipient cell surface by means of a DNA-binding protein. (b) After binding, a nuclease contained at the cell surface degrades one strand of the donor DNA, leaving the other strand to be ferried, by other transformation-specific proteins, to the interior of the cell. (c) A fragment of single-stranded DNA aligns with a homologous stretch on the recipient chromosome. (d) The donor fragment becomes integrated by a process of non-reciprocal recombination. At the next cell division, one daughter cell is a transformant, whilst the other retains the parental genotype

the recombinant genotype. In the latter, there is a stable alteration of the cell's genetic composition, and a new phenotype is expressed in subsequent generations.

Uptake of donor DNA from an unrelated species will not result in transformation; this is due to a failure to locate a homologous sequence and integrate into the host's chromosome, rather than an inability to gain entry to the cell.

Induced competence

Transformation is not thought to occur naturally in *E. coli*, but, if subjected to certain treatments in the laboratory, cells of this species can be made to take up DNA, even from a completely unrelated source. This is done by effecting a state of induced or artificial competence, and by introducing the foreign DNA in a self-replicating *vector* molecule, which does not depend on integration into the host chromosome. As we shall see in the next chapter, this is of enormous significance in the field of genetic engineering.

Box 11.8 Auxotrophic mutants

A type of mutant that has proved to be of great use to the microbial geneticist is the *auxotrophic* mutant. Here, the mutation causes the organism to lack a gene product, usually an enzyme, involved in the synthesis of a nutrient such as an amino acid or vitamin. If the nutrient in question is supplied in the culture medium, the auxotroph can survive quite happily, as its 'handicap' is not exposed. If the nutrient is not provided however, as in a *minimal medium*, the cells would be unable to grow. Thus microbiologists can detect the existence of an auxotrophic mutant by use of selective media.

Conjugation

In 1946, Edward Tatum and Joshua Lederberg (the latter aged only 21 and still a student!) demonstrated a second form of genetic transfer between bacteria. These experiments involved the use of *auxotrophic* mutants (Box 11.8), which have lost the ability to make a particular enzyme involved in the biosynthesis of an essential nutrient. These can be recognised experimentally by their inability to grow on a medium lacking the nutrient in question (Figure 11.26). The results of Tatum and Lederberg's experiments suggested that a process was taking place in bacteria akin to sex in eucaryotes, in which there was a recombination of the cells' genetic material (Figure 11.27). Transformation, as demonstrated by Griffith, could not explain their results, since the addition of a DNA-containing extract of one strain to whole cells of the other did not result in prototroph formation. This was confirmed in an ingenious experiment in which Bernard Davis showed that Tatum and Lederberg's results were only obtained if direct cell-to cell contact was allowed (Figure 11.28).

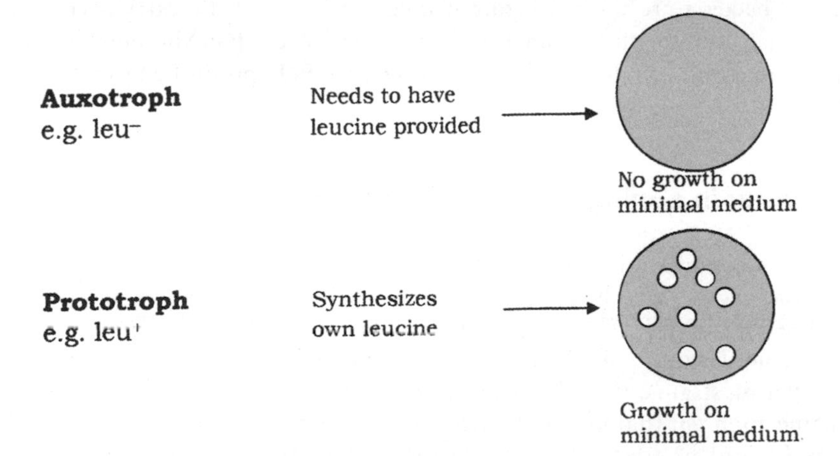

Figure 11.26 Auxotrophic mutants provide useful genetic markers. They are unable to synthesise a particular nutrient, and can be detected by their inability to grow on a minimal medium (one containing only inorganic salts and a carbon source such as glucose)

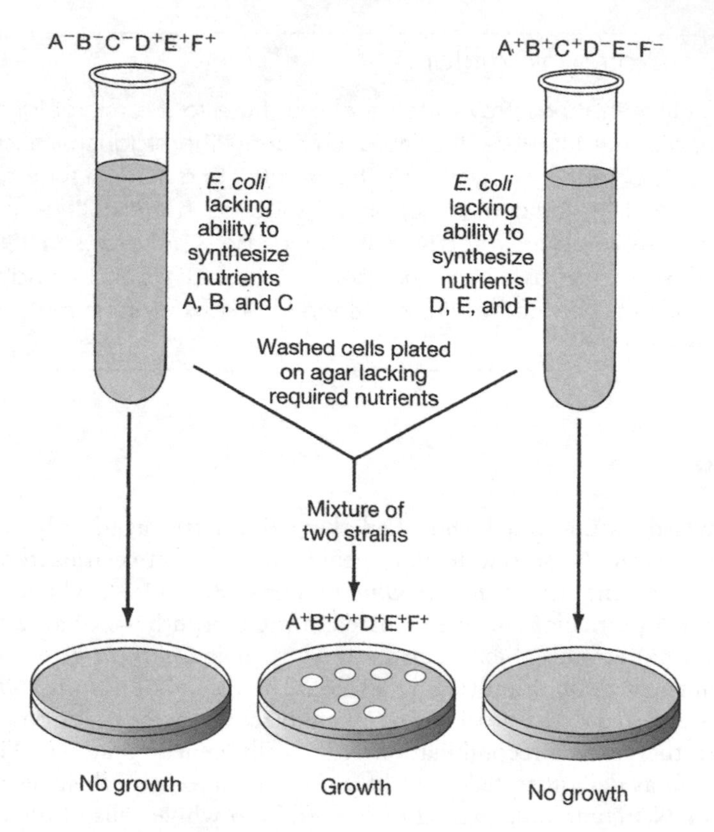

Figure 11.27 Tatum and Lederberg: sexual recombination in bacteria. Two strains with complementary genotypes (auxotrophic for certain genes, prototrophic for others) are each unable to grow on a minimal medium. When the two are mixed, however, colonies are formed, indicating that recombinant bacteria prototrophic for all the genes have been formed. The bacteria are washed before plating out to avoid the carry-over of nutrients from the initial broth to the minimal medium. From Black, JG: Microbiology: Principles and Explorations, 4th edn, John Wiley & Sons Inc., 1999. Reproduced by permission of the publishers

Gene transfer in conjugation is one way only

The process of conjugation was initially envisaged as a fusion of the two partner cells to give a diploid zygote, which subsequently underwent meiosis to give haploid offspring with modified genotypes. The work of William Hayes, however, showed that the development of colonies of recombinant cells was dependant on the survival of only one of the participating strains, the other strain being required only as a donor of DNA.

It became apparent that in *E. coli*, there are two distinct mating types, which became known as F^+ and F^-, depending on whether or not they possessed a plasmid called the *F (fertility) plasmid*. This contains some 30 or 40 genes responsible for its own replication and for the synthesis of a thread-like structure expressed on the cell surface called a *sex pilus*. In a mixture of F^+ and F^- cells, the sex pilus contacts an F^- cell, then

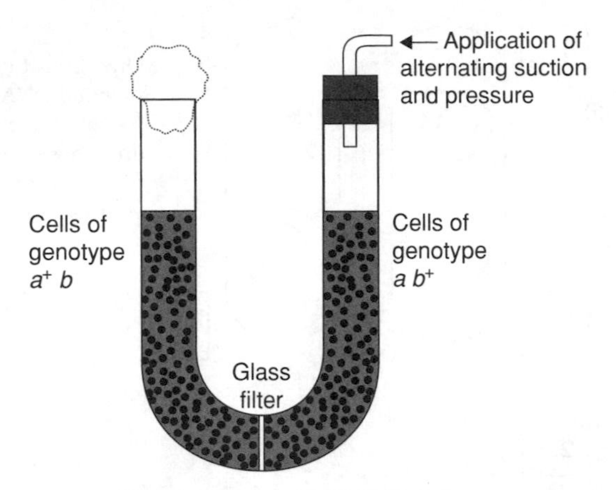

Figure 11.28 The Davis U-tube experiment. Two bacterial strains were placed in two arms of a U-tube, separated by a filter which did not permit the passage of cells. Although suction was used to transfer medium between compartments, no recombinants resulted. The results provided direct evidence that cell-to-cell contact is essential for conjugation to occur

contracts, to pull the two cells together. A single strand of plasmid DNA is then passed across a channel made between the two cells, and enters the F^- cell (Figure 11.29). This then serves as a template for the production of the complementary second strand, and similarly the single strand left behind in the donor cell is replicated, leaving us with a double-stranded copy of the F plasmid in both cells. By acquiring the F plasmid, the recipient F^- cell has been converted to F^+.

When conjugation involves a form of donor cell called *Hfr* (high frequency of recombination), genes from the main bacterial chromosome may be transferred. In these, the F plasmid has become integrated into the main bacterial chromosome; and thus loses its ability to replicate independently (Figure 11.30). It behaves just like any other part of the chromosome, although of course it still carries the genes for conjugation and pilus formation. When conjugation occurs, a single strand of Hfr DNA is broken within the F sequence, and is transferred in a linear fashion, carrying behind it chromosomal DNA. Recombination in the recipient cell results in the replacement of a homologous segment of DNA by the transferred material, which is then faithfully replicated in subsequent generations. If transfer is uninterrupted, a process which takes around 100 minutes in *E. coli*, eventually the remaining stretch of F plasmid will enter the F^- cell, bringing up the rear. The fragile nature of the pilus means, however, that transfer is rarely complete, and only a limited portion of the bacterial genome is transferred. This means that those F^- cells receiving DNA from Hfr cells usually remain F^-, unlike those in a cross with F^+ cells, because the remainder of the F sequence is not transferred. It soon became clear that this phenomenon afforded a great opportunity to determine the relative positions of genes on the bacterial chromosome. This was done by *interrupted mating* experiments, in which the time allowed for conjugation is deliberately limited by mechanical breakage of the sex pili, and correlated with the phenotypic traits transferred to the recipient cells (Figure 11.31). By these means a *genetic map* of the bacterial chromosome could be developed.

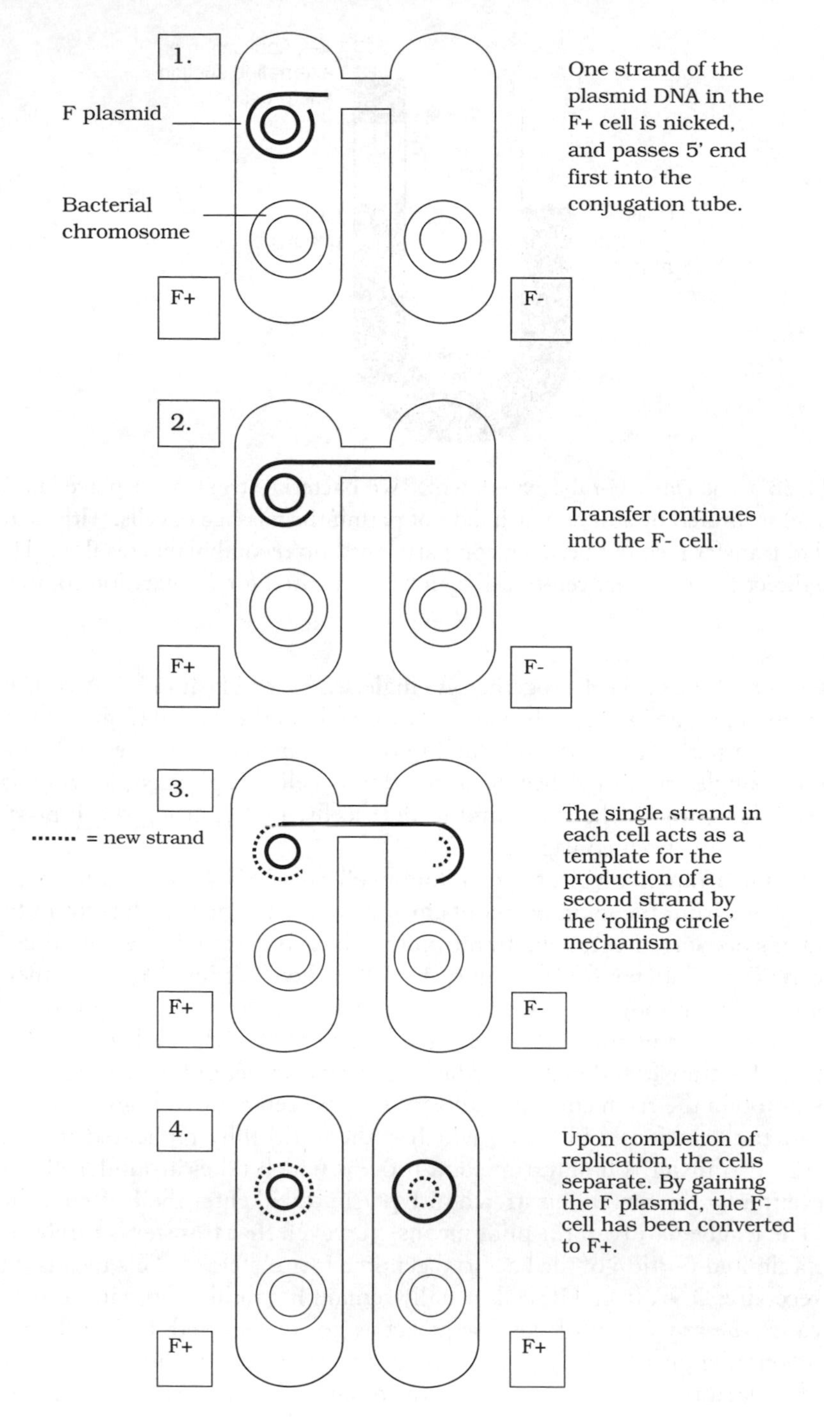

Figure 11.29 Conjugation in bacteria. A single-stranded copy of the F plasmid passes across a conjugation tube from an F^+ to an F^- cell. Both this and the copy left behind act as templates for their own replication, leaving both cells with a complete F plasmid. This means that the recipient cell is converted from F^- to F^+.

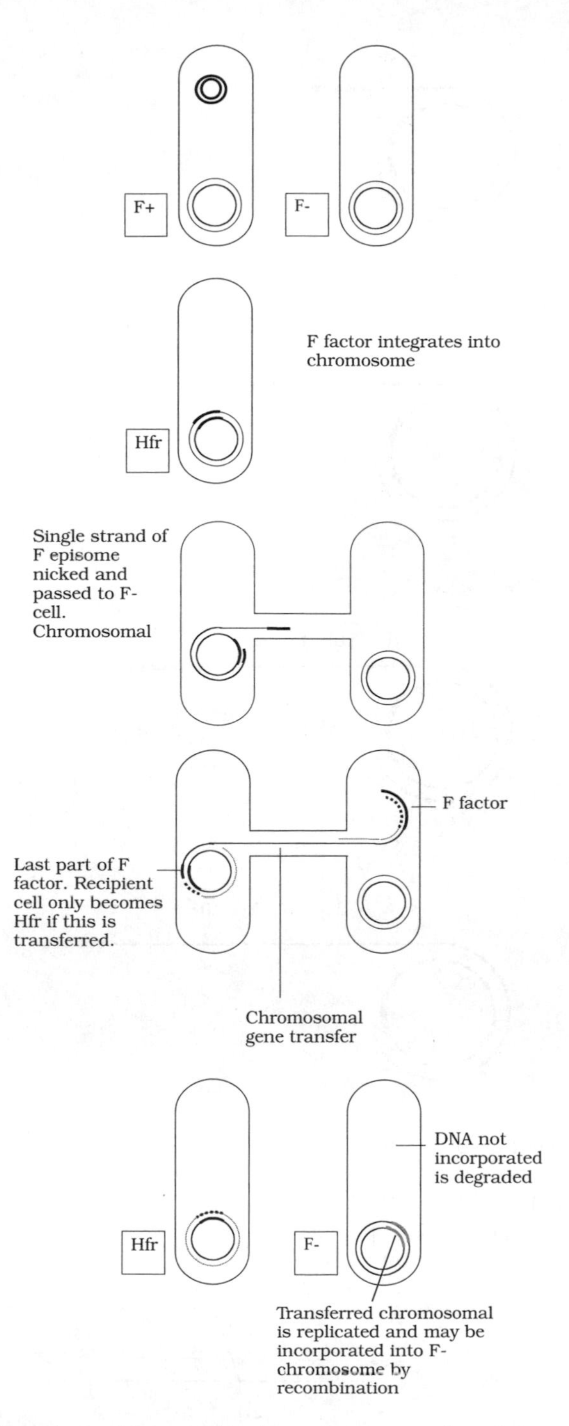
F+
F-
F factor integrates into chromosome
Hfr
Single strand of F episome nicked and passed to F- cell. Chromosomal
F factor
Last part of F factor. Recipient cell only becomes Hfr if this is transferred.
Chromosomal gene transfer
DNA not incorporated is degraded
Hfr
F-
Transferred chromosomal is replicated and may be incorporated into F- chromosome by recombination

Figure 11.30 Conjugation with an Hfr cell results in transfer of chromosomal genes. Hfr cells are formed by the integration of the F factor into the bacterial chromosome. During conjugation, transfer begins part of the way along the F episome, and continues with chromosomal DNA. The amount of chromosomal material transferred depends on how long conjugation is able to proceed. Conjugation may be followed by recombination of transferred chromosomal material with its homologous sequence on the recipient cell's chromosome

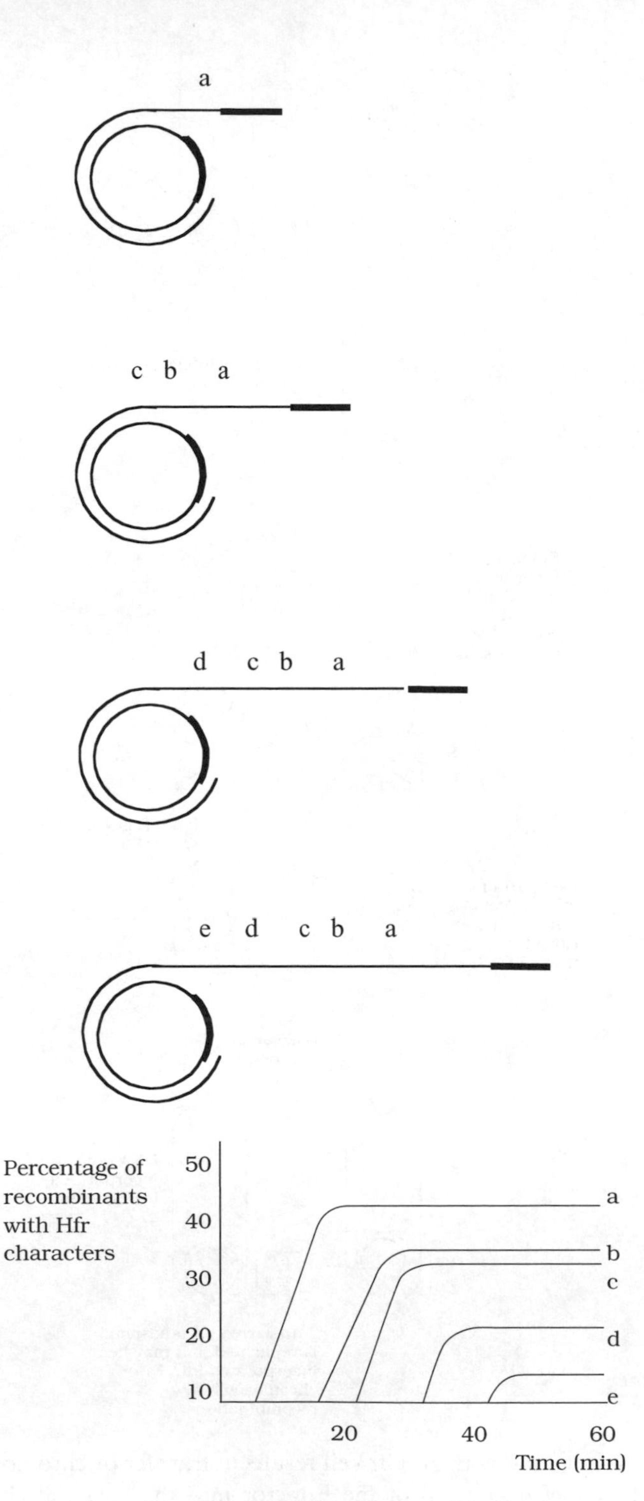

Figure 11.31 In interrupted mating experiments, the conjugation tube is broken after different time periods. The time at which different genes are transferred reflects their relative positions on the bacterial chromosome. By plating out onto selective media, the order in which the different genes are transferred can be determined. The graph shows that the first genes to enter the F^- cell are those present in the highest proportion of recombinants

Box 11.9 It's that man again!

Not content with going down in history as the man who first demonstrated conjugation in bacteria, Joshua Lederberg was also, in 1952, one of the co-discoverers of transduction, along with Norton Zinder. Lederberg was to become a dominant figure in microbial genetics for over half a century. He was even responsible for coining the term 'plasmid'!

The integration of the F plasmid into the bacterial chromosome is reversible; thus Hfr cells can revert to F^+. Excision of the integrated plasmid is not always precise, and sometimes a little chromosomal DNA is removed too. When this happens, the plasmid, and the cell containing it, are called F′ ('F prime'); transfer of the plasmid to an F^- cell takes with it the extra DNA from the host chromosome. The recipient genome thus becomes partially diploid (*merodiploid*), because it has its own copy, plus the 'guest' copy of certain genes.

Transduction (Box 11.9)

In the third form of genetic transfer in bacteria, bacteriophages act as carriers of DNA from one cell to another. In order to appreciate the way in which this is done, it is necessary to recall the sequence of events in phage replication cycles discussed in the previous chapter (see Figure 10.11).

Generalised transduction occurs in virulent phages, that is, those with a lytic life cycle. Sometimes, the enzymes responsible for packaging phage DNA into its protein coat package instead similarly sized fragments of degraded chromosomal DNA (Figure 11.32). Despite containing the wrong DNA, this *transducing phage* particle is still infective, since this is dependant on its protein element. Thus following infection of another bacterial cell, the DNA can be incorporated by recombining with the homologous segment in the recipient cell. Since any chromosomal fragment can be mistakenly packaged in this way (as long as it finds an area of homology), all genes are transferred at a similar (low) frequency.

Specialised transduction results in a much higher efficiency of transfer for specific genes, however it is limited to genes having a particular chromosomal location. Recall from Chapter 10 that in lysogenic life cycles, the phage DNA is integrated into the host chromosome, and later, perhaps after many rounds of cell division, excised again before re-entering a lytic cycle. If this excision does not happen precisely, some of the adjoining chromosomal DNA, carrying a gene or two, may be incorporated into the phage particle (we saw a similar mechanism in the case of F′ plasmid formation). Upon infecting another cell, the transduced genes would undergo recombination and become incorporated into the recipient's chromosome (Figure 11.33). Although limited to genes in the vicinity of the lysogenic phage's integration, this is a highly efficient form of transfer, since the genes become stably integrated into the host cell.

Transduction experiments, like those involving conjugation, can be used to determine the relative positions of genes on a bacterial chromosome.

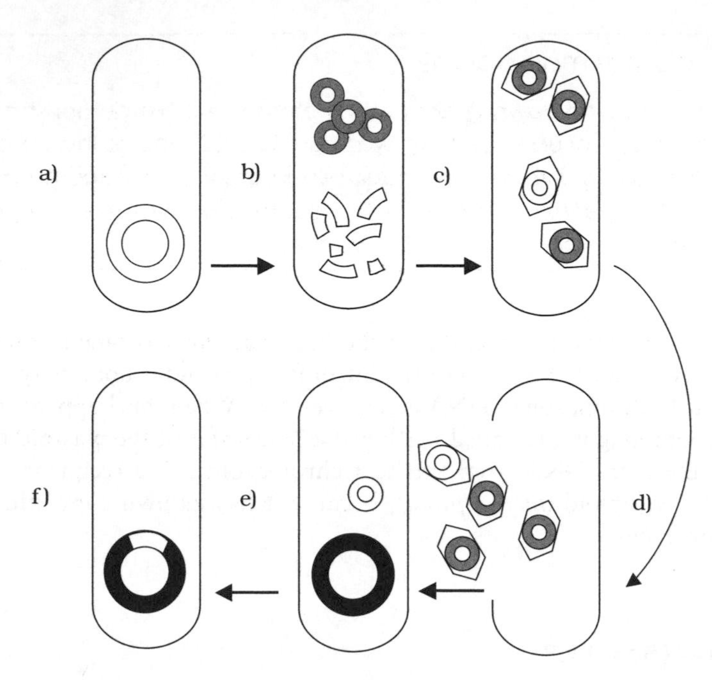

Figure 11.32 Generalised transduction. During the lytic cycle of a phage, the host DNA is degraded (b), and a fragment may be mistakenly packaged into a newly synthesised phage particle (c). Upon infecting a new host cell, the transducing phage releases its DNA (e); although unable to replicate, this can undergo recombination with a homologous sequence on the host chromosome (f)

Transposable elements

An unusual type of genetic transfer which takes place *within* an individual cell involves sequences of DNA called *transposable elements*. One type is known as an *insertion sequence* (IS), a relatively short piece of chromosomal or plasmid DNA which contains a gene for the enzyme *transposase* (Figure 11.34). This recognises, cuts and re-ligates the insertion sequence anywhere in the bacterial genome. In so doing, it may interrupt a gene sequence, and thereby cause a mutation. Unlike recombination events, no homology is required between the transposable element and the point at which it inserts. This relocation of a transposable element from one place in the genome to another is termed *conservative transposition*. In *replicative transposition*, the element remains in its original position and a copy is made and inserted elsewhere in the genome. Insertion sequences are flanked by inverted sequences some 9–41bp in length, which are thought to be essential for the recognition of the sequence by the transposase.

Transposable elements that also carry genes other than those required for transposition, such as genes for antibiotic resistance or toxins, are known as *transposons*.

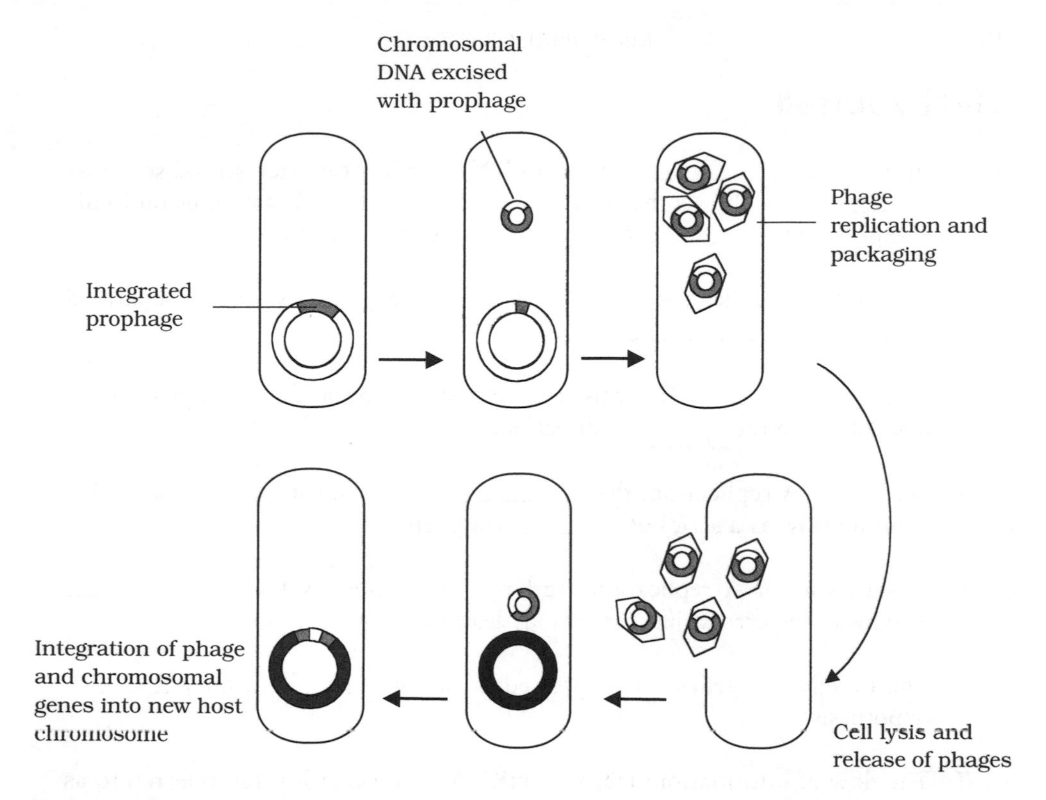

Figure 11.33 Specialised transduction. During the replication cycle of a lysogenic bacteriophage, phage DNA is incorporated into the host chromosome (see Figure 10.11). When a lytic cycle resumes and the phage DNA is excised, it may take with it an amount of surrounding chromosomal DNA. This is packaged into phage particles and infects new host cells, where it is integrated into the bacterial chromosome. Only genes surrounding the site of phage integration may be transduced in this way

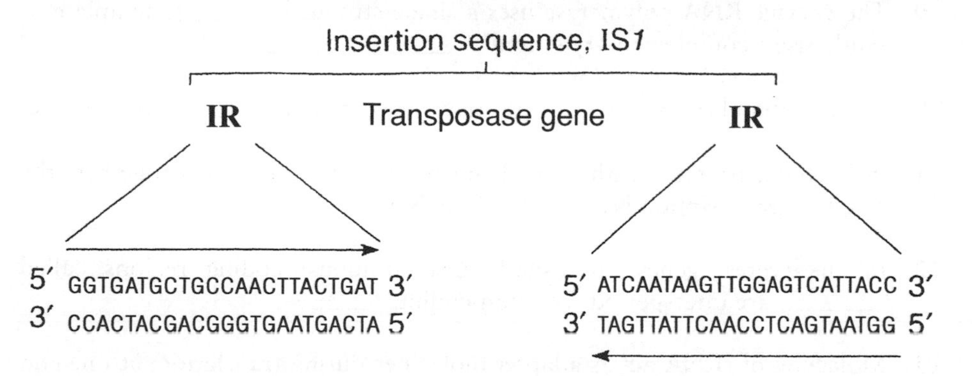

Figure 11.34 Transposable elements. The insertion sequence IS*1* of *E. coli* is 768bp in length and is flanked by 23bp inverted repeat sequences. The IS contains the gene for a transposase enzyme, which catalyses the movement of the insertion sequence from one location to another. By integrating at random points in the genome where there is no sequence homology, IS sequences may disrupt functional genes and give rise to mutations

Test yourself

1 In the ________-________ model of DNA replication, each strand serves as a ________ for the synthesis of a ________ stand. Each daughter molecule comprises one ________ strand and one newly synthesised strand.

2 During DNA replication, the point where the two strands become separated is called the ________ ________.

3 ________ ________ III forms a second strand by adding complementary nucleotides in the ________ direction.

4 During DNA replication, the ________ ________ must be synthesised discontinuously, as a series of ________ fragments.

5 Mistakes in DNA replication are largely corrected by the cell's ________ enzymes. Any errors that persist may lead to ________.

6 The function of genes was expressed in the one ________, one, ________ hypothesis.

7 The flow of information: DNA → mRNA → protein is often referred to as the ________ ________ of biology.

8 In the genetic code, many amino acids are encoded by more than one triplet sequence; the code is therefore said to be ________. Three of the ________ triplet combinations do not code for an amino acid, but instead serve as ________ codons.

9 The enzyme RNA polymerase uses a single-stranded ________ template to synthesise a complementary strand of ________ ________.

10 Transcription begins at a ________ sequence, situated ________ of the gene.

11 In bacteria, proteins with related functions may be encoded together; the result of transcription is a ________ mRNA.

12 In eukaryotes, genes are usually discontinuous; coding regions called ________ are interspersed with non-coding ________.

13 Molecules of tRNA act as adapter molecules during translation; at one end they have a three-base ________ complementary to a triplet codon, and at the other end carry the corresponding ________ ________.

14 Lactose acts as an ________ for the three genes that make up the *lac* operon. It neutralises the effects of a ________ protein encoded by the *I* gene.

15 The *trp* operon contains five genes involved in the synthesis of ________. The presence of this substance activates a ________, which prevents transcription of the operon by binding to the ________ sequence.

16 A ________ mutation alters the ________ ________ of a gene and will change the sense of the encoded message. Such a mutation arises through ________ or ________ of DNA.

17 A ________ mutation changes a normal codon into a ________ codon, and results in the premature termination of translation.

18 ________ ________ such as 5-bromouracil mimic the structure of normal nucleotide bases and become ________ into the DNA structure.

19 The ________ test is used to assess the mutagenicity of a substance.

20 In Griffith's famous experiment, ________ ________ cells of the virulent S-strain appeared to pass on the ability to synthesise a capsule to the non-virulent R-strain. Griffith coined the term ________ ________ for the factor responsible.

21 Transformation only occurs between related cells as it depends on the donor DNA finding a ________ sequence on the host chromosome with which to ________.

22 ________ ________ experiments can be used to map the order in which genes on a bacterial chromosome are transferred by ________.

23 In ________ transduction, chromosomal genes close to the point of integration of the ________ may be excised along with it and be transferred to another host cell.

24 In ________ transduction, fragments of ________ DNA are mistakenly packaged into phage coats and can be transferred to another bacterial cell.

25 ________ ________ are sequences of DNA that can move from one location on a chromosome to another.

12

Microorganisms in Genetic Engineering

Introduction

In the last 30 years or so, there has been a revolution in the field of genetics, which has had a profound effect on virtually every other area of biology. This has been due to the development of new techniques that have enabled scientists to analyse and manipulate DNA in a quite unprecedented way. Genetically modified crops, DNA 'fingerprinting' and gene therapy are just three of the many applications made possible by these advances. The subject of 'genetic engineering' is too huge to be discussed here in detail, and indeed it extends into areas far beyond the remit of this book. In this chapter, however, we shall examine some of the ways that microorganisms have contributed to the genetic revolution. As we shall see, their role in the development of new techniques of DNA manipulation since the 1970s has been just as important as their earlier contribution to the elucidation of the structure, role and replication of DNA several decades earlier.

The beginnings of genetic engineering can be said to date from the discovery, in the late 1960s, of a class of bacterial enzymes called *restriction endonucleases* (REs). These are enzymes that cleave DNA into pieces by making breaks in the sugar-phosphate backbone; in nature, they serve to destroy any foreign DNA that may enter the cell. They do not cut the DNA in a random fashion, however; their unique usefulness to the molecular biologist lies in the fact that they break the DNA *in a precise and reproducible manner*. They do this by cutting only at specific *recognition sites*, sequences of typically four to six nucleotides (Figure 12.1). Thus, under favourable conditions, a particular RE will digest a given piece of DNA into an identical collection of fragments, time after time. In the ensuing years, many hundreds of restriction endonucleases have been discovered, many of which recognise different specific sequences, providing biologists with a hugely versatile tool for the manipulation of DNA, often likened to a pair of molecular 'scissors'. Not long after REs were first isolated, they were used to create the first man-made *recombinant DNA* molecule (Figure 12.2). This involved cutting

> Restriction endonucleases do not destroy the host bacterium's own DNA, because certain nucleotides in the recognition sequence are modified by *methylation*. The REs are unable to cleave the DNA at methylated sites.

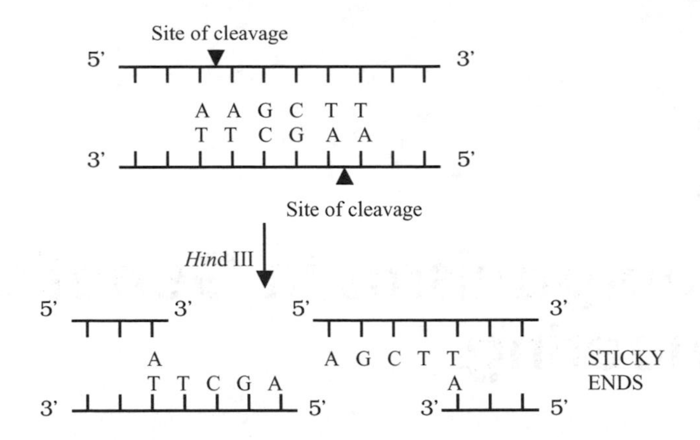

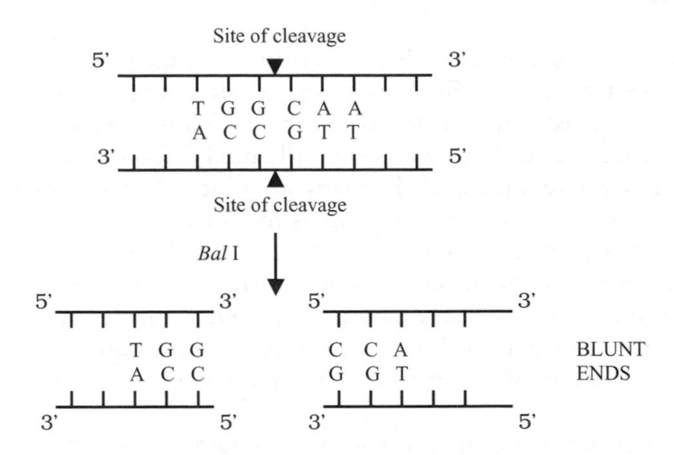

Figure 12.1 Restriction endonucleases fragment DNA molecules by breaking the sugar–phosphate backbone within a specific sequence of nucleotides. Depending on the site of cleavage, the fragments so produced may be blunt-ended or 'sticky'-ended

fragments of DNA from different sources, then using another enzyme, *DNA ligase* to join them together, a process facilitated by using fragments with *compatible 'sticky' ends*. Remember from Chapter 11 that A always pairs with T and C with G; because of this, complementary sequences that come into contact with one another will 'stick' together. DNA, it seems, is DNA, wherever it comes from; consequently DNA from plants, animals, bacteria or viruses can be joined together to create novel sequences undreamed of by Mother Nature.

Of course, a single molecule of our newly recombinant DNA is not much use to us. The important breakthrough came with the development of *cloning* – the ability to produce huge numbers of copies of a given molecule. To do this, two further things are needed: a carrier DNA molecule called a *vector*, and a *host cell* in which it can be replicated.

Cloning is the production of multiple copies of a specific DNA molecule. The term is also used to describe the production of genetically identical cells or even organisms.

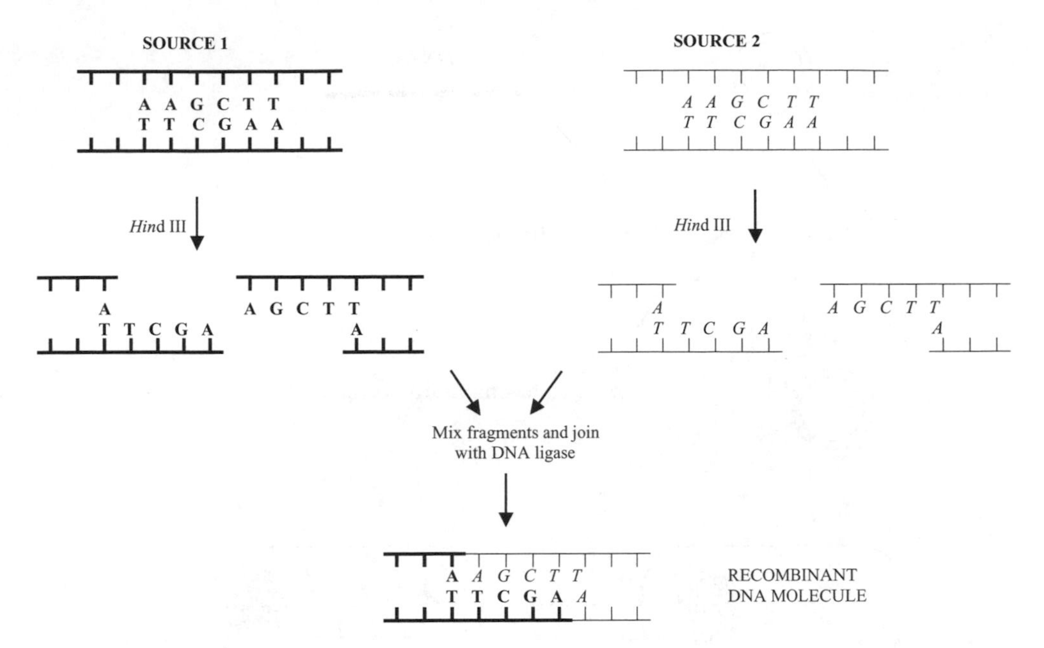

Figure 12.2 DNA from different sources can be joined together. 'Sticky'-ended restriction fragments from one DNA source have single-stranded sequences that are compatible with fragments produced from another source by the same RE. Compatible base pairing attracts the fragments together and the join is made more permanent by the action of DNA ligase

Figure 12.3 shows the main steps of a cloning protocol:

- 'donor' DNA and vector are digested with an RE to provide compatible sticky ends
- a fragment of donor DNA is spliced into the vector molecule
- the recombinant vector gains entry to a host cell (*e.g. E. coli*)
- the vector replicates inside the cell, making further copies of the inserted DNA
- host multiplication results in the formation of a clone of cells, all containing the same recombinant plasmid – we now have millions of copies of our donor DNA 'insert'. A collection of such clones is called a *DNA library*.

Let us look at role of vectors in a little more detail. The main features required of a cloning vector are:

> A vector is a self-replicating DNA molecule used in gene cloning. The sequence to be cloned is inserted into the vector, and replicated along with it.

- *it must be capable of replicating autonomously inside a host cell* – when it does so, any DNA it carries will also be replicated. Vectors make multiple copies of themselves inside the host cell.
- *it must be relatively small* – to facilitate manipulation and entry into a host cell, vectors must not exceed a certain size.

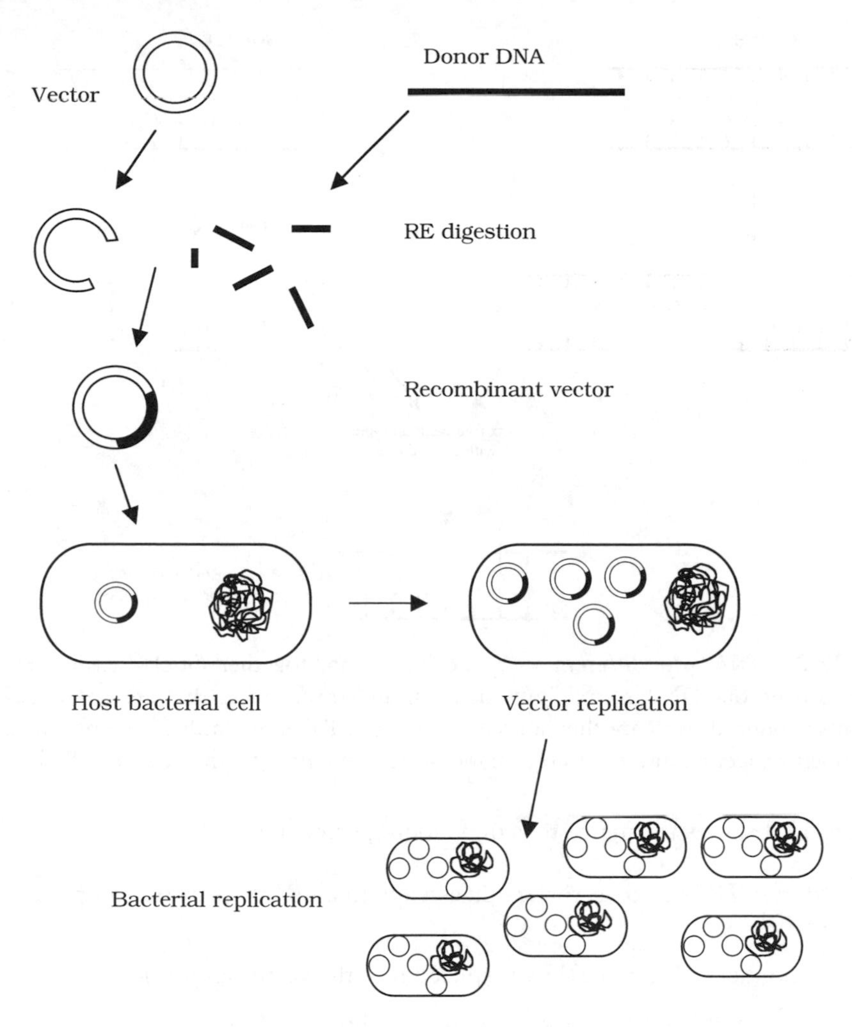

Figure 12.3 Gene cloning. DNA fragments obtained by restriction digestion can be spliced into a similarly digested vector molecule and transformed into a host bacterial cell. See the text for details

- *it must carry a selectable marker* – since only a proportion of host cells will take up the vector, there must be a means of differentiating them from those that do not. A common way to do this is to use a vector that carries a gene that confers resistance to an antibiotic such as ampicillin. When bacterial cells are plated out on a medium containing the antibiotic, only those that have taken up the vector will be able to form colonies. (The host strain must, of course, normally be susceptible to the antibiotic.)

A selectable marker is a gene that allows cells containing it to be identified by the expression of a recognisable characteristic.

- *it must carry a single copy of RE restriction sites* – in order to accommodate a piece of donor DNA, a vector must be cut by a restriction endonuclease in one place only (Figure 12.3).

Plasmid cloning vectors

Two main types of vector system are used in cloning, those that use *plasmids* and those that use *bacteriophages* (revisit Chapters 3 and 10, respectively, for a reminder of the main features of these). Naturally occurring examples of these are manipulated so that they possess the above properties. A popular vector in the early days of gene cloning was the plasmid pBR322; Figure 12.4 shows how it contains the features described above.

Let us consider now what happens, at a molecular level, when donor DNA is ligated into a plasmid vector (Figure 12.5). Sticky ends of the donor fragment form hydrogen bonds with the exposed compatible ends of the opened up plasmid by complementary base pairing, and DNA ligase consolidates the join. Many plasmids contain engineered sequences called *multiple cloning sites* (MCS); these provide additional flexibility with respect to the restriction fragments that may be accommodated.

The ligation of insert DNA into the cloning site of the plasmid is not the only possible outcome of the procedure described above, however. Unless experimental conditions are carefully controlled (there are ways of doing this), a more likely outcome is that the two compatible ends of the plasmid will simply 'find' each other

A multiple cloning site or *polylinker* is a region of a cloning vector designed to contain recognition sequences for several REs.

The *lacZ′* gene actually only encodes a part of the β-galactosidase enzyme, called the α-peptide. The strain of *E. coli* used as host makes an incomplete version of the enzyme, which lacks this portion. Only if the cells contain the plasmid with the lac Z′ gene can they produce functional β-galactosidase, by α-complementation.

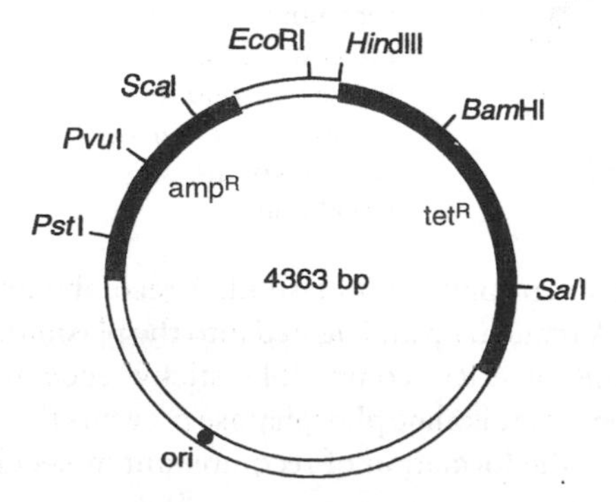

Figure 12.4 Plasmid pBR322. One of the earliest plasmid vectors, pBR322 illustrates the major features required for use in gene cloning: an origin of replication (ori), selectable markers (genes for resistance to ampicillin and tetracycline), and single recognition sites for a number of REs

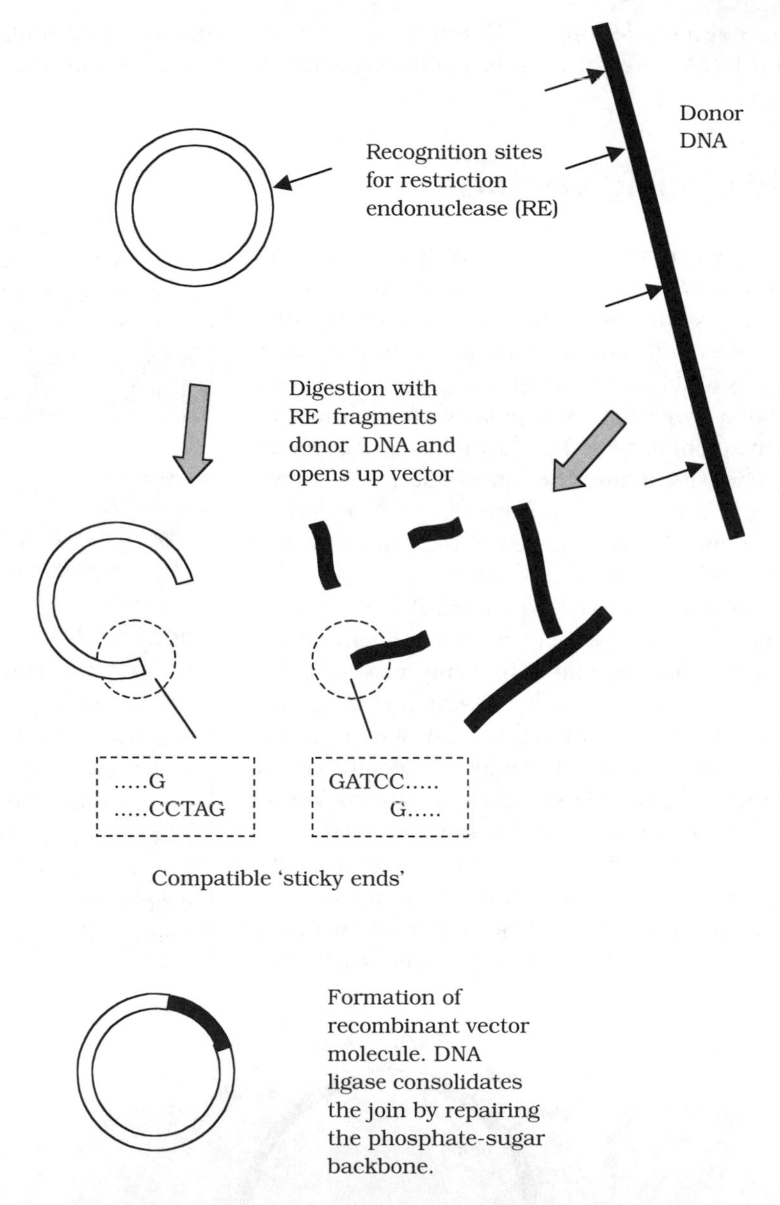
Donor
DNA
Recognition sites
for restriction
endonuclease (RE)
Digestion with
RE fragments
donor DNA and
opens up vector
.....G
.....CCTAG
GATCC.....
G.....
Compatible 'sticky ends'
Formation of
recombinant vector
molecule. DNA
ligase consolidates
the join by repairing
the phosphate-sugar
backbone.

Figure 12.5 Formation of a recombinant plasmid. A recombinant plasmid is formed when a fragment of foreign DNA is taken up and ligated into the plasmid. By cleaving both plasmid and foreign DNA with the same RE, compatible 'sticky' ends are created, facilitating the join. Treatment with the enzyme alkaline phosphatase prevents the cut plasmid ends rejoining together, thereby favouring the formation of recombinant molecules

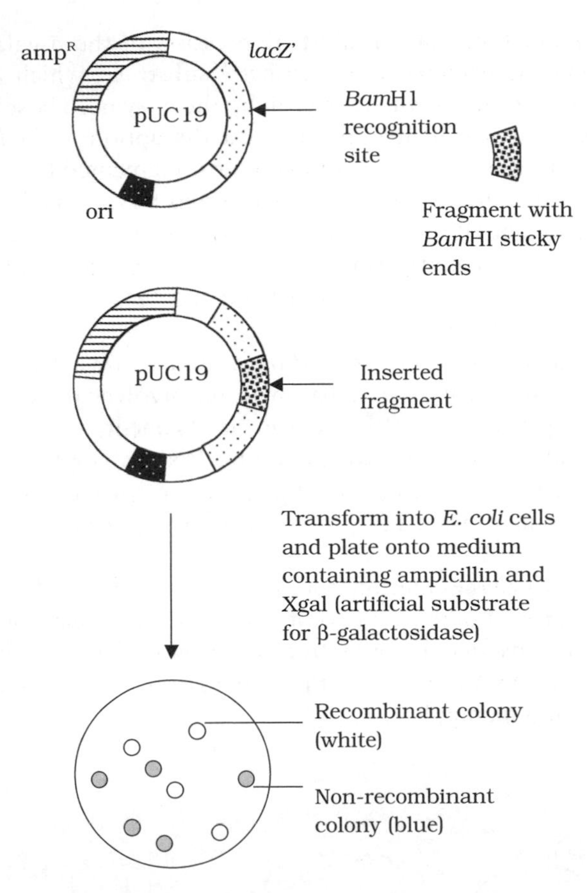

Figure 12.6 Recombinant plasmids can be detected by insertional inactivation. Insertion of foreign DNA is carried out at a site within one of the selectable markers, thus interrupting its gene sequence. Here, a fragment has been inserted into the *Bam*HI site situated within the gene that codes for β-galactosidase. Bacteria transformed with such a plasmid will not produce the functional enzyme, and so can be distinguished from those carrying plasmids with no inserted DNA

again, and rejoin. Since a certain amount of this is inevitable, how are we able to tell the difference between those bacteria that contain a recombinant plasmid (one containing a piece of donor DNA) and those that have taken up a recircularised 'native' plasmid? Since both types will contain the gene for ampicillin resistance, we cannot distinguish them by this means. A strategy commonly used to get around the problem is *insertional inactivation* (Figure 12.6). This clever ploy exploits the fact that we can manipulate DNA, and, for example, insert RE recognition sequences at desired points. If a recognition site occurs in the middle of a gene sequence, and a piece of foreign DNA is inserted at this position, the gene will be interrupted, and unable to produce a functional gene product. In the example shown, the gene *is lacZ'*, necessary for the successful expression of the enzyme β-galactosidase. This will only be expressed in those bacteria that contain plasmids in which the gene has remained uninterrupted, i.e. those that have *not* taken up an insert. Expression of the β-galactosidase can be detected by growing the bacteria on an artificial substrate, which is converted to a coloured (usually blue) product when acted on by the enzyme. Those cells that contain recombinant plasmids are easily identified because disruption of the *lacZ′* gene means that no β-galactosidase is produced, resulting in non-pigmented (white) colonies.

One problem remains. Remember that our inserted DNA was derived from the digestion of total (genomic) DNA from the donor organism; this means that our DNA library will contain recombinant plasmids with a whole range of fragments from that digestion, and not just the specific one that interests us. How are we able to distinguish this fragment from the others?

A technique called *nucleic acid hybridisation* is used to solve the problem. This once again depends on complementary base pairing, and involves the creation of a *probe*, a short length of single-stranded DNA that is complementary to part of the desired 'target' sequence, and therefore able to seek it out. If searching for the right clone can be likened to looking for a needle in a haystack, then the probe is a powerful 'magnet' that makes the task much easier. The probe carries a tag or label, so that its location can be identified (Figure 12.7).

Once we have identified the clone of bacteria containing plasmids with the insert that interests us, we can grow a pure culture of it and then isolate plasmid DNA. Using the same RE as before, the inserted donor DNA can be removed and purified. We now have enough of this specific DNA sequence (a tiny proportion of the donor organism's total genome) to analyse and manipulate.

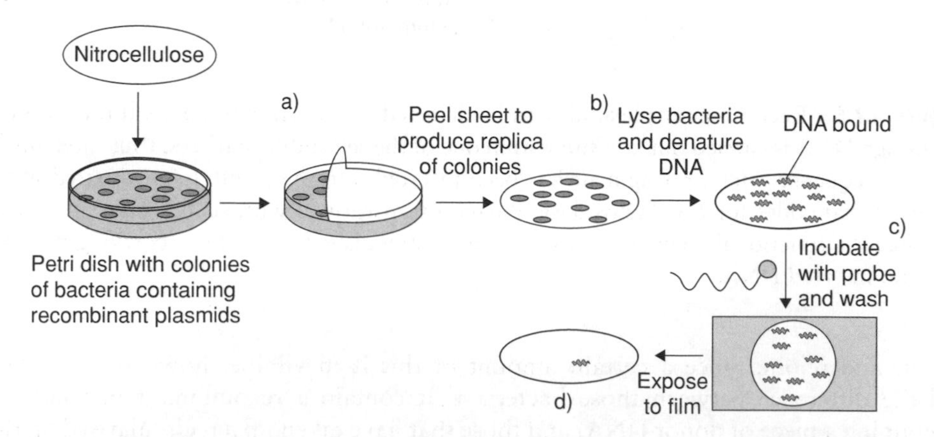

Figure 12.7 Colony probing. (a) A replica of the bacterial colonies is made using a nitrocellulose membrane. (b) Alkali treatment lyses the cells and denatures the DNA, making it single stranded. (c) Following a period of incubation with a radiolabelled probe to allow hybridisation to take place, the membrane is washed to remove any unbound radioactivity. (d) Position of bound radioactivity revealed by autoradiography. Comparison with original master plate reveals the location of colonies carrying specific target sequence. Alternative detection systems such as biotin–streptavidin that avoid the use of radioactivity have become much more commonly used in recent years. From Reece, RJ: Analysis of Genes and Genomes, John Wiley & Sons, 2003. Reproduced by permission of the publishers

Although plasmids are easily isolated and manipulated, their use as cloning vectors is limited by the fact that they tend to become unstable if we attempt to insert much more than about 5kb of foreign DNA. For inserts larger than this, we must turn to other vector systems.

Bacteriophages as cloning vectors

The most commonly used bacteriophage vectors are those based on *phage Lambda* (λ) (Figure 12.8). The genome of λ is 48.5kb in length; it was the first genome to have its entire sequence determined (1982). As drawn conventionally, the genome is linear, and contains 46 genes. At each 5′ end of the double-stranded structure is a 12 base single-stranded sequence known as a *cos* site; these have a complementary base sequence (i.e. they are cohesive, or 'sticky' ends). They can join together to form a double-stranded, circular λ molecule, a conformation that is essential for insertion and integration into the host genome (see Chapter 10).

The naturally occurring phage is unsuitable as a vector, because being relatively large, it contains multiple copies of recognition sites for a number of REs. By using genetic manipulation techniques, however, unwanted sites can be removed, allowing the DNA to be 'opened up' at a single location, and fragments of foreign DNA with complementary sticky ends to be ligated.

A further limitation that must be overcome is that in this form, phage λ can only accommodate about another 3kb of DNA. This is because if the genome exceeds 52kb, it cannot be packaged properly into the protein head to produce viable phage particles. The arrangement of genes on the phage λ genome offers a solution to this problem. It is known that genes encoding specific functions are clustered together, with genes necessary for the lytic replication cycle being found at the ends of the map as it is conventionally drawn (Figure 12.8). It is possible, therefore, to remove much of the central part of the λ genome without affecting its ability to infect and lyse its bacterial host.

Insertion vectors have had some of this non-essential material removed, reducing their genome size to around 42kb, and thus allowing them to carry an insert of up to 10kb. A single-copy restriction site is opened up, and a fragment with complementary sticky ends inserted with the aid of DNA ligase. Vectors such as λZAPII have a multiple cloning site or polylinker, positioned so that it falls within the *lacZ′* gene. This allows insertional

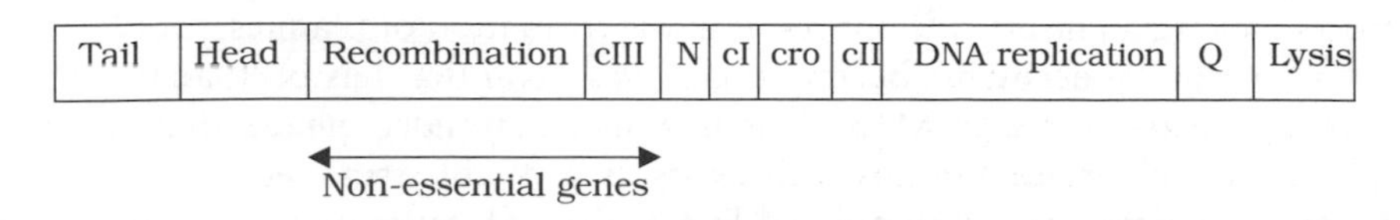

Figure 12.8 Bacteriophage λ. The central region of the λ genome contains genes involved in its lysogenic cycle. This region can therefore be excised, without affecting the ability of the phage to infect *E. coli* via the lytic cycle. The genes at the right-hand end of the genome as shown have a regulatory function, whilst the structural genes are situated at the left

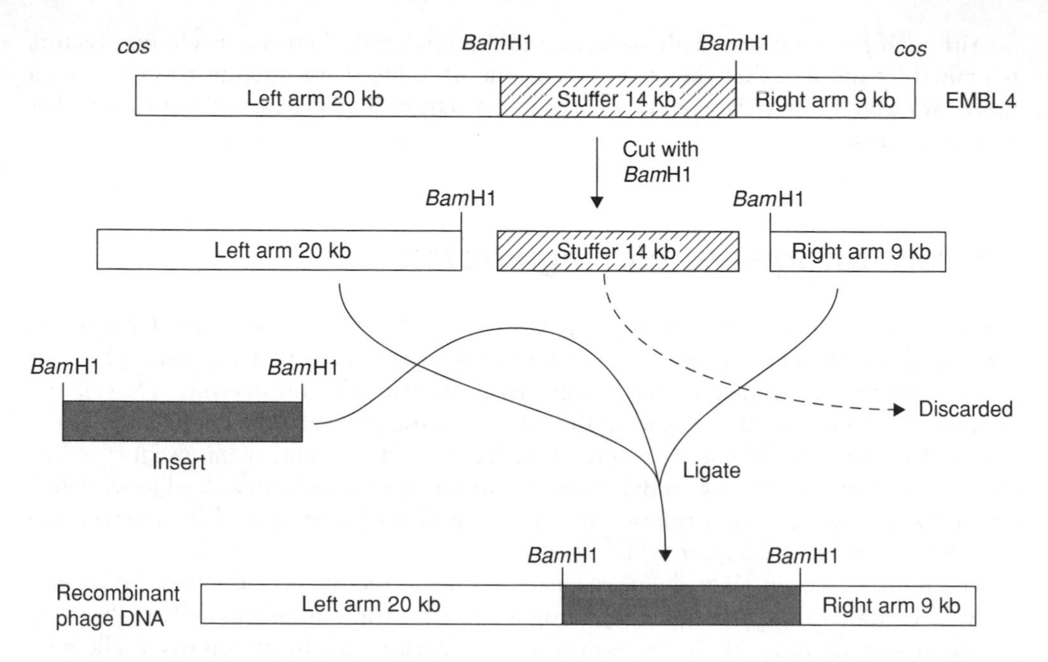

Figure 12.9 Cloning in bacteriophage λ. The removal of non-essential genes in the centre of the genome allows the incorporation of large fragments of foreign DNA. From Dale, JW and von Schantz, M: From Genes to Genomes: Concepts and Applications of DNA Technology, John Wiley & Sons, 2002. Reproduced by permission of the publishers

inactivation with blue/white selection to be used for the detection of recombinants as described above for pUC plasmids.

Replacement vectors (Figure 12.9) are able to accommodate larger inserts, because instead of a single copy of a particular RE site, they have two, one situated either side of a disposable central section known as the 'stuffer' fragment. This can be removed by digestion with the appropriate RE and replaced by insert DNA. The vector λEMBL3, which can accommodate inserts of between 9 and 23kb, has two polylinker sequences containing sites for *Sal*I, *Bam*HI and *Eco*RI flanking the stuffer fragment, allowing a variety of restriction fragments to be inserted.

Recombinants in replacement vector systems can be detected by a method that exploits the limitations in the phage's ability to package DNA. Just as too big a genome cannot be packaged (see above), neither can one that is too small (<37kb); consequently, constructs lacking an insert will not result in the formation of plaques.

Another naturally occurring bacteriophage of *E. coli* that has been adapted for use as a cloning vector is phage M13. This is a single-stranded phage that for part of its replication cycle inside the host cell exists as a double-stranded replicative form, which can be isolated and manipulated like a plasmid. Systems based on M13 have proved popular as a means of obtaining DNA in a single-stranded form for applications such as DNA sequencing using the dideoxy (Sanger) method. M13 vectors contain a multiple cloning site situated within the *lacZ′* gene, allowing blue/white selection of recombinants. The cloning capacity is generally limited to fragments of less than 1.5kb,

Box 12.1 A large number of clones are required to cover an entire genome

The factors that determine the number of clones necessary to make a genomic library are the size of the genome and the average size of each insert. In addition the probability that a given fragment will be present in the library must be set: the higher the probability, the more clones will be necessary.

$$N = \frac{\log(1 - P)}{\log(1 - a/b)}$$

Where: N = number of clones required; a = average insert size; b = total genome size; P = probability that a given fragment will be present.

e.g. Number of clones required for a 95% probability of a given fragment to be present in a genomic library of *E. coli* (genome size 4.8×10^6) with an average insert size of 20kb:

$$\begin{aligned} N &= \log(1 - 0.95)/\log(1 - 4.167 \times 10^{-3}) \\ &= \log 0.05/\log(0.9958) \\ &= 712 \text{ clones} \end{aligned}$$

This may seem a manageable number, but remember that the genome of *E. coli* is very small compared to that of higher organisms (*Homo sapiens*: 3×10^9bp).

but hybrid M13/plasmid vectors (*phagemids*) have been developed that are able to take inserts of as much as 10kb.

As we have seen, a large collection of cloned DNA fragments is called a DNA library. If the original donor DNA comprised the entire genome of an organism, then the collection of clones, which between them contain an entire donor genome, is known as a *genomic library*. We could create a library of the entire *E. coli* genome (total size 4.8×10^6bp) for example, by having just over 700 clones with an average insert size of 20kb (see Box 12.1 for calculation). For a more complex genome, however, we would either need to have very many more clones, or to increase the average fragment size.

For this reason, further types of cloning vectors have been developed, to allow the cloning of larger fragments. *Cosmids* are entirely man-made creations, and have no naturally occurring counterpart. They incorporate beneficial features of both plasmids and phage vectors, and may be capable of accommodating insert fragments of more than 45kb. Cosmids are essentially plasmids that contain the *cos* sites from λ phage, with the plasmid supplying the necessary origin of replication, restriction sites and selectable marker. As we've seen, λ DNA will be packaged into phage heads if the *cos* sites are 37–52kb apart; however the only parts of the DNA recognised by the packaging enzymes are the *cos* sites, so any DNA can be used to fill the intervening sequence.

The recombinant DNA is packaged into phage particles by a process called *in vitro* packaging, and introduced into the host *E. coli*. Lacking phage genes, cosmids do not lead to lysis of the host cells and plaque production, but are instead replicated as if they were plasmids. Growth of host cells on a selective medium allows the detection of transformants, i.e. those that have taken up the cosmid. There is no need to select for recombinants, because non-recombinants are too small to be packaged into the phage heads.

Expression vectors

Sometimes, the aim of a cloning experiment is not just to obtain large amounts of a specific gene, but for the gene to be *expressed*. This involves using the host cell as a sort of 'factory', to manufacture the specific protein encoded by the cloned gene. One of the earliest applications of genetic engineering was the production of human insulin in *E. coli* (Figure 12.10). Insulin is needed in considerable quantities for the treatment of diabetics; for years it was obtained from the pancreas of pigs and cattle, but this had several disadvantages including immunological complications and the risk of viral contamination. Insulin generated by recombinant means is free from these problems. Many proteins can now be produced in this way by microorganisms at a rate several times that of the normal host cell. In order for a gene to be expressed, it must have specific nucleotide sequences around it that act as signals for the host cell's transcription/translation machinery (promoter, ribosomal binding site and terminator – see Chapter 11). Since these sequences differ between, say, humans and *E. coli*, the bacterial RNA polymerase will not recognise the human sequences, and therefore, although a human gene may be cloned in *E. coli* using a simple vector, it will not be expressed. If, however, the human gene could be inserted so that it was under the control of the *E. coli* expression signals, then transcription should take place. Specially designed vectors that provide these signals are called *expression vectors*. The choice of promoter sequence is particularly important; often, a strong (i.e. very efficient) promoter is selected, so as to maximise the amount of protein product obtained. Genes whose protein products are naturally synthesised in abundance are likely to have such promoters. It is often helpful to be able to regulate gene expression; *inducible* promoters can be switched on and off by the presence of certain substances. The *lac* promoter (which controls the *lacZ* gene) is an example of this. *Cassette vectors* have promoter, ribosomal binding site and terminator sequences clustered together as a discrete unit, with a single recognition site for one or more REs being situated downstream of the promoter (Figure 12.11).

The small size of the insulin molecule (and gene) and the size of the potential market made it a prime early candidate for production by recombinant DNA technology. Most insulin used in the treatment of diabetes nowadays is produced in this way. Systems based on *E. coli* have also been used to synthesise other small human proteins with therapeutic potential such as human growth hormone, γ-interferon and tumour necrosis factor (TNF).

Bacteria, however, are not suitable host cells for the production of many other human proteins such as tissue plasminogen activator (TPA) or blood clotting Factor VIII, due to the size and complexity of their genes. This is because many proteins of complex eucaryotes are subject to *post-translational modifications*; this does not occur in procaryotes, so bacteria such as *E. coli* are not equipped with the cellular machinery to make the necessary modifications to human proteins.

The polypeptide products of translation in eucaryotes may require post-translational modification before functional protein is produced. Examples include phosphorylation, acetylation and glycosylation.

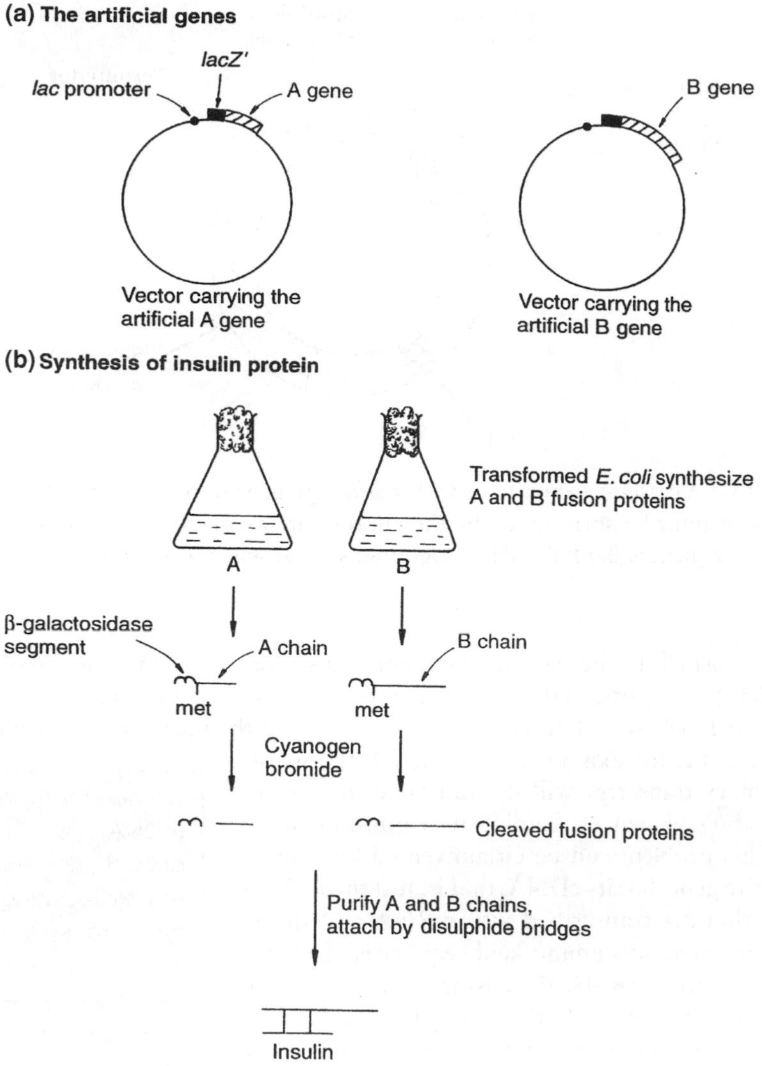

Figure 12.10 Production of recombinant human insulin. Human insulin comprises two short peptide chains, 21 and 30 amino acid residues in length. Because they are so short, the nucleotide sequence of their genes can be predicted from their amino acid sequence, and synthetic genes produced. Each is cloned under the influence of the lac promoter and downstream of part of the *lacZ'* gene. By being produced as fusion proteins with β-galactosidase, the insulin chains are protected against being degraded by the *E. coli* cell. From Brown, TA: Gene Cloning, 3rd edn, Chapman & Hall, 1995. Reproduced by permission of Blackwell Publishing Ltd.

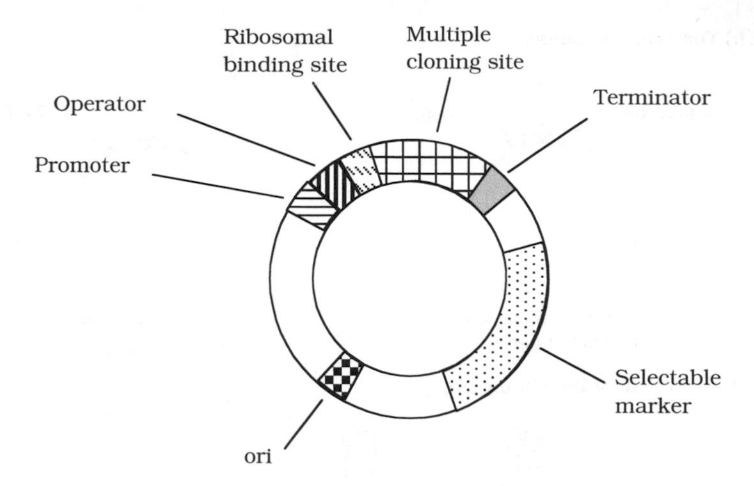

Figure 12.11 A cassette vector. In order for a foreign gene to be transcribed and translated successfully, it must be surrounded by the appropriate expression sequences. In a cassette vector, these sequences flank the RE recognition sites at which the foreign DNA is inserted

Another obstacle to the cloning of such proteins concerns a fundamental difference in the way that procaryotic and eucaryotic systems convert the message encoded in DNA to messenger RNA (see Chapter 11). Procaryotes lack the means to remove introns, so if a human gene, for example, is expressed, the whole of the primary transcript will be translated, instead of just the coding sequences, leading to a non-functional protein. This problem can be circumvented by cloning not the entire gene, but its cDNA, that is, just those DNA sequences that are transcribed into mRNA and subsequently translated into amino acid sequences. This can be done by isolating mRNA, then using reverse transcriptase to make a DNA copy. In the case of proteins such as insulin, the very small size enabled artificial genes to be synthesised, based on their known amino acid sequences.

Complementary DNA (cDNA) is DNA produced using mRNA as a template, using the enzyme reverse transcriptase. It represents only the coding sequences of the parent DNA molecule.

Eucaryotic cloning vectors

A number of problems are associated with the expression of complex eucaryotic genes in bacterial cloning systems, some of which have just been outlined. In order to avoid these, such genes may be expressed in eucaryotic cloning systems, the model system for which is the yeast *Saccharomyces cerevisiae*. This holds many attractions for the experimenter:

- it is safe and easy to handle in the laboratory

Table 12.1 Characteristics of yeast cloning vectors

Plasmid type	Abbreviation	Features
Yeast episomal plasmid	YEp	Contain 2 μm circle (naturally occurring yeast plasmid) origin of replication and whole of pBR322 sequence. Can integrate into yeast chromosomal DNA or replicate independently. High copy number
Yeast integrative plasmid	YIp	Bacterial plasmid containing yeast selectable marker. Stably integrates into yeast chromosome. Unable to exist independently in yeast. Low transformation rate and single copy number
Yeast replicative plasmid	YRp	Carries yeast chromosomal origin of replication (ARS); high copy number; unstable
Yeast centromere plasmid	YCp	Contain centromere sequence – ensures stability, but low (single) copy number

- it grows at higher density than *E. coli*
- it grows much faster than other eucaryotic cells in culture (e.g. cultured mammalian cells)

Several types of vector have been developed for use in yeasts. These are summarised in Table 12.1. The choice of which type to use will depend on the relative importance to a particular application of factors such as yield and long-term stability. Most yeast vectors are *shuttle vectors*, that is, vectors that can be replicated in more than one type of host cell (in this case, yeast and *E. coli*). Initial cloning is more easily done in *E. coli*, then the recombinant vector is transferred to yeast cells, which can carry out functions such as protein folding and glycosylation that are not possible in bacteria. Shuttle vectors must contain selectable markers and an origin of replication (or means of integration) for both cell types.

YACs, BACs and PACs

We saw earlier in this chapter how the use of cosmid vectors extended the size of cloneable insert to around 45kb. Many eucaryotic genes, however, are bigger than this, and so cannot be cloned in a single sequence using such a system. In addition, the ability to clone large DNA fragments is essential for the physical mapping of complex eucaryotic genomes such as that of humans (3×10^9bp). *Yeast artificial chromosomes* (YACs) have been developed to accommodate insert fragments of up to 1Mb (1000kb). They contain key sequences from a

The *centromere* is the central region of the chromosome that ensures correct distribution of chromosomes between daughter cells during cell division, whilst *telomeres* are important in preserving the stability of the tips of each chromosome arm.

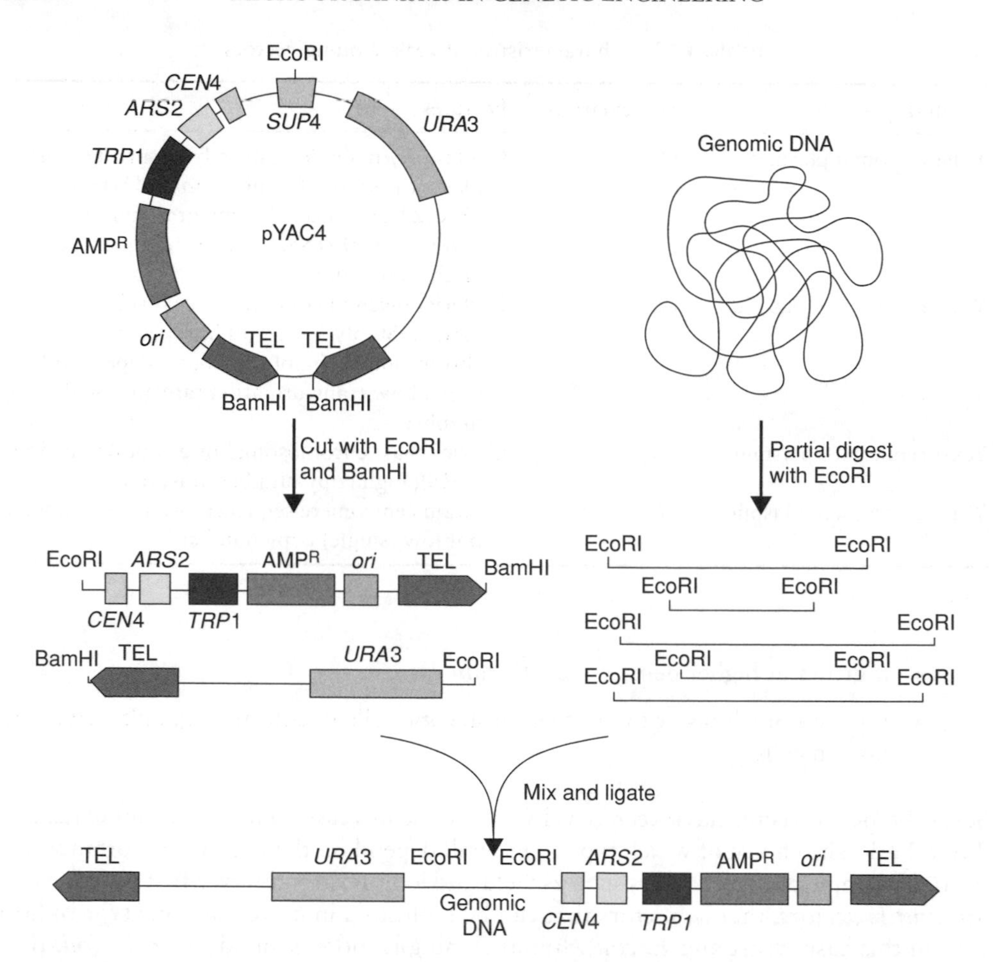

Figure 12.12 Yeast artificial chromosomes (YACs) are used to clone large fragments of DNA. Cleavage with the REs *Bam*HI and *Sna*BI leads to the formation of two arms, between which fragments of foreign DNA of several hundred kb may be ligated. The final construct contains a centromere, telomere sequences and an origin of replication, and is able to act like a natural chromosome. From Reece, RJ: Analysis of Genes and Genomes, John Wiley & Sons, 2003 Reproduced by permission of the publishers

yeast chromosome that ensure its stable replication; these include an origin of replication (the autonomous replication sequence, ARS) the CEN sequence from around the centromere, and the telomeres at the end of the chromosome. When placed in *S. cerevisiae* cells, the presence of these sequences allows the YAC to replicate along with the natural chromosomes.

The fragment to be cloned is inserted between the two arms of the YAC (Figure 12.12). Each arm carries a selectable marker; it is important to have two, to ensure that each construct comprises both a right and a left arm, and not two of the same. Insertional inactivation of a third selectable marker (situated around the point of insertion) allows the detection of recombinants.

Bacterial artificial chromosomes (BACs) allow the cloning of fragments as large as 300kb in length into *E. coli* host cells, although 100–150kb is more routinely used. They are based on the naturally occurring F plasmid of *E. coli*; recall from our description of bacterial conjugation in Chapter 11 that the F plasmid can pick up considerable lengths of chromosomal DNA, when it becomes known as F'. BACs have the advantage over YACs that they are easier to manipulate, and inherently more stable.

Phage P1-derived artificial chromosomes (PACs) are another relatively recently developed class of vector for use in *E. coli*, with a comparable capacity to that of BACs.

Viruses as vectors in eucaryotic systems

Several proteins of clinical or commercial interest are too complex to be expressed using microbial host cells, even eucaryotic ones, and are only properly produced in mammalian systems. Vectors based on animal viruses such as SV40, adenoviruses and vaccinia virus have been successfully developed for use in these. Vaccinia has been particularly valuable in the development of recombinant vaccines.

Viruses of humans such as adenoviruses and retroviruses have also been tested as vectors in the exciting new technique of *gene therapy*, which attempts to ameliorate the effects of genetic disorders by introducing the 'correct' form of the defective gene into the patient's cells. Here it is important to ensure stable integration of the inserted DNA into the host chromosome.

A large virus that infects insects, the baculovirus, has been found to be a highly efficient vector for the large-scale expression of eucaryotic proteins in cultured insect cells. The rate of expression is much higher than in cultured mammalian cells, and the necessary protein folding and post-translational modifications are correctly executed.

Cloning vectors for higher plants

The most important single tool for the genetic engineering of plants is the *Ti plasmid*. This is found naturally in the soil bacterium *Agrobacterium tumefaciens,* which infects plants at wound sites, and leads to a condition called crown gall disease. The important feature of this plasmid is that part of it, called the *T-DNA*, can integrate into the host plant's chromosomes, and be expressed along with host genes (Figure 12.13). Geneticists were quick to spot the potential of the Ti plasmid, replacing tumour-forming genes with foreign genes, and having them expressed in plant tissues. The recombinant *A. tumefaciens* is used to infect protoplasts, which can be regenerated into a whole plant, every cell of which will contain the integrated foreign gene. The Ti plasmid system has been used in the successful transfer of genes for insect- and herbicide-resistance into economically significant crop plants.

Plant viruses have very limited usefulness as vectors. Only the caulimoviruses and the geminiviruses have DNA rather than RNA as their genomic material, and a variety of problems, including instability of inserts and narrow host range, have been encountered with the use of these.

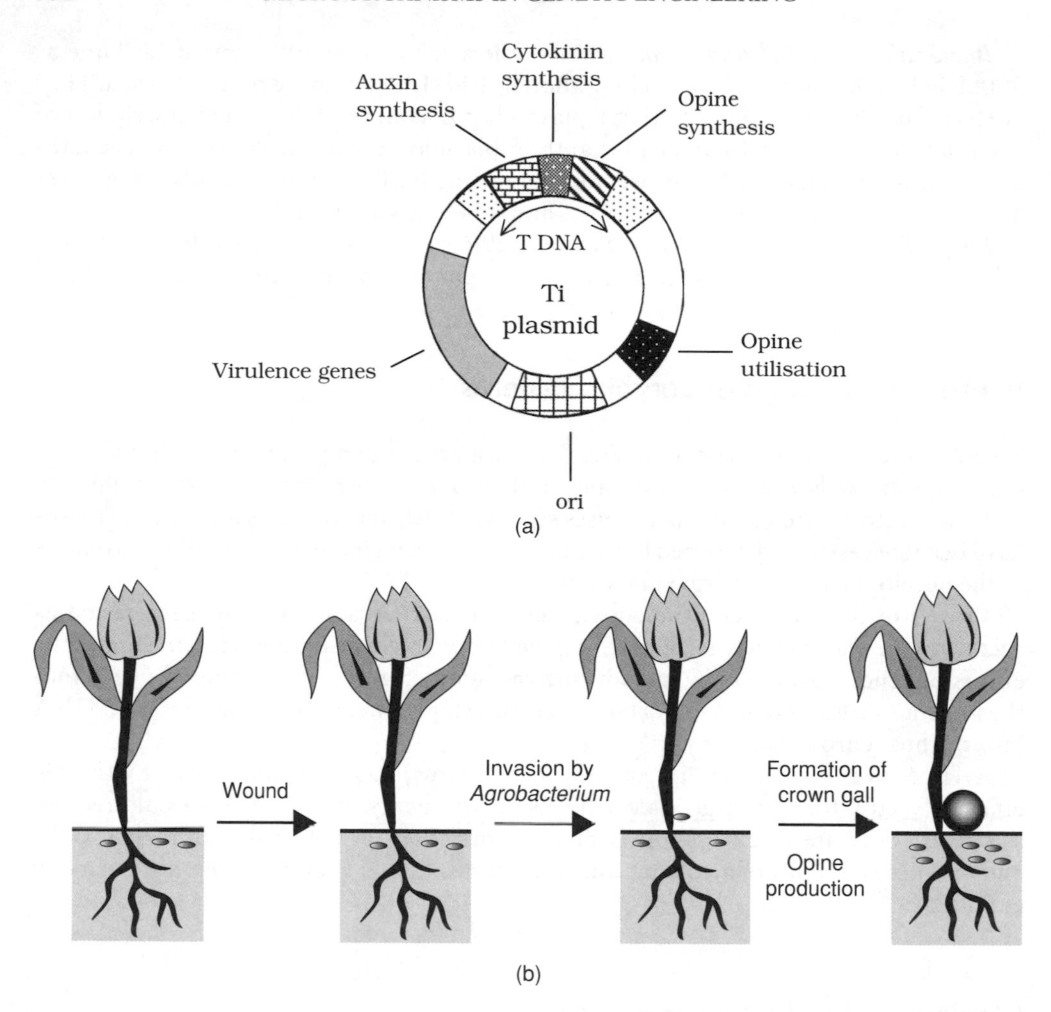

Figure 12.13 *Agrobacterium tumefaciens*. (a) The Ti plasmid. The T-DNA contains genes for tumour production and the synthesis of opines, unusual amino acid derivatives that serve as nutrients for *A. tumefaciens*. (b) Crown gall formation by *A. tumefaciens*. The T-DNA integrates into the DNA of the host, where its genes are expressed. This ability has been exploited in order to transfer foreign genes into plant cells. From Reece, RJ: Analysis of Genes and Genomes, John Wiley & Sons, 2003. Reproduced by permission of the publishers

Genetically engineered insect resistance using *Bacillus thuringiensis*

Destruction of crops by insect pests is a huge problem throughout the world, and the use of chemical insecticides is only partially successful in countering it. The drawbacks to the use of such chemicals are manifold:

- they are often non-specific, so beneficial insects as well as harmful ones are killed

- they are often non-biodegradable, so they can have lasting environmental effects
- aerial spraying only reaches upper leaf surfaces
- resistance develops with continued use.

A form of natural insecticide does exist; it is a crystalline protein produced during sporulation by *Bacillus thuringiensis*. This is highly toxic to insects when converted to its active form by the enzymes of the gut. This *δ-endotoxin* is relatively selective; different strains of *B. thuringiensis* produce different forms of the toxin, which are effective against the larvae of different insects. Surprising as it may seem, the use of the δ-endotoxin as an insecticide was patented a hundred years ago; however, the success of it has been limited due to a variety of practical considerations.

The development of genetic engineering techniques has meant that attention has turned in more recent times to introducing the genes for the δ-endotoxin into the crops themselves, so that they synthesise their own insecticide. This has been achieved with some success, but the problem of the insects building up resistance remains.

As a spin-off from the Human Genome Project, the genetic make-up of many microorganisms has been elucidated. One of these is *Photorhabdus luminescens*, which encodes a toxin lethal to the two species of mosquito responsible for the spread of malaria, and it is hoped that determining the sequence of the gene will lead to applications in insect control.

Polymerase chain reaction (PCR)

First described in the mid-1980s by Kary Mullis, PCR is probably the most significant development in molecular biology since the advent of gene cloning. PCR allows us to amplify a specific section of a genome (for example a particular gene) millions of times from a tiny amount of starting material (theoretically a single molecule!). The impact of this powerful and highly specific process has been felt in all areas of biology and beyond. Medicine, forensic science and evolutionary studies are but three areas where PCR has opened up new possibilities over the last 15 years or so. To appreciate how PCR works, you will need to understand the role of the enzyme DNA polymerase in DNA replication, as outlined in the previous chapter. This is the enzyme, you'll recall, that when provided with single-stranded DNA and a short *primer*, can direct the synthesis of a complementary second strand. Figure 12.14 illustrates the three steps in the PCR process:

PCR selectively replicates a specific DNA sequence by means of *in vitro* enzymatic amplification.

A primer is a short nucleotide sequence that can be extended by DNA polymerase. In PCR, synthetic primers are designed to flank either side of the target sequence.

- *Denaturation*: by heating to 95 °C, the DNA is separated into single strands, providing a template for the DNA polymerase.

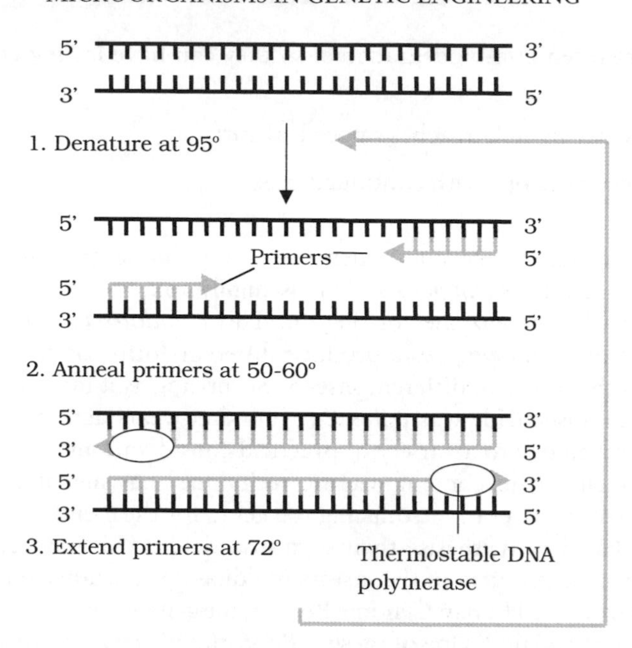

Figure 12.14 The polymerase chain reaction (PCR). Millions of copies of a specific sequence of DNA can be made by PCR. Samples are subjected to repeated cycles of denaturation, annealing of primers and extension by a thermostable DNA polymerase (for details see the text). At the end of the first cycle (shown in the diagram), each of the four strands can serve as a template for replication in cycle two. After 30 cycles, over a billion copies of the target sequence will have been made

- *Primer annealing*: at a lower temperature (typically 50–60 °C, the exact value depends on the primer sequence), short single-stranded primers are allowed to attach. Primers are synthetic oligonucleotides with a sequence complementary to the regions flanking the region we wish to amplify. One primer anneals to each strand, at either side of the target sequence.

- *Polymerase extension*: at around 72 °C, the DNA polymerase extends the primer by adding complementary nucleotides and so forms a second strand.

The net result of this process is that instead of one double-stranded molecule, we now have two. We now come to the key concept of PCR; if we raise the temperature to 95 °C again to start another cycle, we will have not two, but four single-stranded templates to work on, each of which can be converted to the double-stranded form as before. After 20 such cycles we should have, in theory, over a *million* copies of our original molecule! Typical PCR protocols run for 30–35 cycles. All this can be achieved in just a couple of hours; the temperature cycling is carried out by a programmable microprocessor-controlled machine called a *thermal cycler*.

PCR has widespread applications in microbiology, as in other fields of biology, but in addition, microorganisms play a crucial role in the process itself. If you have read the above description of PCR carefully, something should have struck you as being not quite right: how can the DNA polymerase (a protein) tolerate being repeatedly heated to over 90 °C, and what sort of an enzyme can work effectively at 72 °C? The answer is that the DNA polymerase used in PCR comes from thermophilic bacteria such as *Thermus aquaticus*, a species found naturally in hot springs. The optimum temperature for *Taq* polymerase is 72 °C, and it can tolerate being raised to values considerably higher than this for short periods. The availability of this enzyme meant that it was not necessary to add fresh enzyme after each cycle, and the whole process could be automated, a key factor in its subsequent phenomenal success.

Test yourself

1 ________ ________ are enzymes that cut DNA at specific ________ sequences in the ________ ________ backbone.

2 Joining together DNA fragments from different sources produces a ________ molecule.

3 DNA fragments with ________ ________ tend to join together because of their complementary single-stranded overhangs.

4 In order to gain access to a host cell and be replicated, a fragment of DNA must be taken up by a ________ molecule.

5 ________ ________ enable us to detect which host cells have been transformed. A commonly used example is resistance to ________ . By inserting foreign DNA at this point, recombinants can be detected by ________ ________ .

6 Another name for a polylinker is a ________ ________ ________ .

7 A short sequence of labelled single-stranded DNA used to locate a specific target sequence is known as a ________ .

8 The removable central section of the phage Lambda genome is known as a ________ fragment.

9 Bacteriophage ________ is a useful vector if DNA is required in a ________ ________ form.

10 Cloning vectors combining features of plasmids and phage Lambda are called ________ .

11 ________ vectors incorporate foreign DNA at a site that is surrounded by essential signal sequences.

12 *E. coli* is not a suitable host for the expression of complex eucaryotic proteins because it lacks the molecular machinery to carry out ________ modifications such as ________ .

13 ________ DNA contains only those sequences that are involved in coding for a polypeptide product.

14 Yeast plasmids are based on a naturally occurring sequence called the ________ ________ .

15 ________ vectors are able to replicate in both ________ and ________ cells.

16 A virus found to be useful in the expression of eucaryotic genes is the ________ . It is grown in cultured ________ cells.

17 The soil bacterium *A. tumefaciens* contains a plasmid called ________ , which causes crown gall disease. Part of it, the ________ , integrates into host chromosomes.

18 Genes from *Bacillus thuringiensis* coding for ________ have been introduced into plant cells, making them ________ resistant.

19 The polymerase chain reaction (PCR) comprises repeated cycles of the following three steps: ________ , ________ of primers and ________ by ________ DNA polymerase.

20 PCR results in an ________ increase in the number of copies of the target sequence.

Part V

Control of Microorganisms

13

The Control of Microorganisms

An *antiseptic* is a chemical agent of disinfection that is mild enough to be used on human skin or tissues.

Prior to Lister's pioneering work with antiseptics, around four out of every ten patients undergoing an amputation failed to survive the experience, such was the prevalence of infections associated with operative procedures, and yet, just 40 years later, this figure had fallen to just three in a hundred. This is a dramatic demonstration of the fact that in some situations it is necessary for us to destroy, or at least limit, microbial growth, because of the undesirable consequences of the presence of microorganisms, or their products.

Control of microorganisms can be achieved by a variety of chemical and physical methods. *Sterilisation* is generally achieved by using physical means such as heat, radiation and filtration. Agents which destroy bacteria are said to be *bactericidal*. Chemical methods, whilst effective at *disinfection*, are generally not reliable for achieving total sterility. Agents which inhibit the growth and reproduction of bacteria without bringing about their total destruction are described as *bacteriostatic*.

Sterilisation

Sterilisation is the process by which all microorganisms present on or in an object are destroyed or removed.

One of the oldest forms of antimicrobial treatment is that of heating, and in most cases this remains the preferred means of sterilisation, provided that it does not cause damage to the material in question. The benefits of boiling drinking water have been known at least since the 4th century BC, when Aristotle is said to have advised Alexander the Great to order his troops to take this precaution. This of course was many centuries before the existence of microorganisms had been demonstrated or perhaps even suspected.

Sterilisation by heat

Boiling at 100 °C for 10 minutes is usually enough to achieve sterility, provided that organisms are not present in high concentrations; in fact most bacteria are killed at about 70 °C. If, however, endospores of certain bacteria (notably *Bacillus* and *Clostridium*)

Table 13.1 Temperature of steam at different pressures

Pressure		Temperature
(kPa)	(psi)	(°C)
0	0	100
69	10	115
103.5	15	121
138	20	126
172.5	25	130

psi = pounds/in^2.

are present, they can resist boiling, sometimes for several hours. As we saw in Chapter 7, the causative agents of some particularly nasty conditions, such as botulism and tetanus, are members of this group. In order to destroy the heat resistant endospores, heating beyond 100 °C is required, and this can be achieved by heating under pressure in a closed vessel (Table 13.1). A typical laboratory treatment is 15 minutes at a pressure of 103 kpa (15 psi), raising the temperature of steam to 121 °C. This is carried out in an *autoclave*, which is, to all intents and purposes, a large-scale pressure cooker (Figure 13.1). Air is driven out of the system so that the atmosphere is made up entirely of steam; the desired temperature will not be reached if this is not achieved (Figure 13.2). Large loads, or large volumes of liquids, may need a longer treatment time in order for

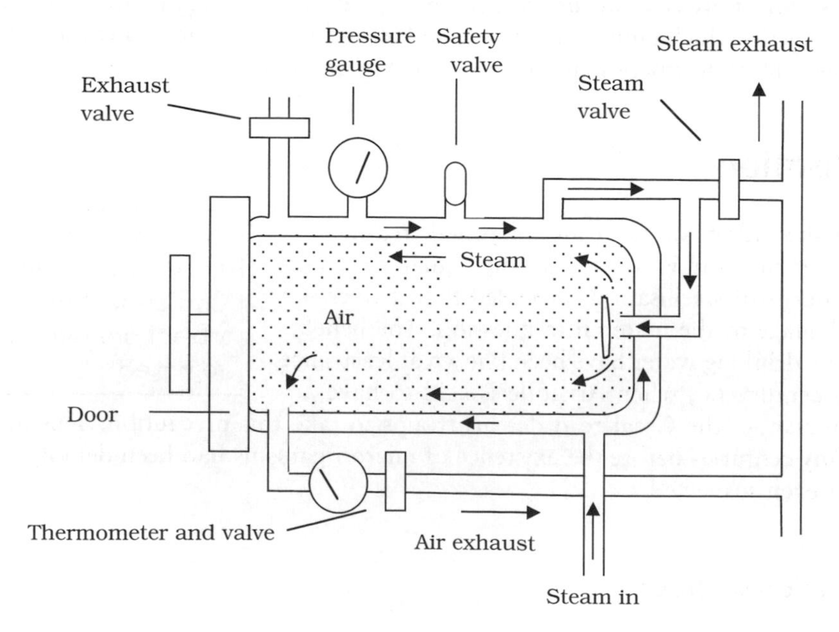

Figure 13.1 The autoclave. Steam enters the chamber, driving air out. As the pressure of the steam reaches 103 kPa (15 psi) the temperature reaches 121 °C, sufficient to kill resistant endospores as well as vegetative cells

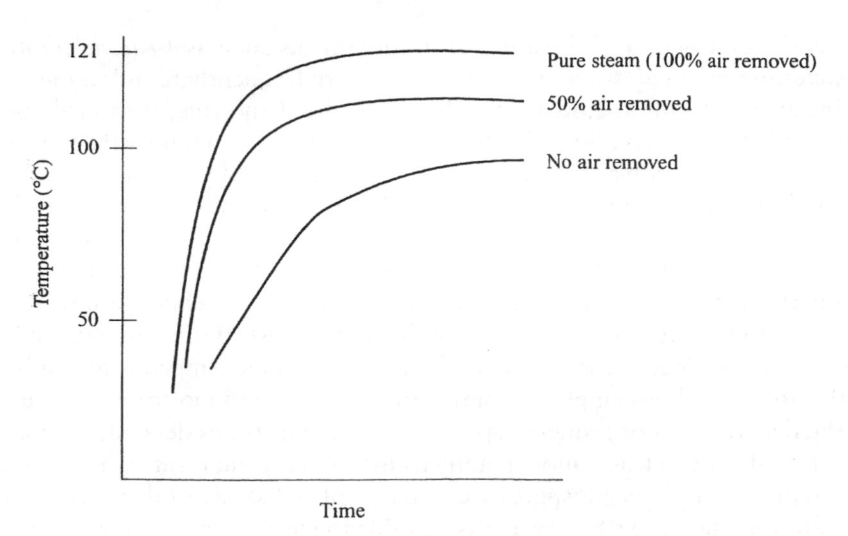

Figure 13.2 An autoclave only reaches maximum temperature in an atmosphere of pure steam. Temperature achieved at 103 kPa (15 psi) in different atmospheres. Any air remaining in the system will reduce the final temperature achieved. From Hardy, SP: Human Microbiology, Taylor and Francis, 2002. Reproduced by permission of Thomson Publishing Services

the heat to penetrate throughout. Modern autoclaves include probes designed to assess the temperature within a load rather than that of the atmosphere. Sometimes, spores of *Bacillus stearothermophilus* are introduced into a system along with the material to be sterilised; if subsequent testing shows that the spores have all been destroyed, it is reasonable to assume that the system has also destroyed any other biological entity present. Special tape which changes colour if the necessary temperature is reached can act as a more convenient but less reliable indicator.

An effect similar to that achieved by autoclaving can be obtained by a method called intermittent steaming or *tyndallisation* (after the Irish physicist John Tyndall, who was one of the first to demonstrate the existence of heat-resistant microbial forms). This is used for those substances or materials that might be damaged by the high temperatures used in autoclaving. The material is heated to between 90 and 100 °C for about 30 minutes on each of three successive days, and left at 37 °C in the intervening periods. Vegetative cells are killed off during the heating period, and during the 37 °C incubation, any endospores that have survived will germinate. Once these have grown into more vegetative cells, they too are killed in the next round of steam treatment. Clearly this is quite a long-winded procedure, and it is therefore reserved for those materials which might be harmed by steam sterilisation.

High temperatures can cause damage to the taste, texture and nutritional value of many food substances and in such instances, it is sufficient to destroy vegetative cells by a process of *pasteurisation* (among his many other achievements, Pasteur demonstrated that the microbial spoilage of wines could be prevented by short periods of heating). Milk was traditionally pasteurised by heating large volumes at 63 °C for 30 minutes, but the method employed nowadays is to pass it over a heat exchanger at 72 °C for 15 seconds (HTST – high temperature, short time). This is not sterilisation as such, but it ensures the destruction of disease-causing organisms such as *Brucella abortus* and *Mycobacterium tuberculosis*, which at one time were frequently found in milk, as well as significantly reducing the organisms that cause food spoilage, thus prolonging the time the milk can be kept. Some protocols exceed these minimum values in order to reduce further the microbial content of the milk. One type of milk on sale in the shops is subjected to more extreme heating regimes; this is 'UHT' milk, which can be kept for several weeks without refrigeration, though many find that this is achieved at some cost to its palatability! It is heated to *u*ltra *h*igh *t*emperatures (150 °C) for a couple of seconds using superheated steam. The product is often referred to as being 'sterilised', but this is not true in the strictest sense. Milk is not the only foodstuff to be pasteurised; others such as beer, fruit juices and ice cream each has its own time/temperature combination.

All the above methods employ a combination of heat and moisture to achieve their effect; the denaturation of proteins, upon which these methods depend, is enhanced in the presence of water. Heat is more readily transferred through water than through air, and the main reason that endospores are so resistant is because of their very low water content. In some situations however, it is possible to employ dry heat, using an oven to sterilise metal instruments or glassware for example. It is a more convenient procedure, but a higher temperature (160–170 °C) and longer exposure time (2 hours) are required. Dry heat works by oxidising ('burning') the cell's components. Microorganisms are quite literally burnt to destruction by the most extreme form of dry heat treatment, incineration. Soiled medical dressings and swabs, for example are potentially hazardous, and are destroyed in this way at many hundreds of degrees Celsius. As we saw in Box 4.1, sterilising the loops and needles used to manipulate microorganisms by means of flaming is a routine part of aseptic procedures.

Sterilisation by irradiation

Certain types of irradiation are used to control the growth of microorganisms. These include both ionising and non-ionising radiation.

The most widely used form of non-ionising radiation is ultraviolet (UV) light. Wavelengths around 260 nm are used because these are absorbed by the purine and pyrimidine components of nucleic acids, as well as certain aromatic amino acids in proteins. The absorbed energy causes a rupture of the chemical bonds, so that normal cellular function is impaired. You will recall from Chapter 11 that UV light causes the formation of *thymine dimers* (Figure 11.21), where adjacent thymine nucleotides on the same strand are linked together, inhibiting DNA replication. Although many bacteria are capable of repairing this damage by enzyme-mediated photoreactivation, viruses are much more susceptible. UV lamps are commonly found in food preparation areas, operating theatres and specialist areas such as tissue culture facilities, where it is important to prevent contamination. Because they are also harmful to humans (particularly the skin and eyes), UV lamps can only be operated in such areas when people are not present. UV radiation has very poor penetrating powers; a thin layer of glass, paper or fabric is able to impede the passage of

The world's biggest UV wastewater treatment plant was opened in Manukau, New Zealand in 2001. Its 8000 UV lamps are able to treat up to 50 000 cubic metres of water per hour.

the rays. The chief application is therefore in the sterilisation of work surfaces and the surrounding air, although it is increasingly finding an application in the treatment of water supplies.

Ionising radiations have a shorter wavelength and much higher energy, giving them greater penetrating powers. The effect of ionising radiations is due to the production of highly reactive free radicals, which disrupt the structure of macromolecules such as DNA and proteins. Surgical supplies such as syringes, catheters and rubber gloves are commonly sterilised employing gamma (γ) rays from the isotope cobalt 60 (^{60}Co).

Gamma radiation has been approved for use in over 40 countries for the preservation of food, which it does not only by killing pathogens and spoilage organisms but also by inhibiting processes that lead to sprouting and ripening. The practice has aroused a lot of controversy, largely due to concerns about health and safety, although the first patent applications for its use date back nearly a hundred years! Although the irradiated product does not become radioactive, there is a general suspicion on the part of the public about anything to do with radiation, which has led to its use on food being only very gradually accepted by consumers. In this respect Europe lags behind the USA, where during the 1990s a positive attitude towards irradiation of food both by professional bodies and the media has led to a more widespread acceptance of the technology. Gamma radiation is used in situations where heat sterilisation would be inappropriate, because of undesirable effects on the texture, taste or appearance of the product. This mainly relates to fresh produce such as meat, poultry, fruit and vegetables. Irradiation is not suitable for some foodstuffs, such as those with a high fat content, where unpleasant tastes and odours result. Ionising radiations have the great advantage over other methods of sterilisation that they can penetrate packaging.

Filtration

Many liquids such as solutions of antibiotics or certain components of culture media become chemically altered at high temperatures, so the use of any of the heat regimes described above is not appropriate. Rather than killing the microorganisms, an alternative approach is simply to isolate them. This can be done for liquids and gases by passing them through filters of an appropriate pore size. Filters used to be made from materials such as asbestos and sintered glass, but have been largely replaced by membrane filters, commonly made of nitrocellulose or polycarbonate (Figure 13.3). These can be purchased ready-sterilised and the liquid passed through by means of pressure or suction. Supplies of air or other gases can also be filter-sterilised in this way. A pore size of 0.22 μm is commonly used; this will remove bacteria plus, of course, anything bigger, such as yeasts; however, mycoplasma and viruses are able to pass through pores of this size. With a pore size 10 times smaller than this, only the smallest of viruses can pass through, so it is important that an appropriate pore size is chosen for any given task. A drawback with all filters, but especially those of a small pore size, is that they can become clogged easily. Filters in general are relatively expensive, and are not the preferred choice if alternative methods are available.

High efficiency particulate air (HEPA) filters create clean atmospheres in areas such as operating theatres and laboratory laminar-flow hoods.

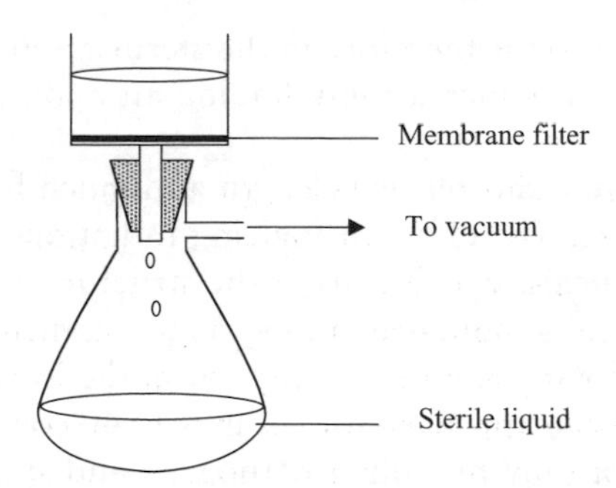

Figure 13.3 Membrane filtration. Membrane filters are used to sterilise heat-labile substances. They are available in a variety of pore sizes, according to the specific application

Sterilisation using ethylene oxide

Generally, chemical methods achieve only disinfection (see below); the use of the gas ethylene oxide, however, is effective against bacteria, their spores and viruses. It is used for sterilising large items of medical equipment, and materials such as plastics that would be damaged by heat treatment. Ethylene oxide is particularly effective in sterilising items such as dressings and mattresses, due to its great powers of penetration. In the food industry, it is used as an antifungal fumigant, for the treatment of dried fruit, nuts and spices. The materials to be treated are placed in a special chamber which is sealed and filled with the gas in a humid atmosphere at 40–50 °C for several hours. Ethylene oxide is highly explosive, so it must be used with great caution; its use is rendered safer by administering it in admixture (10 per cent) with a non-flammable gas such as carbon dioxide. It is also highly toxic, so all items must be thoroughly flushed with sterile air following treatment to remove any trace of it. Ethylene oxide is an alkylating agent; it denatures proteins by replacing labile hydrogens such as those on sulphydryl groups with a hydroxyl ethyl radical (Figure 13.4).

Disinfection

We saw at the beginning of this chapter how sterilisation is an absolute term, implying the total destruction of all microbial life. Disinfection, by comparison, allows the possibility that some organisms may survive, with the potential to resume growth when

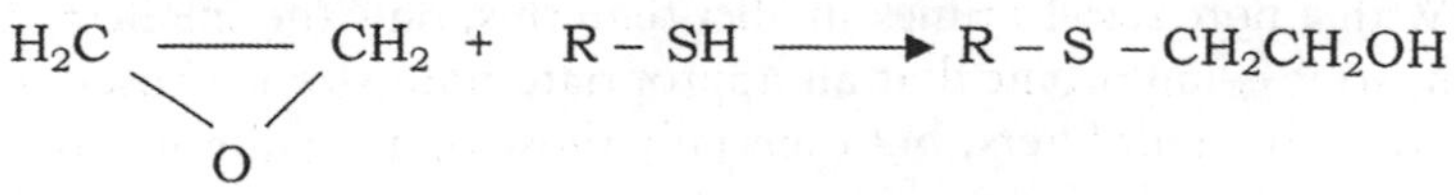

$$\underset{\text{Ethylene oxide}}{H_2C - CH_2 \text{ (bridged by O)}} + R - SH \longrightarrow R - S - CH_2CH_2OH$$

Figure 13.4 Ethylene oxide is an alkylating agent. Like other alkylating agents, ethylene oxide affects the structure of both proteins and nucleic acids. Labile hydrogen atoms such as those on sulphydryl groups are replaced with a hydroxyl ethyl radical

conditions become more favourable. A *disinfectant* is a chemical agent used to disinfect inanimate objects such as work surfaces and floors. In the food and catering industry, especially in the USA, the term *sanitisation* is used to describe a combination of cleaning and disinfection. Disinfectants are incapable of killing spores within a reasonable time period, and are generally effective against a narrower range of organisms than physical means. *Decontamination* is a term sometimes used interchangeably with disinfection, but its scope is wider, encompassing the removal or inactivation of microbial products such as toxins as well as the organisms themselves.

Disinfection is the elimination or inhibition of pathogenic microorganisms in or on an object so that they no longer pose a threat.

The lethal action of disinfectants is mainly due to their ability to react with microbial proteins, and therefore enzymes. Consequently, any chemical agent that can coagulate, or in any other way denature, proteins will act as a disinfectant, and compounds belonging to a number of groups are able to do this.

Alcohols

The antimicrobial properties of ethanol have been known for over a century. It was soon realised that it worked more effectively as a disinfectant at less than 100 per cent concentration, that is, when there was some water present. This is because denaturation of proteins proceeds much more effectively in the presence of water. (Recall that moist heat is more effective than dry heat for the same reason.) It is important, however, not to overdo the dilution, as at low percentages some organisms can actually utilise ethanol as a nutrient! Ethanol and isopropanol are most commonly used at a concentration of 70 per cent. As well as denaturing proteins, alcohols may act by dissolving lipids, and thus have a disruptive effect on membranes, and on the envelope of certain viruses. Both bacteria and fungi are killed by alcohol treatment, but spores are often resistant because of problems in rehydrating them; there are records of anthrax spores surviving in ethanol for 20 years! The use of alcohols is further limited to those materials that can withstand their solvent action.

Alcohols may also serve as solvents for certain other chemical disinfectants. The effectiveness of iodine for example, can be enhanced by being dissolved in ethanol.

Halogens

Chlorine is an effective disinfectant as a free gas, and as a component of chlorine-releasing compounds such as hypochlorite and chloramines. Chlorine gas, in compressed form, is used in the disinfection of municipal water supplies, swimming pools and the dairy industry. Sodium hypochlorite (household bleach) oxidises sulphydryl (–SH) and disulphide (S–S) bonds in proteins. Like chlorine, hypochlorite is inactivated by the presence of organic material. Chloramines are more stable than hypochlorite or free chlorine, and are less affected by organic matter. They are also less toxic and have the additional benefit of releasing their chlorine slowly over a period of time, giving them a prolonged bactericidal effect.

Iodine acts by combining with the tyrosine residues on proteins; its effect is enhanced by being dissolved in ethanol (1 per cent I_2 in 70 per cent ethanol) as tincture of iodine, an effective skin disinfectant. Its use is being superseded by iodophores (Betadine, Isodine), in which iodine is combined with an organic molecule, usually a detergent, to combat bacteria, viruses and fungi, but not spores.

Phenolics

As we saw in Chapter 1, the germicidal properties of phenol (carbolic acid) were first demonstrated by Lister in the middle of the 19th century. Since it is highly toxic, phenol's use in the disinfection of wounds has long since been discontinued, but derivatives such as cresols and xylenols continue to be used as disinfectants and antiseptics. These are both less toxic to humans and more effective against bacteria than the parent compound. Phenol is still used, however, as a benchmark against which the effectiveness of related disinfectants can be measured. The *phenol coefficient* compares the dilution at which the derivative is effective against a test organism with the dilution at which phenol achieves the same result. A phenol coefficient of more than one means that the new compound is more effective than phenol against the organism tested, whereas a value of less than one means that it is not as effective as phenol.

Phenolics act by combining with and denaturing proteins, as well as disrupting cell membranes. Their advantages include the retention of activity in the presence of organic substances and detergents, and their ability to remain active for some time after application; hence their effect increases with repeated use. Familiar disinfectants such as Dettol, Lysol and chlorhexidine (Hibitane, Hibiscrub) are all phenol derivatives. Hexachlorophene (Figure 13.5) is very effective against Gram-positive bacteria such as staphylococci and streptococci, and used to be a component of certain soaps, surgical scrubs, shampoos and deodorants. Its use is now confined to specialist applications in hospitals since the finding that in some cases, prolonged application can lead to brain damage.

Surfactants

Surface active agents or surfactants, such as soaps and detergents, have the ability to orientate themselves between two interfaces to bring them into closer contact

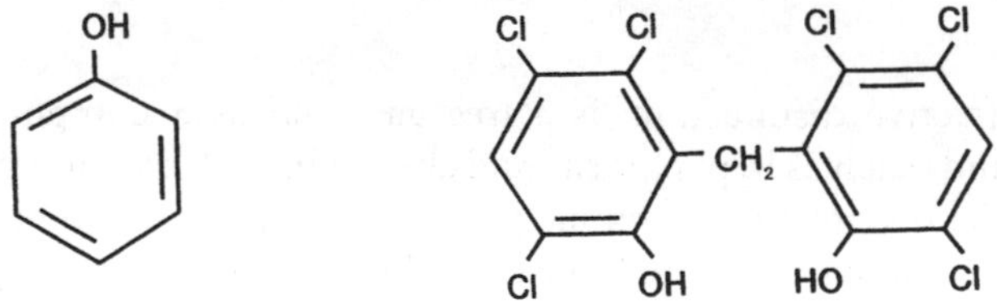

(a) (b)

Figure 13.5 The structure of (a) phenol and (b) hexachlorophene

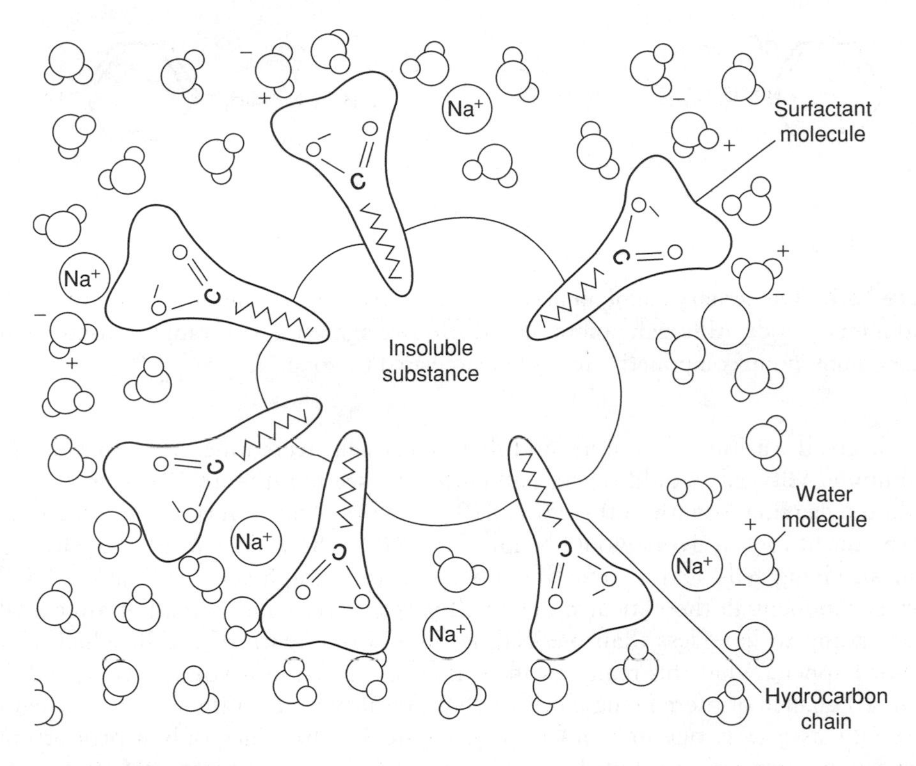

Figure 13.6 The action of surfactants. The long hydrophobic tails of the detergent are able to penetrate an insoluble grease particle. The negatively charged carboxyl group at the other end attracts the positive pole of water molecules, enhancing the water solubility of the grease. From Black, JG: Microbiology: Principles and Explorations, 4th edn, John Wiley & Sons Inc., 1999. Reproduced by permission of the publishers

> A surfactant reduces the tension between two molecules at an interface.

(Figure 13.6). The value of soap has less to do with its disinfectant properties than with ability to facilitate the mechanical removal of dirt and microorganisms. It does this by emulsifying oil secretions, allowing the debris to be rinsed away. Detergents may be *anionic* (negatively charged), *cationic* (positively charged) or non-ionic. Cationic detergents such as quaternary ammonium compounds (ammonium chloride with each hydrogen replaced by an organic group, Figure 13.7) act by combining with phospholipids to disrupt cell membranes and affect cellular permeability.

The kinetics of cell death

When microorganisms are exposed to any of the treatments outlined in the preceding pages, they are not all killed instantaneously. During a given time period, only a certain proportion of them will die. Suppose we had 1000 cells (an unrealistically small number,

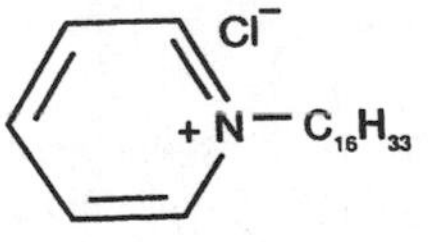

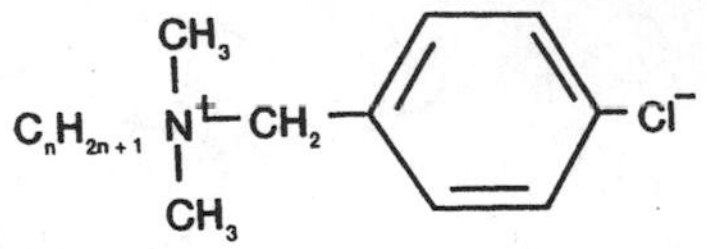

(a) (b)

Figure 13.7 Quaternary ammonium compounds. (a) cetylpyridinium chloride and (b) benzalkonium chloride. Although non-toxic and effective against a wide range of bacteria, quaternary ammonium compounds are easily inactivated by soap

but it keeps the arithmetic simple) and that 10 per cent were killed each minute. After one minute, 900 cells would remain, and after the second minute 10 per cent of these would die, leaving us with 900 − 90 = 810 survivors. After a further minute, another 10 per cent of the survivors would be killed, so 810 − 81 = 729 would be left. A plot of the surviving cells against time of exposure gives a graph such as Figure 13.8. The curve is exponential; theoretically, there will never be zero survivors, but after a while we are going to have less than one cell, let us say one tenth of a cell, which clearly can not happen. What this really means is that in a given unit volume, there will be a one in 10 chance of there being a cell present. Sterility is generally assumed when this figure falls as low as one in a million (see Figure 13.8b). Since only a proportion of the surviving population is killed per unit time, it follows that the more cells you have initially, the longer it will take to eliminate them (Figure 13.9).

The steepness of the slope in Figure 13.8(b) is an indication of the effectiveness of heat sterilisation. The *decimal reduction time* or *D value* is the time needed to reduce the population by a factor of ten (i.e. to kill 90 per cent of the population) using a particular heat treatment. The *D* value applies to a particular temperature; at higher temperatures, the rate of killing is enhanced, and so the *D* value is reduced (Figure 13.10). The increase in temperature required to reduce *D* by a factor of 10 is the *Z* value.

Since in real life, the microbial population is certain to be a mixed one, then the critical factor is the death rate of the most resistant species, that is, the one with the highest percentage of survivors per minute. Sterilisation protocols should therefore be based on the rate of destruction of endospores.

It should be stressed that an effective treatment regime for one organism may be wholly inappropriate for another, since organisms differ in their susceptibility to different agents. Organism A may resist heat treatment better than organism B, but be more sensitive to a particular chemical treatment.

Killing by irradiation

Microorganisms as a group are much more resistant to the effects of ionising radiation than are higher organisms, and some more resistant than others. A plot of log surviving numbers against time of irradiation is shown in Figure 13.11. The decimal reduction value (*D* 10) is analogous to the *D* value used for heat sterilisation (see above).

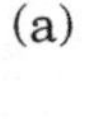

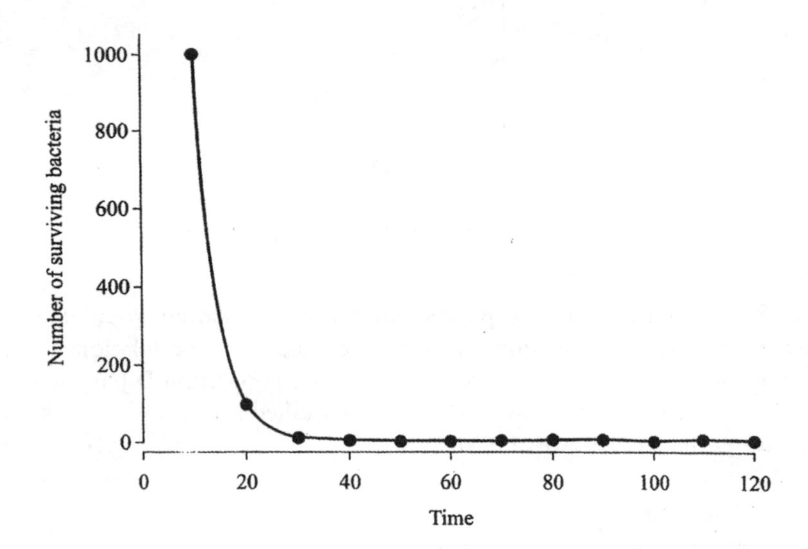

(b)

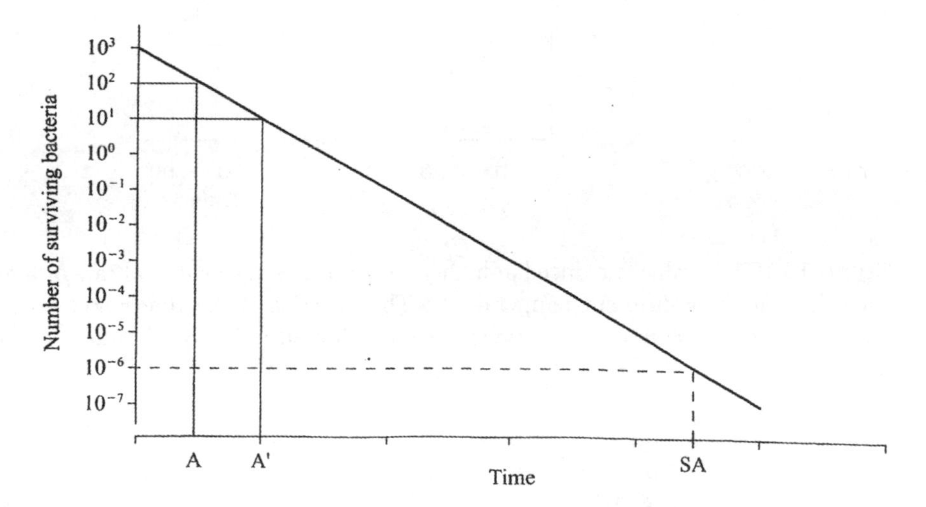

Figure 13.8 The kinetics of cell death. (a) During heat sterilisation, the number of living cells decreases by the same proportion per unit time, giving an exponential curve. (b) When plotted on a logarithmic scale, the decrease in numbers is seen as a straight line, whose slope is a reflection of the rate of killing. The time period between A and A′ is the decimal reduction time (D): the time taken to reduce the population to one-tenth of its size. The total period until the point SA is the sterility assurance value, when there is only a one in a million probability of any cells having survived. From Hardy, SP: Human Microbiology, Taylor and Francis, 2002. Reproduced by permission of Thomson Publishing Services

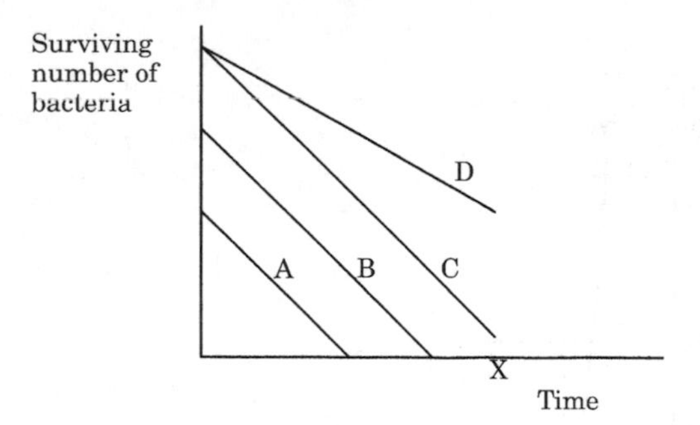

Figure 13.9 Sterilisation time is dependent on starting population. Populations A, B and C all have the same decimal reduction rate, but different starting populations, therefore after a given time they have different numbers of survivors. Population D has the same starting numbers as C, but because a smaller proportion are killed per minute, (i.e. the slope is less steep), a greater number survive after time *X*

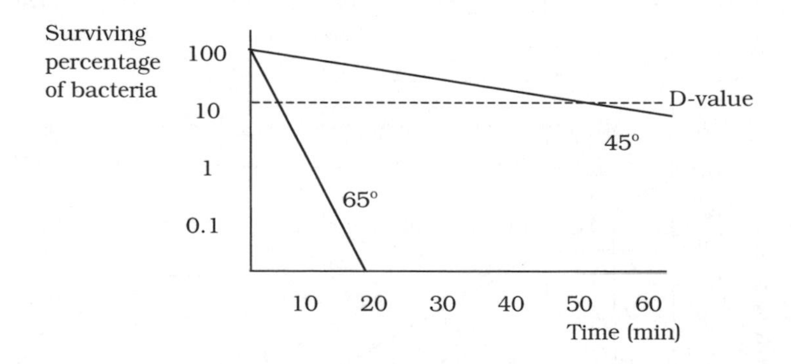

Figure 13.10 *D* value is reduced at higher temperatures. The survival rate of a mesophilic bacterium at three different temperatures. The *D* value is the time taken to reduce the population to one tenth of its starting value: note how it falls from 50 min (45°C) to 5 min (65°C)

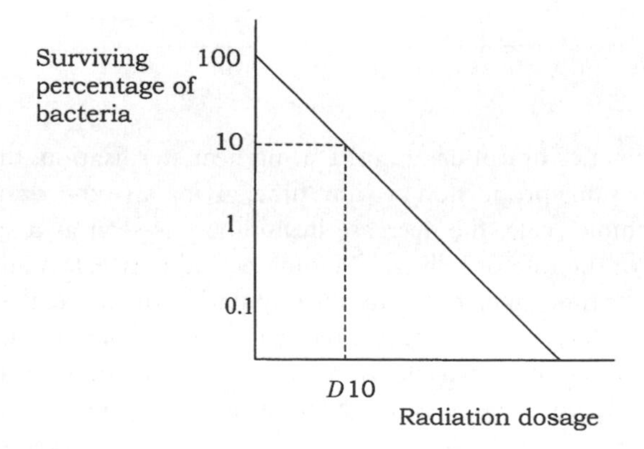

Figure 13.11 Killing by irradiation. The decimal reduction value (*D*10) is the dose of radiation necessary to reduce the population of microorganisms to one-tenth of its value. The value for *D*10 differs greatly between organisms

Test yourself

1 An antimicrobial agent that kills microorganisms is termed ________ one that inhibits their growth is described as ________.

2 Heat treatment at temperatures in excess of 100°C is necessary to ensure the destruction of ________.

3 Steam heated under a pressure of 103 kPa in an autoclave reaches a temperature of ________ °C.

4 During tyndallisation, the material being treated is heated to between 90 and 100°C for about ________ on three consecutive days. In the intervening periods, it is left at ________°C, to allow any surviving ________ to________.

5 The traditional method for the ________ of milk was to heat at ________ for 30 minutes.

6 Moist heat is more effective than dry heat at the same temperature, because water is needed for the ________ of ________.

7 Ultraviolet light is used mainly in the sterilisation of________, because it has very poor ________ power.

8 ________ radiation is used in many countries to irradiate foodstuffs.

9 Liquids may be sterilised by the use of ________ made of polycarbonate or nitrocellulose.

10 ________ is a gas used in the sterilisation of hospital equipment.

11 After disinfection, there is a possibility that some microorganisms will ________.

12 Ethanol is most effective as a disinfectant at a concentration of ________.

13 A number of chemical disinfectants are more effective when dissolved in ________.

14 Chlorine and its derivatives are most widely used as disinfectants of ________.

15 Tincture of ________ is an effective skin disinfectant.

16 The use of carbolic acid in operating theatres, introduced by ________, reduced the number of fatalities due to infections.

17 The efficacy of a disinfectant can be measured as its ________ coefficient.

18 Soaps and detergents are examples of ________.

19 ________ is assumed when it is reckoned to be less than a one in a million chance of a microbial cell being present.

20 The time needed for a bacterial population to be reduced to one tenth at a particular temperature is called the ________.

14

Antimicrobial Agents

In Chapter 1, we saw how by the end of the 19th century, the germ theory of infectious diseases had finally gained widespread acceptance. It is hardly surprising then, that the scientific community at that time should turn its collective mind towards ways of controlling infections and the organisms that cause them. In the following pages we shall discover how the development of chemical agents targeted against pathogenic microorganisms had a dramatic impact on the treatment of infectious diseases in the twentieth century.

The term *chemotherapy* is most closely associated in the minds of most people with the treatment of cancer. In fact the term was first used by Paul Ehrlich to describe any use of a drug or other chemical substance for the treatment of disease; thus, it has much wider terms of reference. In our present discussion, we shall confine ourselves to chemotherapy as it relates to the treatment of infectious diseases. It was Ehrlich who, 100 years ago, observed how certain dyes would stain bacteria but not the surrounding tissues, leading him to formulate the idea of *selective toxicity*, whereby a substance would selectively target harmful microorganisms but leave human tissues undamaged. He tested hundreds of synthetic compounds in the search for his 'magic bullet' before finding, in 1910, an arsenic-containing drug, Salvarsan, which was effective against *Treponema pallidum*, the causative agent of syphilis (Figure 14.1).

It was another 20 years before another significant antimicrobial drug was developed, when the German chemist Gerhard Domagk showed a synthetic dye, Prontosil, to be active against a range of Gram-positive bacteria. The active component of prontosil was shown soon afterwards to be sulphanilamide. In the following decade, numerous derivatives of sulphanilamide were synthesised, many of which were more potent antimicrobial agents than the parent molecule. This class of compounds is known collectively as the *sulphonamides*, or sulfa drugs (Box 14.1). In the years leading up to the Second World War, sulphonamides dramatically improved the mortality rates due to pneumonia and puerperal fever.

Nowadays, sulphonamides have largely been replaced by *antibiotics* because of their side-effects, and because, due to wholesale and indiscriminate use in the early years, bacterial resistance to sulphonamides has become widespread. Some synthetic compounds are still useful as antimicrobial agents, however. Isoniazid is one of the principal agents used in the treatment of tuberculosis. It

A side-effect is an undesirable and unintended effect of a therapeutic treatment on the recipient.

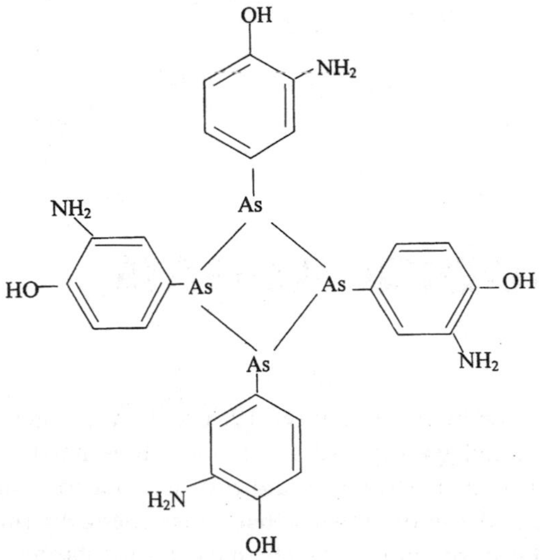

Figure 14.1 Salvarsan. Ehrlich's 'Compound 606' proved to be a very effective treatment for syphilis. The name derives from the facts that it brought *sal*vation to sufferers, and that it contained *ar*senic!

Box 14.1 Sulphonamides - the deadly mimics

Sulphonamides exert their effect by fooling the bacterial cell into thinking they are molecules of *p*-aminobenzoic acid (PABA) because of their similar structures (see below).

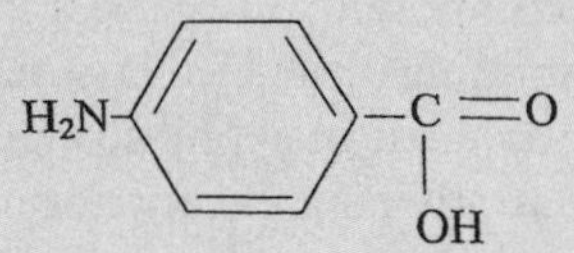

p- aminobenzoic acid (PABA)

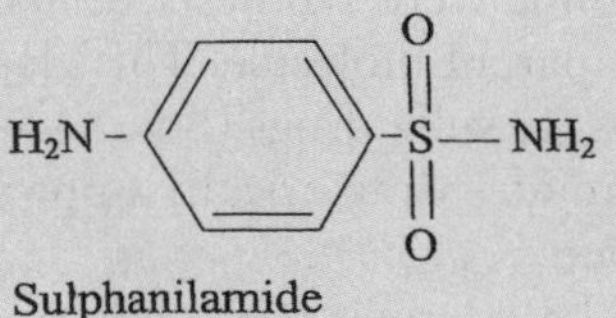

Sulphanilamide

PABA is a precursor of folic acid, which is required by cells as a coenzyme in the synthesis of nucleic acids. The sulphonamide acts as a competitive enzyme inhibitor (see Chapter 6), preventing the synthesis of folic acid, which in turn affects nucleic acid metabolism and leads to cell death. Because of their close structural resemblance to the PABA, the sulphonamides are said to be *structural analogues*; another, equally descriptive name is *antimetabolites*.

is nearly always given in association with another antimicrobial agent because of the high incidence of resistant forms of the mycobacteria that cause the disease. Like the sulphonamides, isoniazid is a structural analogue, and is thought to inhibit the production of mycolic acid in the mycobacterial cell wall (see Chapter 7).

The *quinolones* are synthetic substances related to nalidixic acid, and include ciprofloxacin and norfloxacin. They interfere with nucleic acid synthesis by inhibiting DNA gyrase, the enzyme responsible for unwinding DNA prior to replication (Chapter 11). Quinolones are used in the treatment of urinary infections, as are the *nitrofurans*. These are active against certain fungi and protozoans as well as a range of bacteria.

Antibiotics

The other major breakthrough in the treatment of infectious diseases was of course the discovery of naturally occurring antimicrobial agents, or *antibiotics*. These are metabolites produced by certain microorganisms, which inhibit the growth of certain other microorganisms. As we shall see, the definition has been extended to include semisynthetic derivatives of these naturally occurring molecules. Table 14.1 lists some commonly used antibiotics.

> An antibiotic is a microbially produced substance (or a synthetic derivative) that has antimicrobial properties.

One of the best known of all stories of scientific discovery is that of how Sir Alexander Fleming discovered penicillin in 1928. Before we consider that, however, it is worth noting that a number of treatments for infectious diseases practised over the preceding

Table 14.1 Some antibiotics and their microbial source

Antibiotic	Microbial Source
	Bacteria (Gram-positive)
Bacitracin	*Bacillus subtilis*
Polymixin	*Bacillus polymixa*
	Actinomycetes
Gentamicin	*Micromonospora purpurea*
Actinomycin D	*Streptomyces parvulus*
Erythromycin	*Streptomyces erythreus*
	Streptomyces
Nystatin	*Streptomyces noursei*
Rifamycin	*Streptomyces mediterranei*
Streptomycin	*Streptomyces griseus*
Tetracycline	*Streptomyces rimosus*
Vancomycin	*Streptomyces orientalis*
	Fungi
Cephalosporins	*Cephalosporium acremonium*
Griseofulvin	*Penicillium griseofulvum*
Penicillin	*Penicillium chrysogenum*

Box 14.2 Was Fleming lucky?

In the frequent retelling of the discovery of penicillin, much is made of the role played by chance. It is true that the *Penicillium* mould in Fleming's petri dish was an accidental contaminant, but arguably the real stroke of luck was not so much the fact that the contamination occurred but that it was observed by somebody who immediately recognised its significance.

centuries can, with the benefit of hindsight, be regarded as a form of antibiotic therapy. Many hundreds of years ago for example, the Chinese used mouldy soybean curd in the treatment of boils and South American Indians controlled foot infections by wearing sandals which had become furry with mould! In the late 19th century, Tyndall (see Chapter 13) made the observation that a culture medium cloudy with bacterial growth would clear when mould grew on the surface. Around the same time Pasteur and Joubert demonstrated that cultured anthrax bacilli could be inactivated in the presence of certain other microorganisms from the environment. By the early 1920s the search was on for the isolation of a microbially produced antibacterial agent, and Gratia and Dath isolated a substance from a soil actinomycete which came to be known as *actinomycin*. However, although potent against a number of pathogens, actinomycin is too toxic to be useful therapeutically.

Fleming was also looking for a naturally occurring antimicrobial agent. On one occasion, he noticed that a plate culture of *Staphylococcus aureus* had become contaminated by the growth of a mould; around it were clear areas, where the *S. aureus* did not grow. The mould was subsequently identified as *Penicillium notatum*. and the substance that had diffused through the agar from it, preventing bacterial growth, became known as *penicillin*. Further investigation revealed that broth from a culture of the *Penicillium* mould was inhibitory towards the growth of a number of other Gram-positive pathogens, and remained so even when diluted several hundred times. Critically, when tested on mice, it was, for the most part, harmless.

When it came to purifying the active ingredient and using it *in vivo* however, a number of problems were encountered. The penicillin proved to be impure, only produced in minute amounts, and unstable in the acid conditions of the stomach, thereby limiting its therapeutic potential. After publishing a few papers on the subject, Fleming ceased work on penicillin and it was left to Howard Florey and Ernst Chain in 1939 to take up the challenge of producing it in sufficient quantities and in a pure enough form for therapeutic use. Early work in Oxford had to continue in the United States because of the German air raids in Britain. The American entry into the Second World War in late 1941 meant that the development of penicillin was awarded war project status, giving it greatly added impetus. As a result of their endeavours, the yield of penicillin rose hugely (see Box 14.3), and in 1945, Fleming, Chain and Florey shared the Nobel Prize for their work.

Other antibiotics were also isolated during this period, most notably *streptomycin*, isolated by Selman Waksman from *Streptomyces griseus*, which was to prove so effective against tuberculosis.

In 1942 there was only enough penicillin in the world to treat a few hundred individuals, but by the end of the Second World War production had grown to such an extent that 7 million people a year could be treated. By the mid-1950s, such well-known

Box 14.3 How did Florey and Chain improve the yields of penicillin?

The work of Florey and Chain resulted in pure penicillin being produced on a large scale, suitable for therapeutic use. Among their achievements were:

- isolation of a better penicillin-producing species (*P. chrysogenum,* famously isolated from a mouldy cantaloupe melon in Peoria Illinois!) and selection of mutant strains induced by X-rays and UV irradiation
- development of submerged culture technique, with sterile air being forced through the medium to supply essential oxygen
- improvements in medium composition
- addition of precursors to the medium
- refinements to recovery methods.

antibiotics as *tetracycline*, *chloramphenicol* and *neomycin* had been isolated. The discovery of a few naturally occurring compounds had revolutionised the treatment of infectious diseases.

New antibiotics are still being sought today; in Chapter 17 we discuss the stages in the isolation, testing and development of a putative new antibiotic. Of the thousands isolated so far, only a small proportion have proved to be of any real therapeutic or commercial value. This is because, like the actinomycin mentioned above, most of the substances isolated harm not only bacteria but humans too.

A key prerequisite for any chemotherapeutic agent is selective toxicity. An obvious way of achieving this is for a compound to direct its effect against a metabolic or physiological function found in microbial cells but not in the host. We shall look at some examples of this later in this chapter. Those chemotherapeutic agents which inhibit the same process in host as in pathogen, or which cause harm to the host in some other way, are said to have side-effects. These may include directly toxic effects, hypersensitivity (allergic) reactions or adverse effects on the host's normal resident microflora.

One of the reasons why penicillin was, and continues to be, so successful, was that the target of its action is unique to bacteria, so it its degree of selective toxicity is high.

What other properties should an antibiotic have?

Selective toxicity is the most important single attribute of an antibiotic, but ideally it should also have as many of the following properties as possible:

- antibiotics, like other chemotherapeutic agents, need to be *soluble in body fluids*, in order to exert their effect by penetrating the body tissues. The compound must not be metabolised so quickly that it is excreted from the body before having a chance to act.

- if administered orally, it *must not be inactivated* by the acid environment of the stomach, and must be capable of being absorbed by the small intestine.
- an antibiotic should not have any significant effect on the *resident microflora* of the host.
- it should not be easy for the target pathogen to establish *resistance* against an antibiotic.
- *side-effects* such as *allergic reactions* should be minimal.
- it should be sufficiently stable to have a good *shelf life*, without special storage considerations.

Antibiotics should not produce a *hypersensitivity* (allergic) reaction in the host. This is caused by an extreme response by the host immune system, and is not the same as toxicity.

How do antibiotics work?

All antibiotics have the common property of interfering in some way with a normal, critical function of the target bacterial cell. The most commonly used antibiotics exert their effect by one of the following methods:

1. Inhibition of cell wall synthesis (group I)
2. Disruption of cell membranes (group II)
3. Interference with protein synthesis (group III)
4. Interference with nucleic acid synthesis (group IV)

Table 14.2 lists examples of each group. Those antibiotics belonging to groups I and III are better able to discriminate between procaryotic and eucaryotic cells, and consequently show more selective toxicity and a higher *therapeutic index*.

The therapeutic index provides a measure of the selective toxicity of a chemotherapeutic agent. It is the ratio between the concentration at which the substance causes harm to its host (toxic dose) and that at which it is required to be clinically effective (therapeutic dose). It is therefore desirable for an antibiotic to have a high therapeutic index.

Group I: Inhibitors of cell wall synthesis

The main group which work in this way are the β-lactam antibiotics, so-called because they contain a β-lactam ring in their structure. Included among this group are the penicillins and the cephalosporins. See also Box 14.4.

You may recall from our discussion of bacterial cell wall structure in Chapter 3 that an important factor in the strengthening of the peptidoglycan component of the bacterial cell walls is the cross-linking of chains by transpeptidation. It is this process which is acted on by the β-lactams; they bind irreversibly to the transpeptidase enzyme, forming covalent bonds with a serine

Table 14.2 Some commonly used antibiotic classes

Inhibitors of cell wall synthesis	Penicillins, cephalosporins
Disrupters of cell membranes	Polymixins, polyenes
Inhibitors of protein synthesis	Streptomycin, tetracyclines
Inhibitors of nucleic acid synthesis	Rifamycins

residue within the enzyme's active site. The cell wall continues to form, but becomes progressively weaker as more new, unlinked, peptidoglycan is set down. Since bacteria are generally to be found in a hypotonic environment, as the wall weakens, water enters the cell, leading to swelling and then lysis.

Penicillins The first β-lactam antibiotic to be discovered was benzylpenicillin, or penicillin-G, whose action is restricted to Gram-positive bacteria, because it is unable to penetrate the Gram-negative cell wall. It is effective against Gram-positive bacteria when administered intramuscularly, but cannot be taken by mouth because it is broken down in the acid conditions of the stomach. Another naturally occurring penicillin, penicillin-V, represented an advance inasmuch as it is less acid-labile and can therefore be taken orally. All the penicillins are based on a core structure or nucleus called 6-amino-penicillanic acid (Figure 14.2); extensive research has led to the development of many variants of this, the so-called *semisynthetic penicillins*. These have attached to their nucleus novel side chains not encountered in nature, and have overcome some of the problems inherent in naturally occurring penicillins such as instability and narrow specificity (Figure 14.3).

Semisynthetic penicillins are based on the core structure of the naturally occurring molecule, with the addition of chemically synthesized side chains.

Ampicillin is a semi-synthetic penicillin that has a broader specificity (Box 14.5) than Penicillin G; it is appreciably more effective against Gram-negative bacteria such as *Salmonella* and *E. coli*, its hydrophobic nature making it better able to penetrate their outer membrane. It has the additional benefit of being acid-stable and can therefore be taken orally.

Another drawback to natural penicillins is that they are susceptible to naturally occurring bacterial β-lactamases (also called penicillinases), which breaks a bond in the β-lactam core of the penicillin molecule (Figure 14.4). Sometimes, β-lactam antibiotics

Box 14.4 β-lactam antibiotics have a second mode of action

The β-lactams also act by preventing the natural regulation of enzymes called *autolysins*. These enzymes function by breaking down peptidoglycan in a controlled fashion, causing breaks to allow for the addition of new peptidoglycan as the cell grows, and are normally regulated by naturally occurring inhibitors. The β-lactams neutralise the activity of these inhibitors, leading to further breakdown of the cell wall.

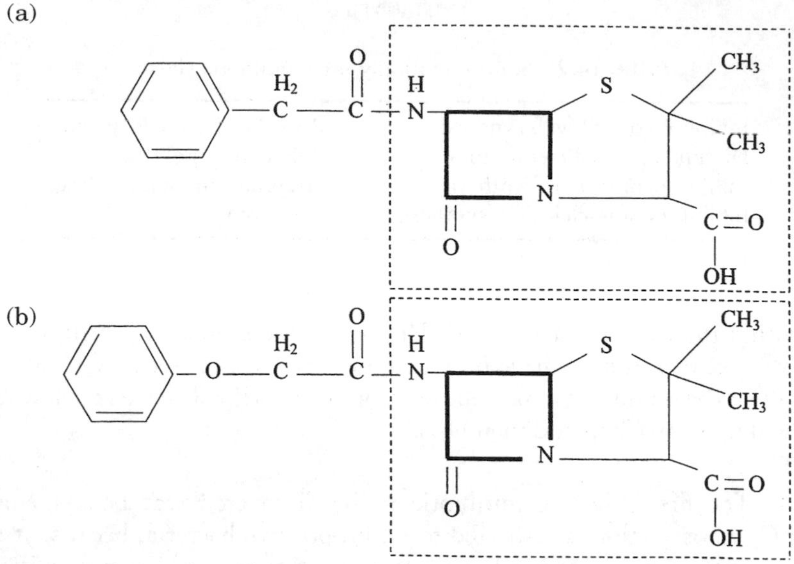

(a)

(b)

Figure 14.2 Naturally occurring penicillins. (a) Penicillin G (benzylpenicillin); (b) penicillin V. The dotted outline covers the 6-aminopenicillanic acid nucleus present in all variants of penicillin. The heavy outline denotes the β-lactam ring

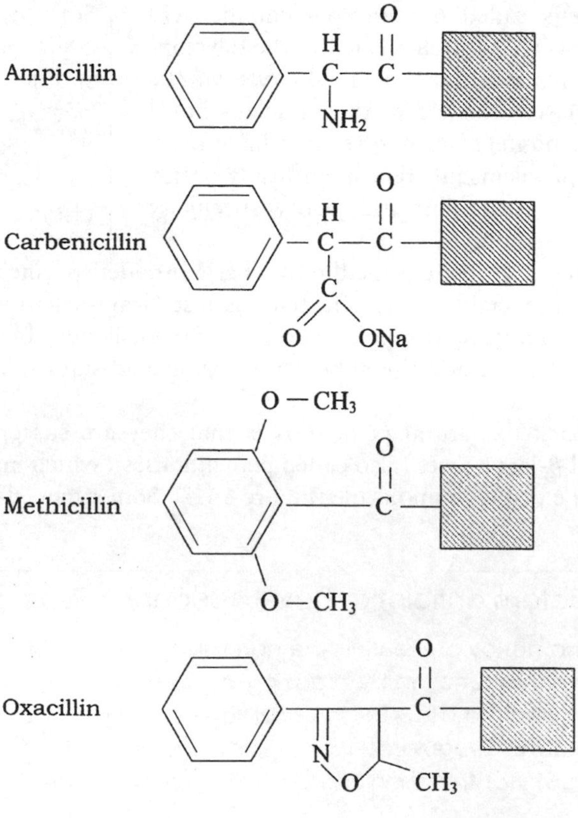

Ampicillin

Carbenicillin

Methicillin

Oxacillin

Figure 14.3 Some important semisynthetic penicillins. The shaded square represents the 6-aminopenicillanic acid nucleus common to all penicillins (see Figure 14.2)

Box 14.5 Broad spectrum or narrow spectrum?

Certain antibiotics, due to the mechanism of their action, are only effective against a few different pathogens, while others can be used successfully against many different kinds. They are said to have, respectively, a *narrow spectrum* and a *broad spectrum* of activity.

On the face of it, all things being equal, you would expect your doctor to choose the antibiotic with the broadest possible spectrum of activity, but this isn't always the wisest option. When the cause of an infection isn't known, it makes sense to hedge one's bets and prescribe a broad-spectrum antibiotic ('whatever it is, this should sort it!'), but this policy is not without its dangers. The drug is likely to kill off many of the host's own resident microflora, which can lead to a *superinfection*, and the development of antibiotic-resistant strains is also made more likely. If the identity of the pathogen is suspected, an appropriate narrow-spectrum drug is to be preferred.

are taken in combination with a β-lactamase inhibitor such as clavulanic acid. This binds to the β-lactamase with a high affinity, preventing it from acting on the antibiotic. Some semisynthetic penicillins such as methicillin and oxacillin are resistant to attack by the β-lactamases that can render certain bacteria resistant to their naturally occurring forms.

Penicillin is not an appropriate treatment for the estimated 1–5 per cent of adults who show an allergic reaction to it; in extreme cases, death from *anaphylactic shock* can result.

An anaphylactic shock is an extreme form of hypersensitivity reaction.

Cephalosporins The cephalosporins, like the penicillins, have a structure based on a β-lactam ring (Figure 14.5). They also exert their effect on transpeptidases, but generally have a broader specificity and are more resistant to the action of β-lactamases. Ceftriaxone, for example, is now used in the treatment of gonorrhoeal infections, caused by penicillin-resistant strains of *Neisseria gonorrhoeae*. In addition, patients who are allergic to penicillin are often treated with cephalosporins. Cephalosporins were first

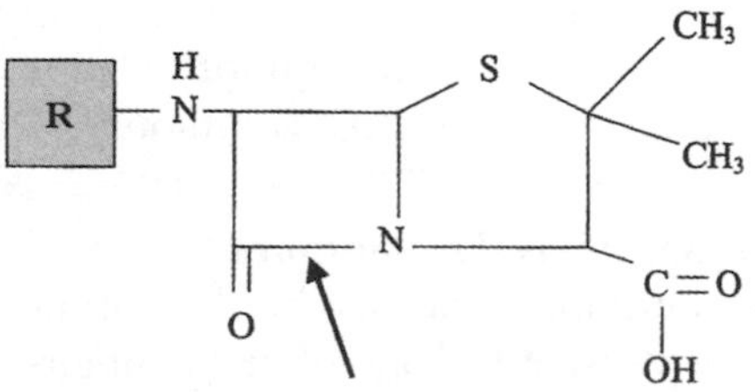

Figure 14.4 Action of β-lactamase on penicillin. A number of bacteria, especially staphylococci, possess the enzyme β-lactamase (penicillinase), which inactivates penicillin by cleavage of the β-lactam ring at the point marked by the arrow

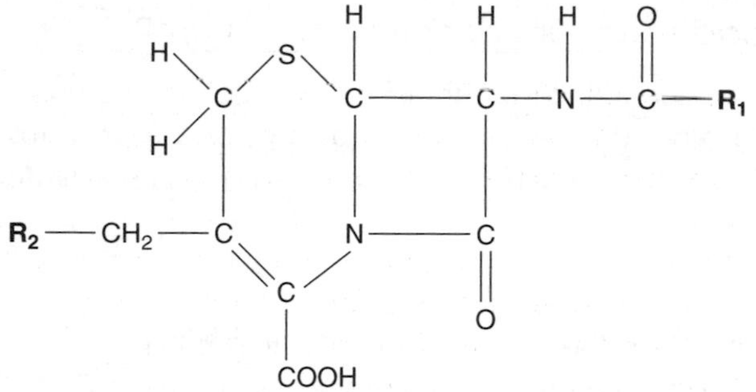

Figure 14.5 Cepahalosporins are based on a nucleus of 7-amino-cephalosoranic acid, which, like the penicillins, features a β-lactam ring (shown as a square). Note that each molecule has two variable side chains

isolated in the late 1940s from a marine fungus called *Cephalosporium acremonium*, and came into general use in the 1960s. So-called second, third and fourth generation cephalosporins have been developed to widen the spectrum of activity to include many Gram-negative organisms, and to keep one step ahead of pathogens developing resistance to earlier versions.

Both penicillins and cephalosporins are also used *prophylactically*, that is, in the prevention of infections, prior to surgery in particularly vulnerable patients.

Other antibiotics that affect the cell wall Carbapenems are β-lactam antibiotics produced naturally by a species of *Streptomyces*. A semisynthetic form, imipenem, is active against a wide range of Gram-positive and -negative bacteria, and is used when resistance to other β-lactams has developed.

Bacitracin and vancomycin are two other antibiotics that exert their effect on the cell wall, but by a different mechanism. Bacitracin is derived from species of *Bacillus* and acts on bactoprenol pyrophosphate, the lipid carrier molecule responsible for transporting units of peptidoglycan across the cell membrane to their site of incorporation into the cell wall (see Chapter 3). Its use is restricted to topical (surface) application, since its use internally can cause kidney damage. Vancomycin is a highly toxic antibiotic with a narrow spectrum of use against Gram-positive organisms such as streptococci and staphylococci. It is particularly important in its use against infections caused by organisms resistant to methicillin and the cephalosporins, such as methicillin-resistant *Staphylococcus aureus* (MRSA) (see Resistance to Antibiotics below). It is not absorbed from the gastrointestinal tract and is therefore most commonly administered intravenously.

Group II: Antibiotics that disrupt cell membranes

Polymixins are a class of antibiotic that act by disrupting the phospholipids of the cytoplasmic membrane and causing leakage of cell contents. Produced naturally by a species of *Bacillus*, polymixins are effective against pseudomonad infections of wounds and burns, often in combination with bacitracin and neomycin (an inhibitor of protein synthesis; see below). Their toxicity makes them unsuitable for internal use. Polyene antibiotics such as amphotericin and nystatin are antifungal agents that act on the sterol components of membranes; they are discussed more fully towards the end of this chapter.

Group III: Inhibitors of protein synthesis

Antibiotics that act by affecting protein synthesis generally have a relatively broad spectrum of action. As we saw in our historical review earlier in this chapter, streptomycin was the first antibiotic that was shown to be effective against Gram-negative organisms. Its discovery in 1943 was particularly welcome since such organisms were unaffected by penicillin or sulphonamides. It proved to be particularly useful in the treatment of tuberculosis, the causative agent of which, *Mycobacterium tuberculosis*, is protected against the effects of penicillin by the waxy layer of mycolic acids in its cell wall.

Streptomycin belongs to a group of antibiotics called *aminoglycosides*, which act by binding to the 30S subunit of the bacterial ribosome, preventing attachment of the 50S subunit to the initiation complex (Figure 14.6). They can thus discriminate between procaryotic (70S) and eucaryotic (80S) ribosomes, and consequently have a relatively high therapeutic index (although not as high as cell wall inhibitors). Other members of this group are gentamicin, kanamycin and neomycin. Like some other 'wonder drugs', streptomycin has proved to have undesirable side-effects; these have led to it being replaced in most applications by safer alternatives. In addition, bacterial resistance to streptomycin is widespread, further diminishing its usefulness. Use of the

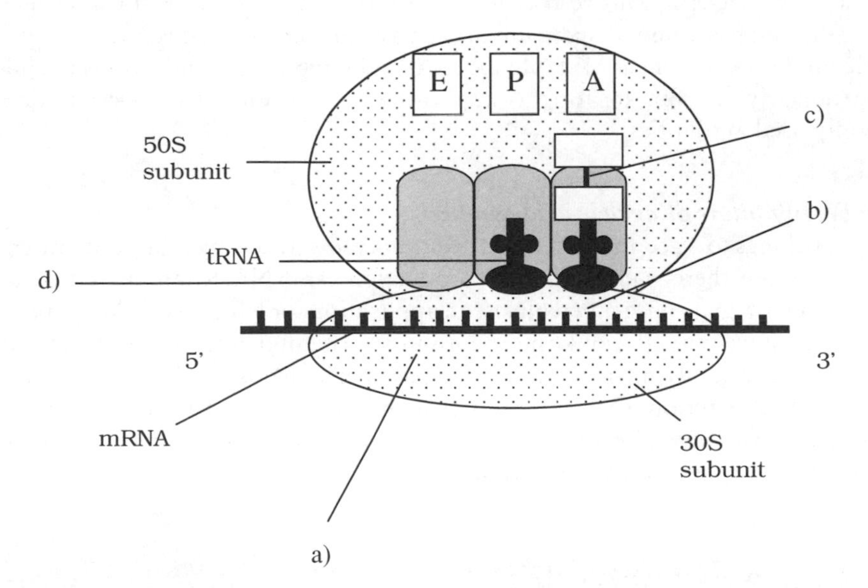

Figure 14.6 Inhibitors of protein synthesis. (a) By binding to the 30S subunit of the bacterial ribosome, aminoglycosides block the attachment of the 50S subunit. This prevents completion of the initiation complex, thus protein synthesis is inhibited. (b) Tetracyclines distort the shape of the 30S subunit, preventing the attachment of the appropriate aminoacyl tRNA. (c) Chloramphenicol inhibits peptidyltransferase and prevents formation of new peptide bonds. (d) Macrolides such as erythromycin bind to the 50S subunit, preventing elongation of the growing peptide chain

aminoglycosides as a group has diminished since the development of later generation cephalosporins and the tetracyclines.

Tetracyclines also work by binding to the 30S ribosomal subunit, preventing the attachment of aminoacyl tRNA, and therefore extension of the peptide chain (Figure 14.6). They are yet another group of antibiotics produced by *Streptomyces* spp. Both natural and semisynthetic tetracyclines are easily absorbed from the intestine, allowing them to be taken orally. Coupled with their broad specificity (the broadest of any antibiotic), this led to inappropriately widespread use in the years following their discovery, sometimes resulting in complications caused by the destruction of the normal resident microflora. Tetracyclines are still used for a number of applications, notably to treat a variety of sexually transmitted diseases.

Two important antibiotics which act on the larger, 50S, subunit of the procaryotic ribosome are erythromycin and chloramphenicol. Both combine with the subunit in such a way as to prevent the assembly of amino acids into a chain (Figure 14.6). Chloramphenicol was the first antibiotic to be discovered with a broad spectrum of activity; it also derives originally from *Streptomyces* spp., but is nowadays produced synthetically. Its use has become severely restricted since it was shown to have some serious side-effects, notably on the bone marrow, but it remains the agent of choice for the treatment of typhoid fever.

Erythromycin is the best known of the *macrolide* group of antibiotics. Unlike chloramphenicol, it has a large hydrophobic molecule and is unable to gain access to most Gram-negative bacteria, thus restricting its spectrum of activity. Erythromycin can be taken orally and has a similar spectrum of activity to penicillin G; it is often used instead of penicillin in the treatment of staphylococcal and streptococcal infections in children. It is particularly appropriate for this application as it is one of the least toxic of all commonly used antibiotics.

Group IV: Inhibitors of nucleic acid synthesis

Rifampin belongs to a group of agents called *rifamycins*. It acts by inhibiting the enzyme RNA polymerase, thereby preventing the production of mRNA. Rifampin is used against the mycobacteria that cause tuberculosis, an application for which its ability to penetrate tissues makes it well suited. Unlike most antibiotics, rifampin interacts with other drugs, often reducing or nullifying their effect. When used in high doses, it has the unusual side-effect of turning secretions such as tears, sweat and saliva, as well as the skin, an orange-red colour. As we have already seen, the quinolone group of synthetic antimicrobial drugs act by disrupting DNA replication.

Resistance to antibiotics

The global increase in resistance to antimicrobial drugs, including the emergence of bacterial strains that are resistant to all available antibacterial agents, has created a public health problem of potentially crisis proportions.

American Medical Association, 1995

Box 14.6 Where did antibiotic resistance come from?

Have genes responsible for antibiotic resistance always existed in nature, or have they arisen since the development and widespread use of antibiotics? The answer, almost certainly, is the former. A sample of an *E. coli* strain, freeze-dried in 1946, was revived many years later, and found to have plasmid-encoded genes for resistance to streptomycin and tetracycline, neither of which were in clinical use until some years after the culture was preserved. It seems likely that bacteria possessed these genes to protect against naturally occurring antibiotics, an idea supported by the fact that R-plasmids have been found in non-pathogenic soil bacteria. Also, resistance to a number of antibiotics has been demonstrated in soil and water bacteria from sources sufficiently remote to be free from anthropogenic influence.

As we have already suggested, the impact of certain antibiotics can be greatly reduced due to the development of resistance by target pathogens (Box 14.6). This represents the greatest single challenge facing us in the fight against infectious diseases at the start of the 21st century. Fleming himself foresaw that the usefulness of penicillin might become limited if resistant forms of pathogens arose.

Not long after penicillin was put into general use, strains of *Staphylococcus aureus* were found which did not respond to treatment, and by 1950 penicillin-resistant *S. aureus* was a common cause of infections in hospitals. A decade later, a semi-synthetic form of penicillin, methicillin, was introduced; this was not affected by the β-lactamase enzymes that inactivated Penicillin G, and was used to treat resistant forms. Within years, however, came the first reports of strains of *S. aureus* that did not respond to methicillin. The incidence of methicillin-resistant *S. aureus* (MRSA) has increased greatly since, and it represents the major source of *nosocomial infections*. In 1980, synthetic fluoroquinolones were introduced to counter the threat of MRSA, but within a year 80 per cent of isolated strains had developed resistance to these too. Vancomycin is regarded as a last-resort treatment for MRSA, for a number of reasons; it has a number of serious side-effects, its widespread use would encourage resistance against it, and it is extremely expensive.

Nosocomial infections are ones that are acquired in hospitals or similar locations. Some 5--10 percent of hospital patients acquire such an infection during their stay. This may prove to be fatal, especially among the elderly and immunocompromised. As well as the human cost, such infections extend the average time spent in hospital and therefore add greatly to the costs of treatment.

A case of vancomycin-resistant *Staphylococcus aureus* (VRSA) emerged in Japan in 1996; a few months later it had reached the USA. This represents a serious threat; some of these strains respond to treatment with a cocktail of antibiotics, but already people have died from untreatable VRSA infections. In 2003, a strain of VRSA was shown to have obtained its vancomycin resistance by cross-species transfer from a strain of *Enterococcus faecalis*.

How does antibiotic resistance work?

We saw earlier in this chapter how antibiotics exert their effects in a variety of ways, so it should come as no surprise that there is no single mechanism of resistance. Resistance may be natural, that is, intrinsic to the microorganism in question, or it may be acquired.

Some bacteria are able to resist antibiotic action by denying it entry to the cell; penicillin G for example is unable to penetrate the Gram negative cell wall. Others can pump the antibiotic back out of the cell before it has had a chance to act, by means of enzymes called *translocases*; this is fairly non-specific, leading to multiple drug resistance. Other bacteria are naturally resistant to a particular antibiotic because they lack the target for its action, for example, mycoplasma do not possess peptidoglycan, the target for penicillin's action.

To avoid the action of an antibiotic, bacteria may be able to use or develop alternative biochemical pathways, so that its effect is cancelled out. Many pathogens can secrete enzymes that modify or degrade antibiotics, causing them to lose their activity; we have already seen that penicillins can be inactivated by enzymatic cleavage of their β-lactam ring. Similarly, chloramphenicol can be acetylated, while members of the aminoglycoside family can be acetylated, adenylated or phosphorylated, all leading to loss of antimicrobial activity.

Mutations may occur which modify bacterial proteins in such a way that they are not affected by antimicrobial agents. You will recall that streptomycin normally acts by binding to part of the 30S subunit on the bacterial ribosome; the actual binding site is a protein called S12. Mutant forms of the S12 gene can lead to a product which still functions in protein synthesis, but loses its ability to bind to streptomycin. Similarly, mutations in transpeptidase genes in staphylococci means they do not bind to penicillin any more, so cross-linking of the cell wall is not inhibited.

How does resistance arise?

Occasionally, mutations occur spontaneously in bacteria, which render them resistant to one antibiotic or another. Usually the mutation leads to a change in a receptor or binding site such as those just described, rendering the antibiotic ineffective. The changes are usually brought about by point mutations (see Chapter 11) occurring at very low frequency on chromosomal DNA. Bacteria can, however, become resistant much more rapidly by acquiring the mutant resistance-causing gene from another bacterium. This is called transmissable antibiotic resistance; it occurs mainly as a result of bacterial conjugation, and is the cause of most of the resistance problems we presently face. Transmissable resistance was first reported in Japan in the late 1950s, when multi-drug resistance in *Shigella* was shown to have been acquired by conjugation with resistant *E. coli* in a patient's large intestine. *E. coli* is known to transfer R (resistance) plasmids to several other gut bacteria including *Klebsiella, Salmonella* and *Enterobacter,* as well as *Shigella*. Whereas chromosomal mutations usually result in a modification to the drug's binding site, genes carried on plasmids code for enzymes which inactivate it, (e.g. β-lactamases) or lead to its exclusion from the cell (translocases).

There is a strong link between the use of a particular antibiotic in a locality and the incidence of resistant bacterial strains. This is because of selective pressure favouring the resistant forms of a bacterium. Fortunately this can, at least in part, be reversed, as several studies have shown, where a more restricted use of certain antibiotics over several years was followed by a reduction in the incidence of resistant bacterial forms.

Antibiotic susceptibility testing

In order to determine the most appropriate antimicrobial agent to use against an infection, it is necessary to determine the susceptibility of the pathogen. There are several ways of doing this, but here we describe the two most commonly employed techniques.

The *tube dilution assay* determines the *minimum inhibitory concentration* (MIC) of the antibiotic, that is, the lowest concentration at which it prevents growth of a given organism. A series of tubes containing increasingly dilute preparations of the antibiotic is introduced into a broth with a standard number of test organisms and incubated. The lowest concentration in the series to show no microbial growth is the MIC (Figure 14.7).

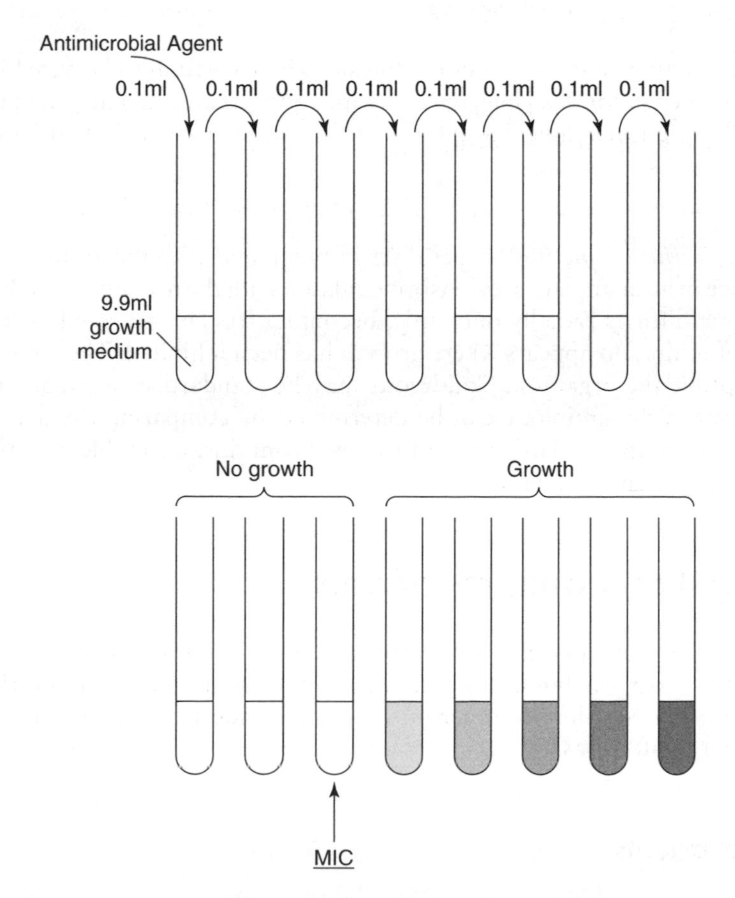

Figure 14.7 Minimal inhibitory concentration (MIC). The test organism is incubated with serially diluted antibiotic. The lowest concentration capable of preventing microbial growth is the MIC

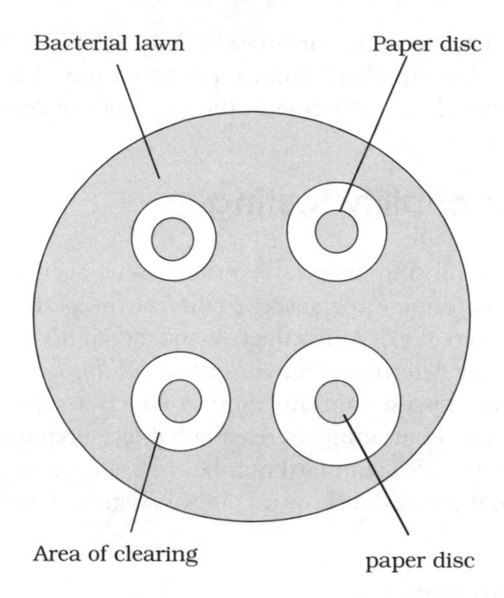

Figure 14.8 Antibiotic testing by disc diffusion. The bacterium to be tested is spread on an agar plate, then paper discs impregnated with appropriate antibiotics are placed on the surface. Following incubation, susceptibility to an antibiotic is indicated by a clear ring surrounding the disc

In the *disc diffusion method*, paper discs impregnated with the antibiotic are placed on the surface of an agar plate previously inoculated with the test organism (Figure 14.8). The antibiotic diffuses radially outwards, becoming less concentrated as it does so. A clear zone of inhibition appears where growth has been inhibited. The larger this is, the more susceptible the organism. Conditions may be standardised so that susceptibility (or otherwise) to the antibiotic can be determined by comparing the diameter of the zone of clearing with standard tables of values. From this, a suitable concentration for therapeutic use can be determined.

Antifungal and antiviral agents

We have focused our attention up to now on those antimicrobial agents which are directed against bacteria, but of course these are not the only cause of infections. Below, we review the relatively limited repertoire of compounds available for the treatment of fungal and viral infections.

Antifungal agents

Fungi are eucaryotes, and therefore are unaffected by those agents which selectively target uniquely procaryotic features such as peptidoglycans and 70S ribosomes. Therein lies the problem: anything that damages fungal cells is likely to damage human cells too.

Polyene antibiotics such as amphotericin and nystatin (both produced by species of *Streptomyces*) act on the sterol components of membranes; their use is limited, because human cells can also be affected by their action (to use a term we learnt earlier in this chapter, they have a low therapeutic index). Nystatin is used topically against *Candida* infections, while amphotericin B is generally used against systemic infections of fungal origin. The latter substance can have a wide range of serious side-effects, but in some cases infections are so severe that the physician is faced with no alternative. Synthetic compounds such as the *imidazoles* have a similar mode of action to the polyenes; they are effective against superficial mycoses (fungal infections of the skin, mouth and urino-genital tract). Griseofulvin, a natural antibiotic produced by a species of *Penicillium,* is another antifungal agent whose use is restricted; it works by interfering with mitosis and not surprisingly has a range of side-effects. Although used to treat superficial infections, it is taken orally.

Antiviral agents

In spite of the looming threat of resistant strains, there is no doubt that antibiotics have been hugely successful in the control of bacterial diseases. We have, however, been a lot less successful when it comes to finding a treatment for diseases caused by viruses; a quick revision of their *modus operandi* (Chapter 10) should make it clear why this is so. Viruses survive by entering a host cell and hijacking its replicative machinery, thus a substance interfering with the virus is likely to harm the host as well. A number of compounds have been developed however, which are able to act selectively on a viral target.

All antiviral agents act by interfering with some aspect of the virus's replication cycle. A number of such compounds have been found, but only a few have been approved for use in humans.

One of the first antiviral agents to be approved for use was amantidine, which inhibits uncoating of the influenza A virus by preventing the formation of acid conditions in the host cell's endocytotic vesicles (see Chapter 10). Its specificity for the virus is due to selective binding to M_2, a matrix protein. Amantidine's efficacy is dependent on administration within the early stages of an infection. It can be administered prophylactically, but may have side-effects.

Most antiviral agents target nucleic acid synthesis, usually by acting as *base analogues*. These are molecules that are incorporated into viral nucleotides instead of the normal deoxynucleosides, disrupting synthesis because DNA polymerase is unable to act on them. The majority of viruses encode their own DNA polymerases, and the base analogues exert their effect by selectively inhibiting these, thus having little effect on that of the host cell. An example is acyclovir, which is an analogue of guanosine; it is converted to the nucleoside triphosphate by the action of thymidine kinase and then in this form acts as a competitive inhibitor of the 'correct' version (Figure 14.9). When the acyclovir nucleotide is incorporated into the viral DNA, there is no attachment point for the next nucleotide, so further elongation of the chain is prevented. Acyclovir exerts

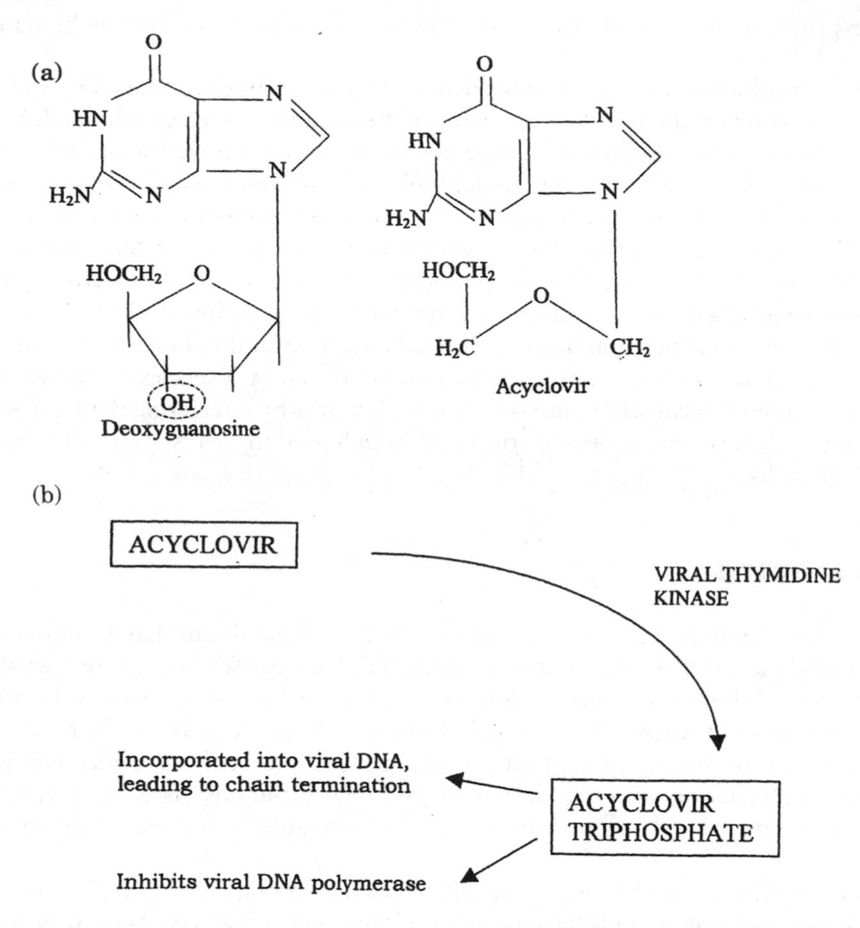

Figure 14.9 Acyclovir inhibits viral DNA synthesis. (a) Acyclovir has a similar structure to the nucleoside deoxyguanosine, but lacks the -OH group (circled) necessary for chain extension. (b) Acyclovir needs to be phosphorylated to become active; virally encoded thymidine kinase is required for this. Acyclovir triphosphate (ACV-T) selectively inhibits viral, but not human, DNA polymerase. Any ACV-T that is incorported into viral DNA acts as a chain terminator

its selective action by having a much higher affinity for the viral polymerase than that of the host. It is used in the treatment of herpes simplex infections; unfortunately, in a scenario echoing our experience with antibiotics, resistant strains of herpes simplex virus have been shown to exist. This can be seen as even more serious than the emergence of antibiotic resistant bacteria, because the choice of alternative antiviral agents is so restricted. Vidarabine and azidothymidine (AZT) are other examples of base analogues. AZT (Retrovir) was one of the first substances shown to have an effect against HIV, which it does by preventing cDNA synthesis by the enzyme reverse transcriptase (see Chapter 10).

Zanamivir was approved by the US Food and Drug Administration (FDA) in 1999. It belongs to a new class of synthetic compounds called *neuraminidase inhibitors*, which act selectively against both influenza A and B viruses. They block the active site of the enzyme neuraminidase, preventing the release of new virus particles from infected cells, hence reducing the spread of the infection. Zanamivir is inhaled as a fine powder directly into the lungs of patients who are in the early stages of infection.

The future

The development and widespread use of antibiotics must rank as the most remarkable of all medical advances made in the 20th century. Overconfident assertions that infectious diseases would soon be a thing of the past, however, now have a hollow ring to them. Viruses have proved to be more difficult to deal with than bacteria, and many viral diseases continue to elude effective treatment. Most alarmingly, the threat of resistant strains casts a shadow over all the past achievements of antibiotics. The major aim of scientists now must be to develop new antibiotics or other therapeutic strategies at a pace greater than that at which bacteria are developing resistance. In 2000, the FDA approved a new synthetic agent shown to be effective against both MRSA and vancomycin-resistant *Enterococcus faecalis*. Linezolid (Zyvox), which works by blocking the initiation of protein synthesis, belongs to a new class of antibiotics called *oxazolidinones*. It is the first new anti-MRSA compound to be introduced for more than 40 years.

Another approach to countering resistant forms is to identify and target the mechanism by which they combat antibiotic therapy. A team at Rockefeller University in New York have identified two genes that enable resistant forms to rebuild their cell walls after antibiotic treatment. By targeting these genes, they hope to restore the potency of a cell wall inhibitor such as penicillin. Perhaps by the time you read this, other, less conventional approaches will have yielded promising results (see Boxes 14.7 and 14.8), but you can be just as sure that bacteria will have new tricks up their sleeves, and that the battle of Man versus microorganisms will continue well into the new millennium.

Box 14.7 Bugs against bugs?

Scientists are always on the lookout for new weapons in the fight against infectious diseases. The emergence of resistant strains of pathogens means that new solutions continually need to be found. One novel line of research is hoping to utilise the bactericidal powers of a defence system used by certain insects. A sap-sucking species has been found that produces substances that interfere with bacterial protein synthesis. Most of these would harm protein synthesis in humans too, but certain peptides appear to be more selective in their action, protecting mice from *E. coli* and *Salmonella* infections. It may seem an unlikely source for a life saver, but then, so was *Penicillium*!

Box 14.8 Bacteriophages: our secret weapon against infections?

If bacteriophages are viruses that infect bacterial cells, why aren't they used in the fight against bacteria that cause infectious diseases? The short answer is: in some places they are! In the years following their discovery, there was considerable enthusiasm in some quarters for the notion that phages might be useful in the treatment of bacterial diseases. A particular attraction of phages as therapeutic agents is that they are extremely selective in their action, targeting one specific cell type. Early studies and trials had mixed results and before too long, antibiotics had revolutionised the treatment of infectious diseases in the West. This led to phage therapy being forgotten for several decades, but its use continued in the Soviet bloc countries, however, where even now phage preparations can be bought over the counter.

Western interest revived in the 1980s in the wake of the upsurge in antibiotic-resistant strains of bacteria, since when phages have attracted attention as possible allies in the control of infectious diseases, including some not responsive to antibiotic therapy. Several American companies have become involved in phage therapy, some in collaboration with the former Soviet state of Georgia, where research has been carried out over several decades.

Particular attention has been paid to veterinary applications, with a view to reducing the amount of antibiotic usage in animals, and it is hoped that one day phages may prove to be the weapon we need to fight antibiotic resistant strains such as MRSA and VRE.

Test yourself

1 Following Fleming's discovery of penicillin, ________ and ________ were largely responsible for developing it and bringing it to large scale production.

2 It is essential for an antimicrobial agent to have ________ toxicity, otherwise host cells will be affected as well as microorganisms.

3 Penicillins and cephalosporins both belong to the ________ group of antibiotics. They bind to the ________ enzyme, preventing ________ of peptidoglycan.

4 ________ penicillins such as ampicillin have been developed to overcome some of the problems inherent in naturally occurring forms, such as ________ and narrow ________.

5 ________ was the first antibiotic to shown to be effective against Gram-negative bacteria.

6 ________ have a particularly broad range of specificity and act by preventing extension of the peptide chain during protein synthesis.

7 Antibiotics that affect protein synthesis target the differences between procaryotic and eucaryotic ________.

8 Aminoglycoside antibiotics inhibit protein synthesis by preventing attachment of the ________ ________ subunit.

9 ________ is a member of the macrolide group of antibiotics, and is often used instead of penicillin in the treatment of staphylococcal and streptococcal infections in children.

10 Rifamycins affect mRNA production by inhibiting the enzyme ________ ________.

11 One of the antibiotic resistant organisms most frequently found in hospitals is MRSA (________ ________ *Staphylococcus aureus*). Infections originating in such environments are termed ________ infections.

12 ________ is used against antibiotic-resistant infections; however bacteria have been found that are also resistant to this agent.

13 Some bacteria are able to pump antibiotics out of the cell before they have had time to act. They are able to do this by means of enzymes called ________.

14 Antibiotic resistance acquired through bacterial conjugation is described as ________ resistance.

15 Genes for enzymes that inactivate antibiotics are usually carried on ________.

16 The lowest concentration of an antibiotic that prevents growth of a microorganism is described as its ________ ________ ________.

17 In the disc diffusion method, ________ of ________ develop as the antibiotic diffuses outwards.

18 The usefulness of polyene antibiotics in fungal infections is limited because human cells may also be affected. Such compounds are said to have a low ________ ________.

19 The antiviral agent acyclovir works by acting as a ________ ________. It inhibits viral ________ ________.

20 ________ has been used as a therapeutic agent in the treatment of AIDS. It acts by inhibiting viral ________ ________.

Part VI

Microorganisms in the Environment

15

Microbial Associations

We have stressed on more than one occasion during the course of this book that microorganisms in nature do not exist as pure cultures but alongside numerous other organisms, microbial or otherwise, with which they may have to compete in the never-ending struggle for survival. In a number of cases, this coexistence may extend beyond merely sharing the same environmental niche; some microorganisms form a close physical association with another type of organism, from which special benefits may accrue for one or both parties. Such associations are termed collectively *symbiosis* ('living together') (Table 15.1). Three general forms of symbiotic relationship may be defined:

> Symbiosis is sometimes taken to mean a relationship between different organisms from which both participants derive benefit. We use the term in its broader sense, as described in the text.

- *Parasitism*: an association from which one partner derives some or all of its nutritional requirements by living either in or on the other (the *host*), which usually suffers harm as a result.
- *Mutualism:* an association from which both participants derive benefit. The relationship is frequently obligatory, that is, both are dependent upon the other for survival. Non-obligatory mutualism is sometimes called *protocooperation*.
- *Commensalism:* an association from which one participant (the commensal) derives benefit, and the other is neither benefited nor harmed. The relationship is not usually obligatory.

Microorganisms may be associated with plants, animals or other types of microorganism in any of these types of symbiosis (Tables 15.2–15.4).

Microbial associations with animals

Termites are insects belonging to the order Isoptera that are found particularly in tropical regions. Their famous ability to destroy trees and wooden structures such as buildings and furniture is due to a resident population of flagellated protozoans in their hindgut, which are able to break down cellulose. Termites lack the enzymes necessary to do

Table 15.1 Types of symbiotic association

Association	Species A	Species B
Mutualism	+	+
Protocooperation	+	+
Commensalism	–	+
Parasitism	x	+

Participants in symbiosis may derive benefit, harm or neither from the association. + denotes benefit, x denotes harm and – denotes neither.

this, and would thus starve to death if the protozoans were not present. In return, they are able to provide the anaerobic conditions required by the protozoans to ferment the cellulose to acetate, carbon dioxide and hydrogen. The acetate is then utilised as a carbon source by the termites themselves.

In addition to the protozoans, anaerobic bacteria resident in the hindgut also play an important role in the metabolism of the termites. Acetogenic and methanogenic species compete for the carbon dioxide and hydrogen produced by the protozoans. The former contribute more acetate for the termite to use, whilst the latter produce significant amounts of methane. Some methanogens exist as endosymbionts within the protozoans.

> The total global amount of methane production by termites is comparable to that generated by ruminants.

In other types of termite, no resident population of cellulose digesters is present. Instead, the termite ingests a fungus, which provides the necessary cellulolytic enzymes.

Another example of a host's staple diet being indigestible without the assistance of resident microorganisms is provided by the brightly coloured African bird the honey guide. The honey guide eats beeswax, and relies on a two-stage digestion process by bacteria (*Micrococcus cerolyticus*) and yeast (*Candida albicans*) to render it in a usable form.

At the bottom of the deepest oceans, around geothermal vents, live enormous (two metres or more) tube worms belonging to the genus *Riftia*. These lack any sort of digestive system, but instead contain in their body cavity a tissue known as the *trophosome*. This comprises vascular tissue plus cells packed with endosymbiotic bacteria. These are able to generate ATP and NADPH by the oxidation of hydrogen sulphide generated by

Table 15.2 Microorganism–animal associations

Microorganism	Animal	Type of relationship
Anaerobic bacteria	Ruminants	Mutualism
Flagellated protozoans	Termites	Mutualism
Sulphur-oxidising bacteria	*Riftia* (marine tube worm)	Mutualism
Luminescent bacteria	Fish, molluscs	Mutualism
Bacteria, yeasts	Honey guide (bird)	Mutualism
Fungus	Leaf cutter ants	Protocooperation
Resident bacteria of skin, large intestine, etc.	Humans	Commensalism

Table 15.3 Microorganism–plant associations

Microorganism	Plant	Type of relationship
N_2-fixing bacteria	Legumes	Mutualism
Mycorrhizal fungi	Various	Mutualism
Agrobacterium tumefaciens	Various	Parasitism (Crown gall disease)
Acremonium (fungus)	Grass	Mutualism

volcanic activity and fix carbon dioxide via the Calvin cycle, providing the worm with a supply of organic nutrients. Hydrogen sulphide is transported to the trophosome from the worm's gill plume by a form of haemoglobin present in its blood (Figure 15.1).

Warm-blooded animals such as humans play host in their lower intestinal tract to vast populations of bacteria. Although some of these are capable of producing useful metabolites such as vitamin K, most live as *commensals*, neither benefiting nor harming their host. It could be argued, however, that the very presence of the resident intestinal microflora acts as an important defence against colonisation by pathogens, thus making the association more one of mutualism.

A number of bacteria, viruses, fungi, protozoans and even algae act as pathogens in animals, and cause millions of human deaths every year. A detailed description of these falls outside of the scope of this introductory text, however examples of diseases caused by each group are described in Chapters 7 to 10.

Microbial associations with plants

The roots of almost all plants form mutualistic associations with fungi, known as *mycorrhizae*, which serve to enhance the uptake of water and mineral nutrients, especially phosphate, by the plants. The beneficial effect of a mycorrhizal association is particularly noticeable in soils with a poor phosphorus content. In return, the plant supplies reduced carbon in the form of carbohydrates to the fungi. Unlike other plant–microorganism interactions that occur in the rhizosphere, mycorrhizal associations involve the formation of a distinct,

A mycorrhiza is a mutually beneficial association between plant roots and a species of fungus.

The rhizosphere is the region around the surface of a plant's root system.

Table 15.4 Microorganism–microorganism associations

Microorganism	Microorganism	Type of relationship
Fungi	Alga/blue green	Mutualism (Lichen)
Amoeba, flagellates	Methanogenic archaea	Mutualism

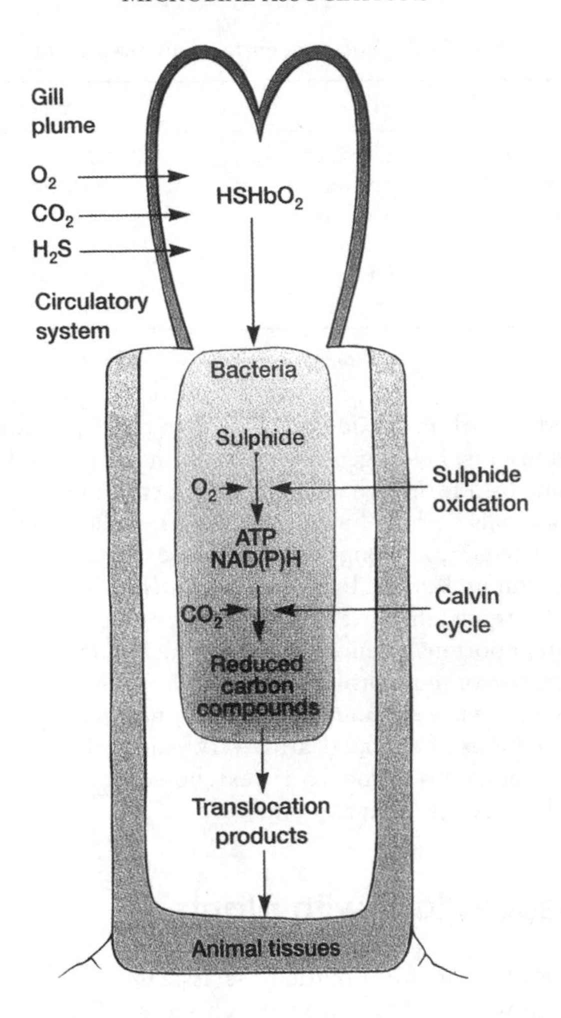

Figure 15.1 Symbiosis in *Riftia*, the giant tube worm. Found in deep-sea hydrothermal vents, *Riftia* acts as host to sulphur-oxidising bacteria. Energy and reducing power derived from sulphide oxidation are used to fix CO_2 via the Calvin cycle and provide the worm with organic carbon. From Prescott, LM, Harley, JP & Klein, DA: Microbiology 5th edn, McGraw Hill, 2002. Reproduced by permission of the publishers

Box 15.1 Once bitten, twice shy

The fungus *Acremonium* derives reduced carbon compounds and shelter by living within the tissues of the grass *Stipa robusta*, and in return deters animals from grazing on it. It does this by producing various alkaloids which, ingested in sufficient amounts, are powerful enough to send a horse to sleep for several days! The horse clearly does not relish the experience, as it avoids the grass thereafter. The *Acremonium* passes to future generations through the seeds, so the relationship between plant and fungus is perpetuated. The nickname of 'sleepy grass' is self-explanatory!

Figure 15.2 Endomycorrhizae. Section through a plant root colonised by an endomycorrhizal fungus. Note the spreading, 'tree-like' arbuscles

integrated structure comprising root cells and fungal hyphae. In *ectomycorrhizae* the plant partner is always a tree; the fungus surrounds the root tip, and hyphae spread between (but do not enter) root cells. In the case of the more common *endomycorrhizae*, the fungal hyphae actually penetrate the cells by releasing cellulolytic enzymes. Arbuscular mycorrhizae are found in practically all plant types, including 'lower' plants (mosses, ferns). They form highly branched *arbuscules* within the root cells that gradually lyse, releasing nutrients into the plant cells (Figure 15.2). In contrast to pathogenic fungi, mycorrhizal fungi are often rather non-specific in their choice of 'partner' plant.

An interesting example of mutualistic association concerns the endophytic (='inside plant') fungus *Acremonium* (Box 15.1).

The ability of crop plants to thrive is frequently limited by the supply of available nitrogen; although there is a lot of it in the atmosphere, plants are unable to utilise it, and instead must rely on an inorganic supply (both naturally-occurring and in the form of fertilisers). As we saw in Chapter 7, however, certain bacterial species are able to 'fix' atmospheric nitrogen into a usable form. Some of these, notably *Rhizobium* spp. form a mutualistic relationship with leguminous plants such as peas, beans and clover, converting nitrogen to ammonia, which the legume can incorporate into amino acids. In return, the bacteria receive a supply of organic carbon, which they can use as an energy source for the fixation of nitrogen.

The free-living *Rhizobium* enters the plant via its root hairs, forming an infection thread and infecting more and more cells (Figure 15.3). Normally rod-shaped, they proliferate as irregularly-shaped *bacteroids*, densely packing the cells and causing them to swell, forming *root nodules*.

Root nodules are tumour likc growths on the roots of legumes, where nitrogen fixation takes place.

Rhizobium requires oxygen as a terminal electron acceptor in oxidative phosphorylation, but as you may recall from Chapter 7, the nitrogenase enzyme, which fixes the nitrogen, is sensitive to oxygen. The right microaerophilic conditions are maintained

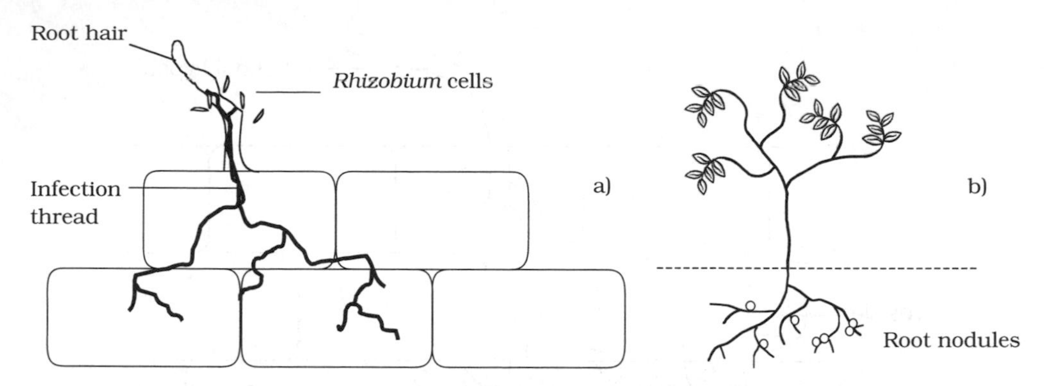

Figure 15.3 Nitrogen-fixing bacteria form root nodules in legumes. (a) *Rhizobium* cells attach to root hairs and penetrate, forming an infection thread and spreading to other root cells. (b) Root nodule formation.

by means of a unique oxygen-binding pigment, *leghaemoglobin*. This is only synthesised by means of a collaboration between both partners. Nitrogen fixation requires a considerable input of energy in the form of ATP (16 molecules for every molecule of nitrogen), so when ammonia is in plentiful supply the synthesis of the nitrogenase enzyme is repressed.

Farmers have long recognised the value of incorporating a legume into a crop rotation system; the nodules left behind in the soil after harvesting the crop appreciably enhance the nitrogen content of the soil.

Legumes are not the only plants able to benefit from the nitrogen-fixing capabilities of bacteria. The water fern *Azolla*, which grows prolifically in the paddy fields of southeast Asia, has its nitrogen supplied by the blue-green bacterium *Anabaena*. When the fern dies, it acts as a natural fertiliser for the rice crop. *Anabaena* does not form root nodules, but takes up residence in small pores in the *Azolla* fronds. Nitrogen fixation takes place in heterocysts, specialised cells whose thick walls slow down the rate at which oxygen can diffuse into the cell, providing appropriate conditions for the oxygen-sensitive nitrogenase.

> Unlike higher plants, ferns do not possess true roots, stems and leaves. The structure equivalent to a leaf is called a *frond*.

The alder tree (*Alnus* spp.) is able to grow in soils with poor nitrogen content due to its association in root nodules with the nitrogen-fixing actinomycete *Frankia*. The filamentous *Frankia* solves the problem of nitrogenase's sensitivity to oxygen by compartmentalising it in thick-walled vesicles at the tips of its hyphae, which serve the same function as the heterocysts of *Anabaena*.

Many microorganisms, particularly bacteria and yeasts, are to be found living as harmless commensals on the surface structures (leaves, stem, fruits) of plants.

> Organisms that grow on the surface of a plant are called *epiphytes*. They frequently live as commensals.

Plant disease may be caused by viruses, bacteria, fungi or protozoans. These frequently have an impact on humans, especially if the plant affected is a commercially important crop. Occasionally the effect on a human

Table 15.5 Some microbial diseases of plants

Causative agent	Type of microorganism	Host	Disease
Heterobasidion	Fungus	Pine trees	Heart rot
Ceratocystis	Fungus	Elm trees	Dutch elm disease
Puccinia graminis	Fungus	Wheat	Wheat rust
Phytophthora infestans	Water mould	Potato	Potato blight
Erwinia amylovera	Bacterium	Apple, pear tree	Fire blight
Pseudomonas syringae	Bacterium	Various	Chlorosis
Agrobacterium	Bacterium	Various	Crown gall disease
Tobacco mosaic virus	Virus	Tobacco	Tobacco mosaic disease

population can be catastrophic, as with the Irish famine of the 1840s brought about by potato blight. A number of microbial diseases of plants are listed in Table 15.5.

We have already encountered the soil bacterium *Agrobacterium tumefaciens* in Chapter 12, where we saw how it has been exploited as a means of genetically modifying plants. *A. tumefaciens* is useful for introducing foreign DNA because it is a natural pathogen of plants, entering wounds and causing crown gall disease, a condition characterised by areas of uncontrolled growth, analogous to tumour formation in animals. This proliferation is caused by the expression within the plant cell of genes that encode the sequence for enzymes involved in the synthesis of certain plant hormones. The genes are carried on the T-DNA, part of an *A. tumefaciens* plasmid, which integrates into a host chromosome. Also on the T-DNA are genes that code for amino acids called opines. These are of no value to the plant, but are utilised by the *A. tumefaciens* as a food source.

Microbial associations with other microorganisms

A lichen is a mutually beneficial association between a fungus and an alga or cyanobacterium (blue-green).

The most familiar example of mutualism between microorganisms is that of *lichens*, which comprise a close association between the cells of a fungus (usually belonging to the Ascomycota) and a photosynthetic alga or cyanobacterium. Although many different fungal species may take part in lichens, only a limited number of algae or cyanobacteria do so. Lichens are typically found on exposed hard surfaces such as rocks, tree bark and roofs, and grow very slowly at a rate of a millimetre or two per year. They often occupy particularly harsh environments, from the polar regions to the hottest deserts. The photosynthetic partner usually exists as a layer of cells scattered among fungal hyphae (Figure 15.4). Often unicellular, it fixes carbon dioxide as organic matter, which the heterotrophic fungus absorbs and utilises. The fungal member provides anchorage and supplies inorganic nutrients and water, as well as protecting the alga from excessive exposure to sunlight.

Although lichens are tolerant of extremes of temperature and water loss, they have a well-known sensitivity to atmospheric pollutants such as the oxides of nitrogen and

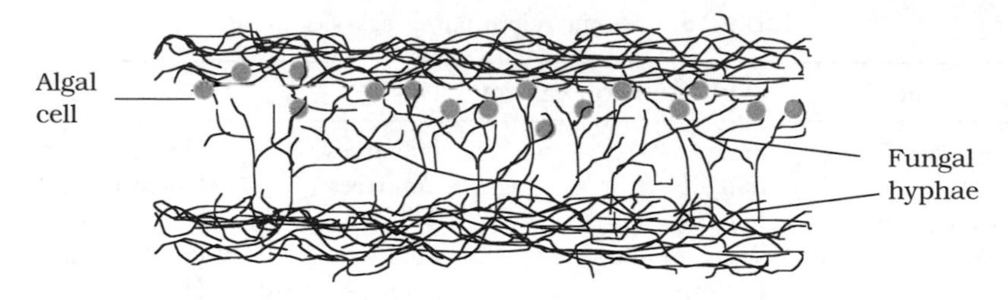

Figure 15.4 Algae and fungi combine to form lichens. Algal cells are embedded in among the fungal hyphae just below the surface, where light is able to penetrate. Organic carbon and oxygen produced by photosynthesis are used by the fungus, whilst it provides water, minerals and shelter for its algal partner

sulphur. Their presence in an urban setting is therefore a useful indicator of air quality. Lichens were used for many years as a source of brightly coloured dyes for the textile industry; they are also used in the perfume industry. The dye used in litmus paper is derived from a lichen, belonging to the genus *Roccella*.

It should be stressed that lichens are not just a mixture of fungal and algal cells. They are distinctive structures with properties not possessed by either of their component species. Indeed, the relationship between the two partners of a lichen is so intimate that the composite organisms are given taxonomic status. Many thousands of species of lichen have been identified.

As noted in the definitions at the beginning of this chapter, a mutualistic relationship need not always be as intimate and essential to both partners as it is in a lichen. The sulphate-reducing bacteria *Desulfovibrio*, for example, can obtain the sulphate and organic substrates it needs from the photosynthetic purple sulphur bacterium *Chromatium*, which in turn receives the carbon dioxide and hydrogen sulphide it requires (Figure 15.5). This protocooperation is not essential to either bacterium, however, as each can satisfy its requirements by alternative means. In another bacterial relationship,

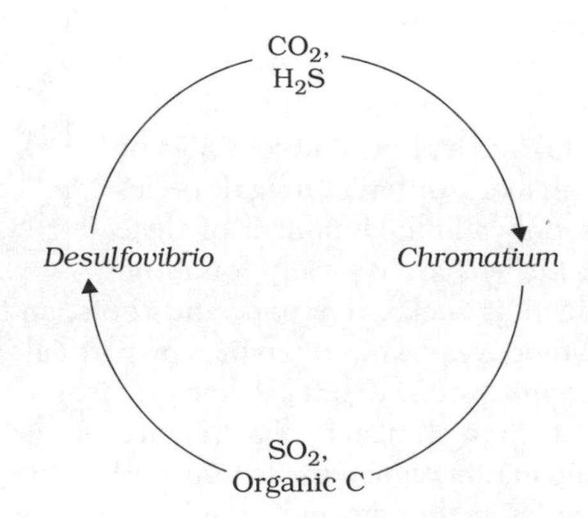

Figure 15.5 Protocooperation in bacteria. The sulphate-reducing *Desulfovibrio* and the sulphur-oxidising *Chromatium* can supply each other with the raw materials required for energy production. Neither is completely dependent on the association

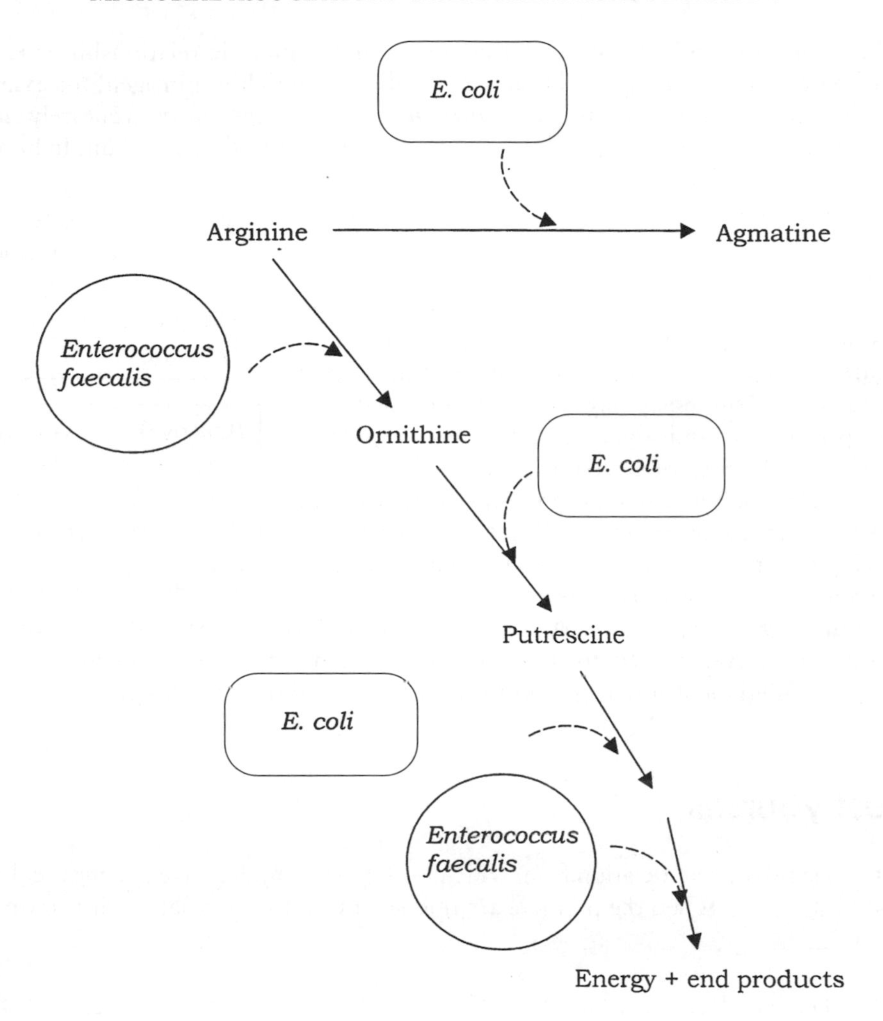

Figure 15.6 Protocooperation can make available an otherwise unutilisable substrate. Individually, neither *Enterococcus faecalis* nor *E. coli* is able to utilise arginine; however, working together they can convert it into putrescine, which can then be metabolised further by either organism to produce energy

the gut bacteria *E. coli* and *Enterococcus faecalis* cooperate to utilise arginine. As shown by Figure 15.6, neither can usefully metabolise this amino acid on its own; however, neither is dependent on the reaction.

Commensal relationships, in which one partner benefits and neither suffer, are common among microorganisms; such associations are rarely obligatory. A common basis for microbial commensalism is for one partner to benefit as a coincidental consequence of the normal metabolic activities of the other. Thus, one may excrete vitamins or amino acids that can be utilised by the other, or a facultative anaerobe may assist its obligate anaerobe neighbour by removing oxygen from the atmosphere, thus providing the conditions for the latter to grow.

The nature of a role played by an organism in a symbiotic relationship may alter according to the prevailing conditions. In a soil already rich in nitrogen, for example, the legume derives no benefit from the *Rhizobium*, which is then more accurately classed as a parasite, as it continues to utilise organic carbon produced by the plant. In humans, harmless gut symbionts such as *E. coli* can become *opportunistic pathogens*, and cause infections if introduced to an inappropriate site such as a wound or the urinary tract.

An *endoparasite* fully enters its host and lives inside it. An *ectoparasite* attaches to the outside of the host.

A limited number of microorganisms exist by living parasitically inside another. Viruses (Chapter 10) are all obligate endoparasites that form an association with a specific host. This host may be microbial: bacteria, fungi, protozoans and algae all act as hosts to their own viruses. Viruses of bacteria are termed *bacteriophages* and are only able to replicate themselves inside an actively metabolising bacterial cell (see Chapter 10 for bacteriophage replication cycles). An unusual bacterium belonging to the genus *Bdellovibrio* also parasitises bacteria, but as it does not properly enter the cell, is more properly thought of as an ectoparasite or even a predator (see Chapter 7). Other non-viral parasitism involves bacteria or fungi on protozoans, and fungi on algae and on other fungi.

Bdellovibrio itself may play be parasitised by bacteriophages; this is known as *hyperparasitism*!

Test yourself

1 A symbiotic association from which both participants derive benefit is called ________. When the partners are not reliant on the association, it is termed ________.

2 Flagellated ________ are required by termites to digest the ________ component of wood.

3 Bacteria in the trophosome of the tube worm *Riftia* generate energy by oxidation of ________ ________.

4 Most of the resident bacteria in our gut live as ________.

5 ________ comprise an association between plant root cells and a fungus.

6 The area surrounding a plant's roots is called the ________.

7 The bacterium *Rhizobium* has the unusual ability to ________ ________.

8 In return for acting as host for *Rhizobium*, the cells of a legume's roots receive a supply of ________ ________.

9 Inside root cells, *Rhizobium* takes on an irregular, swollen form called a ________.

10 In a *Rhizobium*-legume association, the pigment ________ maintains ________ at a low level, in order that the ________ enzyme can function.

11 Microorganisms that live as commensals on the surface of plants are termed ________.

12 Potato blight is caused by a fungus belonging to the genus ________.

13 In a lichen, the fungal partner derives ________ ________ from its photosynthetic partner.

14 The sensitivity of lichens to gases such as SO_2 makes them a good indicators of ________ ________.

15 *Escherichia coli* and *Enterococcus faecalis* form a ________ partnership to enable them to metabolise arginine.

16

Microorganisms in the Environment

Living organisms, together with their physical surroundings, make up an ecosystem.

At various points in this book we have referred to the different environments in which particular microorganisms are to be found. Like other living organisms, they live as part of *ecosystems*, and we shall consider the three main types of ecosystem – terrestrial, freshwater and marine – later in this chapter.

First, however, we must turn our attention again to the subject of energy relations in living things. In Chapter 4, we looked at the different ways in which microorganisms can derive and utilise energy from various sources. We now need to put these processes into a global perspective. All organisms may be placed into one of three categories with respect to their part in the global flow of energy:

- *(Primary) Producers*: autotrophs that obtain energy from the sun or chemical sources (e.g. green plants, photosynthetic bacteria, chemolithotrophic bacteria). They use the energy to synthesise organic material from carbon dioxide and water.
- *Consumers*: heterotrophs that derive energy through the consumption of other organisms (producers or other consumers). They may serve as a link between the primary producers and the decomposers.
- *Decomposers*: organisms that break down the remains and waste products of producers and consumers, obtaining energy and releasing nutrients, including CO_2, that can be reused by the producers.

Natural systems exist in a balance; carbon and all the other elements that make up living things are subject to repeated *recycling*, so that they are available to different organisms in different forms. Think back to Chapter 6, in which we discussed how algae, green plants and certain bacteria capture light energy, then use it to synthesise organic carbon compounds from carbon dioxide and water. What happens to all this organic carbon? It does not just accumulate, but is recycled by other living things, which convert it back to carbon dioxide by respiration. This can be seen in its simplest form in Figure 16.1. Many other elements such as sulphur, nitrogen, and iron are similarly changed from one form to another in this way, by a cyclic series of reactions. Microorganisms are responsible for most of these reactions, oxidising and reducing the elements according

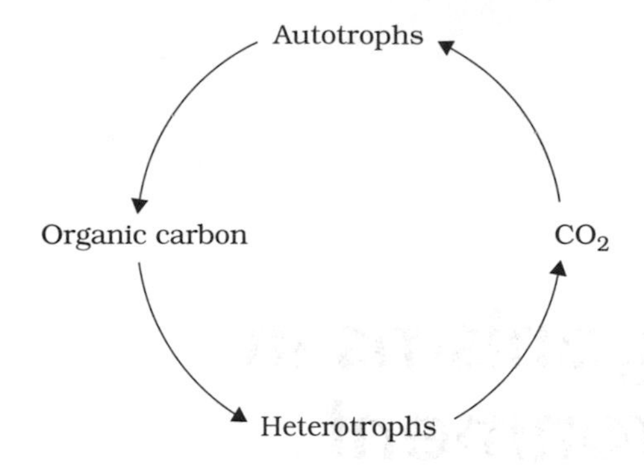

Figure 16.1 The carbon cycle. Autotrophs fix CO_2 as an organic compound, which heterotrophs convert back to CO_2. The recycling of carbon satisfies the requirements of both nutritional types

to their metabolic needs. The continuation of life on Earth is dependent on the cycling of finite resources in this way.

The carbon cycle

> The carbon cycle is the series of processes by which carbon from the environment is incorporated into living organisms and returned to the atmosphere as carbon dioxide.

A more detailed scheme of the carbon cycle is shown in Figure 16.2. Both aerobic and anaerobic reactions contribute to the cycle. The numbers in parentheses in the following description refer to those in Figure 16.2.

Atmospheric CO_2 is fixed into organic compounds by plants, together with phototrophic and chemoautotrophic microorganisms (**1**). The organic compounds thus synthesised undergo cellular respiration and CO_2 is returned to the atmosphere (**2**). The carbon may have been passed along a food chain to consumers before this occurs. Carbon dioxide is also produced by the decomposition of dead plant, animal and microbial material by heterotrophic bacteria and fungi.

Methanogenic bacteria produce methane from organic carbon or CO_2 (**3**, **4**). This in turn is oxidised by methanotrophic bacteria; carbon may be incorporated into organic material or lost as CO_2 (**5**, **6**).

The nitrogen cycle

Nitrogen is essential to all living things as a component of proteins and nucleic acids. Although elemental nitrogen makes up three quarters of the Earth's atmosphere, only a handful of life forms are able to utilise it for metabolic purposes. These are termed nitrogen-fixing bacteria, and incorporate the nitrogen into ammonia (Figure 16.3, reaction 1):

$$N_2 + 8e^- + 8H^+ + 16ATP \longrightarrow 2NH_4{}^+ + 16ADP + 16Pi$$

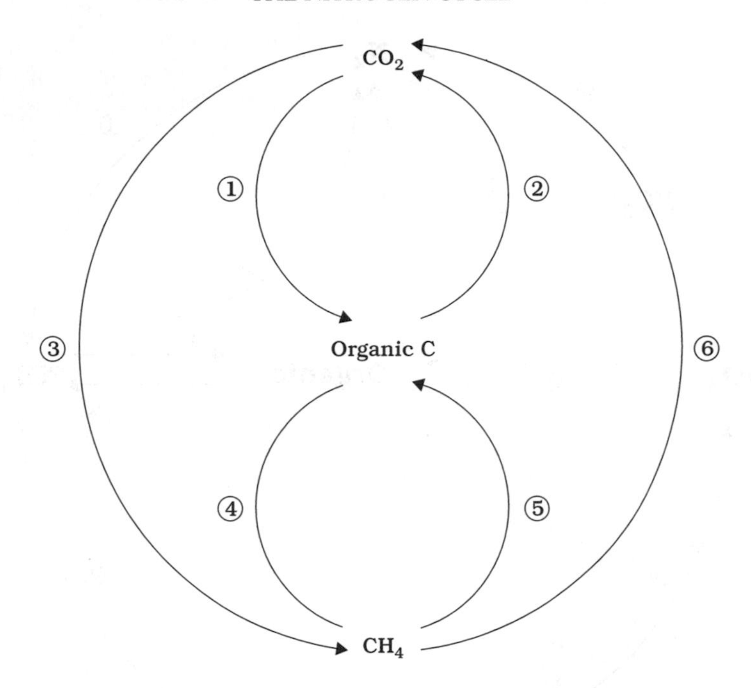

Figure 16.2 The carbon cycle – a closer look. Carbon is circulated as one of three forms, carbon dioxide, methane and organic compounds. Different organisms are able to utilise each form for their own metabolic requirements, converting it in the process to one of the others. Numbered arrows refer to reactions described in the text

The nitrogenase enzyme complex responsible for the reaction is very sensitive to oxygen, and is thought to have evolved early in the Earth's history, when the atmosphere was still largely oxygen-free. Many nitrogen-fixing bacteria are anaerobes; those that are not have devised ways of keeping the cell interior anoxic. *Azotobacter* species, for example, utilise oxygen at a high rate, so that it never accumulates in the cell, inactivating the nitrogenase. Many cyanophytes (blue-greens) carry out nitrogen fixation in thick-walled heterocysts which help maintain anoxic conditions.

Some nitrogen-fixing bacteria such as *Rhizobium* infect the roots of leguminous plants such as peas, beans and clover, where they form nodules and form a mutually beneficial association (see Chapter 15).

The process by which microorganisms convert organic matter to an inorganic form is termed mineralisation.

Ammonia produced by nitrogen fixation is assimilated as amino acids, which can then form proteins and feed into pathways of nucleotide synthesis (2). Organic nitrogen in the form of dead plant and animal material plus excrement re-enters the environment, where it undergoes *mineralisation* (3) at the hands of a range of microorganisms, involving the deamination of amino acids to their corresponding organic acid. This process of mineralisation may occur aerobically or anaerobically, in a wide range of microorganisms, e.g.:

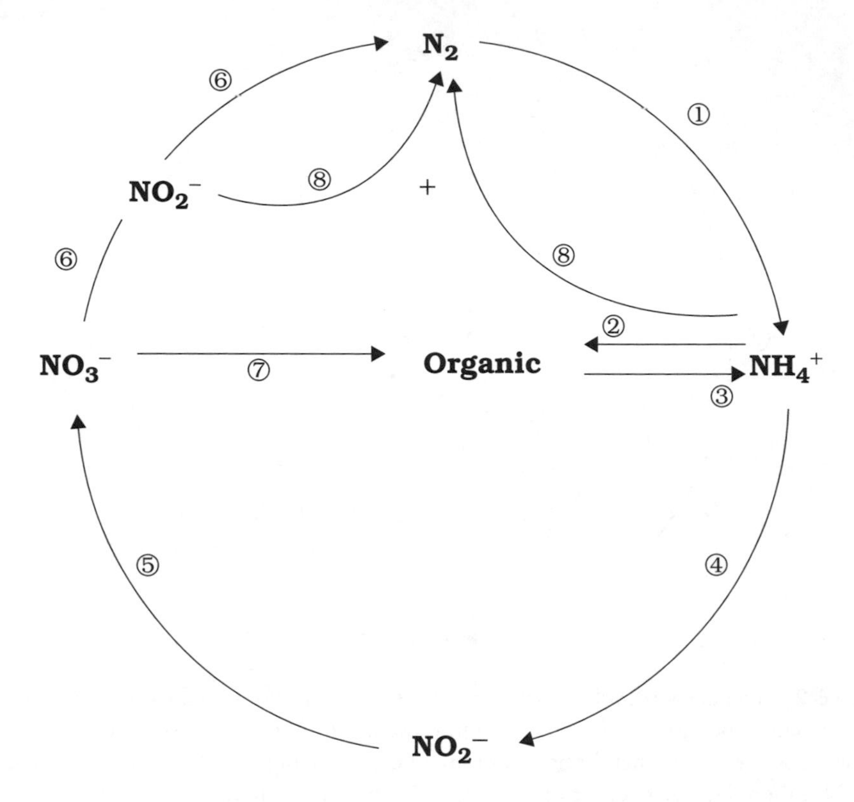

Figure 16.3 The nitrogen cycle. See the text for further details of reactions. Numbered arrows refer to reactions described in the text

$$2\ \underset{\text{Alanine}}{\begin{matrix} CH_3 \\ | \\ HC\text{-}NH_3^+ \\ | \\ COO^- \end{matrix}} + O_2 \xrightarrow{\textbf{Deamination}} 2\ \underset{\text{Pyruvate}}{\begin{matrix} CH_3 \\ | \\ C{=}O \\ | \\ COO^- \end{matrix}} + 2\ NH_4^+$$

The process of *nitrification*, by which ammonia is oxidised stepwise firstly to nitrite and then to nitrate, involves two different groups of bacteria (4, 5).

$$NH_4^+ \longrightarrow NO_2^-$$
$$NO_2^- \longrightarrow NO_3^-$$

The nitrate thus formed may suffer a number of fates. It may act as an electron acceptor in anaerobic respiration, becoming reduced to nitrogen via a series of intermediates including nitrite (6). This process of *denitrification* occurs in anaerobic conditions such as waterlogged soils. Alternatively, it can be reduced once again to ammonia and thence converted to organic nitrogen (7).

Denitrification is the reduction, under anaerobic conditions, of nitrite and nitrate to nitrogen gas.

A final pathway of nitrogen cycling has only been discovered in recent years. It is known as *anammox* (anaerobic ammonia oxidation), and is carried out by members of a group of Gram-negative bacteria called the Planctomycetes (see Chapter 7). The reaction, which can be represented thus:

Anammox is the formation of nitrogen gas by the anaerobic oxidation of ammonia and nitrite.

$$NH_4^+ + NO_2^- = N_2 + 2H_2O \quad (8)$$

has considerable potential in the removal of nitrogen from wastewater.

The sulphur cycle

Sulphur is found in living organisms in the form of compounds such as amino acids, coenzymes and vitamins. It can be utilised by different types of organisms in several forms; Figure 16.4 shows the principal components of the sulphur cycle.

In its elemental form, sulphur is unavailable to most organisms; however, certain bacteria such as *Acidithiobacillus* are able to oxidise it to sulphate (1), a form that can be utilised by a much broader range of organisms (see Chapter 7):

$$2S + 3O_2 + 2H_2O \longrightarrow H_2SO_4$$

Powdered sulphur is often added to alkaline soils in order to encourage this reaction and thereby reduce the pH.

Sulphate-reducing bacteria convert the sulphate to hydrogen sulphide gas (2) using either an organic compound or hydrogen gas as electron donor:

$$8H^+ + SO_4^{2-} \longrightarrow H_2S + 2H_2O + 2OH^-$$

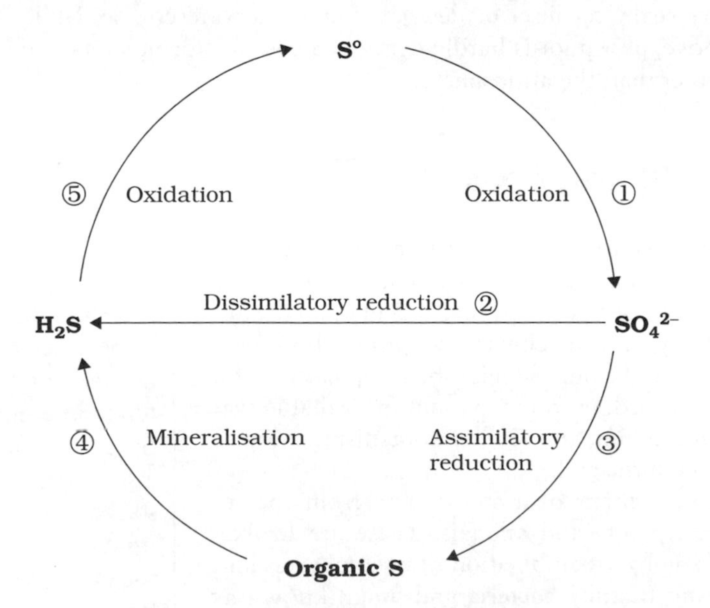

Figure 16.4 The sulphur cycle. See the text for further details of reactions. Numbered arrows refer to reactions described in the text

These bacteria are obligate anaerobes, and the process is termed *dissimilatory* sulphate reduction.

Plants are also able to utilise sulphate, incorporating it into cellular constituents such as the amino acids methionine and cysteine (3) (*assimilatory* sulphate reduction).

When the plants die, these compounds are broken down, again with the release of hydrogen sulphide (4) (see mineralisation, above).

$$\begin{array}{c} CH_2HS \\ | \\ HC\text{-}NH_3^+ + H_2O \\ | \\ COO^- \end{array} \longrightarrow \begin{array}{c} CH_3 \\ | \\ C{=}O + NH_4^+ + H_2S \\ | \\ COO^- \end{array}$$

Cysteine Pyruvate

Green and purple photosynthetic bacteria and some chemoautotrophs use hydrogen sulphide as an electron donor in the reduction of carbon dioxide, producing elemental sulphur and thus completing the cycle (5):

$$H_2S + CO_2 \longrightarrow (CH_2O)_n + S^0$$

Phosphorus

Phosphorus exists almost exclusively in nature as phosphate; however, this is cycled between soluble and insoluble forms. This conversion is pH-dependent, and if phosphate is only present in an insoluble form, it will act as a limiting nutrient. This explains the sudden surge in the growth of plants, algae and cyanobacteria when a source of soluble phosphate (typically fertiliser or detergent) enters a watercourse. Unlike the elements discussed above, phosphorus hardly exists in a gaseous form, so its main 'reservoir' is in the sea rather than the atmosphere.

The microbiology of soil

In the following section it will be necessary to generalise, and treat soil as a homogeneous medium. In fact, it is no such thing; its precise make-up is dependent upon the underlying geology, and the climatic conditions both past and present. In addition, the microbial population of a soil will vary according to the amount of available water and organic matter, and different organisms colonise different strata in the soil.

The organic content of a soil derives from the remains of dead plants and animals. These are broken down in the soil by a combination of invertebrates and microorganisms (mainly bacteria and fungi) known as the *decomposers*. Their action results in the release of

> The *topsoil* is the top few centimetres of a soil, characterised by its high content of organic material in various stages of decomposition. It is distinguished from the succeeding layers underneath it, termed the *subsoil*, *parent layer* and *bedrock*.

substances that can be used by plants and by other microorganisms. Much organic material is easily degraded, while the more resistant fraction is referred to as *humus*, and comprises lignin together with various other macromolecules. The humus content of a soil, then, is a reflection of how favourable (or otherwise) conditions are for its decomposition; the value usually falls between 2 and 10 per cent by weight. The inorganic fraction of a soil derives from the weathering of minerals. Microorganisms may be present in soils in huge numbers, mostly attached to soil particles. Their numbers vary according to the availability of suitable nutrients. Bacteria (notably actinomycetes) form the largest fraction of the microbial population, together with much smaller numbers of fungi, algae and protozoans. Published values of bacterial numbers range from overestimates (those that do not distinguish between living and dead cells) and underestimates (those that depend on colony counts and therefore exclude those organisms we are not yet able to grow in the laboratory – 99 per cent according to some experts!). Suffice to say that many millions (possibly billions) of bacteria may be present in a single gram of topsoil. In spite of being present in such enormous numbers, microorganisms only represent a minute percentage of the volume of most soils. Fungi, although present in much smaller numbers than bacteria, form a higher proportion of the soil biomass, due to their greater size. The majority of soil microorganisms are aerobic heterotrophs, involved in the decomposition of organic substrates; thus, microbial numbers diminish greatly the further down into the soil we go, away from organic matter and oxygen. The proportion of anaerobes increases with depth, but unless the soil is waterlogged, they are unlikely to predominate.

Humus is the complex organic content of a soil, comprising complex materials that remain after micribial degradation.

Other factors affecting microbial distribution include pH, temperature, and moisture. Broadly speaking, neutral conditions favour bacteria, while fungi flourish in mildly acidic conditions (down to about pH 4), although extremophiles survive well outside these limits. Actinomycetes favour slightly alkaline conditions. Bacterial forms occurring commonly in soils include *Pseudomonas*, *Bacillus*, *Clostridium*, *Nitrobacter* and the nitrogen-fixing *Rhizobium* and *Azotobacter*, as well as cyanobacteria such as *Nostoc* and *Anabaena*. Commonly found actinomycetes include *Streptomyces* and *Nocardia*. As we have noted elsewhere, actinomycetes are notable for their secretion of antimicrobial compounds into their surroundings. This provides an example of how the presence of one type of microorganism in a soil population can influence the growth of others, forming a dynamic, interactive ecosystem. In addition, bacteria may serve as prey for predatory protozoans, and secondary colonisers may depend on a supply of nutrients from, for example, cellulose degraders. Important fungal genera common in soil include the familiar *Penicillium* and *Aspergillus*; these not only recycle nutrients by breaking down organic material, but also contribute to the fabric of the soil, by binding together microscopic soil particles. Soil protozoans are mostly predators that ingest bacteria or protists such as yeasts or unicellular algae. All the major forms of protozoans may be present (flagellates, ciliates and amoebas), moving around the water-lined spaces between soil particles. Algae are of course phototrophic, and are therefore to be found mostly near the soil surface, although it will be recalled from Chapter 9 that some forms are capable of heterotrophic growth, and may thus survive further down.

The surface of soil particles is a good natural habitat for the development of *biofilms*, complex structures comprising microbial cells held together in a polysaccharide matrix. The microorganisms themselves produce the polysaccharide, which also allows the passage of nutrients from the environment. Biofilms can form on almost any surface, and are often to be found in rapidly flowing waters. Biofilms may be beneficial (e.g. wastewater treatment – see below) or harmful (e.g. infections resulting from growth in catheters) to humans.

Although we have emphasised the importance of organic matter in soil ecosystems, microorganisms may also be found growing on or even within rocks. The growth of such organisms, together with the action of wind and rainfall, contribute to the weathering of rocks.

The microbiology of freshwater

The microbial population of freshwater is strongly influenced by the presence or absence of oxygen and light. A body of water such as a pond or lake is stratified into zones (Figure 16.5), each having its own characteristic microflora, determined by the

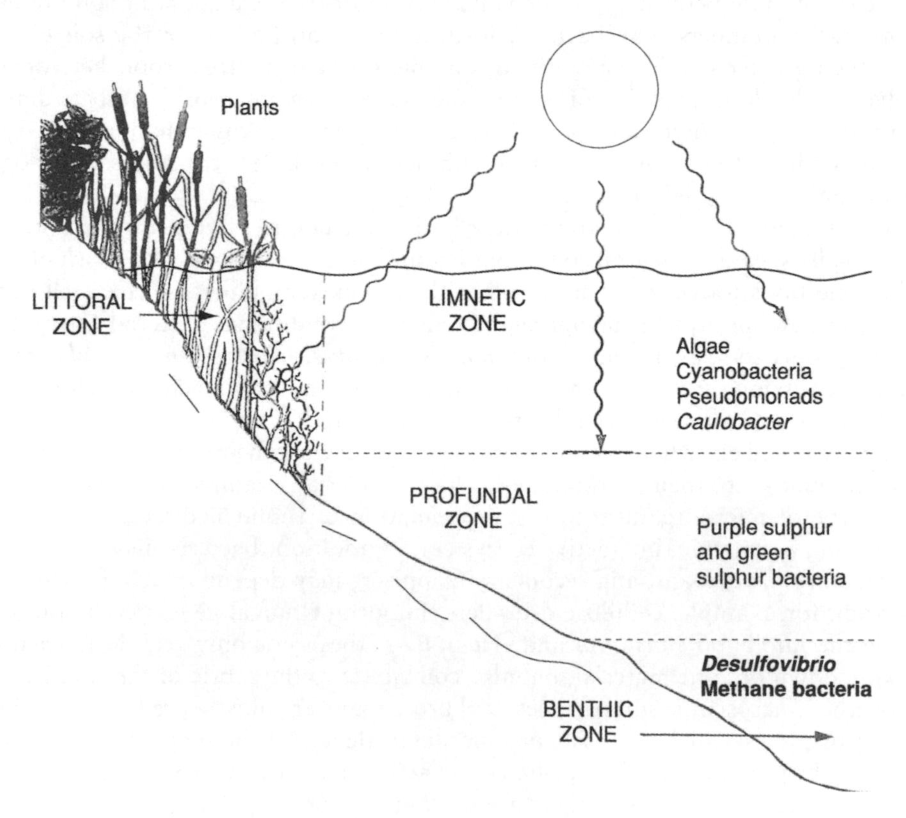

Figure 16.5 Vertical zonation in a lake or pond. Representative organisms are indicated for each zone. From Black, JG: Microbiology: Principles and Explorations, 4th edn, John Wiley & Sons Inc., 1999. Reproduced by permission of the publishers

availability of these factors. The *littoral* zone is the region situated close to land where the water is sufficiently shallow for sunlight to penetrate to the bottom. The *limnetic* zone occupies the same depth, but is in open water, away from the shore. The *profundal* zone occupies deeper water, where the sun is unable to penetrate, and finally the *benthic* zone comprises the sediment of mud and organic matter at the bottom of the pond or lake.

Oxygen is poorly soluble in water (9 mg/l at 20 °C), so its availability is often a limiting factor in determining the microbial population of a body of water. Oxygen availability in lakes and ponds is closely linked to oxygenic photosynthesis and therefore, indirectly, to the penetrability of light. Phototrophs such as algae and blue-greens are limited to those regions where light is able to penetrate. Oxygen is absent or very limited in the benthic zone, where anaerobic forms such as the methanogenic bacteria are to be found. Another factor influencing microbial populations is the organic content of the water; if this is high, the growth of decomposers will be encouraged, which will in turn deplete the oxygen. This is much less of an issue in rivers and streams, where physical agitation of the water generally ensures its continued oxygenation.

The temperature of freshwater ecosystems ranges between extremes (0–90 °C), and microorganisms may be found throughout this range.

Microorganisms play a central role in the purification of wastewaters, a topic we shall examine in more later in this chapter.

The microbiology of seawater

The world's oceans cover some 70 per cent of the Earth's surface and have a fairly constant salt content of 3.5 per cent (w/v). The depth to which light can penetrate varies, but is limited to the first 100 metres or so. A world of permanent darkness exists at greater depths, however in spite of the absence of photosynthesis, oxygen is often still present. This is because the generally low levels of mineral nutrients in seawater limit the amount of primary production, and therefore heterotrophic activity. At extreme depths, however, anoxic conditions prevail.

Compared to freshwater habitats, marine ecosystems show much less variability in both temperature and pH, although there are exceptions to this general rule. A more pertinent issue in marine environments is that of pressure; this increases progressively in deeper waters, and at 1000 metres reaches around 100 times normal atmospheric pressure. Concomitant with this increase in pressure is a decrease in temperature and nutrients. Surprisingly, however, certain members of the Archaea have been isolated even from these extreme conditions.

In contrast to terrestrial ecosystems, where plants are responsible for most of the energy fixation via photosynthesis, marine primary production is largely microbial, in the shape of members of the *phytoplankton*. As we have seen, such forms are restricted to those zones where light is able to penetrate. Also found here may be protozoans and fungi that feed on the phytoplankton. Because of the high salt concentration of seawater, the bacteria that are typically found in such environments

> Phytoplankton is a collective term used to describe the unicellular photosynthesisers, which include cyanobacteria, dinoflagellates, diatoms and single-celled algae.

differ from those in freshwater. In the last decade or so, the presence of *ultramicrobacteria* has been detected in marine ecosystems at relatively high densities; these are around one-tenth of the size of 'normal' bacteria. Marine bacteria are of necessity halophilic. Anaerobic decomposing bacteria inhabit the benthic zone, carrying out reactions similar to those that occur in freshwater sediments, whilst the profundal zone is largely free of microbial life.

Ultramicrobacteria are bacteria that are much smaller than normal forms, and can pass through a 0.22 μm filter. They may represent a response to reduced nutrient conditions.

Detection and isolation of microorganisms in the environment

As we emphasised in the last chapter, microorganisms rarely, if ever, exist in nature as pure cultures but rather form mixed populations. Methods are required, therefore, for the detection and isolation of specific microbial types from such mixtures. The traditional method of isolation is the use of an enrichment culture, as described in Chapter 4. As examples, aerobic incubation with a supply of nitrite would assist in the isolation of nitrifying bacteria such as *Nitrobacter* from mud or sewage, whilst a minimal medium containing $FeSO_4$ at pH 2 would encourage the isolation of *A. ferrooxidans* from a water sample.

We now know however that there are many types of microorganism in the environment that have so far resisted all attempts to culture them in the laboratory (often referred to as *viable but non-culturable*). The use of modern molecular techniques has helped us to identify the existence of a much broader range of bacteria and archaea than had previously been thought to exist. The extreme sensitivity of such methods means that we are able to demonstrate the presence of even a single copy of a particular bacterium in a mixed population. One such technique is called *fluorescence in situ hybridisation* (FISH). This uses a probe comprising a short sequence of single-stranded DNA or RNA that is unique to a particular microorganism, attached to a fluorescent dye. The microorganisms are fixed to a glass slide and incubated with the probe. The rules of base pairing in nucleic acids mean that the probe will seek out its complementary sequence, and cells carrying this sequence can be visualised under a fluorescence microscope. The most commonly used 'target' is ribosomal RNA, since this shows sequence variation from one microbial type to another, and because there are multiple copies within each cell, providing a stronger response. The polymerase chain reaction (PCR, Chapter 12) is another valuable tool in the identification of specific nucleic acid sequences. Other methods, not dependent on DNA, include the use of fluorescence-labelled *antibodies* raised against specific microorganisms.

Antibodies are proteins produced by the immune systems of higher animals in response to infection by a foreign organism; their main characteristic is their extreme specificity, thus they can be used to locate a specific protein.

Beneficial effects of microorganisms in the environment

The central role played by microorganisms in the recycling of essential elements on a global scale has already been stressed in this chapter. Many of their natural activities are exploited by humans for their own benefit. Some form the basis of industrial processes such as those used in the food and drink industries and are considered in Chapter 17, while the application of others is essentially environmental. Notable among these is the harnessing of natural processes of *biodegradation* to treat the colossal volumes of liquid and solid wastes generated by our society. These are reviewed briefly in the following section.

Biodegradation is the term used to describe the natural processes of breakdown of matter by microorganisms.

Solid waste treatment: composting and landfill

We in the modern Western world are often described as living in a 'throwaway society'. On average, each of us generates around 2 tonnes of solid waste material per year, and all of this must be disposed of in some way! Most of it ends up in landfill sites, huge holes in the ground where refuse is deposited to prevent it being a hazard. The non-biodegradable components (metals, plastics, rubble, etc.) remain there more or less indefinitely; however, over a period of time biodegradable material (food waste, textiles, paper, etc.) undergoes a decomposition process. The rate at which this happens is dependent on the nature of the waste and the conditions of the landfill, but could take several decades. Aerobic processes give way to anaerobic ones and a significant result of the latter is the generation of methane. Modern landfill sites incorporate systems that remove this to prevent it being a fire/explosion hazard, and may put it to good use as a fuel source.

Many householders separate organic waste items such as vegetable peelings and grass cuttings and use them to make compost. This practice, apart from providing a useful gardening supplement, also substantially reduces the volume of material that has to be disposed of by other means (see above). We have already mentioned the role of microorganisms in the recycling of carbon in the biosphere; these same processes serve to degrade the organic waste, especially the cellulose, resulting in a considerable reduction of the bulk. Fungi and bacteria, particularly actinomycetes, break down the organic matter to produce CO_2, water and humus, a relatively stable organic end product. Compost is not really a fertiliser, since its nitrogen content is not high, but it nevertheless provides nutrients to a soil and generally helps to improve its condition. Composting is carried out on a large scale by local authorities using the waste generated in municipal parks and gardens.

Wastewater treatment

The aim of wastewater treatment is the removal of undesirable substances and hazardous microorganisms in order that the water may safely enter a watercourse such as a river or stream. Further purification procedures are required before it can be used as drinking water. Wastewater treatment is fundamental to any developed society, and greatly reduces the incidence of waterborne diseases such as cholera. Wastewater may come from domestic or commercial sources; highly toxic industrial effluents may require pre-treatment before entering a water treatment system. Sewage is the term used to describe liquid wastes that contain faecal matter (human or animal).

The effectiveness of the treatment process is judged chiefly by the reduction of the wastewater's *biochemical oxygen demand* (BOD). This is a measure of the amount of oxygen needed by microorganisms to oxidise its organic content. A high BOD leads to the removal of oxygen from water, a certain indicator of pollution.

Wastewater treatment usually occurs in stages, the first of which (primary treatment) is purely physical, and involves the removal of floating objects followed by sedimentation, a process that removes up to a third of the BOD value. Secondary treatment involves microbial oxidation, leading to a substantial further reduction in BOD. This may take one of two forms, both of which are aerobic, the traditional *trickling filter* and the more recent *activated sludge* process (Figure 16.6). In the former, the wastewater is passed slowly over beds of stones or pieces of moulded plastic. These develop a biofilm comprising bacteria, protozoans, fungi and algae, and the resulting treated water has its

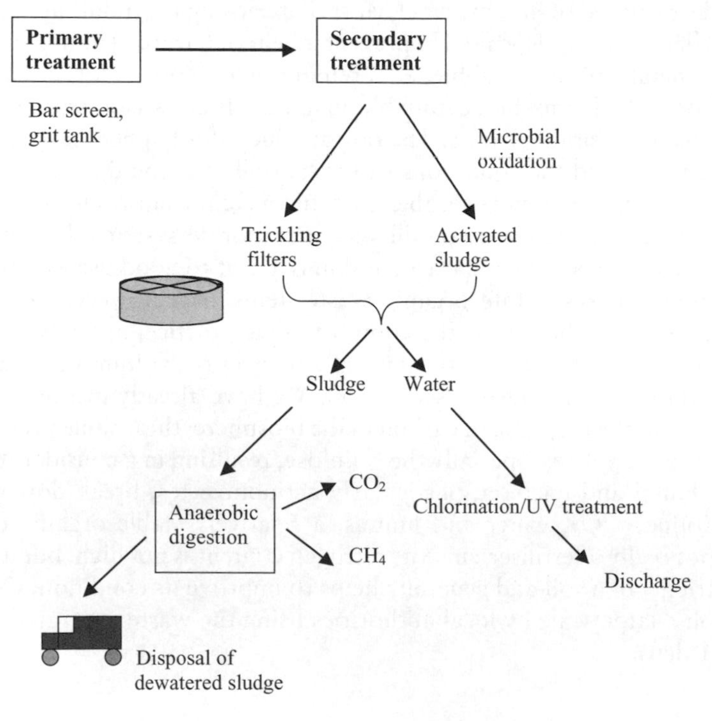

Figure 16.6 Wastewater treatment. Wastewater treatment achieves a reduction of the biochemical oxygen demand of the water by primary (physical) and secondary (biological) treatment

BOD reduced by some 80–85 per cent. Activated sludge plants achieve an even higher degree of BOD reduction. Here the wastewater is aerated in tanks that have been seeded with a mixed microbial sludge. The main component of this is the bacterium *Zoogloea*, which secretes slime, forming aggregates called *flocs*, around which other microorganisms such as protozoans attach. Some of the water's organic content is not immediately oxidised, but becomes incorporated into the flocs. After a few hours' residence in the tank, the sludge is allowed to settle out, and the treated water passes out of the system. Before being discharged to a watercourse, it is treated with chlorine to remove any pathogenic microorganisms that may remain.

The principal operating problem encountered with activated sludge is that of *bulking*. This is caused by filamentous bacteria such as *Sphaerotilus natans*; it results in the sludge not settling properly and consequently passing out with the treated water.

Both secondary treatment processes result in some surplus sludge, which undergoes *anaerobic digestion*, resulting in the production of methane and CO_2. The methane can be used as a fuel to power the plant, and any remaining sludge is dewatered and used as a soil conditioner. Care must be taken in this context, however, that the sludge does not have a high heavy metal content.

Bioremediation

Perhaps the biggest problem facing the developed world at the start of the twenty first century is that of pollution of the environment. Our dependence on the products of the chemical industries has resulted in the production of vast amounts of toxic waste material. One way of dealing with such (mostly organic) waste is to encourage the growth of bacteria and fungi that are able to oxidise the pollutants, a process known as *bioremediation*. Elsewhere in this book we have seen how microorganisms are able to utilise an enormous range of organic compounds as carbon sources; the Gram-negative bacterium *Burkholderia cepacia* can use over 100 such compounds. Many organisms can metabolise not only naturally occurring substances, but also synthetic ones, making them valuable allies in the process of bioremediation. Often the most effective microorganisms to use are those found living naturally at the contaminated site, since they have demonstrated the ability to survive the toxic effects of the pollutant, although in other cases specially adapted or genetically modified bacteria may be introduced (*bioaugmentation*). Examples of the use of microorganisms include the treatment of toxic waste sites, chemical spills, pesticides in groundwater and oil spills. One of the first successful applications of bioremediation came in the aftermath of the *Exxon Valdez* disaster in 1989 when thousands of tons of crude oil were released off the coast of Alaska. Depending on the circumstances, bioremediation procedures may occur

> Bioremediation is the use of biological processes to improve a specific environment, such as by the removal of a pollutant.

> Bioaugmentation is the deliberate introduction of specific microorganisms to an environment in order to assist in bioremediation.

in situ, or the contaminated soil or water may be removed to a specialist facility for treatment.

Harmful effects of microorganisms in the environment

The natural processes of bioconversion that are so important in the global recycling of elements may have unwanted consequences for humans. Prominent among these is acid mine drainage, a frequently encountered problem in mining regions. Bacterial oxidation of mineral sulphides, particularly the ubiquitous iron pyrite, leads to the release of a highly acidic leachate into streams and rivers. This also contains dissolved metals, including ferric iron. When it mixes with stream water, the pH is raised sufficiently for the iron to precipitate as unsightly orange ferric hydroxides, blanketing the stream bed and wiping out plant and animal life. The main culprits in the formation of acid mine drainage are sulphur-oxidising bacteria, notably *Acidithiobacillus ferrooxidans*; as we shall see in the final chapter, under controlled conditions this same organism can also provide economic benefits to the mining industry by extracting valuable metals from low grade ores.

Another area in which environmental microorganisms can have detrimental effects is that of *biodeterioration*, whereby economically important materials such as wood, paper, textiles, petroleum and even metals and concrete may be subject to damage by a range of microorganisms, mainly fungi and bacteria.

> Biodeterioration is the damage caused to materials of economic importance due to biological (mainly microbial) processes.

The most important microorganisms in the biodeterioration of wood are members of the Basidiomycota. Wood is only susceptible to fungal attack when its moisture level reaches around 30 per cent. The major component of wood that is subject to microbial attack is cellulose, although some forms can also degrade lignin. There are two main forms of rot; *white rot*, which involves the degradation of lignin as well as cellulose, and *brown rot*, in which the lignin is unaltered. The dry rot fungus *Serpula lacrymans* produces thick strands of hyphae called *rhizomorphs*, which it uses to conduct water and nutrients from damper areas. These are very strong, and able to travel over brickwork and masonry barriers. *S. lacrymans* is able to generate water as a metabolic end product and thus, once established, is able to grow even on dry wood. Dry rot flourishes in areas of static dampness such as badly ventilated, uninhabited properties.

Because cellulose is also an important component of paper and textiles, its breakdown is clearly of great economic importance. Degradation by fungi, and, to a lesser extent, bacteria, results in a loss of strength of the material in question. The paper-making process provides warm, wet conditions rich in nutrients, ideal for microbial growth, which can clog up machinery and discolour the finished product. A variety of biocides are used in an effort to minimise microbial contamination.

The discoloration referred to above raises the point that biodeterioration of a material need not necessarily affect its physical or chemical make-up; aesthetic damage can lessen the economic value of a material by altering its appearance. The blackening of shower curtains by moulds growing on surface detritus, familiar to generations of students, is another example of this!

Test yourself

1 Microorganisms are essential in the ________ of elements in the biosphere.

2 ________ organisms fix carbon dioxide into ________ compounds. These are ________ by heterotrophs and carbon dioxide is returned to the ________.

3 Nitrogen-fixing bacteria need to keep the interior of their cells oxygen-free because of the sensitivity of the ________ towards oxygen.

4 The release of inorganic nutrients from organic compounds as a result of microbial metabolism is termed ________.

5 Powdered ________ is added to alkaline soils in order to encourage the activity of ________ ________ bacteria, which leads to a reduction in pH.

6 Certain photosynthetic bacteria are able to use hydrogen sulphide as a ________ of ________ in the reduction of carbon dioxide.

7 ________ often acts as the limiting nutrient in an ecosystem, because it is only available in an insoluble form.

8 The ________ are the major group of bacteria found in a typical soil.

9 ________ are complex populations of microorganisms held together within a polysaccharide matrix.

10 The microbial population of a aquatic habitat is influenced by the availability of ________ and ________.

11 The profundal zone is the name given to deep waters characterised by the absence of ________.

12 In the benthic zone, the microbial population comprises mostly ________ forms.

13 A high organic content in a body of water encourages the growth of ________, which may lead to a depletion of available ________.

14 Unicellular photosynthesisers such as the ________ and ________ are collectively termed ________.

15 Molecular techniques such as FISH (________ ________ ________ ________) enable the detection of microorganisms it is not possible to culture by conventional means.

16 The effectiveness of a wastewater treatment regime may be judged by the reduction in the ________ ________ ________ (________).

17 The two forms of secondary wastewater treatment are ________ ________ and ________ ________.

18 In anaerobic digestion, organic matter is converted into ________ and ________ ________.

19 The principal component of wood that is susceptible to biodeterioration is ________. However the white rot fungi can also degrade ________.

20 The dry rot fungus *Serpula lacrymans* is able to conduct water via thick strands of hyphae called ________.

Part VII

Microorganisms in Industry

17
Industrial and Food Microbiology

Many aspects of our everyday lives are influenced in some way by microorganisms. In previous chapters we have noted their role in our environment, as well as their ability both to cause and cure infectious diseases. In addition, they are responsible for the production of much of what we eat and drink, synthesise industrially useful chemicals, and can even extract precious metals from the Earth (Table 17.1). In this chapter we shall look at some of the ways in which the activities of microorganisms have been harnessed for the benefit of humans, and developed on an industrial scale. The first applications of biotechnology many thousands of years ago were in the production of food and drink, so it is here that we shall begin our survey.

Microorganisms and food

To the general public, the association of microorganisms and food conjures up negative images of rotten fruit or food poisoning. On reflection, many people may remember that yeast is involved in bread and beer production, but how many realise that microorganisms play a part in the manufacture of soy sauce, pepperoni and even chocolate? In the following pages, we shall look at the contribution of microorganisms to the contents of our shopping baskets before considering one of the negative associations referred to above, the microbial spoilage of food.

The production of foodstuffs as a result of microbial fermentation reactions predates recorded history. The accidental discovery that such foods were less susceptible to spoilage than fresh foods must have made them an attractive proposition to people in those far-off days. Of course, until relatively recent times, nothing was known of the part played by microorganisms, so the production of beer, cheese and vinegar would not have been the carefully controlled processes that are used today. Indeed, it was only with the development of isolation techniques towards the end of the nineteenth century (recall Chapter 1), that it became possible to use pure cultures in food production for the first time. The fermentation of foodstuffs, hitherto an art, became a science.

Table 17.1 Some applications of microorganisms

Food products
Alcoholic drinks
Dairy products
Bread
Vinegar
Pickled foods
Mushrooms
Single-cell protein
Products from microorganisms
Enzymes
Amino acids
Antibiotics
Citric acid
Mining industries
Metal extraction
Desulphurisation of coal
Agriculture
Microbial pesticides
Environment
Bioremediation
Wastewater treatment
Nitrogen fixation
Composting

Alcoholic fermentations

Red wine is made from red (black) grapes only. Both white and red grapes may be used in white wine, but red grapes must first be deskinned.

There is evidence that alcoholic drinks, including beer and wine, were being produced thousands of years before the Christian era, making them among the earliest known examples of the exploitation of microorganisms by humans. Ethanol results from the fermentation process, because the conversion of sugar to carbon dioxide and water is incomplete:

$$C_6H_{12}O_6 \longrightarrow 2CH_3CH_2OH + 2CO_2$$

In dry wines, most of the sugar is fermented to alcohol; in sweet varieties, some sugar remains unfermented.

Although, in principle, wine can be made from almost any fruit juice with a high sugar content, the vast majority of commercially produced wines derive from the fermentation of the sugar present in grapes (Figure 17.1). Such fermentation reactions may be initiated by yeasts naturally found on the grape skin; however the results of such fermentations are erratic and may be unpalatable. In commercial winemaking the *must* (juice) resulting from the crushed grapes is treated with sulphur dioxide to kill off the natural microflora, and then inoculated with the yeast *Saccharomyces cerevisiae*, variety *ellipsoideus*. Specially developed strains are

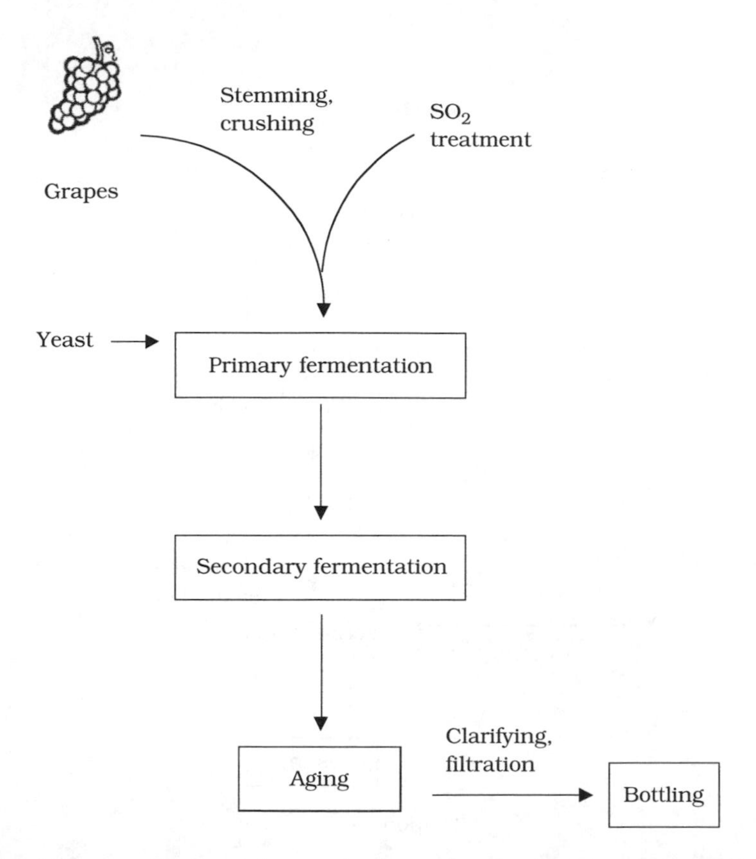

Figure 17.1 Wine production. The production of red and white wines differ in certain details, but share the main steps of crushing the grapes, fermenting their sugar content into alcohol and ageing to allow flavour development

used, which produce a higher percentage of alcohol (ethanol) than naturally occurring yeasts. Fermentation proceeds for a few days at a temperature of 22–27 °C for red wines (lower for whites), after which the wine is separated from the skins by pressing. This is followed by ageing in oak barrels, a process that may last several months, and during which the flavour develops. *Malolactic* fermentation is a secondary fermentation carried out on certain types of wine. Malic acid, which has a sharp taste, is converted to the milder lactic acid, imparting smoothness to the wine.

$$\underset{\text{Malic acid}}{COOH\text{-}H_2OC\text{-}H_2C\text{-}COOH} \longrightarrow \underset{\text{Lactic acid}}{CH_3\text{-}H_2OC\text{-}OOH} + CO_2$$

A secondary product of malolactic fermentation is diacetyl, which imparts a 'buttery' flavour to the wine. Spirits such as brandy and rum result from the products of a fermentation process being concentrated by *distillation*. This gives a much higher alcohol content than that of wines.

Most wines have an alcohol content of around 10–20 per cent. Fortified wines have extra alcohol added.

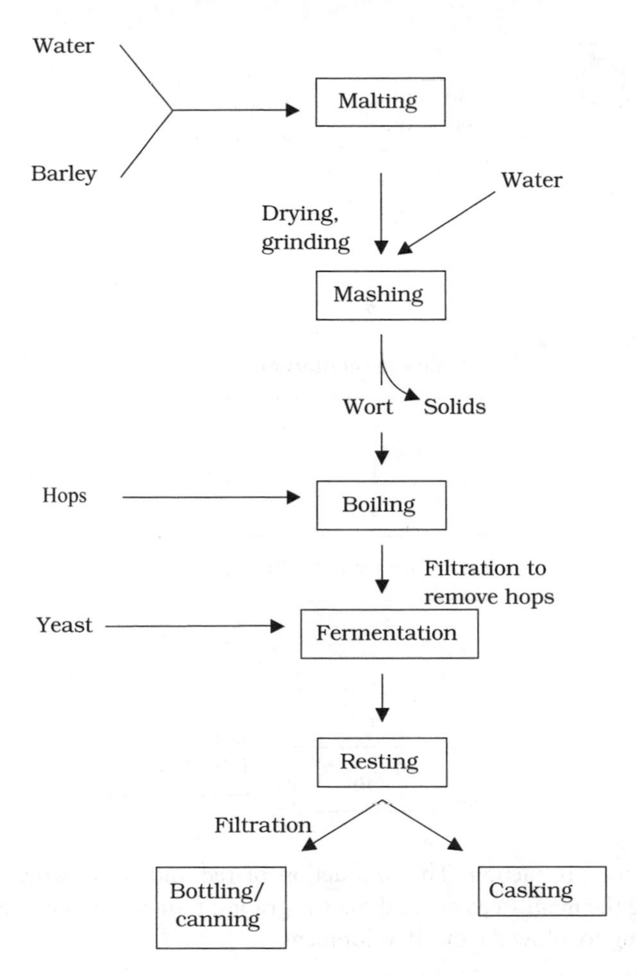

Figure 17.2 Beer production. The early stages serve to convert the carbohydrate present in the grain into a form that can be fermented by the yeast. The details shown refer to the production of a light, lager-type beer

Malting is the process whereby grain is soaked in water to initiate germination and activate starch-digesting enzymes.

Mashing releases soluble material from the grain in preparation for fermentation.

Beer is produced by the fermentation of barley grain. The procedure varies according to the type of beer, but follows a series of clearly defined steps (Figure 17.2). Grain, unlike grapes, contains no sugar to serve as a substrate for the yeast, so before fermentation can begin, it is soaked in water and allowed to germinate. This stimulates the production of the enzymes necessary for the conversion of starch to maltose (*'malting'*). An additional source of starch may be introduced during the next stage, *mashing*, in which the grains are ground up in warm water, and further digestion takes place. The liquid phase or *wort* is drained off and hops are added. They impart flavour and colour to the finished product and also possess antimicrobial properties, thereby helping to prevent contamination. The mixture is boiled, inactivating the enzymes, precipitating proteins and killing off any microorganisms. In the next stage, the wort is filtered and transferred to the fermentation vessel where yeast is introduced.

Two species of yeast are commonly used in the brewing process, both belonging to the genus *Saccharomyces*. *S. cerevisiae* is mainly used in the production of darker beers such as traditional English ales and stouts, whereas *S. carlsbergensis* (no prizes for guessing where this one was developed!) gives lighter coloured, less cloudy, lager-type beers. Cells of *S. cerevisiae* are carried to the surface of the fermentation by carbon dioxide bubbles (top fermenters), while *S. carlsbergensis* cells form a sediment at the bottom (bottom fermenters). 'Spent' yeast may be dried, and used as an animal food supplement.

Hops are the dried flowers of a type of vine. They were originally added to beer because of their antimicrobial properties, but were soon found to improve the flavour too! Hops are added to the wort after boiling.

Fermentation takes about a week to complete, at a temperature appropriate for each type of yeast (*S. carlsbergensis* prefers somewhat lower temperatures than *S. cerevisiae*). Following fermentation, the beer is allowed to age or 'rest' for some months in the cold. Beers destined for canning or bottling are filtered to remove remaining microorganisms.

Beers typically have an alcohol content of around 4 per cent. Small amounts of other secondary products such as amyl alcohol and acetic acid are also produced, and contribute to the beer's flavour. 'Light' or low-carbohydrate beers are produced by reducing the levels of complex carbohydrates. The yeast do not possess the enzymes necessary to cope with these branched molecules, so a supplement of debranching enzymes may be added to aid their breakdown.

Dairy products

Milk can be fermented to produce a variety of products, including butter, yoghurt and cheese (Figure 17.3). In each case, acid produced by the fermentation process causes coagulation or curdling of the milk proteins.

In cheese-making, this coagulation is effected by the addition of the protease *rennin,* or by the action of lactic acid bacteria (especially *Streptococcus lactis* and *S. cremoris*). Coagulation allows the separation of the semisolid *curd* from the liquid *whey*. The subsequent steps in the cheese-making process depend on the specific type of cheese (Table 17.2). Following separation, the curd of most cheeses is pressed and shaped, removing excess liquid and firming the texture. During the ripening process, salt is often added, and flavour develops due to continuing microbial action on the protein and fat components of the cheese. In some cases, a fresh inoculation of microorganisms is made at this point, such as the addition of *Penicillium* spores to Camembert and Brie. The length of the ripening period varies from a month to more than a year according to type, with the harder cheeses requiring the longer periods.

Rennin is traditionally derived from the stomachs of calves, but nowadays is more frequently the product of genetically engineered bacteria, allowing its consumption even by strict vegetarians.

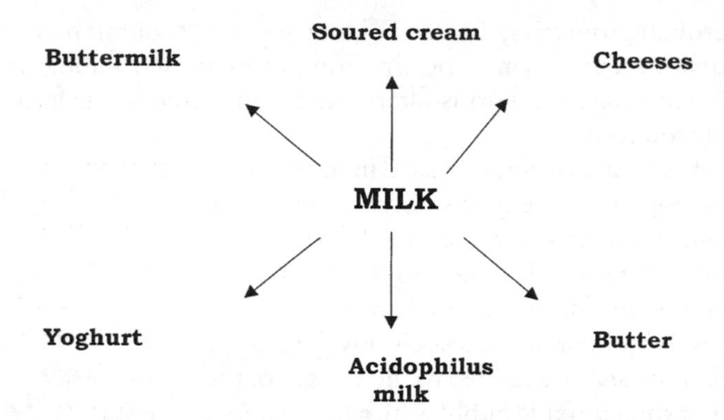

Figure 17.3 Fermented dairy products. Fermentation is initiated by the inoculation of a starter culture of lactic acid bacteria to convert lactose to lactic acid. Heterolactic fermenters such as *Leuconostoc* are added when aromatic flavouring compounds such as diacetyl are required

Yoghurt is another milk derivative. Thickened milk is exposed to the action of two bacteria, *Streptococcus thermophilus* and *Lactobacillus bulgaricus*, both of which ferment lactose present in milk into lactic acid. In addition, *L. bulgaricus* contributes aromatics responsible for imparting flavour to the yoghurt.

Other dairy products, such as soured cream and buttermilk, are also produced by means of the fermentative properties of species of streptococci and lactobacilli.

Bread

The biological agent responsible for bread production is yeast. In fact baker's yeast and brewer's yeast are just different strains of the same species, *Saccharomyces cerevisiae*. In breadmaking, aerobic, rather than anaerobic conditions are favoured, so sugar present in the dough is converted all the way to carbon dioxide rather than to alcohol. It is this

Table 17.2 Types of cheese

Soft	Semisoft	Semihard	Hard
Unripened	Roquefort	Cheddar	Parmesan
Mozzarella	Stilton	Emmentaler	Pecorino
Cottage			
Ripened			
Camembert			
Brie			

Cheeses are classified according to their texture. Unripened cheeses are those that have not undergone the ageing or ripening process, during which additional flavours develop.

Table 17.3 Fermented food products

Product	Source	Fermentative microorganisms
Sauerkraut	Cabbage	*Leuconostoc brevis, Lactobacillus plantarum*
Olives	Olives	*Leuconostoc brevis, Lactobacillus plantarum, Lactobacillus brevis*
Pepperoni	Ground beef, pork	*Lactobacillus plantarum, Pediococcus pentosaceus*
Pickles	Cucumber	*Leuconostoc brevis, Lactobacillus plantarum, Pediococcus* spp.
Soy sauce	Soybean curd	*Aspergillus oryzae, Saccharomyces rouxii, Pediococcus soyae*
Tempeh*	Soybean	*Rhizopus oligosporus*
Kombucha	Tea	*Gluconobacter, Saccharomyces*, etc.
Sake	Rice	*Aspergillus oryzae, Saccharomyces cerevisiae*
Vinegar	Wine, cider	*Acetobacter, Gluconobacter*
Cocoa, chocolate	Cacao beans	*Saccharomyces cerevisiae, Candida rugosa, Acetobacter, Geotrichum*

* Tempeh is a solid fermented soya bean cake that forms an important part of the diet of many Indonesians.

that causes the bread to rise. Any small amount of ethanol that may be produced is evaporated during the baking process.

Many other popular foodstuffs are the result of microbial fermentation processes (see Table 17.3). These include vinegar, soy sauce and sauerkraut. *Silage* is animal fodder made from the fermentation of grass and other plant material by the action of lactic acid bacteria.

Microorganisms *as* food

As we have seen in the previous section, a number of microorganisms are involved in the production of food products. Others, however, *are* foodstuffs! Perhaps the most obvious of these are mushrooms, the stalked fruiting bodies of certain species of basidiomycete (see Chapter 8), notably *Agaricus bisporus*. These are grown in the dark at favourable temperatures, in order to stimulate the production of fruiting bodies. Another fungus, *Fusarium* forms thc basis of Quorn™, a processed mycoprotein that has been used as a meat substitute for some years in the UK. Whereas mushrooms are grown as agricultural products, Quorn™ must be produced under highly regulated sterile conditions. Other microbial food sources include certain algae (seaweed), which form an important part of the diet in some parts of the world, and bacteria and yeast grown in bulk as *single-cell protein* (SCP) for use as a protein-rich animal food supplement. The cyanobacterium *Spirulina* has been collected from dried-up ponds in parts of central Africa for use as a food supplement since time immemorial and is now available at health stores in the West.

The microbial spoilage of food

We have described in previous chapters the nutritional versatility of microorganisms and their role in the global recycling of carbon. Unfortunately for us, fresh foods such as meats, fruit and vegetables provide a rich source of nutrients, which a wide range of heterotrophic microorganisms find just as attractive as we do. Certain microbial types are associated with particular foodstuffs, depending on their chemical composition and physical factors such as pH and water content. Acidic foods such as fruits, for example, tend to favour the growth of fungi rather than bacteria.

Often, spoilage organisms come from the same source as the food, for example soil on vegetables, or meat exposed to intestinal contents following slaughter. Others are introduced as contaminants during transport, storage or preparation. Among the most commonly found spoilage organisms are a number of human pathogens, including *Pseudomonas, Salmonella, Campylobacter* and *Listeria*. Thus, although microbial spoilage may merely lead to foodstuffs being rendered unpalatable, it can also result in serious and even fatal illness ('food poisoning'). Whilst observable changes to foodstuffs are only likely after the microbial population has reached a considerable size, food poisoning can result from the presence of much smaller numbers of contaminants.

Some foodstuffs are more susceptible to spoilage than others: fresh items such as meat, fish, dairy produce and fruit and vegetables are all highly perishable. Foods such as rice and flour, on the other hand, are much more resistant, because having no water content they do not provide suitable conditions for microbial growth. Drying is one of a number of methods of food preservation, all designed to prevent growth of microorganisms by making conditions unfavourable. Other methods include heating/canning, drying, pickling, smoking and, in many countries, irradiation.

Microorganisms in the production of biochemicals

Many products of microbial metabolism find an application in the food and other industries. These include amino acids, steroids, enzymes and antibiotics (Table 17.4). Microbial growth conditions are adjusted so that production of the metabolite in question takes place at an optimal rate. Often an unnaturally high rate of production is achieved by the use of a mutated or genetically engineered strain of microorganism, or by manipulating culture conditions to favour excess metabolite production.

The development of a microbial means of producing acetone was vital to the allied effort in the First World War. Acetone was a crucial precursor in explosives manufacture and the demands of war soon outstripped supply by traditional methods. The problem was solved when Chaim Weismann isolated a strain of *Clostridium acetobutylicum* that could ferment molasses to acetone and butanol (another industrially useful product). Nowadays, acetone is made more cheaply from petrochemicals.

Microbially produced amino acids are used in the food industry, in medicine and as raw materials in the chemical industry. The one produced in the greatest quantities by far is glutamic acid (in excess of half a billion tonnes per year), with most of it ending up as the flavour enhancer monosodium glutamate. The amino acids aspartic acid and phenylalanine are components of the artificial sweetener aspartame and are also synthesised on a large scale.

Table 17.4 Some industrial applications of microbially produced enzymes

Industry	Enzyme	Application
Food & drink	Rennin	Cheese manufacture
	Lipase	
	Pectinase	Fruit juice production
		Coffee bean extraction
	Amylase	Improved bread dough quality
		Haze removal in beer
	Amylase	Fructose syrup production
	Glucoamylase	
	Glucose isomerase	
Animal feed	Amylase	Improved digestibility
	Cellulase	
	Protease	
Detergent	Protease	Stain and grease removal
	Lipase	
	Amylase	
	Cellulase	Fabric softener
Paper	Cellulase	Pulp production
Textile	Cellulase	'Stone-washed' jeans
Leather	Protease	Dehairing, softening, fat
	Lipase	removal
Molecular biology	*Taq* polymerase	Polymerase chain reaction

A number of organic acids are produced industrially by microbial means, most notably citric acid, which has a wide range of applications in the food and pharmaceutical industries. This is mostly produced as a secondary metabolite by the large-scale culture of the mould *Aspergillus niger*.

Certain microorganisms serve as a ready source of *vitamins*. In many cases these can be synthesised less expensively by chemical means; however, riboflavin (by the mould *Ashbya gossypii*) and vitamin B_{12} (by the bacteria *Propionibacterium shermanii* and *Pseudomonas denitrificans*) are produced by large-scale microbial fermentation. Microorganisms play a partial role in the production of ascorbic acid (vitamin C). Initially, glucose is reduced chemically to sorbitol, which is then oxidised by a strain of *Acetobacter suboxydans* to the hexose sorbose. Chemical modifications convert this to ascorbic acid (Figure 17.4).

Enzymes of fungal and bacterial origin have been utilised for many centuries in a variety of processes. It is now possible to isolate and purify the enzymes needed for a specific process and the worldwide market is currently worth around a billion pounds. The most useful industrial enzymes include proteases, amylases, lipases

If you wear contact lenses, you have probably used an enzyme preparation to remove protein deposits.

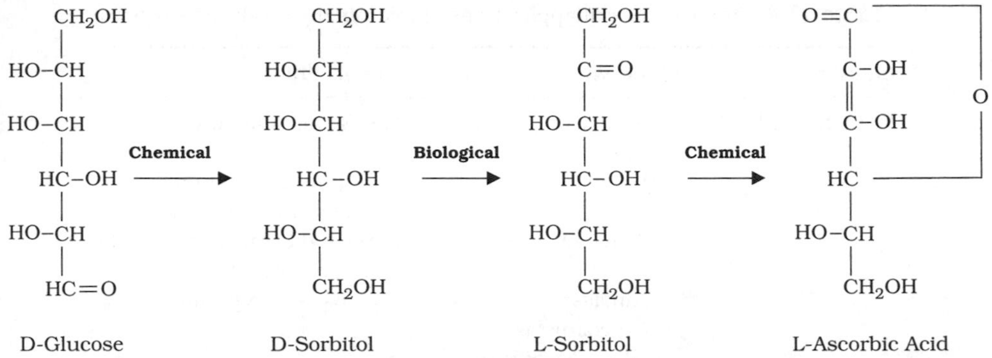

D-Glucose —Chemical→ D-Sorbitol —Biological→ L-Sorbitol —Chemical→ L-Ascorbic Acid

Figure 17.4 Ascorbic acid is produced by a combination of chemical and microbial reactions Most steps in the synthesis of ascorbic acid are purely chemical, but the conversion of sorbitol to sorbose is carried out by the sorbitol dehydrogenase enzyme of *Acetobacter suboxydans*, with NAD^+ acting as an electron acceptor

and pectinase (Table17.4). Some applications of enzymes are listed in Table 17.5, and two examples are briefly described below.

Syrups and modified starches are used in a wide range of foodstuffs, including soft drinks, confectionery and ice cream, as well as having a wealth of other applications. Different enzymes or combinations of enzymes are used to produce the desired consistencies and physical properties. High fructose corn syrup (HFCS) is a sweetener used in a multitude of food products. It is some 75 per cent sweeter than sucrose and has several other advantages. HFCS is a mixture of fructose, dextrose (a form of glucose) and disaccharides, and is produced by the action of a series of three enzymes on the starch (amylose and amylopectin) of corn (maize). Alpha amylase hydrolyses the internal α-1, 4-glycosidic bonds of starch, but is not able to degrade ends of the chain. The resulting di- and oligosaccharides are broken down to the monomer glucose by the action of glucoamylase, then finally glucose isomerase converts some of the glucose to its isomer, fructose.

Enzymes have been added to cleaning products such as washing powders, carpet shampoos and stain removers since the 1960s, and this remains one of the principal

Table 17.5 Commercially useful products of microbial metabolism

Product	Use
Amino acids:	
Glutamic acid	Flavour enhancer
Lysine	Animal feed additive
Aspartic acid & phenylalanine	Artificial sweetener (aspartame)
Citric acid	Antioxidant, flavour enhancer, emulsifier
Enzymes	Numerous – see Table 17.4
Antibiotics	Treatment of infectious diseaases
Vitamins	Dietary supplements
Steroids	Anti-inflammatory drugs, oral contraceptives

industrial applications of enzymes. *Proteases* are the most widely used enzymes in this context; working in combination with a surfactant, they hydrolyse protein-based stains such as blood, sweat and various foods. Greasy and oily stains present a different challenge, made all the more difficult by the move towards lower washing temperatures. The inclusion of *lipases* aids the removal of stains such as butter, salad dressing and lipstick, while *amylases* deal with starch-based stains such as cereal or custard. The food and detergent industries between them account for around 80 per cent of all enzyme usage.

We have already seen in Chapter 14 that antibiotics are now produced on a huge scale worldwide. Figure 17.5 outlines the stages in the isolation, development and production of an antibiotic.

Isolating an antibiotic from a natural source is not all that difficult, but finding a new one that is therapeutically useful is another matter. Initially, the antimicrobial properties of a new isolate are assessed by streaking it across an agar plate, then inoculating a range

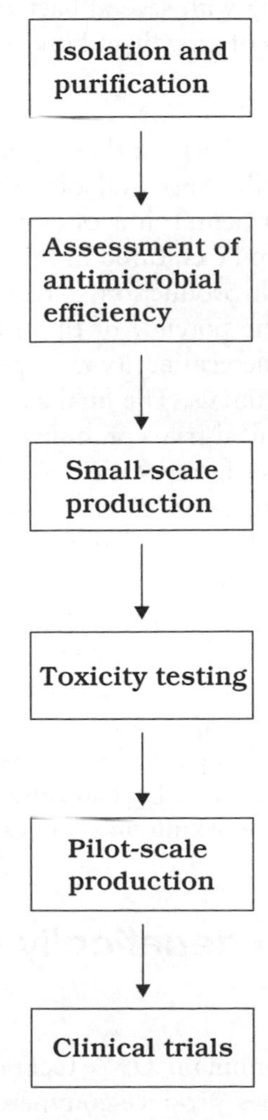

Figure 17.5 Stages in the isolation and development of an antibiotic. See the text for details

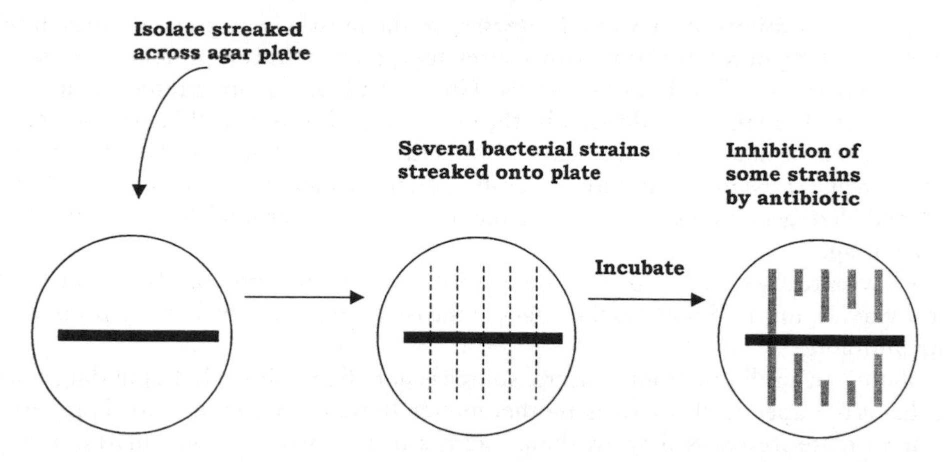

Figure 17.6 Assessing the antimicrobial properties of an antibiotic. Candidate antibiotic is streaked onto an agar plate along with several bacterial isolates. Following incubation, areas of clearing indicate inhibition of growth and thus susceptibility to the antibiotic

of bacteria at right angles (Figure 17.6). As the antibiotic diffuses through the agar, it will inhibit growth of any susceptible species. Isolates that still show potential are then grown up in a laboratory scale fermenter; it is essential for commercial culture that the antibiotic-producing organism can be cultured in this way.

Before committing to large-scale production, exhaustive further tests must be carried out on two fronts: to ascertain the potency of the preparation and the breadth of its antimicrobial spectrum, and to determine its *therapeutic index* (see Chapter 14) by carrying out toxicity testing on animals. The final stages of development involve pilot-scale production, followed by clinical trials on human volunteers.

When an antibiotic or any other fermentation product finally goes into production, it is cultured in huge stirred fermenters or *bioreactors*, which may be as large as 200 000 litres. A typical stirred fermenter has impellers for mixing the culture, an air line for aeration and microprocessor-controlled probes for the continuous monitoring and regulation of temperature, pH and oxygen content (Figure 17.7). Cultures with a high protein content may also have an antifoaming agent added. The process of *scale-up* is a complex operation, and not simply a matter of growing the microorganism in question in ever-larger vessels. Factors such as temperature, pH, aeration, must all be considered at the level of the individual cell if scale-up is to be successful. Fermenters are usually made from stainless steel, which can withstand heat sterilisation; the economic consequences of microbial contamination when working on such a large scale can be immense.

Products derived from genetically engineered microorganisms

In Chapter 12 we saw how recombinant DNA technology can be used to genetically modify microorganisms so that they produce commercially important proteins such as human insulin. This is done by incorporating the gene for the desired protein into an

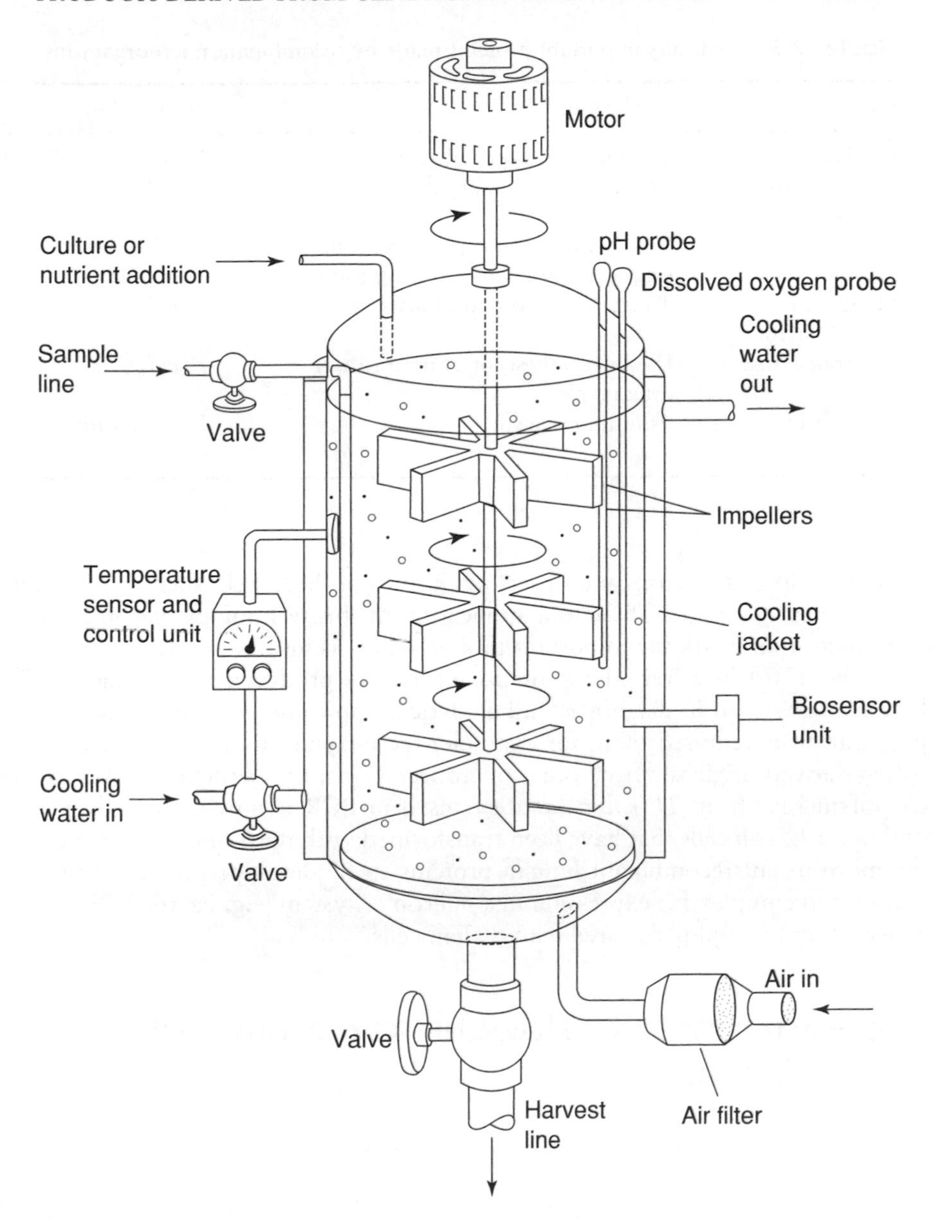

Figure 17.7 A continuous flow stirred tank reactor. Parameters such as pH and concentrations of specific metabolites are closely monitored to ensure the maintenance of optimum conditions. Outlets allow for the collection of samples during fermentation as well as the collection of cells and medium at the conclusion of the reaction. Addition and collections are carried out under aseptic conditions. From Prescott, LM, Harley, JP & Klein, DA: Microbiology 5th edn, McGraw Hill, 2002. Reproduced by permission of the publishers

Table 17.6 Medically important proteins made by recombinant microorganisms

Protein	Application	Produced in
Insulin	Treatment of Type 1 diabetes	*E. coli*
Human growth hormone	Treatment of pituitary dwarfism	*E. coli*
Hepatitis B vaccine	Vaccination of susceptible personnel e.g. healthcare workers, drug users	*Saccharomyces cerevisiae*
Epidermal growth factor	Treatment of wounds, burns	*E. coli*
Acyltransferase	Used in synthesis of ovarian cancer drug taxol	*E. coli*
Endostatin	Antitumour agent	*Pichia pastoris* (yeast)

appropriate cloning vector, and inserting it into a host cell such as *E. coli* or *Saccharomyces cerevisiae*. The initial application of this technology was in the microbial production of medically important proteins such as insulin and epidermal growth factor (Table 17.6), however other proteins may also be produced by these means. These include enzymes used in diagnostic and analytical applications, where a higher purity of preparation is required than, for example, the enzymes used in detergents. These are often derived originally from other microorganisms; for example the thermostable DNA polymerase from *Thermus aquaticus* used in PCR is now commonly made by recombinant *E. coli* cells that have been transformed with the *T. aquaticus* gene. Many of the more recent recombinant human proteins to be developed for therapeutic use have been too complex for expression in a microbial system (e.g. Factor VIII), so it has been necessary to employ cultured mammalian cells.

Microorganisms in wastewater treatment and bioremediation

These applications of microbial processes in an environmental context are discussed in Chapter 16.

Microorganisms in the mining industry

An unexpected application for microorganisms is to be found in the mining industry. Acidophilic bacteria including *Acidithiobacillus ferrooxidans* are increasingly being used to extract valuable metals, notably copper, from low-grade ores that would not be worth working by conventional technologies. You may recall from Chapter 16 that *A. ferrooxidans* is the organism largely responsible for the phenomenon of acid mine drainage; by carrying out the same reactions in a different context, however, it can be put to a beneficial use. Tailings, that is, mineral waste with a low metal content, are

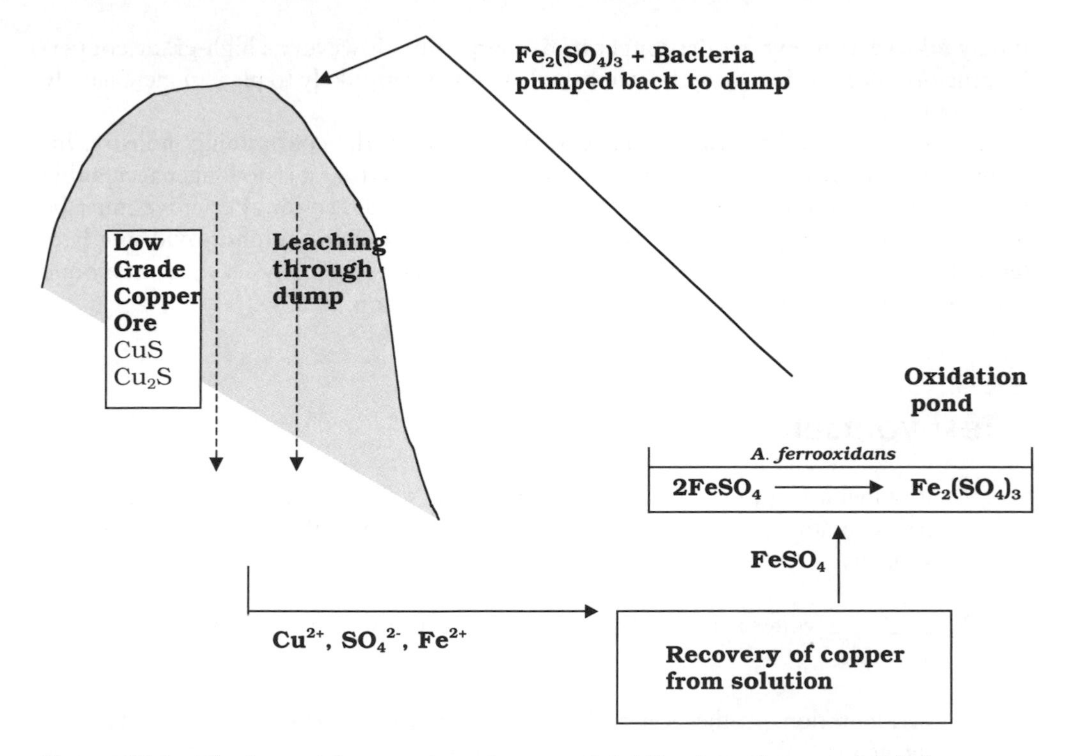

Figure 17.8 The bacterial extraction of copper. Solubilisation of copper sulphides occurs by a combination of direct (biological) and indirect (chemical) leaching. The ferric iron necessary for chemical oxidation is produced by bacterial oxidation of ferrous iron in the oxidation pond

gathered in huge tips and acidified water sprinkled over them (Figure 17.8), stimulating the growth of indigenous bacteria. Bacterial oxidation results in soluble copper sulphate leaching from the tip and being collected for copper extraction. This bacterial action is known as *direct bioleaching*, but if you follow the process in Figure 17.8, you will see that *A. ferrooxidans* has not finished yet! This remarkable organism can also oxidise iron from its ferrous to ferric form; the resulting ferric sulphate is a potent oxidising agent, which, when recycled to the tip, carries out *indirect* (chemical) *bioleaching*, and so the cycle continues. *A. ferrooxidans* has a number of other unusual features, which enable it to survive in this hostile environment; it thrives in acidic conditions ($pH < 2.0$), and has an unusually high tolerance of metal ions such as copper.

Bacteria are also involved in the extraction of other metals such as uranium and gold; the methodologies differ slightly, but still involve the conversion of an insoluble compound to a soluble one. It is only in the last 20 years or so that the economic possibilities of *biohydrometallurgy* have been realised, and now a significant proportion of the world's copper and other metals is produced in this way. The method is inexpensive but rather slow;

Other metals, including zinc, nickel uranium and gold have also been extracted with the aid of bacteria.

it may take years to extract the copper from a large tip. However as high-grade copper-bearing ores become increasingly scarce, bioleaching seems likely to play an increasingly important role.

Sulphur-oxidising bacteria also have a role to play in the coal mining industry. Increased environmental awareness in many countries means that it is no longer acceptable to burn off the sulphur content of coal as sulphur dioxide, so an alternative must be found. One possibility is the *biodesulphurisation* of coal, using sulphur-oxidising bacteria to remove the sulphur before combustion. Whilst technically feasible, economic considerations mean that this has not yet been widely adopted.

Test yourself

1 In winemaking, juice from the crushed grapes (known as the ________) is treated with ________ ________ to remove any naturally occuring microorganisms.

2 ________ wines result from converting most of the sugar from the grapes to alcohol.

3 Fermentations in the beer and wine industries are carried out by yeasts of the genus ________.

4 Hops were originally added to beers because of their ________ properties.

5 The alcohol content of an average beer is around ________ per cent.

6 Coagulation of milk using rennet allows separation into ________ and ________.

7 ________ cheeses, such as Parmesan, require the longest periods of ________.

8 The bacteria ________ ________ and ________ ________ are involved in yoghurt making, converting lactose to ________ ________.

9 Quorn™ is a meat substitute derived from the fungus ________.

10 The end-product of sugar fermentation in bread making differs from that in alcoholic fermentations beacuse it takes place in ________ conditions.

11 ________ and ________ are two human pathogens found as spoilage organisms in foodstuffs.

12 Foodstuffs with a low ________ content are less susceptible to microbial attack.

13 ________ produced microbially from by a species of *Clostridium*, was important in explosive manufacture during the First World War.

14 The three amino acids produced in greatest quantities for use in the food industry are ________ ________, ________ ________ and ________.

15 Citric acid is widely used in the food industry; it is produced by the mould ________ ________.

16 Starch is converted to ________ ________ ________ ________ by the action of the enzymes α-amylase, glucoamylase and glucose isomerase.

17 The enzymes most widely used in cleaning products are ________ and ________.

18 Microbial cultures for the commercial production of antibiotics are grown in large-scale ________.

19 ________ ________ is able to convert copper from its insoluble ________ to its soluble ________. It can also convert iron from its ________ to its ________ from.

20 The biggest disadvantage of using biological methods of metal extraction is that the process is ________.

Glossary

Acid–fast stain a procedure for assessing the ability of an organism to retain hot carbol fuchsin stain when rinsed with acidic alcohol.

Acidophilic ('acid-loving'): a term applied to organisms that show optimal growth in acid conditions ($pH<5.5$).

Activated sludge treatment a method of wastewater treatment involving aeration in tanks that have been seeded with a mixed microbial sludge.

Activation energy the energy required to initiate a chemical reaction.

Active site the part of an enzyme involved in binding its substrate; the site of catalytic action.

Active transport an energy-requiring process in which a substance is transported against an electrochemical gradient.

Adenosine triphosphate (ATP) the principal compound used for the storage and transfer of energy in cellular systems.

Aerobe an organism that grows in the presence of molecular oxygen, which it uses as a terminal electron acceptor in aerobic respiration.

Aerotolerant anaerobe an anaerobe that is able to tolerate the presence of oxygen, even though it does not use it.

Aetiology the cause or origin of a disease.

Aldose a sugar molecule that contains an aldehyde group.

Alga a photosynthetic, eucaryotic plant-like organism. It may be unicellular or multicellular.

Alternation of generations a pattern of sexual reproduction that includes both haploid and diploid mature individuals.

Ames test a test to assess the mutagenicity of a substance.

Amino acid an organic acid bearing both amino and carboxyl groups. The building block of proteins.

Amphipathic having a polar region at one end and a nonpolar region at the other.

Anabolism metabolic reactions involved in the synthesis of macromolecules, usually requiring an input of energy.

Anaerobe an organism that grows in the absence of molecular oxygen.

Anammox the formation of nitrogen gas by the anaerobic oxidation of ammonia and nitrite.

Angstrom unit one ten billionth (10^{-10}) of a metre.

Anisogamy the fusion of unequally sized gametes.

Anoxygenic photosynthesis a form of photosynthesis in which oxygen is not generated; found in the purple and green photosynthetic bacteria.

Antibiotic a microbially produced substance (or a synthetic derivative) that has antimicrobial properties.

Antibody a protein of high binding specificity produced by the immune systems of higher animals in response to infection by a foreign organism.

Anticodon the three-nucleotide sequence carried by a tRNA, that base-pairs with its complementary mRNA codon.

Antigenic shift a process by which major variations in viral antigens occur.

Antiseptic a chemical agent of disinfection that is mild enough to be used on human skin or tissues.

Archaea a group of prokaryotes that diverged from all others (see **Bacteria**) at an early stage in evolution, and that show a number of significant differences from them. One of the three domains of life.

Ascospore a haploid spore produced by members of the Ascomycota.

Ascus a sac-like structure possessed by members of the Ascomycota that contains the ascospores.

Aseptic technique a set of practical measures designed to prevent the growth of unwanted contaminants from the environment.

Assimilatory sulphate reduction the incorporation of inorganic sulphate into sulphur-containing amino acids.

Autoclave an appliance that uses steam under pressure to achieve sterilisation.

Atomic mass the average of the mass numbers of an element's different isotopes, taking into account the proportions in which they occur.

Atomic number the number of protons in the nucleus of an element.

Autotroph an organism that can derive its carbon from carbon dioxide.

Auxotroph a mutant that lacks the ability to synthesise an important nutrient such as an amino acid or vitamin, and must therefore have it provided (c.f. **prototroph**).

Axenic culture a pure culture containing one type of organism only, and completely free from contaminants.

Bacillus (pl. bacilli): a rod-shaped bacterium.

Bacteria procaryotes other than the Archaea (q.v.). Less formally, the term is frequently used to describe all procaryotes.

Bactericidal causing the death of bacteria.

Bacteriophage a virus whose host is a bacterial cell.

Bacteriostatic inhibiting the growth of bacteria, but not necessarily killing them.

Basidiocarp the fruiting body of members of the Basidiomycota.

Basidiospore a haploid spore produced by members of the Basidiomycota.

Basidium a club-shaped structure carrying the basidiospores in members of the Basidiomycota.

Batch culture a microbial culture grown in a closed vessel with no addition of nutrients or removal of waste products.

Benthic zone the sediment of mud and organic matter at the bottom of a pond or lake.

Beta-lactamase an enzyme that breaks a bond in the β-lactam core of the penicillin molecule.

Bioaugmentation the deliberate introduction of specific microorganisms into an environment in order to assist in bioremediation.

Biochemical oxygen demand (BOD) the amount of oxygen needed by microorganisms to oxidise the organic content of a water sample.

Biodegradation the natural processes of breakdown of organic matter by microorganisms.

Biodeterioration the damage caused to materials of economic importance due to biological (mainly microbial) processes.

Biofilm a complex system of microorganisms and their surrounding polysaccharide matrix.

Biomass the total amount of cellular material in a system.

Bioreactor a fermentation vessel for the controlled growth of microoganisms.

Bioremediation the use of biological processes to improve a specific environment, such as by the removal of a pollutant.

Calvin cycle a pathway for the fixation of carbon dioxide, used by photosynthetic organisms and some chemolithotrophs.

Capsid the protein coat of a virus particle.

Capsomer a protein subunit of a viral capsid.

Capsule a clearly defined polysaccharide layer surrounding the cell of certain procaryotes.

Carcinogen an agent capable of causing cancer.

Catabolism those reactions that break down large molecules, usually coupled to a release of energy.

Catabolite repression the mechanism by which the presence of a preferred nutrient (generally glucose) has the effect of preventing the synthesis of enzymes that metabolise other nutrients.

Central Dogma of biology the proposal that information (= genetic) flow in organisms is in one direction only, from DNA to RNA to protein.

Centromere the central region of the chromosome that ensures correct distribution of chromosomes between daughter cells during cell division.

Chemotroph an organism that obtains its energy from chemical compounds.

Chloroplast a chlorophyll-containing organelle found in photosynthetic eucaryotes.

Chromosome a nuclear structure on which most of a eucaryotic cell's genetic information (DNA) is carried, in association with specialised proteins called histones. The **nucleoid** of procaryotes (q.v.) is also often referred to as a chromosome.

Cilium (pl. cilia) a short motile hair-like structure found on the surface of some eucaryotic cells.

Citric acid cycle see **Tricarboxylic acid cycle.**

Clamp connection a mechanism, unique to members of the Basidiomycota, for ensuring the maintenance of the dicaryotic state.

Cloning the production of multiple copies of a specific DNA molecule. The term is also used to describe the production of genetically identical cells or even organisms.

Coccus (pl. cocci) a spherically-shaped bacterium.

Codon a sequence of three nucleotide bases that corresponds to a specific amino acid.

Coenocytic containing many nuclei within a single plasma membrane.

Coenzyme a loosely-bound organic cofactor that influences the activity of an enzyme.

Cofactor a non-protein component of an enzyme (often a metal ion) essential for its normal functioning.

Commensal an organism that lives in or on another organism, deriving some benefit from the association but not harming the other party.

Commensalism an association between two species from which one participant (the commensal) derives benefit, and the other is neither benefited nor harmed.

Competence (of a bacterial cell): the state of being able to take up naked DNA from outside the cell.

Compound a substance comprising the atoms of two or more elements.

Conidiophore an aerial hypha that bears conidia.

Conidium (conidiospore) an asexual spore, found in members of the Ascomycota and Actinomycetes. Often forms chains.

Conjugation a process of genetic transfer involving intimate contact between cells and direct transfer of DNA across a sex pilus.

Consumer a heterotroph that derives energy from the consumption of other organisms.

Continuous culture a microbial culture in which nutrient concentrations and other conditions are kept constant by the addition of fresh medium and the removal of old.

Contractile vacuole a fluid-filled vacuole involved in the osmoregulation of certain protists.

Cosmid a hybrid cloning vector capable of accommodating inserts of up to 50kb.

Covalent bond a bond formed by the sharing of a pair of electrons between atoms.

Cyanobacteria a group of mostly unicellular procaryotes that carry out oxygenic photosynthesis. Commonly known as the blue-greens.

Decimal reduction time (D value) the time needed to reduce a cell population by a factor of ten (i.e. to kill 90% of the population) using a particular heat treatment.

Decomposer an organism that breaks down the remains and waste products of producers and consumers.

Defined medium a medium whose precise chemical composition is known.

Denitrification the reduction, under anaerobic conditions, of nitrite and nitrate to nitrogen gas.

Diauxic growth a form of growth that has two distinct phases due to one carbon source being used preferentially before a second.

Dicaryon a structure formed by two cells whose contents, but not nuclei, have fused.

Differential medium a medium that allows colonies of a particular organism to be differentiated from others growing in the same culture.

Differential stain a stain that employs two or more dyes to distinguish between different cellular structures or cell types.

Dimorphic existing in two distinct forms.

Dioecious having male and female reproductive structures on separate individuals.

Diploid having two sets of chromosomes.

Disaccharide a carbohydrate formed by the joining of two monosaccharides.

Disc diffusion method a method for assessing the antimicrobial properties of a substance.

Disinfection the elimination or inhibition of pathogenic microorganisms in or on an object so that they no longer pose a threat.

Dissimilatory sulphate reduction the reduction of sulphate to hydrogen sulphide by obligate anaerobes, using either an organic compound or hydrogen gas as electron donor.

DNA library a collection of cloned DNA fragments.

Domain the highest level of taxonomic grouping.

Eclipse period the stage of a viral replication cycle in which no complete viral particles are present in the infected host cell.

Ecosystem the organisms of a particular habitat, together with their inanimate surroundings.

Ectoparasite a parasite that attaches to the outside of its host.

Electron a subatomic particle carrying a negatively charge.

Electron transport chain a series of donor/acceptor molecules that transfer electrons from donors (e.g. NADH) to a terminal electron acceptor (e.g. O_2).

Embden Meyerhof pathway see **Glycolysis.**

Endoparasite a parasite that fully enters its host and lives inside it.

Endoplasmic reticulum a tubular network found in the cytoplasm of eucaryotic cells.

Endospore a highly resistant spore found within certain bacteria.

Enrichment culture a culture that uses a selective medium to encourage the growth of an organism present in low numbers.

Entner-Doudoroff Pathway an alternative pathway for the oxidation of glucose, producing a mixture of pyruvate and glyceraldehyde-3-phosphate.

Enzyme a cellular catalyst (usually protein), specific to a particular reaction or group of reactions.

Epiphyte an organism that grows on the surface of a plant.

Eucaryote an organism whose cells contain a true nucleus and membrane-bound organelles such as mitochondria and endoplasmic reticulum.

Excision repair a repair mechanism in which damaged sections of DNA are cut out and replaced.

Exon a coding region of a gene. c.f. **intron**

Expression vector a vector that allows the transcription and translation of a foreign gene inserted into it.

Facilitated diffusion the transport of molecules across a membrane with the help of carrier proteins. Transport takes place down a concentration gradient, and energy is not required.

Facultative anaerobe an organism that can grow in the absence of oxygen, but utilises it when available.

Fastidious (of an organism): unable to synthesise a range of nutrients and therefore having complex requirements in culture.

Fatty Acid a long-chain hydrocarbon chain with a carboxyl group at one end.

Feedback inhibition a control mechanism whereby the final product of a metabolic pathway acts as an inhibitor to the enzyme that catalyses an early step (usually the first) in the pathway.

Fermentation a microbial process by which an organic substrate (usually a carbohydrate) is broken down without the involvement of oxygen or an electron transport chain, generating energy by substrate-level phosphorylation

Flagellum a long hair-like extracellular structure associated with locomotion. Found in both procaryotes and eucaryotes, although each has its own distinctive structure.

F plasmid a plasmid containing genes that code for the construction of the sex pilus, across which it is transferred to a recipient cell.

Frameshift mutation a mutation that results in a change to the reading frame, and thus an altered sequence of amino acids downstream of the point where it occurs.

Fungi a kingdom of non-photosynthetic eucaryotes characterised by absorptive heterotrophic nutrition.

Gamete a haploid reproductive cell arising from meiosis. Fuses with another gamete to form a diploid zygote.

Gametophyte the haploid, gamete-forming stage in a life cycle with alternation of generations.

Gene a sequence of DNA that usually encodes a polypeptide.

Generalised transduction the transfer of a gene from one bacterial cell to another as a result of being inadvertently packaged into a phage particle.

Generation (doubling) time the time taken for a population of cells to double in time under specific conditions.

Genetic code the 64 triplet codons and the corresponding amino acids or termination sequences.

Genome the complete genetic material of an organism.

Genotype the genetic make-up of an organism.

Gluconeogenesis a series of reactions by which glucose is synthesised from compounds such as amino acids and lactate.

Glycolysis a series of reactions by which glucose is oxidised to two molecules of pyruvate with the synthesis of two molecules of ATP.

Glycosidic linkage a covalent linkage formed between monosaccharides.

Golgi apparatus an organelle of eucaryotes, comprising flattened membranous sacs.

Gram stain a differential stain that divides bacteria into Gram-positive (purple) or Gram-negative (pink). The reaction depends on the constitution of the organism's cell wall.

Group translocation a form of active transport in which a solute is modified to prevent its escape from the cell.

Haploid having only one set of chromosomes.

Haloduric able to tolerate high salt concentrations.

Halophilic showing a requirement for moderate to high salt concentrations.

Heterotroph an organism that must use one or more organic molecules as its source of carbon.

Hexose a six-carbon sugar.

Hexose monophosphate shunt see **Pentose phosphate pathway**

Hfr cell a bacterial cell that has a transferred F plasmid integrated into its chromosome.

Histone a basic protein found associated with DNA in eucaryotic chromosomes.

Horizontal transfer the transfer of genetic material between members of the same generation.

Human genome project an international effort to map and sequence all the DNA in the human genome. The project has also involved sequencing the genomes of several other organisms.

Hydrogen bond a relatively weak bond that forms between covalently bonded hydrogen and any electronegative atom, most commonly oxygen or nitrogen.

Hydrophilic ('water-loving'): having an affinity for water.

Hydrophobic ('water-hating'): repelled by water.

Hypha a thread-like filament of cells characteristic of fungi and actinomycetes.

Immersion oil a viscous oil used to improve the resolution of a light microscope at high power.

In vitro = 'in glass', i.e. outside of the living organism, in test tubes etc.

In vivo = 'in life', i.e. within the living organism.

Inoculum the cells used to 'seed' a new culture.

Insertion vector a cloning vector based on phage λ that has nonessential genes removed.

Insertional inactivation the insertion of foreign DNA within the gene sequence of a selectable marker in order to inactivate its expression.

Interrupted mating a technique in which the order of gene transfer during conjugation is determined by terminating transfer at different time intervals.

Intron a non-coding sequence within a gene. c.f. **exon**

Ion an atom or group of atoms that carries a charge due to losing or gaining one or more electrons.

Isogamy the fusion of morphologically identical motile gametes.

Isotope forms of an element, having the same number of protons and electrons but differing in the number of neutrons

Karyogamy the fusion of nuclei from two different cells.

Ketose a sugar molecule that contains a ketone group.

Kinetoplast a specialised structure within the mitochondria of certain flagellated protozoans.

Koch's postulates a set of criteria, proposed by Robert Koch, which must be satisfied in order to link a specific organism to a specific disease.

Krebs cycle see **Tricarboxylic acid cycle.**

Latent period the period in a viral replication cycle between infection of the host and release of newly synthesised viral particles.

Latent virus a virus that remains inactive in the host for long periods before being reactivated.

Lichen a symbiotic association of a fungus (usually an ascomycete) and an alga or cyanophyte (blue-green).

Limnetic zone a zone of a body of water away from the shore, and where light is able to penetrate.

Lipase an enzyme that digests lipids.

Lithotroph an organism that uses inorganic molecules as a source of electrons.

Littoral zone the region of a body of water situated close to land where the water is sufficiently shallow for sunlight to penetrate to the bottom.

Lysogeny a form of bacteriophage replication in which the viral genome is integrated into that of the host and is replicated along with it. See **prophage**.

Lytic cycle a process of viral replication involving the bursting of the host cell and release of new viral particles.

Magnetosome a particle of magnetite (iron oxide) found in certain bacteria, allowing them to orient themselves in a magnetic field.

Malting the stage in beer-making in which grain is soaked in water to initiate germination and activate starch-digesting enzymes.

Mashing the stage in beer-making in which soluble material is released from the grain in preparation for fermentation.

Mass number the combined total of protons and neutrons in the nucleus of an element.

Meiosis a form of nuclear division in diploid eucaryotic cells resulting in haploid daughter nuclei.

Merodiploid a genome that is partly haploid and partly diploid.

Mesophile an organism that grows optimally at moderate temperatures (20–45°).

Messenger RNA (mRNA) a form of RNA that is synthesised as a complementary copy of the template strand of DNA.

Methanotroph a bacterium capable of using methane as a carbon and energy source.

Michaelis constant (K_m) the concentration of substrate in an enzyme reaction that results in a rate of reaction equal to one half of the maximum.

Mineralisation the breakdown of organic matter to carbon dioxide and inorganic compounds.

Minimum inhibitory concentration (MIC) the lowest concentration of an antimicrobial substance that prevents growth of a given organism.

Mismatch repair a DNA repair system that replaces incorrectly inserted nucleotides.

Missense mutation a mutation that alters the sense of the message encoded in the DNA, resulting in an incorrect amino acid being incorporated.

Mitochondrion a spherical to ovoid organelle found in eucaryotes. The site of ATP generation via TCA cycle and oxidative phosphorylation.

Mitosis a form of nuclear division in eucaryotic cells, resulting in daughter nuclei each with the same chromosome complement as the parent.

Monoecious having reproductive structures of both sexes on the same individual.

Monosaccharide the simplest from of carbohydrate molecule.

Most probable number (MPN) a statistical method of estimating microbial numbers in a liquid sample based upon the highest dilution able to support growth.

mRNA see **Messenger RNA**

Mutagen a chemical or physical agent capable of inducing mutations.

Mutation any heritable alteration in a DNA sequence. It may or may not have an effect on the phenotype.

Mutualism an association between two species from which both participants derive benefit.

Mycelium a tangled mass of branching hyphae.

Mycorrhiza a mutualistic relationship between a fungus and a plant root.

Mycosis a disease caused by a fungus.

Neutron a subatomic particle carrying neither positive nor negative charge.

Nitrification the two-step process by which ammonia is oxidised to nitrite and then to nitrate.

Nonsense codon see **stop codon.**

Nonsense mutation a mutation that results in a 'stop' codon being inserted into the mRNA at the point where it occurs, and the premature termination of translation.

Nosocomial infection an infection that is acquired in hospital or other healthcare setting.

Nuclear membrane the double membrane surrounding the nucleus of a eucaryotic cell.

Nucleocapsid the genome of a virus and its surrounding protein coat.

Nucleoid another name for the bacterial chromosome. The site of most of a procaryotic cell's DNA.

Nucleolus a discrete region of the eucaryotic nucleus, where ribosomes are assembled.

Nucleotide the building block of nucleic acids, comprising a pentose sugar, a nitrogenous base and one or more phosphate groups.

Nucleus the central, membrane-bound structure in eucaryotic cells that contains the genetic material. Also, the region of an atom that contains the protons and neutrons.

Obligate anaerobe an organism that is incapable of growth in the presence of oxygen.

Okazaki fragments discontinuous fragments of single-stranded DNA synthesised complementary to the lagging strand during DNA replication.

Oncogene a gene associated with the conversion of a cell to a cancerous form.

Oogamy the fusion of a small, motile sperm cell and a larger, immobile egg cell.

Operon a group of related genes under the control of a single operator sequence.

Organotroph an organism that uses organic molecules as a source of electrons.

Outer membrane the outermost part of the Gram-negative cell wall, comprising phospholipids and lipopolysacchride.

Oxidation a chemical reaction in which an electron is lost.

Oxygenic photosynthesis a form of photosynthesis in which oxygen is produced; found in algae, cyanobacteria (blue greens) and also green plants.

Parasitism an association between two species from which one partner derives some or all of its nutritional requirements by living either in or on the other (the host), which usually suffers harm as a result.

Pasteurisation a mild heating regime used to destroy pathogens and spoilage organisms present in food and drink, especially milk.

Pathogen an organism with the potential to cause disease.

Pellicle a semi-rigid structure composed of protein strips found surrounding the cell of certain unicellular protozoans and algae.

Pentose a five-carbon sugar.

Pentose phosphate pathway a secondary pathway for the oxidation of glucose, resulting in the production of pentoses that serve as precursors for nucleotides.

Peptide bond the bond formed between the amino group of one amino acid and the carboxyl group of another.

Peptidoglycan a polymer comprising alternate units of N-acetylmuramic acid and N-acetylglucosamine that forms the major constituent of bacterial cell walls.

Phagemid a hybrid cloning vector, comprising elements of plasmid and phage.

Phagocytosis the ingestion and digestion of particulate matter by a cell, a process unique to eucaryotes.

Phenol coefficient a measure of the efficacy of a disinfectant against a given organism, compared to that of phenol.

Phenotype the observable characteristics of an organism.

Phospholipid an important constituent of all membranes, comprising a triacylglycerol in which one fatty acid is replaced by a phosphate group.

Phosphorylation the addition of a phosphate group.

Photophosphorylation the synthesis of ATP using light energy.

Photoreactivation a DNA repair mechanism involving the light-dependent enzyme DNA photolyase.

Photosynthesis a process by which light energy is trapped by chlorophyll and converted to ATP, which is used to drive the synthesis of carbohydrate by reducing CO_2.

Phototroph an organism that is able to use light as its source of energy.

Phylogenetic relating to the evolutionary relationship between organisms.

Phytoplankton the collective term used to describe the unicellular photosynthesisers, including cyanobacteria, dinoflagellates, diatoms and single-celled algae.

Pilus (pl. pili) a short, hair-like appendage found on surface of procaryotes and assisting with attachment. Specialised **sex pili** (q.v.) are involved in bacterial conjugation.

Plankton the floating microscopic organisms of aquatic systems.

Plasma membrane the membrane that surrounds a cell.

Plasmid a small, self-replicating loop of extrachromosomal DNA, found in bacteria and some yeasts. Specially engineered forms are used as vectors in gene cloning.

Plasmodium a mass of protoplasm containing several nuclei and bounded by a cytoplasmic membrane.

Plasmogamy the fusion of the cytoplasmic content of two cells.

Plasmolysis the shrinkage of the plasma membrane away from the cell wall, due to osmotic loss of water from the cell.

Point mutation a mutation that involves the substitution of one nucleotide by another.

Polar having unequal charge distribution, caused by unequal sharing of atoms.

Polymerase chain reaction (PCR) a technique that selectively replicates a specific DNA sequence by means of *in vitro* enzymatic amplification.

Polypeptide a chain of many amino acids joined together by peptide bonds.

Polyribosome (polysome) a chain of ribosomes attached to the same molecule of mRNA.

Polysaccharide a carbohydrate polymer of monosaccharide units.

Primary producer an autotroph that obtains energy from the sun or chemical sources.

Primer a short sequence of single-stranded DNA or RNA required by DNA polymerase as a starting point for chain extension.

Prion a self-replicating protein responsible for a range of neurodegenerative disorders in humans and mammals

Procaryote an organism lacking a true nucleus and membrane-bound organelles.

Profundal zone the deepest part of a body of water, where the sun is unable to penetrate

Promoter a sequence upstream of a gene, where RNA polymerase binds to initiate transcription.

Prophage the DNA of a temperate phage that has integrated into the host genome. It remains inactive whilst in this form.

Prostheca a stalked structure formed by an extension of the cell wall and plasma membrane of certain bacteria.

Prosthetic group a non-polypeptide component of a protein, such as a metal ion or a carbohydrate.

Protease an enzyme that digests proteins.

Protista a eucaryotic kingdom, comprising mostly unicellular organisms.

Protocooperation a form of mutualistic relationship that is not obligatory for either partner.

Proton a positively-charged subatomic particle.

Protoplast a cell that has had its cell wall removed.

Prototroph the normal, nonmutant form of an organism.

Protozoa a group of single celled eucaryotes with certain animal-like characteristics.

Pseudomurein a modified form of peptidoglycan found in some archaean cell walls.

Pseudopodium a projection of the plasma membrane of the amoebas that causes the cell to change shape and allows movement.

Psychrophile an organism that grows optimally at low temperatures ($<15^\circ$).

Psychrotroph an organism that is able to tolerate low temperatures, but grows better at more moderate values.

Reading frame the way in which a sequence of nucleotides is read in triplets, depending on the starting point.

Real image an image that can be projected onto a flat surface such as a screen.

Recombinant DNA DNA that comprises material from more than one source.

Recombination any process that results in new combinations of genes.

Redox potential (E_O) the tendency of a compound to lose or gain electrons.

Reduction a chemical reaction in which an electron is gained.

Refractive index the ratio between the velocity of light as it passes through a substance and its velocity in a vacuum.

Regulatory gene a gene whose protein product has an effect in controlling the expression of other genes.

Replacement vector a cloning vectors based on phage λ, in which a central section of nonessential DNA is replaced by the insert DNA.

Replication fork a Y-shaped structure formed by the separating strands of DNA during replication.

Repressor protein a protein that prevents transcription of a gene by binding to its operator.

Resolution the capacity of an optical instrument to discern detail.

Restriction endonuclease an enzyme of microbial origin that cleaves DNA at a specific nucleotide sequence.

Reverse transcriptase an enzyme found in retroviruses that can synthesise DNA from an RNA template.

Rhizosphere the region around the surface of a plant's root system.

Ribosomal RNA (rRNA) a form of RNA that forms part of the structure of ribosomes.

Ribosome an organelle made up of protein and RNA, found in both procaryotes and eucaryotes. The site of protein synthesis.

rRNA see **Ribosomal RNA**

Saprobe an organism that feeds on dead and decaying organic materials. Previously termed saprophyte.

Saturated fatty acid a fatty acid that has only single covalent bonds between adjacent carbon atoms. (c.f. unsaturated)

Secondary metabolite a substance produced by a microorganism after the phase of active growth has ceased.

Selectable marker a gene that allows cells containing it to be identified by the expression of a recognisable characteristic.

Selective medium a medium that favours the growth of a particular organism or group of organisms, often by suppressing the growth of others.

Semi-conservative replication the process of DNA replication by which each strand acts as a template for the synthesis of a new complementary strand. Each resultant double stranded molecule thus comprises one original strand and one new one.

Septate separated by septa or cross-walls.

Sex pilus a narrow extension of the bacterial cell, through which genetic material is transferred during conjugation.

Shuttle vector a cloning vector that can replicate in both bacterial and yeast host cells.

Silent mutation a mutation that has no effect on the amino acid encoded by the triplet.

Single-cell protein (SCP) bacteria and yeast grown in bulk for use as a protein-rich food supplement.

Slime layer a diffuse and loosely attached polysaccharide layer surrounding the cell of certain procaryotes.

Specialised transduction the transfer of a limited selection of genes due to imprecise excision of a prophage in a lysogenic infection cycle.

Sporangiophore a specialised aerial hypha that bears the sporangia.

Sporangium a structure inside which spores develop.

Spore a resistant, non-motile reproductive cell.

Sporophyte the diploid, spore-forming stage in a life cycle with alternation of generations.

Sporozoite a motile infective stage of members of the Sporozoa that gives rise to an asexual stage within the new host.

Sterilisation the process by which all microorganisms present on or in an object are destroyed or removed.

Steroid a member of a group of lipids based on a four-ring structure.

Stop codon use definition below.

Substrate level phosphorylation the synthesis of ATP by the direct transfer of a phosphate group from a phosphorylated organic compound to ADP.

Symbiosis a close physical association between two species from which special benefits may accrue for one or both parties.
The term is sometimes used specifically to describe such a relationship in which both parties derive benefit (see **mutualism**).

Tautomerism the ability of a molecule such as a nucleotide base to exist in two alternative forms.

Taxonomy the science of naming and classifying living (and once-living) organisms.

Temperate phage a bacteriophage with a lysogenic replication cycle.

Terminator a sequence of DNA that indicates transcription should stop.

stop codons one of the three triplet sequences (also called nonsense codons) that indicate translation should stop.

Thallus a simple vegetative plant body showing no differentiation into root, stem and leaf.

Therapeutic index a measure of the selective toxicity of a chemotherapeutic agent.

Thylakoid a photosynthetic membrane found in chloroplasts or free in the cytoplasm (in cyanobacteria). It contains photosynthetic pigments and components of the electron transport chain.

Total cell count a method that enumerates all cells, living and dead.

Trace element an element required in minute amounts for growth.

Transamination a reaction involving the transfer of an amino group from one molecule to another.

Transcription the process by which single-stranded mRNA is synthesised from a complementary DNA template.

Transduction the bacteriophage-mediated transfer of genetic material between bacteria.

Transformation the uptake of naked DNA from the environment and its integration into the host genome.

Transposable elements a sequence of DNA that is able to relocate to another position on the genome.

Transition a mutation in which a purine replaces a purine or a pyrimidine replaces a pyrimidine.

Translation the process by which the message encoded in mRNA is converted into a sequence of amino acids.

Transfer RNA (tRNA) a form of RNA that carries specific amino acids to the site of protein synthesis.

Transversion a mutation in which a purine replaces a pyrimidine, or a pyrimidine replaces a purine.

Triacylglycerol a lipid formed by the joining of three fatty acids to a molecule of glycerol.

Tricarboxylic acid cycle a series of reactions that oxidise acetate to CO_2, generating reducing power in the form of NADH and $FADH_2$ for use in the electron transport chain. Also known as citric acid cycle or Krebs cycle.

Triose a three-carbon sugar.

tRNA see **Transfer RNA**

Tyndallisation a form of intermittent steam sterilisation.

Undefined or complex medium a medium whose precise chemical composition is not known.

Unsaturated fatty acid a fatty acid that contains one or more double bonds between adjacent carbon atoms.

Vaccination inoculation with a vaccine to provide protective immunity.

Vaccine a preparation of dead or inactivated living pathogens or their products used to provide protective immunity.

Vector a self-replicating DNA molecule used in gene cloning. The sequence to be cloned is inserted into the vector, and replicated along with it.

Viable cell count a method that enumerates only those cells capable of reproducing to form a visible colony.

Virtual image an image that has no physical existence in space, and cannot be projected onto a screen.

Virion a complete, intact viral particle

Viroid a plant pathogen that comprises only ssRNA and does not code for protein product.

Virulent phage a bacteriophage with a lytic replication cycle.

Virus a submicroscopic, noncellular parasite, comprising protein and RNA or DNA.

Wildtype the normal, nonmutant form of an organism or gene.

Wobble the degree of flexibility allowed in the third base in a codon when pairing with tRNA. The wobble hypothesis explains how a single tRNA can pair with more than one codon.

Yeast artificial chromosome (YAC) a cloning vector able to accommodate inserts of several hundred kb in size.

Z value the increase in temperature required to reduce the D value (q.v.) by a factor of ten.

Zoonosis a disease normally found in animals, but transmissible to humans under certain circumstances.

Zoospore a flagellated, asexual spore.

Zygospore a thick-walled resistant diploid structure formed by members of the Zygomycota.

Appendix

Solution to Numerical Questions

Question 2.5: 1000 ($10 \times 10 \times 10$)

Question 2.15: 61 cytosine residues; 286 hydrogen bonds (61×3 for GC and 52×2 for AT)

Further Reading

The declared intention of *Essential Microbiology* is to serve as an introduction to microbiology for students who do not intend to specialise in the subject. It would please the author greatly, however, if the book stimulated readers to deepen their knowledge and understanding by referring to other, more specialist texts. Below are listed a number of suggested sources for further reading; the list is by no means comprehensive, and the reader may well find other texts to suit his or her individual needs. Although titles are listed under the heading of a particular chapter, many will of course also be equally useful in the context of other chapters. Journal articles have not been included, with the exception of a few review-type articles from readily accessible titles such as *Scientific American*.

General microbiology

For a general overview of the world of microorganisms and their applications, there can be no better starting point than John Postgate's classic *Microbes and Man*, 4th edition (2000), published by Cambridge University Press. Those wishing to study microbiology in greater depth may consider one of the many excellent general microbiology texts available; the following are well-established examples that embrace the whole field of microbiology, and cover many topics in greater depth than is possible or indeed desirable in the present book.

Black JG (2004) *Microbiology: Principles and Explorations*, 6th edn. John Wiley & Sons, Chichester.

Lederberg, J (ed) (2000): Encyclopaedia of Microbiology, 2nd edition. Academic Press.

Madigan MT, Martinko JM and Parker J (2003) *Brock Biology of Microorganisms*, 10th edn. Prentice Hall Inc., Englewood Cliffs, NJ.

Prescott LM, Harley JP and Klein DA (2004) *Microbiology*, 6th edn. McGraw-Hill, New York.

Singleton P and Salisbury D (2002): *Dictionary of Microbiology and Molecular Biology*, 3rd edition. John Wiley & Sons.

Tortora GJ, Funke BR and Case CL (2004) *Microbiology: An Introduction,* 8th edition. Benjamin Cummings, San Francisco, CA.

Chapter 1

Beck RW (2000) *A Chronology of Microbiology in Historical Context.* American Society of Microbiology.

De Kruif P (2002) *Microbe Hunters*. Harcourt Publishers, New York.
Lagerqvist U (2003) *Pioneers of Microbiology and the Nobel Prize*. World Scientific Publishing, London.
Murphy DB (2001) *Fundamentals of Light Microscopy and Electronic Imaging*. John Wiley & Sons, Chichester.

Chapter 2

Hames DB, Hooper NM and Houghton JD (2000): *Instant Notes in Biochemistry*, 2nd edn. Bios Scientific Publishers, Oxford.
Lewis R and Evans W (2001) *Chemistry*, 2nd edn. Palgrave, Basingstoke.

Chapter 3

Cooper GM and Hausman RE (2004) *The Cell: A Molecular Approach*, 3rd edn. Sinauer Associates Inc, Sunderland, MA.
Singleton P (2004) *Bacteria in Biology, Biotechnology and Medicine*, 6th edn. John Wiley & Sons, Chichester.

Chapter 4

Atlas, RM (2004) *Handbook of Microbiological Media*, 3rd edn. CRC Press, Boca Raton, FL.
Barrow GI and Feltham RKA (eds) (2004) *Cowan and Steel's Manual for the Identification of Medical Bacteria*, 3rd edn. Cambridge University Press, Cambridge.
Cappuccino JG and Sherman N (2005) *Microbiology: A Laboratory manual*, 7th edn. Benjamin Cummings, San Francisco, CA.
Collins CH, Lyne PM, Grange JM and Falkinham J (2004) *Microbiological Methods*, 8th edn. Hodder Arnold, London.

Chapter 5

Isaac S and Jennings D (1995) *Microbial Culture*. Bios Scientific Publishers, Oxford.
Madigan MT and Marrs BL (1997) Extremophiles. *Scientific American* **276**, 82–7.
Moat AG, Foster JW and Spector MP (2002) *Microbial Physiology*, 4th edn. John Wiley & Sons, Chichester.

Chapter 6

Gilbert HF (2000) *Basic Concepts in Biochemistry: A Student's Survival Guide*, 2nd edition. McGraw-Hill, New York.
Matthews HR, Freedland R and Miesfeld RL (1997) *Biochemistry: A Short Course*. John Wiley & Sons, Chichester.
Zubay G (1998) *Biochemistry*, 4th edn. McGraw-Hill, New York.

Chapter 7

Garrity GM (ed.) (2001) *Bergey's Manual of Systematic Bacteriology*, 2nd edn, Vol 1: The Archaea and the Deeply Branching and Phototrophic Bacteria. Springer-Verlag, Berlin.

Garrity GM (ed.) (2005) *Bergey's Manual of Systematic Bacteriology*, 2nd edn, Vol 2: The Proteobacteria. Springer-Verlag, Berlin.

Lengeler JW, Drews G and Schlegel HG (eds) (1998) *Biology of the Prokaryotes*. Blackwell Science, Oxford.

Chapter 8

Alexopoulos CJ, Mims CW and Blackwell M (1996) *Introductory Mycology*, 4th edn. John Wiley & Sons, Chichester.

Carlile MJ, Watkinson SC and Gooday GW (2001) *The Fungi*, 2nd edn. Academic Press, London.

Deacon JW (1997) *Modern Mycology*, 3rd edn. Blackwell Scientific, Oxford.

Jennings DH and Lysek G (1999) *Fungal Biology*, 2nd edn. Bios Scientific Publishers, Oxford.

Chapter 9

Bhamrah HS and Juneja K (2002) An *Introduction to the Protozoa*, 2nd edn. Anmol Publications, New Delhi.

Graham LE and Wilcox LW (2000) *Algae*. Prentice Hall, Englewood Clifs, NJ.

Van den Hoek C, Mann DG and Jahns HM (1996) *Algae: An Introduction to Phycology*. Cambridge University Press, Cambridge.

Chapter 10

Cann AJ (2001) *Molecular Virology*, 3rd edn. Academic Press, London.

Dimmock MJ, Easton AJ and Leppard KN (2001) *Introduction to Modern Virology*, 5th edn. Blackwell Science, Oxford.

Wagner EK and Hewlett MJ (2003) *Basic Virology*, 2nd edn. Blackwell Science, Oxford.

Chapter 11

Brown TE (1998) *Genetics: A Molecular Approach*, 3rd edn. Chapman & Hall, London.

Calladine, CR and Drew HR (2004) *Understanding DNA: The Molecule and How it Works*, 3rd edn. Academic Press, London

Dale JW and Park SF (2004) *Molecular Genetics of Bacteria*, 4th edn. John Wiley & Sons, Chichester.

Klug WS and Cummings MR (2005) *Essentials of Genetics*, 5th edn. Prentice Hall, Englewood Cliffs, NJ.

Miller RV (1998) Bacterial gene swapping in nature. *Scientific American* **278**, 67–71.

Watson JD, Baker TA, Bell SP, Gann A, Levine M and Losick R (2004) *Molecular Biology of the Gene*, 5th edn. Benjamin Cummings, San Francisco, CA.

Chapter 12

Alcamo IE (1999) *DNA Technology: The Awesome Skill*, 2nd edn. Academic Press, London.
Brown TE (2001) *Gene Cloning and DNA Analysis: An Introduction*, 4th edn. Blackwell Science, Oxford.
Dale JW and von Schantz M (2002) From Genes to Genomes: Concepts and Applications of DNA Technology. John Wiley & Sons, Chichester.
Kreuzer H and Massey A (2000) *Recombinant DNA and Biotechnology: A Guide For Students*, 2nd edn. American Society for Microbiology.

Chapter 13

Block SS (2000) *Disinfection, Sterilisation and Preservation*, 5th edn. Lippincott, Williams & Wilkins, London.
Russell AD (ed) (1998) *Principles and Practice of Disinfection, Preservation and Sterilisation.* Blackwell Science, Oxford.

Chapter 14

Brown K (2004) *Penicillin Man: Alexander Fleming and the Antibiotic Revolution.* Sutton Publishing, Stroud.
Franklin TJ (1999) *Biochemistry and Molecular Biology of Antimicrobial Drug Action*, 5th edn. Springer Verlag, Berlin.
Greenwood D (ed) (2000) *Antimicrobial Chemotherapy*, 4th edn. Oxford University Press, Oxford.

Chapter 15

Nash TH (ed.) (1996) *Lichen Biology*. Cambridge University Press, Cambridge.
Paracer S and Ahmadjian V (2000) *Symbiosis: An Introduction to Biological Associations*, 2nd edn. Oxford University Press, Oxford.

Chapter 16

Alexander, M (1999) *Biodegradation and Bioremediation*, 2nd edn. Academic Press, London.
Atlas RM and Bartha R (1998) *Microbial Ecology: Fundamentals and Applications*, 4th edn. Benjamin Cummings, San Francisco, CA.
Bitton G (1999) *Wastewater Microbiology*, 2nd edn. John Wiley & Sons, Chichester.
DeLong EF (1997) Marine microbial diversity: the tip of the iceberg. *Trends in Biotechnology* **15**, 203–7.
Mitchell R (ed) (1993) *Environmental Microbiology*. John Wiley & Sons, Chichester.

Chapter 17

Adams MR and Moss MO (2000) *Food Microbiology*, 2nd edn. Royal Society of Chemistry, London.

Demain AL (2000) Microbial Technology. *Trends in Biotechnology* **18**, 26–31.

Garbutt, J (1997) *Essentials of Food Microbiology*. Hodder Arnold, London.

Hornsey IS (1999) *Brewing*. Royal Society of Chemistry, London.

Ratledge C and Kristiansen B (2001) *Basic Biotechnology*, 2nd edn. Cambridge University Press, Cambridge.

Smith JE (2004) *Biotechnology*, 4th edn. Cambridge University Press, Cambridge.

Index

The suffix 't' indicates an entry in a table; 'f' indicates a footnote